Springer Hydrogeology

The Springer Hydrogeology series serves as a comprehensive platform for advancing knowledge and fostering collaboration in the diverse and multidisciplinary field of hydrogeology. This book series seeks to cater to a wide audience, including researchers, educators, practitioners, policymakers, and students who are passionate about understanding and solving groundwater-related challenges. It aims to provide high-quality, peer-reviewed monographs, edited volumes, textbooks, and conference proceedings to disseminate cutting-edge research, practical insights, and innovative approaches in hydrogeology and related disciplines.

The series covers the entire spectrum of hydrogeological topics, including but not limited to: groundwater science and management, isotope hydrogeology, hydrogeological modeling, groundwater resources and systems, hydrogeochemistry, ecohydrology and groundwater-energy-food nexus, cryohydrogeology, urban and agricultural hydrogeology, emerging technologies in hydrogeology, climate change impacts on groundwater and groundwater-dependent ecosystems, and geotechnical and environmental aspects.

The Springer Hydrogeology series is aligned with the United Nations Sustainable Development Goals (SDGs), particularly SDG 6 (Clean Water and Sanitation) and SDG 13 (Climate Action). By publishing works that address sustainable groundwater use, pollution mitigation, climate resilience, and integrated groundwater resource management, the series supports efforts to achieve equitable and sustainable access to groundwater resources for current and future generations. Contributions that explore the intersection of hydrogeology with other SDGs—such as those on biodiversity, sustainable cities, and food security—are also strongly encouraged.

Dimple · Chaitanya Baliram Pande · Jitendra Rajput
Editors

Surface Water and Groundwater Quality for Sustainable Utilization

Advanced Methods and Technology

Springer

Editors
Dimple
College of Agriculture, Kishangarh Bas
Sri Karan Narendra Agriculture University
Jobner, Rajasthan, India

Jitendra Rajput
Division of Agricultural Engineering
ICAR-Indian Agricultural Research Institute
New Delhi, Delhi, India

Chaitanya Baliram Pande
Department of Civil Engineering, Faculty of Science, Technology and Architecture (FoSTA)
Manipal University Jaipur
Jaipur, Rajasthan, India

ISSN 2364-6454 ISSN 2364-6462 (electronic)
Springer Hydrogeology
ISBN 978-981-96-8188-4 ISBN 978-981-96-8189-1 (eBook)
https://doi.org/10.1007/978-981-96-8189-1

This Springer imprint is published by the registered company Springer Nature Singapore Pte Ltd.
The registered company address is: 152 Beach Road, #21-01/04 Gateway East, Singapore 189721, Singapore

Preface

The proposed book, *Surface Water and Groundwater Quality for Sustainable Utilization: Advanced Methods and Technology*, offers a comprehensive journey into the complex and critical realm of water quality assessment and management. Spanning a wide array of interconnected topics, this book presents a holistic view of the importance of water quality assessment for both drinking and irrigation, laying the foundation for an in-depth exploration of advanced methods, technologies, and policies aimed at safeguarding this invaluable resource. The following chapters provide a detailed overview of the multifaceted aspects of water quality.

This book contributes to the global effort to secure safe and high-quality water resources for current and future generations, making it an essential reference for anyone engaged in water quality assessment and managed. Water is a fundamental resource that sustains life, agriculture, industry, and ecosystems. However, increasing pollution, climate change, and over-extraction have significantly affected surface and groundwater quality, raising concerns about sustainable water utilization.

The book begins with an exploration of effective approaches to drinking water purification (Chapter "Effective Approaches Towards Drinking Water Purification"), discussing traditional and advanced treatment methods. This is followed by a detailed analysis of contamination in surface water and groundwater (Chapter "Contamination of Surface Water and Groundwater Quality"), covering natural and anthropogenic sources of pollution. To build a foundation, physical, chemical, and biological parameters of water quality (Chapter "Physical, Chemical, and Biological Parameters of Surface Water and Groundwater Quality") are explained, providing a basis for understanding water assessment.

With the increasing use of artificial intelligence and machine learning, soft computing methods for water quality analysis (Chapter "Soft Computing Methods for Surface Water and Groundwater Quality Analysis") offer innovative solutions for data processing and predictive modeling. Traditional approaches remain crucial, and the book presents chemical analysis methods (Chapter "Chemical Analysis Method for Surface and Ground Water Quality Assessment") and physical analysis methods (Chapter "Physical Analysis Methods for Surface Water and Ground Water Quality Quantification") for assessing water quality parameters accurately.

Technological advancements in remote sensing techniques for water quality assessment (Chapters "Remote Sensing and Google Earth Engine for Surface and Groundwater Quality Assessment" and "Integrating Remote Sensing and GIS for Surface and Groundwater Quality Assessment") have revolutionized monitoring capabilities, providing real-time data for large-scale analysis. The impact of poor water quality on crop productivity (Chapter "Influence of Poor Water Quality on Crop Productivity and Quality of Produce") and soil fertility (Chapter "Effects of Poor Irrigation Water Quality on Soil Fertility and Its Mitigation") is discussed, along with mitigation strategies for sustainable agriculture.

Agricultural resilience requires informed decision-making, and crop selection and management strategies for poor irrigation water (Chapter "Crop Selection Criteria for Different Quality of Irrigation Water and Mitigation Techniques") help optimize farming under adverse conditions. A growing concern in water safety is the emergence of new contaminants; unseen threats from emerging contaminants (Chapters "Crop Selection and Management Strategies for Poor Quality Irrigation Water" and "Unseen Threats: The Impact of Emerging Contaminants on Surface Water and Groundwater Quality") highlight their potential risks to ecosystems and human health.

Climate change is altering hydrological cycles, impacting surface and groundwater quality (Chapter "Unseen Threats: The Impact of Emerging Contaminants on Surface and Groundwater Quality") and necessitating policy reforms, as covered in water quality development policies under climate change (Chapter "Advances in Surface Water and Groundwater Quality Monitoring and Analysis Techniques"). To address future challenges, machine learning-based forecasting models (Chapter "Climate Change's Impacts on Surface Water and Groundwater Quality") enhance water quality predictions, aiding decision-making.

Regional studies provide valuable insights, and the book discusses water budgeting and groundwater utilization assessment in Western Uttar Pradesh, India (Chapter "Surface Water and Groundwater Quality Development Policy Under Climate Changes"), offering a case study for sustainable management. Predictive capabilities are further explored in surface and groundwater quality modeling using machine learning (Chapter "Predictive Modeling of Surface and Groundwater Quality with Machine Learning Techniques"), leveraging big data for enhanced predictions.

Monitoring advancements are essential, and new techniques in water quality monitoring and analysis (Chapter "Surface Water and Groundwater Quality Forecasting Using Machine Learning Models") present modern methods for comprehensive assessments. Finally, sustainable agricultural practices require adaptation, and crop selection criteria for different irrigation water qualities and mitigation techniques (Chapter "Water Budgeting and Groundwater Utilization Assessment in Western Uttar Pradesh (India)") offer practical guidance for farmers facing water quality challenges.

By integrating scientific research, technological innovations, and sustainable strategies, this book serves as a valuable resource for researchers, policymakers, and practitioners working towars improved water management. We hope this book provides useful insights and fosters a deeper understanding of surface water and groundwater quality for a sustainable future.

Jobner, India — Dimple
Jaipur/Pune, India — Chaitanya Baliram Pande
New Delhi, India — Jitendra Rajput

Contents

About the Editors

Dr. Dimple is currently working as an Assistant Professor (Agricultural Engineering), College of Agriculture, Kishangarh Bas, a constituent college of Sri Karan Narendra Agriculture University, Jobner, Jaipur, Rajasthan, India. She has received her Ph.D. from MPUAT, Udaipur, India. She has received her M.Tech. (Soil and Water Conservation Engineering) from Maharana Pratap University of Agriculture and Technology (MPUAT), Udaipur, Rajasthan, India, and B.Tech. (Agricultural Engineering) degree from Dr. Annasaheb Shinde College of Agriculture Engineering, Mahatma Phule Krishi Vidyapeeth Rahuri, Maharashtra, India. She had received prestigious fellowships including ICAR-NTS, UGC, and DST-INSPIRE during UG, PG, and Ph.D. degree programs, respectively. She is a reviewer for several scientific journals of international repute, and she is an editorial board member of *Discover Water Journal*. She is an awardee of University Gold Medal from MPUAT, Udaipur, Rajasthan, for outstanding performance during M.Tech. degree program. Her research interest includes micro irrigation system, groundwater quality, and land and water resources management using Geo-spatial tools and soft computing techniques. Dr. Dimple has participated and presented a research paper at 12th international conference organized by the Thai Society of Agriculture Engineeirng-2019, Thailand in person. She has published more than 15 research paper in national and international journals and participated/presented papers in various conferences. She has published book chapters in national and international edited books and a few popular articles in the leading national magazines.

Dr. Chaitanya Baliram Pande completed, his Ph.D. in environment science at Sant Gadge Baba Amravati University, India, and an M.Sc. in geoinformatics at Amravati University in 2011. He has more than 13 years of teaching, research, and industry experience. He is included in the 2% Stanford scientist list, 2024 and 2025. He is a Reviewer for several scientific journals of international repute, and he is an Editorial Board Member in the *American Journal of Agricultural and Biological Sciences*. He has published 120+ research papers, 2 textbooks, 10 edited books, 19 conference papers, and 25+ book chapters. Altogether, they have received more than 6381 citations as per google scholar. His research interests include remote sensing, GIS, air pollution, Google Earth Engine, machine learning, watershed management, hydrogeology, hydrological modeling, drought monitoring, land use and land cover analysis, groundwater quality, urban planning, hydro-geochemistry, groundwater modeling, geology, hyperspectral remote sensing, and GIS applications in natural resources and watershed management, and environmental monitoring and assessment.

Dr. Jitendra Rajput is currently working as Scientist at Division of Agricultural Engineering, Indian Council of Agricultural Research-Indian Agricultural Research Institute (ICAR-IARI), New Delhi, India. He has received B.Tech. (Agricultural Engineering) degree from Central Agricultural University, Imphal, Manipur, in 2012, M.Tech. (Irrigation Water Management Engineering) degree from Maharana Pratap University of Agriculture and Technology (MPUAT), Udaipur, Rajasthan, in 2014 and Doctor of Philosophy (Ph.D.) degree in Water Science and Technology specialization from ICAR-IARI, New Delhi, India in 2023. He is recipient of prestigious ICAR Fellowships *viz.*, NTS, JRF, and SRF during UG, PG, and Ph.D. degree programs, respectively. He is an awardee of Jain Irrigation Medal from MPUAT, Udaipur, Rajasthan for outstanding performance during M.Tech. degree program. He is recipient of Prof. B. V. Ramana Rao Best Paper Award—2024 by the Association of Agrometeorologists. He has previously worked as Assistant Professor and Subject Matter Specialist in the specialization of Soil and Water Engineering at Dr. Rajendra Prasad Central Agricultural University (DRPCAU), Pusa, Bihar. His research interest includes design of micro-irrigation system, canal irrigation system performance assessment, climate change, groundwater management, groundwater carbon footprint, and land and water resources management using geo-spatial tools and soft computing/AI techniques. Dr. Rajput has published 56 national and international research articles. Also, he has published 20 book chapters in national and international edited books. Dr. Rajput has published two books with national publishers. He is the Editorial Board Member of two springer journals, viz., *Discover Water* and *Discover Sustainability* and reviewer of several journals of Elsevier, Springer, International Water Association, MDPI, Taylor and Francis, Frontiers, and other international publishers.

Effective Approaches Towards Drinking Water Purification

K. S. Ambili, Usha K. Aravind, and C. T. Aravindakumar

Abstract Clean water is a fundamental requirement for the establishment of life and there is an ever-increasing demand for safe and pure water. Water pollution and water scarcity are two major challenges facing the mankind nowadays. These challenges may possibly persist or will get worst in upcoming periods due to immediate population growth, industrialization, improving life standards etc. Water pollution, often driven owing to the impact of recalcitrant organic substances and hazardous inorganic compounds, poses an ample threat to public health and the ecosystems that sustain us. Drinking water treatment is very critical as it ensures water is meeting the quality standards for human consumption by eliminating contaminants. The multi-step process of drinking water treatment is meticulously designed to target various pollutants and adhere to regulatory standards, ensuring the water meets quality requirements. This chapter mainly deals with the different novel processes for drinking water treatments and thus raises awareness about drinking water treatment techniques for ensuring public health and safety. This knowledge can also drive the adoption of advanced technologies and sustainable practices, leading to improved water quality and reduced health risks. Additionally, increased awareness can empower people to engage in local water management initiatives and contribute to maintaining safe and clean water sources.

Keywords Drinking water · Water purification · Quality standards · Advanced treatment methods

---

K. S. Ambili · C. T. Aravindakumar (✉)
Inter-University Instrumentation Centre (IUIC), Mahatma Gandhi University (MGU), Kottayam, Kerala, India
e-mail: cta@mgu.ac.in

K. S. Ambili
e-mail: ambiliks@mgu.ac.in

U. K. Aravind
School of Environmental Studies, Cochin University of Science & Technology (CUSAT), Kochi, Kerala, India
e-mail: uka@cusat.ac.in

C. T. Aravindakumar
School of Environmental Sciences, Mahatma Gandhi University (MGU), Kottayam, Kerala, India

Dimple et al. (eds.), *Surface Water and Groundwater Quality for Sustainable Utilization*, Springer Hydrogeology, https://doi.org/10.1007/978-981-96-8189-1_1

1 Introduction

Drinking water pollution is a significant global issue that arises when contaminants, including chemicals, pathogens, and pollutants, infiltrate sources of water categorized for human ingestion. This has a profound impact on human well-being and ecological system, as water infected with pollutants might result in substantial health challenges, including intestinal disorders, neurological syndromes, and chronic conditions like cancer (Lin et al. 2022). Environmentally, polluted water can disrupt ecosystems, harming aquatic life and reducing biodiversity (Feckler et al. 2023). According to 2019 World Health Organization (WHO) projections, by the year 2025, up to fifty percentages of global inhabitants could face difficulties in accessing sufficient water, highlighting a critical issue that needs urgent attention (Kundu et al. 2024). In today's age of heightened environmental awareness, there is an unprecedented urgency to address and mitigate environmental challenges. Degradation in water quality can arise from both natural and human-involved activities. Natural sources comprise microbial activity, geological factors, and natural disasters. Human activities, such as industrial operations, agriculture, improper waste disposal, and inadequate sewage systems, also contribute to contamination. Pollutants can include microbiological pathogens, chemicals, heavy metals, pesticides, and pharmaceuticals, all of which threaten water safety and purity (Akhtar et al. 2021; Kristanti et al. 2022; Pot et al. 2022; Singh et al. 2024).

The widespread presence of harmful substances such as chlorinated biphenyl compounds, polycyclic aromatic hydrocarbons, micro/nano plastics, pharmaceuticals, various pesticides, and persistent organic pollutants (POPs) across global aquatic systems represents a substantial warning to both water quality and ecosystem health (Chen et al. 2024; Duttagupta et al. 2020). Our water resources are particularly strained due to unsustainable usage practices and the improper disposal of industrial waste. It was estimated that significant quantity of effluents from municipal and industrial sources is released into the environment untreated worldwide, causing significant harm to both human health and ecosystems (Chen et al. 2024; Lin et al. 2022). Pollution from industries like petrochemical refining, construction, and chemical manufacturing endangers water quality and ecosystems. Similarly, pollution from agricultural runoff impacts the environment and human health by degrading aquatic ecosystems and compromising water quality. It comes from diverse sources like fertilizers, pesticides, and sediment, both point and diffuse. As the global population grows and food demand rises, addressing this pollution becomes even more urgent (Albou et al. 2024).

Pharmaceutical industries, with their multi-batch and bulk production processes, often release pharmaceuticals into the environment due to improper disposal and use. Facilities involved in chemical synthesis and fermentation are major contributors, generating significant waste during manufacturing and maintenance stages (Mohammed et al. 2021). These sources uphold the release of various POPs in to the environment, where they can gather in the food chain and source long-term risks to human health and environment. When persistent organic pollutants enter the environment, they pose significant risks to all entities in the nature. Hazardous anions and heavy metals are also harmful pollutants in drinking water, known for their toxic and carcinogenic effects (John et al. 2018). The quantity of these contaminants can sometimes rise to dangerous levels, diverging markedly from their typical environmental concentrations. Pollutants entering lakes, estuaries, and rivers heighten the risk to drinking water, highlighting that clean, safe water is a fundamental human right and essential to our existence (Bashir et al. 2020).

Addressing this issue requires robust monitoring, effective treatment methods, stringent regulations, and public awareness-raising programs to safeguard clean and accessible drinking water for all. This chapter deals with the different conventional and novel processes for drinking water treatments which refine water quality, ensuring it meets the drinking water standards.

2 Quality Criterions for Drinking Water

Drinking water quality standards are guidelines or regulations established by governmental agencies or international organizations to confirm that water intended for human consumption is safe and free from harmful contaminants. These standards help safeguard public health by setting limits on various pollutants and ensuring that water treatment processes meet stringent safety criteria (Dinka 2018). It typically specifies acceptable levels of various substances and parameters that could affect human health if detected in potable water. Ensuring the safety of drinking water is vital to prevent and control waterborne diseases (Shayo et al. 2023). WHO develops guidelines through extensive global collaboration, involving WHO member states, international agencies and national authorities. These guidelines are crafted in consultation with the expert advisory panel of WHO, reflecting a comprehensive approach to safeguarding public health. The parameters include microbial parameters, chemical parameters, physical parameters and radiological parameters. Microbial parameters focus on the detection of harmful microbes such as viruses, bacteria, and protozoa, which are able to cause waterborne diseases. Physical parameters deal with the observable characteristics of water, including turbidity, colour, odour, and taste, which can indicate contamination or the presence of pollutants. Chemical parameters involve the measurement of various chemical contaminants, comprising heavy metals, disinfection by-products, herbicides, pharmaceuticals, pesticides, and other industrial chemicals. Radiological parameters address the standards for radioactive substances like radium, uranium, and radon, which may occur naturally in groundwater or surface water. Acceptable range of some common contaminants in drinking water by WHO as per the guideline 2022 are given in Table 1. Adhering to these standards is vital for safeguarding public health, and regulatory agencies continually review and update these guidelines based on new scientific research and emerging health concerns to ensure safe drinking water for communities worldwide.

Table 1 Acceptable range of some common contaminants in drinking water by WHO as per the guideline 2022

Microbial parameters	Parameter	Guideline value
	E coli	Must be untraceable in 100 ml sample
Physical parameters	Taste	Unobjectionable
	Odour	Unobjectionable
	Total dissolved solids	>600 mg/l
Chemical parameters	pH	6.5–8.5
	Total hardness	500 mg/l
	Chlorine	5 mg/l
	Sodium	50 mg/l
	Arsenic	0.01 mg/l
	Fluoride	1.5 mg/l
	Chromium	0.05 mg/l
	Se	0.04 mg/l
	Uranium	0.03 mg/l
	Manganese	0.08 mg/l
	Nitrate	50 mg/l
	Nitrite	3 mg/l
	2,4-D (pesticide)	0.03 mg/l
	Cd	0.003 mg/l
	Hg	0.006 mg/l
	Benzene	0.01 mg/l
	Dichloromethane	%1.%2 g/l

Ancient civilizations demonstrated remarkable ingenuity in water purification process to guarantee safe drinking water. They commonly employed methods such as boiling to kill pathogens, thereby improving microbiological water quality (Bansal and Asthana 2019). Early filtration techniques included the use of sand, charcoal, and cloth to remove impurities. Additionally, settling was a prevalent method where water was left in containers to allow sediments to settle at the bottom. The ancient people harnessed the power of solar purification by exposing water to sunlight in clear containers to enhance its clarity. They also practiced natural coagulation by using plant materials and specific types of clay to help aggregate and remove impurities. These methods reflect the resourcefulness of ancient cultures in managing water quality long before modern technologies were developed (Adeniyi 2020). These methods reflect the ingenuity of ancient societies in addressing the challenge of providing clean drinking water.

3 Drinking Water Purification Techniques

Water purification is a crucial process that involves multiple stages to ensure the water is fit for ingestion. The preliminary stage focuses on the removal of large particles and debris through physical methods such as screening and sedimentation. This step addresses the coarse contaminants and prepares the water for more detailed treatment (Saravanathamizhan and Perarasu 2021). The primary stage involves more refined processes like coagulation and flocculation, where chemicals are added to aggregate smaller particles into larger clumps, which are later filtered out or settled through sedimentation (Bruno et al. 2020). Natural coagulants include plant-based materials (potato starch, guar gum, cactus latifaria extract etc.), animal-based materials (chitosan) and microorganism-based materials (xanthan gum). Apart from this there are chemical coagulants which are divided into three types such as metallic salts (e.g., Alum, Ferric chloride), prehydrolyzed salts (e.g., Polyaluminium sulfate), and synthetic cationic polymers (e.g., Polyethyleneimine) (El-taweel et al. 2023). The coagulants were mixed with water, which neutralizes the charge on the particles, causing them to clump together. This process forms larger aggregates known as "flocs." Following coagulation, the flocculation stage involves gentle mixing to promote the growth of these flocs into even larger clusters. This result in a substantial increase in the size and weight of the particles, which are later filtered out or settled through sedimentation (Badawi et al. 2023). Aeration is another primary treatment to reduce the levels of large organic particles and some biological contaminants. This includes systems where aeration helps to settle out solids or prepare the water for further treatment. Studies indicated that aeration effectively increases dissolved oxygen levels and pre-chlorination during pollutant removal and lowers the need for coagulants (Derakhshan et al. 2023). These primary treatments processes effectively reduce turbidity and improve the quality of water by preparing it for more advanced purification stages.

Secondary purification in water treatment mainly relies on biological processes to further degrade organic matter and pollutants that remain after the primary stage. This treatment technique uses activated sludge systems, trickling filters, sequencing batch reactors (SBR), rotating biological contactors (RBC), oxidation ponds and constructed wetlands to further degrade organic matter and pollutants. In an activated sludge system, microorganisms are introduced into aeration tanks where they consume organic pollutants, effectively degrading them into less hazardous, simpler substances (Ahansazan et al. 2014). Trickling filters, on the other hand, involve passing water over a bed of microbial-covered media, where the microorganisms metabolize the organic pollutants as the water flows through. In rotating biological contactors similar to trickling filters microorganisms covered rotating discs facilitate break down pollutants. The RBC is valued for its simple design, reliable operation, and energy-efficient performance making it a favoured choice in several wastewater treatment applications (Waqas et al. 2023). SBRs treat wastewater in a single tank through batch processing, which involves filling, aerating, settling, and decanting. This approach ensures effective biological treatment and produces

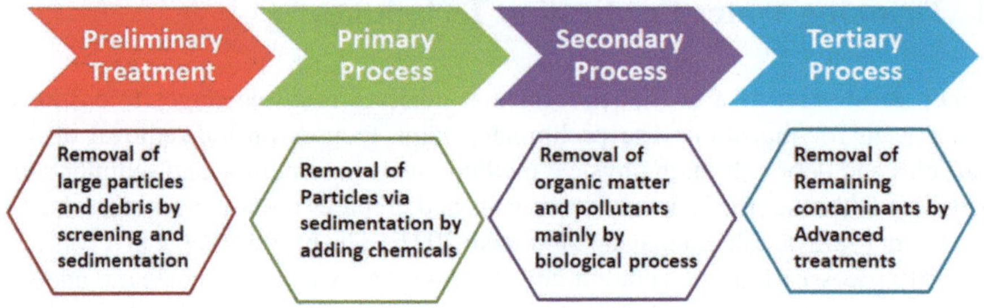

Fig. 1 Different stages of water purification

high-quality effluent. Modifications to the traditional SBR like including sludge granulation, use of biofilm carriers, and the integration of adsorbents are well-documented and enhance the system's performance (Singh et al. 2022). Oxidation ponds or stabilization ponds are large, shallow ponds where sunlight and microorganisms break down organic matter. The ponds use natural treatment through biological and solar processes for waste water treatment (Butler et al. 2017). Constructed wetlands mimic natural wetlands by using plants, soil, and microorganisms to treat wastewater through chemical, physical and biological processes (Retta et al. 2023). Aeration is crucial for these biological processes which provide the oxygen needed for aerobic microorganisms to decompose organic material in the wastewater. All the mentioned methods enhance water quality by significantly reducing biochemical oxygen demand (BOD), chemical oxygen demand (COD) and other contaminants, ensuring that the water is further purified before undergoing tertiary treatment (Abdelfattah et al. 2023).

Tertiary purification provides advanced treatment, including methods like advanced oxidation, membrane filtration, or disinfection with chlorine or ultraviolet light, to remove any remaining contaminants and ensure the water meets the highest safety standards. Each stage is integral to delivering pure and potable water (Ahmed et al. 2021). Figure 1 gives an overview of the primary, secondary, and tertiary treatment processes. The upcoming session provides a detailed description of various tertiary treatment processes used for drinking water purification.

4 Advanced Methods for Drinking Water Purification

Sophisticated drinking water purification methods have evolved significantly to effectively tackle a broad spectrum of contaminants and ensure that the water remains safe, clean, and suitable for consumption (Kekes et al. 2023). Advanced methods include disinfection, which effectively kills bacteria, viruses, and other pathogens. Filtration is another method which targets chlorine, volatile organic compounds, and unpleasant tastes and odors. Additional procedures, including ion-exchange, adsorption, microalgal processes, and bioelectrochemical systems, improve resource recovery by effectively recycling valuable materials and supporting sustainability (Nawaz and Sengupta 2021; Kekes et al. 2023). Advanced biological processes and advanced oxidation processes are also employed in water purification because of their broad applicability and the significant benefits they offer. These techniques are often integrated into multi-stage systems for enhanced purification, confirming that water intended for drinking meets obstinate safety standards and is free from harmful impurities. The aforementioned advanced methods are discussed in detail below.

4.1 Adsorption

Adsorption is a tertiary treatment process used to further purify water by removing contaminants through the adhesion of molecules to a solid surface, typically activated carbon. In adsorption process, water passes through a bed of adsorbent materials such as activated carbon which attract and hold onto a broad spectrum of impurities, including organic materials, chlorine, heavy metals and many more. The surface area and chemical properties of the adsorbent material influence its adsorption efficiency. This method is particularly useful for removing trace contaminants and improving the odour and taste of water. Adsorption is a key component in achieving finest water for drinking and industrial applications, as it complements other purification methods by targeting specific pollutants that may remain after earlier treatment stages (Worch 2012). Adsorption is reversible, with desorption being the process that releases the adsorbate. Effective adsorbent material requires desorption to free up surface sites for new reactants. The general adsorption mechanism is illustrated in Fig. 2. Extensive research is underway to develop effective adsorbents for water purification.

A variety of adsorbents have been used to eliminate pollutants from water among which low-cost options like biomass-based activated carbon, natural clay and agricultural waste peels are commonly used for water processing and wastewater handling (Bilal et al. 2022). Chitosan is a key material for adsorption due to its amino and hydroxyl groups, which enable it to interact with various pollutants like dyes, metals, and pharmaceuticals (Suyambulingam et al. 2023). Apart from this, emerging materials such as nanomaterials and composites, dendrimers, hydrogel nanocomposites, and boron nitride nanomaterials are studied for their water purification efficiency (Gonzalez-Ortiz et al. 2020; Hussain et al. 2024; Kumar et al. 2023a) Several reports are there regarding the development of carbon dots, carbon nanotubes and nanofibers functioned as adsorbents, owing to their unique properties (Huang et al. 2015b; Bhattacharjee et al. 2023; Elamin et al. 2021). Reports indicate that metal–organic frameworks have also been used for water decontamination. (Chaemchuen et al. 2018). Magnetic spinel ferrite materials are gaining attention for pollutant removal on account of their extensive surface area and magnetic properties, which simplify the separation process. (Salih and Mahmood 2023). Apart from these mentioned materials, graphene and graphene derivatives, graphitic carbon nitride based adsorbents,

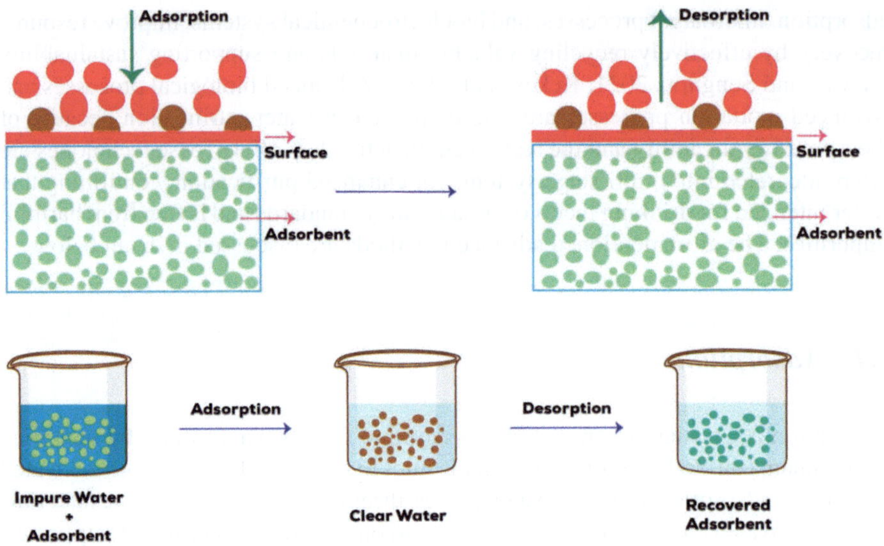

Fig. 2 General adsorption mechanism

metal oxide centred nanocomposites and many more are under study and development for eliminating both organic and inorganic pollutants from surface/drinking water (Jiang et al. 2015; Fronczak 2020; Ghosh et al. 2022; Singh et al. 2018).

4.2 Membrane Distillation

Water distillation is an effective and oldest treatment process that purifies water by removing contaminants through evaporation and condensation (Kocza and Mizsey 2010). In this process, water is heated up to its boiling point, causing pure water to evaporate and leave behind impurities, such as inorganic salts, metals and a range of organic compounds. The steam is then cooled and condensed back into liquid form, resulting in highly purified water. Distillation is particularly useful for removing dissolved solids and microbial contaminants, making it ideal for producing high-quality water from sources that may be heavily polluted or brackish. Apart from this, membrane distillation (MD) is a well-known thermal separation process that leverages a hydrophobic membrane to separate pure water from raw water on account of heating (Camacho et al. 2013). In MD process, the feed solution is heated, causing water vapour to infuse through the membrane while non-volatile solutes and contaminants are retained. The membrane's hydrophobic nature prevents liquid water from passing through, ensuring that only vapour moves across the used MD membrane to the cooler side, where it condenses back to pure water. This method is particularly advantageous for desalination and wastewater treatment because it functions effectively at low temperatures and pressures, which enhances its energy efficiency and makes it ideal for treating high salinity or complex feed solutions (Mohammad Mahdi et al. 2012).

Common polymers for MD membrane materials include PVDF, PTFE, PP, and PE owing to their hydrophobic nature. However, they also face issues like low productivity, high costs, and fouling, which reduce purification efficiency. The membranes must possess high wetting resistance, good permeability and low fouling tendency for effective MD process (Yan et al. 2021). The integration of nano materials into MD membranes has been explored by many researchers to tackle the mentioned three key technical challenges. In membrane distillation, the liquid contacts the membrane's hydrophobic side, preventing liquid flow and forming a liquid/vapor interface (Farid et al. 2024). Different MD systems include:

1. **Direct Contact Membrane Distillation (DCMD)**: In DCMD, two sides of the membrane are directly interacting with the feed and permeate solutions. The water vapour permeates through the membrane and it condenses in a cold infiltrate chamber present inside the module. This configuration is favoured for its simplicity and is ideal for applications where water is the primary feed component (Ameen et al. 2020; Ansari et al. 2021). The stand-alone DCMD process struggles with wetting issues at high feed temperatures. However, coupling DCMD with fractionation or ozonation pretreatment can effectively treat effluents like combined textile waste (Zhang et al. 2021). In a solar-integrated Photovoltaic thermal-DCMD model system reported by Maqbool et al. the intake water temperature for direct contact membrane distillation is powered by photovoltaic thermal output. This integration proves to be both cost-effective and technologically advantageous (Maqbool et al. 2024). DCMD is widely used in industries such as chemicals, food, pharmaceuticals, and environmental protection, with a key application in desalination to produce high-purity water from seawater or brackish sources (Ameen et al. 2020).
2. **Air-Gap Membrane Distillation (AGMD)**: AGMD process incorporates an air gap between the permeation side and the membrane, which aids in removing trace volatile components and reducing heat loss. AGMD avoids the significant reduction in water flux along the module that is experienced with DCMD (Ansari et al. 2022). It was reported that new membrane distillation processes, including material gap, permeate gap, and conductive gap variations, have evolved from AGMD. Combining AGMD systems with various renewable energy sources offers a promising avenue for developing advanced hybrid systems (Shahu and Thombre 2019; Ugarte et al. 2024).
3. **Sweeping Gas Membrane Distillation (SGMD)**: SGMD typically uses an inert gas to sweep vapour from the membrane and is effective for volatile components. In SGMD, an inert gas transports the permeated vapour to the permeate side, where a condenser then converts the vapour back into water. SGMD uses a sweeping gas that enhances permeate flux and mass transfer coefficient (Safi et al. 2020). It was reported that SGMD achieves faster mass transfer rates and has lower conduction heat loss than DCMD (Cao et al. 2024). SGMD offers a

compact and efficient alternative for ammonia removal, potentially replacing air stripping and reducing energy consumption in wastewater treatment plants (Hua et al. 2022).

4. **Vacuum Membrane Distillation (VMD)**: In VMD method vacuum is applied to the permeation side, facilitating the removal of volatile organic components. Vacuum membrane distillation uses a vacuum to evaporate water from brine, with the vapour traveling through the membrane and condensing in a separate condenser. Compared to direct contact membrane distillation, VMD significantly reduces heat loss due to thermal conduction and achieves higher water vapour flux with a high vacuum on the permeate side. Additionally, VMD minimizes temperature polarization since the vapour is quickly removed (Yan et al. 2021; Farid et al. 2024). Recent studies have investigated vacuum membrane distillation (VMD) for desalination using hollow fiber membranes in a shell-side feed configuration (Sun et al. 2014).

Developing dense MD membranes for wastewater treatment faces several challenges, including balancing mechanical integrity with permeate flux, limited long-term studies with real wastewater, stability under harsh conditions, complex fabrication, and the use of harsh chemicals (Prado de Nicolás et al. 2023). Research aims to address these issues and develop MD systems integrated with other technologies to advance the field. Membrane distillation offers high purity and the potential for renewable energy integration, enhancing sustainability. Studies have shown that MD is effective for treating various wastewater types, including desalination brine, textile, radioactive, and oily wastewater (Yan et al. 2021).

4.3 Membrane Filtration

Membrane filtration is an advanced water purification technique that employs semi-permeable membranes to effectively remove impurities from water. This process involves forcing water through a membrane with very fine pores that permit clean water to pass across whereas trapping contaminants on one side (Cevallos-Mendoza et al. 2022). There are several types of membrane filtration, each with specific applications and pore sizes. Figure 3 illustrates the primary membrane filtration techniques, categorized by pore size and transmembrane pressure including microfiltration, ultrafiltration, nanofiltration, and reverse osmosis. Additionally, this section covers forward osmosis, a technique that relies on a concentration gradient rather than transmembrane pressure (Aziz et al. 2024; Shehata et al. 2023).

1. **Microfiltration (MF)**: This is the unpolished form of membrane filtration, with pore sizes extending from 0.1 to 10 μm and typically uses a transmembrane pressure ranging from 0.1 to 10 bar. Microfiltration is effective at removing larger particles such as sediments, bacteria, and some protozoa. It is commonly used as a pre-treatment technique to protect more sensitive filtration stages or to improve the clarity of water (Hussein et al. 2023; Fatima et al. 2024). The MF

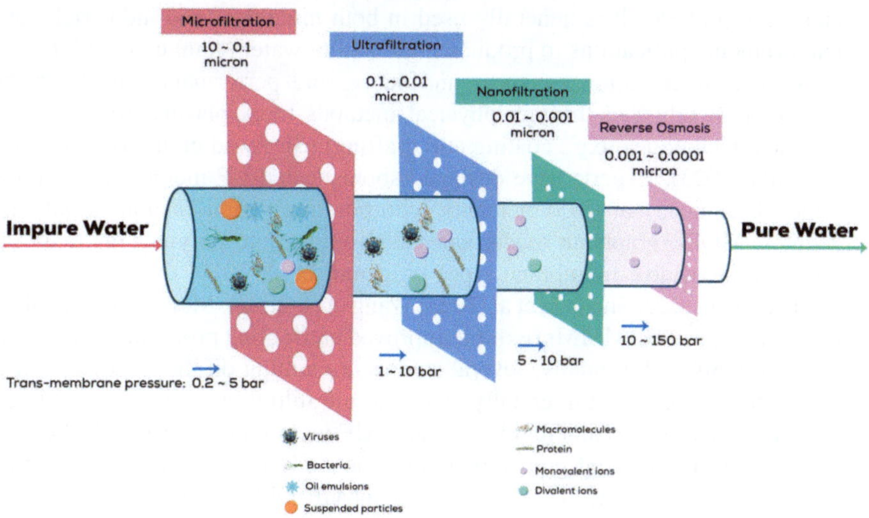

Fig. 3 The primary membrane filtration techniques

membranes are mainly prepared of polymers like polypropylene, polytetrafluoroethylene, polycarbonate, polyvinylidene fluoride, polyimide and polyethylene (Alsoy Altinkaya 2024; Gul et al. 2021). Membrane fouling is a major challenge in membrane separation processes (Gul et al. 2021). Recently, there has been rising concern in modifying membranes to resist fouling and bacterial growth. Addition of silver oxide (AgO) on polyamide MF membranes, integration of cellulose nano whiskers into PAN nanofibers were done to enhance their permeability, antifouling property and antibacterial properties (Kacprzyńska-Gołacka et al. 2020; Ma et al. 2012). Furthermore, creating composite conductive MF membranes by incorporating stainless steel mesh and applying an external electric field to control fouling was reported (Huang et al. 2015a). Zhao et al. found that using of granular activated carbon (1.0 g/L for one hour) improved microfiltration flux by 15–20% and reduced irreversible fouling from algal organic matter (Zhao et al. 2023). Researchers have investigated ceramic membranes from affordable materials like kaolin and natural zeolite, which are especially effective for treating oily wastewater due to their superior hydrophilic properties (Hubadillah et al. 2018; Saja et al. 2018; Hubadillah et al. 2016). Jafari et al. developed two-layer tubular ceramic membranes specifically for treating oily wastewater (Jafari et al. 2020). Studies suggests that coupling of microfiltration with other technologies can effectively decontaminate water from various pollutants (Wang et al. 2020).

2. **Ultrafiltration (UF)**: Ultrafiltration membranes have reduced pore sizes compared to that of MF membranes, normally ranges between 0.01 and 0.1 μm. UF can remove smaller particles, including viruses, larger organic molecules,

and some proteins. It is generally used in both municipal and industrial water management applications to produce high-quality water (Aani et al. 2020).

Surface modification enhances antifouling and performance in membrane engineering, achieved through physical methods (e.g., plasma irradiation) or chemical techniques (e.g., coating and grafting) (Sh Awad et al. 2021; Ibrahim and Hilal 2023). Experts have designed sophisticated UF membranes to boost removal efficiency of toxic elements, incorporating various nanomaterials like zeolites, MOFs, graphene oxide, and metal oxides into polymer matrices and fabricate innovative membrane types (Algamdi et al. 2019; Chan et al. 2024; Hackett et al. 2022; Shoshaa et al. 2023; Wang et al. 2022). Many reported mixed matrix membranes (MMMs) exhibit improved antifouling properties and greater hydrophilicity. Additionally, several studies are aiming on the improvement in the performance of commercially available ultrafiltration membranes. Murcia et al. modified the commercially available UF membranes (GR60PP, GR70PP, GR80PP) using reduced grapheme oxide The study indicate that for enhanced performance with reduced graphene oxide, the GR60PP membrane is best, while the GR80PP excels with the native membrane. A customised PVDF ultrafiltration membrane, blended with a graphene oxide-polyvinyl alcohol-sodium alginate (GO-PVA-NaAlg) hydrogel (HG) and polyvinylpyrrolidone (PVP), has been developed, showing enhanced filtration performance, hydrophilicity, porosity, mean pore size, and surface roughness (Ghobadi and Hemmati 2023).

3. **Nanofiltration (NF)**: Nanofiltration membranes have even smaller pores, ranging from 1 to 10 nm and a transmembrane pressure of 5–10 bars. NF is proficient in removing divalent ions (calcium, magnesium etc.), small contaminants as well as organic compounds. This makes it useful for softening water and removing specific contaminants that are not addressed by microfiltration or ultrafiltration (Charcosset 2016). NF softens water by removing calcium and magnesium ions without adding sodium, unlike ion exchange units (Maroufi and Hajilary 2023).

   Organic NF membranes, prepared using polymers such as polysulfones, cellulose acetate, and polyvinylidene fluoride, struggle with high selectivity due to their moderately large inter-chain free volume (Aburideh et al. 2021; Islam et al. 2023; Ounifi et al. 2020; Ziemann et al. 2022). In contrast, inorganic NF membranes use materials like metal oxides, ceramics and zeolites, but face challenges in creating thin, defect-free films due to their rigid pore structures (Fard et al. 2018; Gordon et al. 2012). Organic-inorganic hybrids combine polymers with inorganic components like metals, metal oxides, or carbon. Thin film composite membranes created by applying a dense active layer onto a porous support are the most widely used type of membranes for nanofiltration. Layer-by-layer (LBL) assembly method is crucial for crafting NF membranes with adjustable charge, pore size, and high water permeability. These membranes also offer excellent anti-fouling properties, enables them ideal selection for resource recovery and seawater desalination (Chen et al. 2023). Recently, Wang et al. fabricated a nano filtration membrane via the inkjet printing of polyethyleneimine and grapheme oxide layers on a polyketone substrate. It is then cross-linked with glutaraldehyde which gave long term filtration stability (Wang et al. 2023).

Advances in NF technology include improvements through interfacial polymerization and nanoparticle incorporation. Various techniques are being utilized and under development to boost membrane selectivity and combat fouling issues of NF membranes (Choi et al. 2024; Galanakis et al. 2012)

4. **Reverse Osmosis (RO)**: Reverse osmosis employs membranes with extremely fine pores, typically less than 1 nm and transmembrane pressure of 10–150 bars. RO is highly effective at removing a broad spectrum of contaminants and widely used for producing potable water from seawater or brackish water. RO treatment is typically used in water treatment practise that requires high-purity water production (Charcosset 2016; Zhai et al. 2022). Generally, RO permeate needs post-treatment for remineralization as well as reconditioning to comply with drinking water regulations and enhance taste (Zhai et al. 2022). Commercially available RO membranes are primarily made from cellulose acetate and aromatic polyamides. Additional polymers like PVDF (polyvinylidene fluoride), PTFE (polytetrafluoroethylene), and polysulfone, as well as their surface-modified, composite, or blended forms, are indeed being developed and researched for reverse osmosis (RO) treatment. These materials are explored for their potential to enhance membrane performance, durability, and efficiency in water purification. Similar to NF membranes, various organic, inorganic, and hybrid membranes are actively being developed and researched for RO processes.

   Recent study revealed that combining Riverbank Filtration (RBF) with reverse osmosis in a one-step RO process significantly reduces the energy and chemical requirements of the water treatment process. RBF acts as a natural pre-treatment stage that successfully eradicates a range of toxins from river water before it undergoes RO filtration. This pre-treatment can significantly lower the load on the RO system, resulting in reduced energy consumption during the RO process and a decrease in the need for chemical treatments. This makes the overall water treatment process more sustainable as well as improves the efficiency and durability of the RO system, leading to better water quality with a smaller environmental footprint (Zhai et al. 2022).

5. **Forward Osmosis**: Forward osmosis (FO) is a water purification procedure that utilizes the natural osmotic potential between two solutions to drive water through a semi-permeable membrane, from a dilute solution to a more concentrated solution (Korenak et al. 2017). Unlike reverse osmosis, which requires external pressure, FO leverages the osmotic gradient created by a high-concentration draw solution to extract water (Wang and Liu 2021). This method is energy-efficient and less prone to fouling, making it advantageous for applications including sewage treatment, desalination, and food processing. The draw solution is then regenerated to recover the extracted water, further enhancing the process's sustainability. The FO process results in a diluted draw solution rather than clean water. To get clean water, FO usually requires a secondary step like RO to recover and purify the solution (Mohammadifakhr et al. 2020).

Overall, membrane filtration is a versatile and powerful tool in water treatment, used across various sectors to ensure clean and safe water for consumption, industrial processes, and environmental protection. However, it also requires careful management of membrane fouling, which can occur when contaminants accumulate on the membrane surface, reducing its performance. Regular cleaning and maintenance are essential to ensure finest operation and endurance of the membranes.

4.4 Ion Exchange

Ion exchange process is a water reclamation method that purifies water by eradicating dissolved ions and contaminants. Water passes through a resin bed that exchanges undesirable ions including calcium, magnesium, and heavy metals, with benign ions like sodium or hydrogen. This process reduces water hardness, removes specific contaminants, and enhances water quality, making it essential for meeting high standards in industrial, pharmaceutical, and high-purity drinking water applications (Alvino et al. 1980). The basic principle involves transferring ions between a liquid and a resin with charged sites. Cation exchange replaces positively charged ions with other cations bound to the resin, while anion exchange replaces negatively charged ions (e.g., chloride and sulfate) with other anions (Gregory and Dhond 1972).

In this method, water with undesirable ions flows through a resin column that swaps these ions for more benign ones like sodium. As the resin becomes saturated with unwanted ions, its regeneration will carry out using a concentrated solution, such as sodium chloride, to restore its capacity. This results in softer, cleaner water with fewer impurities. Ion exchange is a versatile method for water softening, deionization, and contaminant removal, though it's cost and environmental impact must be considered. Extensive global research is exploring the potential effects of ion exchange resins. Combining ion exchange with other water treatment methods can yield high-quality purified water. For instance, Martins et al. demonstrated that coupling electro oxidation with ion exchange resin effectively removes persistent micropollutants (Martins et al. 2020). Recent advancements in water treatment include in-situ ion exchange electrocatalysis combined with biological processes, providing a extremely efficient substitute to traditional electroplating methods (Feng et al. 2019). This method is valuable for separation in various sectors and offers high recovery and low contamination, enabling complex separations like trace metal enrichment and ultrapure water production.

4.5 Electrodeionization

Electrodeionization (EDI) is an advanced tertiary treatment process that combines ion exchange with electrical current to purify water (Santos et al. 2024). Among the three stages of EDI, first phase involves the ion exchange amongst the resin and solution. Water flows through ion-exchange membranes and resins within an electrochemical cell. Second stage encompasses the application of an electric field to move ions through the resin which drives the migration of ions through the membranes and separating them from the water. The final stage includes the electrolytic renewal of the resin to produce ultrapure water (Wei et al. 2024).

Ion exchange resins (IERs) are solid materials with ionic groups that swap ions in solvents. Cation exchange resins (CERs) have structures like R–C, where R is a polymeric resin. These resins can contain mobile counter cations (e.g., H^+, Na^+) and include strong acids (e.g., sulfonic–SO_3H) or weak acids (e.g., phenolic–C_6H_5OH). Anion exchange resins (AERs) feature R–A bonding, with R as a polymeric resin that carries tightly bound cations, resulting in a net positive charge stabilized by counter anions. Strongly basic AERs include quaternary ammonium bases (NR_3OH) and hydroxides (OH), while weakly basic AERs have primary and secondary amino groups (–NH_2 and –NHR). Common anode materials for EDI include platinum, platinized titanium grids, ruthenium dioxide-coated titanium, tantalum oxide, and iridium oxide-coated titanium, while stainless steel and platinized titanium grids are typical for cathodes (Khan et al. 2023). Shen et al. compared graphene composite electrodes with conventional platinum-plated titanium ones in membrane-free EDI systems. They found that graphene composites offer a larger surface area and potentially lower oxygen evolution potential, leading to more effective copper removal from wastewater (Shen et al. 2022). Recent advancements in ion exchange resins have significantly improved water treatment processes. Innovative materials such as phosphonic acid-functionalized resins target and remove heavy metals efficiently. Nanocomposites, including graphene-enhanced resins, offer superior capacity and performance. Sustainable bio-based resins, derived from chitosan, address heavy metal contamination with eco-friendly solutions. High-performance macro reticular resins facilitate rapid ion exchange, and specialized resins effectively target contaminants like pharmaceuticals. Additionally, regenerable resins with high stability are designed to endure harsh conditions and frequent use that optimizes water treatment efficiency.

EDI is highly effective for producing ultrapure water in critical applications like semiconductor manufacturing and pharmaceuticals, where conventional methods fall short. It offers efficiency, continuous operation, and sustainability without needing chemical regenerants. The process reduces energy use, polarization, and pH issues, and works well with low-concentration flows in both batch and continuous modes. However, its performance is dependent on feed water quality (Khan et al. 2023).

4.6 Disinfection

Water disinfection, as a tertiary treatment process implemented immediately before discharge in wastewater treatment plants, guarantees that the treated water is safe for ingestion and devoid of harmful microorganisms (Kumar et al. 2023b). This stage involves advanced techniques such as chlorination, ultraviolet (UV) light, or ozone treatment to eliminate any remaining pathogens that might not have been removed during earlier purification stages (Kong et al. 2021; González et al. 2023) Chlorination involves adding gaseous chlorine or chlorine compounds (chloramine, chlorine dioxide, sodium hypochlorite etc.) to kill bacteria, viruses, and protozoa. Chlorine works by altering the chemical structure of bacterial enzymes, effectively inactivating them and preventing bacterial growth and survival. Chlorine dioxide, due to its potent oxidizing power, is equally or more effective than chlorine in eliminating bacteria by disrupting their enzymatic systems or protein synthesis. UV disinfection uses ultraviolet light to damage the DNA of microorganisms, rendering them inactive. The disinfection efficiency can be impacted by suspended particles, their sizes, and microorganism concentrations in the case of UV induced disinfection method. Ozone treatment, on the other hand, introduces ozone gas into the water to effectively oxidize and neutralize contaminants. Ozone is a highly reactive oxidizing agent with disinfection efficiency that exceeds that of chlorine (Kong et al. 2021). Peracetic acid is a viable substitute to waste water chlorination, achieving disinfection through the formation of acetic acid (Kitis 2004). Either the active oxygen generation or reactive hydroxyl radicals formation is responsible to break the cell wall membrane. Another study revealed that combining peracetic acid with UV radiation resulted in highly effective disinfection (Gori and Caretti 2008).

Although disinfection is a vital water treatment method, it also possess some drawbacks. While UV radiation damages cellular nucleic acids and other components, microorganisms can sometimes retain metabolic functions and develop DNA repair mechanisms over time (González et al. 2023). The disinfection processes can pose health risks due to the formation of disinfection by-products (DBPs) like trihalomethanes, which are associated with major health issues (Li and Mitch 2018). Du et al. examined different disinfection techniques for drinking water and identified health hazards associated with exposure to disinfection by-products through various pathways. Their analysis also indicated that disinfection using chlorine dioxide and sodium hypochlorite poses lower health risks compared to other methods (Du et al. 2021). Research has shown that optimizing disinfectant combinations can simultaneously manage both pathogens and disinfection by-products. For instance, using ozone for primary disinfection alongside chloramines to maintain residuals in the distribution system effectively inactivates pathogens while reducing the generation of regulated trihalomethanes and haloacetic acids (Krasner et al. 2006). Further research is focussed on designing methods that can provide an additional layer of protection, ensuring that the drinking water encounters high safety criterions and is consumable for individuals.

4.7 Advanced Oxidation Processes

Advanced Oxidation Processes (AOPs) are exceptionally effective in drinking water treatment, utilizing hydroxyl radicals to eliminate a broad spectrum of contaminants. These sophisticated techniques achieve high water quality, especially where conventional methods are insufficient. AOPs generate hydroxyl radicals through various chemical and physical processes, which effectively degrade both organic and inorganic pollutants. However, their implementation demands careful attention to costs, operational complexity, and potential by-products to ensure optimal performance and safety. The leading methods for AOP include ozone based AOP, UV-based AOP, together with Fenton and Photo-Fenton methods (Mishra et al. 2017).

1. **Ozone-Based AOPs**
   Various ozone combined processes, such as O_3/H_2O_2, O_3/catalysts, and O_3/filtration, have been well-documented for the exclusion of trace organic impurities (Liu et al. 2021). Ozone/Hydrogen Peroxide (O_3/H_2O_2) process involves the simultaneous use of ozone and H_2O_2) to generate hydroxyl radicals. Ozone reacts with hydrogen peroxide to generate these radicals, which then oxidize the contaminants (Mishra et al. 2017). In the O_3/catalysts process, reported catalysts include a range of materials such as metal oxides, carbon nanotubes and MOFs among others. Solid catalysts, particularly metal oxides, enhance catalytic ozonation by facilitating electron transfer and reducing activation energy due to their active d-electronic structures (Zhang et al. 2020). Similarly, ozone reactor combined with ultrafiltration membrane gave complete removal of viruses and bacteria.
2. **UV-Based AOPs**
   UV based AOP mainly includes UV/Ozone (UV/O_3) and UV/Hydrogen peroxide (UV/H_2O_2) process. In UV/ozone process, UV light is used to decompose ozone, producing hydroxyl radicals that further react with organic pollutants. Whereas in (UV/H_2O_2) process, ultraviolet light is used to decompose hydrogen peroxide that in turn generates hydroxyl radicals. Studies on UV-based AOPs indicate that these techniques can inform the engineering design and operation of cutting-edge water treatment systems (Li et al. 2023).
3. **Fenton and Fenton-Like Reactions**
   Fenton reaction involves the reaction of H_2O_2 with ferrous iron (Fe^{2+}) to form hydroxyl radicals. The process is operative in treating a variety of organic contaminants and is often used in conjunction with other treatments. Fenton-like reactions use different metal catalysts other than ferrous ion to generate hydroxyl radicals (Suhan et al. 2021). Recent research indicates that combining Fenton and Fenton-like processes with photoirradiation, sonolysis, electrolysis, and ultrasonication enhances contaminant removal from wastewater (Xiao et al. 2024). Disinfection process can also be conducted using Fenton like processes (Gosselin et al. 2013).

AOPs enhance water quality by removing trace organic contaminants and disinfection by-products often missed by traditional methods. In advanced treatment systems, AOPs ensure that reclaimed water complies with stringent safety standards. They also proliferates the biodegradability of industrial run-off before biological treatment. Combining AOPs with biological methods and pre-treatments like active sludge can address the limitations of traditional methods, supporting safe wastewater reuse while balancing performance, cost, and safety.

4.8 New Frontiers in AOPs Towards Water Purification

The practical and economic aspects of AOPs in water treatment include their effectiveness, cost, and scalability. Although AOPs are efficient in degrading contaminants, they require specialized equipment and chemicals, which can be expensive. Field-scale case studies show that despite high initial and operational costs, the benefits of improved water quality and regulatory compliance often justify the investment. Successful applications in industries like textiles and pharmaceuticals demonstrate that AOPs, when well-integrated into existing systems, can effectively address complex contamination challenges.

Recent advancements showcase several key developments. Techniques like improved Photo-Fenton processes are boosting hydroxyl radical production, which enhances the breakdown of persistent pollutants such as pesticides and pharmaceuticals. For instance, Spain's Barcelona Water Treatment Plant has successfully employed this technology to manage complex contaminants (Rodríguez-Villanueva and Sauri 2021). The Pittsburgh Water and Sewer Authority integrates AOPs with activated sludge systems to better remove pharmaceutical residues and emerging contaminants, leading to improved water quality. Innovations are making AOPs more cost-effective and energy-efficient. Real-world implementations of AOPs, such as the Singapore NE Water Project, highlight their effectiveness in large-scale operations. This project uses AOPs in its advanced treatment processes to produce high-quality reclaimed water that meets stringent safety standards (Ghernaout et al. 2019). AOPs are effective at removing emerging contaminants that conventional treatments often miss. The Oslo Water Treatment Plant in Norway uses AOPs to address contaminants like personal care products, ensuring high safety standards for treated water (Venkatesh and Brattebø 2011). The AOPs in Modular Units project in Germany illustrates how compact, modular systems can be integrated into existing infrastructure to efficiently handle specific contamination issues. Table 2 illustrates more examples where AOPs are implemented/under process in water treatment plants worldwide (Hübner et al. 2024).

In summary, advancements in AOPs are setting new benchmarks for drinking water purification, enhancing contaminant removal, reducing costs, and improving energy efficiency, leading to safer and cleaner drinking water worldwide.

5 Monitoring System

Water quality monitoring is vital for ensuring high-grade drinking water for every individual. Traditional lab-based methods are time-consuming and impractical for remote locations. Advances like portable testing kits and microwave techniques now offer more efficient solutions. (Grzegorczyk et al. 2022). IoT (Internet of Things) sensors offer high-accuracy remote water monitoring, achieving 95% accuracy in pH, turbidity, TDS, and temperature, versus 85% with traditional methods (de Camargo

Table 2 Illustrates more examples where AOPs are implemented/under process in water treatment plants worldwide

Country	AOP technology used	Efficiency
Spain	Electrochemical AOP	Treat water contaminated with pharmaceuticals and personal care products
Japan	Visible-light-active photocatalysis using nitrogen-doped TiO_2	Treat water contaminated with pesticides
Netherlands	Hybrid AOP system that combines ozone and UV light	Treat micropollutants and pharmaceuticals
U.S	Sonochemical reactor	Remove organic contaminants and pathogens
France	Photo-Fenton process	Remove pesticides and organic pollutants
Australia	Cold plasma reactor	Treat drinking water affected by industrial contaminants
Canada	AOPs combined with biofilm reactors	Treat water with high levels of organic and microbial contaminants
South Korea	Nanomaterial-enhanced photocatalytic system using graphene oxide	Degrade emerging contaminants
India	Solar-powered photocatalytic system	Treat groundwater contaminated with heavy metals and organic pollutants

et al. 2023). AI techniques like machine learning and deep learning further boost accuracy, though it depends on training data, metric complexity, and monitoring frequency. Owing to the widespread application, Geographic Information Systems (GIS) are employed for spatial data analysis and managing water resources (Fakheri 2024).

In developed countries, drinking water quality monitoring systems are highly advanced, using real-time sensors and automated sampling to track parameters like pH, turbidity, and contaminants. Data is analyzed centrally with advanced software for immediate alerts and responses, ensuring high water safety standards. In developing countries, while systems often rely on basic sensors and manual testing due to limited resources, innovations such as portable testing kits and mobile technology are improving monitoring capabilities. Despite these advancements, continued support and infrastructure development are essential for effective and sustainable water quality management (Zainurin et al. 2022).

6 Conclusion

Ensuring safe drinking water involves understanding water quality, employing effective treatment technologies, and utilizing robust monitoring systems. This chapter covered essential water quality parameters, various treatment methods from traditional to advanced, and the role of both sophisticated and practical monitoring approaches. Various treatment technologies, from traditional methods to advanced techniques like reverse osmosis and membrane filtration, advanced oxidation process were also discussed. The real-world application of AOPs, combined with other technologies, has demonstrated their effectiveness in drinking water purification systems worldwide. This practical integration highlights the significant potential of advanced treatments in improving water quality. Accordingly integrating innovative technologies with strategic monitoring is crucial for improving water safety and sustainability, ultimately protecting public health and securing water resources for the future.

References

Abdelfattah A, Ramadan H, Elsamahy T et al (2023) Multifaced features and sustainability of using pure oxygen in biological wastewater treatment: a review. J Water Process Eng 53. https://doi.org/10.1016/j.jwpe.2023.103883

Aburideh H, Tigrine Z, Zioui D et al (2021) Cellulose chemistry and technology. Cellul Chem Technol 55(5–6):697–704. https://doi.org/10.35812/cellulosechemtechnol.2021.55.59

Adeniyi AO (2020) A comparative review of the traditional and modern methods of water treatment. J Adv Biol Biotechnol. https://doi.org/10.9734/jabb/2020/v23i1130191

Ahansazan B, Afrashteh H, Ahansazan N et al (2014) Activated sludge process overview. Int J Environ Sci Dev 5(1). https://www.ijesd.org/papers/455-P20003.pdf

Ahmed SF, Mofijur M, Nuzhat S et al (2021) Recent developments in physical, biological, chemical, and hybrid treatment techniques for removing emerging contaminants from wastewater. J Hazard Mater 416. https://doi.org/10.1016/j.jhazmat.2021.125912

Akhtar N, Syakir Ishak MI, Bhawani SA et al (2021) Various natural and anthropogenic factors responsible for water quality degradation: a review. Water (Switzerland) 13(19). https://doi.org/10.3390/w13192660

Albou EM, Abdellaoui M, Abdaoui A et al (2024) Agricultural practices and their impact on aquatic ecosystems—a mini-review. Ecol Eng Environ Technol 25(1). https://doi.org/10.12912/27197050/175652

al Aani S, MustafaTN, Hilal N (2020) Ultrafiltration membranes for wastewater and water process engineering: a comprehensive statistical review over the past decade. J Water Process Eng 35. https://doi.org/10.1016/j.jwpe.2020.101241

Algamdi MS, Alsohaimi IH, Lawler J et al (2019) Fabrication of graphene oxide incorporated polyethersulfone hybrid ultrafiltration membranes for humic acid removal. Sep Purif Technol 223. https://doi.org/10.1016/j.seppur.2019.04.057

Alsoy Altinkaya S (2024) A review on microfiltration membranes: fabrication, physical morphology, and fouling characterization techniques. Front Membr Sci Technol 3. https://doi.org/10.3389/frmst.2024.1426145

Alvino WM (1980) Stability of ion-exchange resins. 1. Resistance to osmotic shock. Ind Eng Chem Product Res Dev 19(2):271–276. https://doi.org/10.1016/0167-6989(86)90008-5

Ansari A, Galogahi FM, Thiel Dv et al (2022) Downstream variations of air-gap membrane distillation and comparative study with direct contact membrane distillation: a modelling approach. Desalination 526. https://doi.org/10.1016/j.desal.2021.115539

Ansari A, Kavousi S, Helfer F et al (2021) An improved modelling approach for the comprehensive study of direct contact membrane distillation. Membranes 11(5). https://doi.org/10.3390/membranes11050308

Aziz S, Mazhar AR, Ubaid A et al (2024) A comprehensive review of membrane-based water filtration techniques. Appl Water Sci 14(8). Springer Science and Business Media Deutschland GmbH. https://doi.org/10.1007/s13201-024-02226-y

Badawi AK, Salama RS, Mostafa MMM (2023) Natural-based coagulants/flocculants as sustainable market-valued products for industrial wastewater treatment: a review of recent developments. RSC Adv 13(28). https://doi.org/10.1039/d3ra01999c

Bashir I, Lone FA, Bhat RA et al (2020) Concerns and threats of contamination on aquatic ecosystems. In: Bioremediation and biotechnology: sustainable approaches to pollution degradation. https://doi.org/10.1007/978-3-030-35691-0_1

Bhattacharjee T, Konwar A, Boruah JS et al (2023) A sustainable approach for heavy metal remediation from water using carbon dot based composites: a review. J Hazard Mater Adv 10. https://doi.org/10.1016/j.hazadv.2023.100295

Bilal M, Ihsanullah I, Younas M et al (2022) Recent advances in applications of low-cost adsorbents for the removal of heavy metals from water: a critical review. Sep Purif Technol 278. https://doi.org/10.1016/j.seppur.2021.119510

Bruno P, Campo R, Giustra MG et al (2020) Bench scale continuous coagulation-flocculation of saline industrial wastewater contaminated by hydrocarbons. J Water Process Eng 34. https://doi.org/10.1016/j.jwpe.2020.101156

Butler E, Hung YT, Suleiman Al Ahmad M et al (2017) Oxidation pond for municipal wastewater treatment. Appl Water Sci 7(1). https://doi.org/10.1007/s13201-015-0285-z

Camacho LM, Dumée L, Zhang J, Li J et al (2013) Advances in membrane distillation for water desalination and purification applications. Water (Switzerland) 5(1). https://doi.org/10.3390/w5010094

Cao Y, Taghvaie Nakhjiri A, Ghadiri M (2024) Numerical evaluation of sweeping gas membrane distillation for desalination of water towards water sustainability and environmental protection. Sci Rep 14(1). https://doi.org/10.1038/s41598-024-54061-5

Cevallos-Mendoza J, Amorim CG, Rodríguez-Díaz JM et al (2022). Removal of contaminants from water by membrane filtration: a review. Membranes 12(6). https://doi.org/10.3390/membranes12060570

Chaemchuen S, Wang JC, Gilani AG et al (2018) Metal-organic frameworks applied for water purification. Resour Eff Technol 1:1–16. https://doi.org/10.18799/24056537/2018/1/177

Chan MK, Tan SJ, Yeow AT et al (2024) Zeolite-based poly(vinylidene fluoride) ultrafiltration membrane: characterization and molecular weight cut-off estimation with support vector regression modelling. Membranes 14(4). https://doi.org/10.3390/membranes14040091

Charcosset C (2016) Ultrafiltration, microfiltration, nanofiltration and reverse osmosis in integrated membrane processes. In: Basile A, Charcosset C (eds) Integrated membrane processes, pp 1–24. https://doi.org/10.1002/9781118739167.ch1

Chen J, Xu S, Tang CY et al (2023) Stability of layer-by-layer nanofiltration membranes in highly saline streams. Desalination 555. https://doi.org/10.1016/j.desal.2023.116520

Chen L, Bi T, Lizundia E et al (2024) Biomass waste-assisted micro(nano)plastics capture, utilization, and storage for sustainable water remediation. Innovation 5(4). https://doi.org/10.1016/j.xinn.2024.100655

Choi J, Choi K, Kwon YS et al (2024). Ultrathin organosiloxane membrane for precision organic solvent nanofiltration. Nat Commun 15(1). https://doi.org/10.1038/s41467-024-47115-9

de Camargo ET, Spanhol FA, Slongo JS et al (2023) Low-cost water quality sensors for IoT: a systematic review. Sensors 23(9). https://doi.org/10.3390/s23094424

Derakhshan A, Rezaei Kalantari R, Farzadkia M et al (2023) The effect of biological treatment methods on the concentration of carbonaceous pollutants in the slaughterhouse wastewater: a systematic review. Case Stud Chem Environ Eng 8. https://doi.org/10.1016/j.cscee.2023.100451

Dinka MO (2018) Safe drinking water: concepts, benefits, principles and standards. In: Water challenges of an urbanizing world. https://doi.org/10.5772/intechopen.71352

Duttagupta S, Mukherjee A, Bhattacharya A et al (2020) Wide exposure of persistent organic pollutants (PoPs) in natural waters and sediments of the densely populated Western Bengal basin, India. Sci Total Environ 717. https://doi.org/10.1016/j.scitotenv.2020.137187

Du Y, Liang Z, Ban J et al (2021) Cumulative health risk assessment of disinfection by-products in drinking water by different disinfection methods in typical regions of China. Sci Total Environ 770:144662. https://doi.org/10.1016/j.scitotenv.2020.144662

Elamin MR, Elzupir AO, Abdulkhair BY (2021) Synthesis and characterization of functionalized carbon nanofibers for efficient removal of highly water-soluble dextromethorphan and guaifenesin from environmental water samples. Environ Nanotechnol Monit Manag 15. https://doi.org/10.1016/j.enmm.2020.100397

El-taweel RM, Mohamed N, Alrefaey KA et al (2023) A review of coagulation explaining its definition, mechanism, coagulant types, and optimization models; RSM, and ANN. Current Res Green Sustain Chem 6. https://doi.org/10.1016/j.crgsc.2023.100358

Fakheri S (2024) IoT-based river water quality monitoring. https://doi.org/10.22105/cand.2023.166518

Farid MU, Kharraz JA, Sun J et al (2024) Advancements in nanoenabled membrane distillation for a sustainable water-energy-environment nexus. Adv Mater 36(17). https://doi.org/10.1002/adma.202307950

Fard AK, McKay G, Buekenhoudt A et al (2018) Inorganic membranes: preparation and application for water treatment and desalination. Materials 11(1). https://doi.org/10.3390/ma11010074

Fatima F, Fatima S, Du H et al (2024) An evaluation of microfiltration and ultrafiltration pretreatment on the performance of reverse osmosis for recycling poultry slaughterhouse wastewater. Separations 11(4). https://doi.org/10.3390/separations11040115

Feckler A, Wolfram J, Schulz R et al (2023) Reducing pollution to levels not harming biodiversity and ecosystem functions: a perspective on the post-2020 global biodiversity framework. Current Opin Environ Sci Health 35. https://doi.org/10.1016/j.coesh.2023.100495

Feng Y, Yang S, Xia L et al (2019) In-situ ion exchange electrocatalysis biological coupling (i-IEEBC) for simultaneously enhanced degradation of organic pollutants and heavy metals in electroplating wastewater. J Hazard Mater 364. https://doi.org/10.1016/j.jhazmat.2018.10.068

Fronczak M (2020) Adsorption performance of graphitic carbon nitride-based materials: current state of the art. J Environ Chem Eng 8(5). https://doi.org/10.1016/j.jece.2020.104411

Galanakis CM, Fountoulis G, Gekas V (2012) Nanofiltration of brackish groundwater by using a polypiperazine membrane. Desalination 286. https://doi.org/10.1016/j.desal.2011.11.035

Ghernaout D, Elboughdiri N, Alghamdi A (2019) Direct potable reuse: the Singapore NE water project as a role model. Open Access Libr J 6. https://doi.org/10.4236/oalib.1105980

Ghobadi Moghadam A, Hemmati A (2023) Improved water purification by PVDF ultrafiltration membrane modified with GO-PVA-NaAlg hydrogel. Sci Rep 13(1). https://doi.org/10.1038/s41598-023-35027-5

Ghosh N, Das S, Biswas G, Haldar PK (2022) Review on some metal oxide nanoparticles as effective adsorbent in wastewater treatment. Water Sci Technol 85(12):3370–3395. IWA Publishing. https://doi.org/10.2166/wst.2022.153

Gonzalez-Ortiz D, Salameh C, Bechelany M et al (2020) Nanostructured boron nitride–based materials: synthesis and applications. Mater Today Adv 8. https://doi.org/10.1016/j.mtadv.2020.100107

González Y, Gómez G, Moeller-Chávez GE, Vidal G (2023) UV disinfection systems for wastewater treatment: emphasis on reactivation of microorganisms. Sustainability (Switzerland) 15(14). Multidisciplinary Digital Publishing Institute. https://doi.org/10.3390/su151411262

Gori R, Caretti C (2008) Experimental study on municipal and industrial reclaimed wastewater refinement for agricultural reuse. Water Sci Technol 58(1):217–223. https://doi.org/10.2166/wst.2008.651

Gordon N, Lin X, Lee E et al (2012) Experimental investigation of inorganic nanofiltration membranes prepared by atomic layer deposition and sol-gel methods for saltless water softening. Int J Environ Technol Manag 16(1–2). https://doi.org/10.1504/IJETM.2013.050684

Gregory J, Dhond RV (1972) Wastewater treatment by ion exchange. Water Res 6(6). https://doi.org/10.1016/0043-1354(72)90183-2

Grzegorczyk M, Pogorzelski S, Rochowski P (2022) Towards a novel class of photoacoustics-based water contamination sensors. J Environ Chem Eng 10(3). https://doi.org/10.1016/j.jece.2022.107983

Gul A, Hruza J, Yalcinkaya F (2021) Fouling and chemical cleaning of microfiltration membranes: a mini-review. Polymers 13(6). https://doi.org/10.3390/polym13060846

Gosselin F, Madeira LM, Juhna T, Block JC (2013) Drinking water and biofilm disinfection by Fenton-like reaction. Water Res 47(15):5631–5638. https://doi.org/10.1016/j.watres.2013.06.036

Hackett C, Abolhassani M, Greenlee LF et al (2022) Ultrafiltration membranes functionalized with copper oxide and zwitterions for fouling resistance. Membranes 12(5). https://doi.org/10.3390/membranes12050544

Hua J, Straub AP, Karanikola V (2022) Ammonia recovery with sweeping gas membrane distillation: energy and removal efficiency analysis. ACS ES&T Eng. https://doi.org/10.1021/acsestengg.1c00294

Huang J, Wang Z, Zhang J et al (2015a) A novel composite conductive microfiltration membrane and its anti-fouling performance with an external electric field in membrane bioreactors. Sci Rep 5. https://doi.org/10.1038/srep09268

Huang ZN, Wang XL, Yang DS (2015) Adsorption of Cr(VI) in wastewater using magnetic multiwall carbon nanotubes. Water Sci Eng 8(3), https://doi.org/10.1016/j.wse.2015.01.009

Hubadillah SK, Harun Z, Othman MHD et al (2016) Effect of kaolin particle size and loading on the characteristics of kaolin ceramic support prepared via phase inversion technique. J Asian Ceram Soc 4(2). https://doi.org/10.1016/j.jascer.2016.02.002

Hubadillah SK, Othman MHD, Matsuura T et al (2018) Fabrications and applications of low cost ceramic membrane from kaolin: a comprehensive review. Ceram Int 44(5). https://doi.org/10.1016/j.ceramint.2017.12.215

Hussain E, Ahtesham A, Shahadat M et al (2024) Recent advances of clay/polymer-based nanomaterials for the treatment of environmental contaminants in wastewater: a review. J Environ Chem Eng 12(3). https://doi.org/10.1016/j.jece.2024.112401

Hübner U, Spahr S, Lutze H et al (2024) Advanced oxidation processes for water and wastewater treatment—guidance for systematic future research. Heliyon 10(9). https://doi.org/10.1016/j.heliyon.2024.e30402

Hussein TK, Jasim NA, Al-Madhhachi AST (2023) The performance of microfiltration using hydrophilic and hydrophobic membranes for phenol extraction from a water solution. ChemEngineering 7(2). https://doi.org/10.3390/chemengineering7020026

Ibrahim Y, Hilal N (2023) Enhancing ultrafiltration membrane permeability and antifouling performance through surface patterning with features resembling feed spacers. NPJ Clean Water 6(1). https://doi.org/10.1038/s41545-023-00277-3

Islam MD, Uddin FJ, Rashid TU, Shahruzzaman M (2023) Cellulose acetate-based membrane for wastewater treatment—a state-of-the-art review. Mater Adv 4(18):4054–4102. Royal Society of Chemistry. https://doi.org/10.1039/d3ma00255a

Jafari B, Abbasi M, Hashemifard SA (2020) Development of new tubular ceramic microfiltration membranes by employing activated carbon in the structure of membranes for treatment of oily wastewater. J Clean Prod 244. https://doi.org/10.1016/j.jclepro.2019.118720

Jiang Y, Wang WN, Liu D et al (2015) Engineered crumpled graphene oxide nanocomposite membrane assemblies for advanced water treatment processes. Environ Sci Technol 49(11):6846–6854. https://doi.org/10.1021/acs.est.5b00904

John Y, David VE, Mmereki D (2018) A comparative study on removal of hazardous anions from water by adsorption: a review. Int J Chem Eng 2018. Hindawi Limited. https://doi.org/10.1155/2018/3975948

Kacprzyńska-Gołacka J, Kowalik-Klimczak A, Woskowicz E et al (2020) Microfiltration membranes modified with silver oxide by plasma treatment. Membranes 10(6):1–11. https://doi.org/10.3390/membranes10060133

Kekes T, Tzia C, Kolliopoulos G (2023) Drinking and natural mineral water: treatment and quality–safety assurance. Water (Switzerland) 15(13). https://doi.org/10.3390/w15132325

Khan ZU, Moronshing M, Shestakova M et al (2023) Electro-deionization (EDI) technology for enhanced water treatment and desalination: a review. Desalination 548. https://doi.org/10.1016/j.desal.2022.116254

Kitis M (2004) Disinfection of wastewater with peracetic acid: a review. Environ Int 30(1):47–55. https://doi.org/10.1016/S0160-4120(03)00147-8

Kong J, Lu Y, Ren Y et al (2021) The virus removal in UV irradiation, ozonation and chlorination. In: Water cycle, vol 2. KeAi Communications Co., pp 23–31. https://doi.org/10.1016/j.watcyc.2021.05.001

Kocza K, Mizsey P (2010) New area for distillation: wastewater treatment. Periodica Polytech, Chem Eng 54(1):41–45. https://doi.org/10.3311/pp.ch.2010-1.06

Korenak J, Basu S, Balakrishnan M, Hélix-Nielsen C, Petrinic I (2017) Forward osmosis in wastewater treatment processes. Acta Chim Slov 64(1):83–94. https://doi.org/10.17344/acsi.2016.2852

Krasner SW, Weinberg HS, Richardson SD, Pastor SJ, Chinn R, Sclimenti MJ, Onstad GD, Thruston AD (2006) Occurrence of a new generation of disinfection byproducts. Environ Sci Technol 40(23):7175–7185. https://doi.org/10.1021/acs.est.7b05440

Kristanti RA, Hadibarata T, Syafrudin M et al (2022) Microbiological contaminants in drinking water: current status and challenges. Water, Air, Soil Pollut 233(8). https://doi.org/10.1007/s11270-022-05698-3

Kumar N, Gusain R, Pandey S et al (2023) Hydrogel nanocomposite adsorbents and photocatalysts for sustainable water purification. Adv Mater Interfaces 10(2). Wiley. https://doi.org/10.1002/admi.202201375

Kumar M, Shekhar S, Kumar R et al (2023) Drinking water treatment and associated toxic byproducts: concurrence and urgence. In: Environmental pollution, vol 320. https://doi.org/10.1016/j.envpol.2023.121009

Kundu D, Dutta D, Joseph A et al (2024) Safeguarding drinking water: a brief insight on characteristics, treatments and risk assessment of contamination. Environ Monitor Assess 196(2). Springer Science and Business Media Deutschland GmbH. https://doi.org/10.1007/s10661-024-12311-z

Li J, Zhang Z, Xiang Y et al (2023) Role of UV-based advanced oxidation processes on NOM alteration and DBP formation in drinking water treatment: a state-of-the-art review. Chemosphere 311(1):136870. https://doi.org/10.1016/j.chemosphere.2022.136870

Li XF, Mitch WA (2018) Drinking water disinfection byproducts (DBPs) and human health effects: multidisciplinary challenges and opportunities. Environ Sci Technol 52(4):1681–1689. https://doi.org/10.1021/acs.est.7b05440

Lin L, Yang H, Xu X (2022) Effects of water pollution on human health and disease heterogeneity: a review. Front Environ Sci 10. Frontiers Media S.A. https://doi.org/10.3389/fenvs.2022.880246

Liu Z, Demeestere K, Van Hulle S (2021) Comparison and performance assessment of ozone-based AOPs in view of trace organic contaminants abatement in water and wastewater: a review. J Environ Chem Eng 9(4):105599. https://doi.org/10.1016/j.jece.2021.105599

Ma H, Burger C, Hsiao BS, Chu B (2012) Nanofibrous microfiltration membrane based on cellulose nanowhiskers. Biomacromol 13(1):180–186. https://doi.org/10.1021/bm201421g

Maqbool F, Soomro MI, Kumar L, Harijan K (2024) Modeling and simulation of direct contact membrane distillation system integrated with a photovoltaic thermal for electricity and freshwater production. Front Energy Res 12. https://doi.org/10.3389/fenrg.2024.1344214

Maroufi N, Hajilary N (2023) Nanofiltration membranes types and application in water treatment: a review. Sustain Water Resour Manag 9(5). https://doi.org/10.1007/s40899-023-00899-y

Martins VL, Ogden MD, Jones MR, Trowsdale SA, Hall PJ, Jensen HS (2020) Opportunities for coupled electrochemical and ion-exchange technologies to remove recalcitrant micropollutants in water. Sep Purif Technol 239. https://doi.org/10.1016/j.seppur.2020.116522

Mishra NS, Reddy R, Kuila A, Ran A, Mukherjee P, Nawaz A, Pichiah S (2017) A review on advanced oxidation processes for effective water treatment. Current World Environ 12(3). https://doi.org/10.12944/CWE.12.3.02

Mohammad Ameen NA, Ibrahim SS, Alsalhy QF, Figoli A (2020) Highly saline water desalination using direct contact membrane distillation (DCMD): experimental and simulation study. Water (Switzerland) 12(6). https://doi.org/10.3390/W12061575

Mohammadifakhr M, Grooth J de, Roesink HDW et al (2020) Forward osmosis: a critical review. Processes 8(4). https://doi.org/10.3390/PR8040404

Mohammad Mahdi A, Shirazi AK, Shirazi MJA (2012) Direct contact membrane distillation for sea water desalination. Desalin Water Treat 49(1–3):368–375. https://doi.org/10.1080/19443994.2012.719466

Mohammed SA, Kahissay MH, Hailu AD (2021) Pharmaceuticals wastage and pharmaceuticals waste management in public health facilities of Dessie town, North East Ethiopia. PLoS ONE 16(10). https://doi.org/10.1371/journal.pone.0259160

Nawaz T, Sengupta S (2021) Wastewater: novel treatment technologies and source for epidemiological studies. In: Ahuja S (ed) Handbook of water purity and quality, 2nd edn, pp 293–337. https://doi.org/10.1016/B978-0-12-821057-4.00015-X

Ounifi I, Ursino C, Santoro S et al (2020) Cellulose acetate nanofiltration membranes for cadmium remediation. J Membr Sci Res 6(2):226–234. https://doi.org/10.22079/JMSR.2020.120669.1336

Pot EJ, Milakovic M, Chaumot A et al (2022) Pharmaceutical pollution of the world's rivers. https://doi.org/10.1073/pnas.2113947119/-/DCSupplemental

Prado de Nicolás A, Molina-García Á, García-Bermejo JT et al (2023) Desalination, minimal and zero liquid discharge powered by renewable energy sources: current status and future perspectives. Renew Sustain Energy Rev 187. https://doi.org/10.1016/j.rser.2023.113733

Retta B, Coppola E, Ciniglia C et al (2023) Constructed wetlands for the wastewater treatment: a review of Italian case studies. Appl Sci (Switzerland) 13(10). https://doi.org/10.3390/app13106211

Rodríguez-Villanueva P, Sauri D (2021) Wastewater treatment plants in Mediterranean Spain: an exploration of relations between water treatments, water reuse, and governance. Water (Switzerland) 13(12). https://doi.org/10.3390/w13121710

Safi NN, Ibrahim SS, Zouli N et al (2020) A systematic framework for optimizing a sweeping gas membrane distillation (SGMD). Membranes 10(10):1–18. https://doi.org/10.3390/membranes10100254

Salih SJ, Mahmood WM (2023) Review on magnetic spinel ferrite (MFe_2O_4) nanoparticles: from synthesis to application. Heliyon 9(6). https://doi.org/10.1016/j.heliyon.2023.e16601

Santos AB dos, Giacobbo A, Rodrigues MAS, Bernardes AM (2024) Electrodeionization for wastewater reuse in petrochemical plants. Water (Switzerland) 16(3). https://doi.org/10.3390/w16030401

Saja S, Bouazizi A, Achiou B et al (2018) Elaboration and characterization of low-cost ceramic membrane made from natural Moroccan perlite for treatment of industrial wastewater. J Environ Chem Eng 6(1) https://doi.org/10.1016/j.jece.2017.12.004

Saravanathamizhan R, Perarasu VT (2021) Chapter 6 - Improvement of biodegradability index of industrial wastewater using different pretreatment techniques. In: Shah MP, Sarkar A, Mandal

S (eds) Wastewater treatment, pp 103–136. https://doi.org/10.1016/B978-0-12-821881-5.00006-4

Sh Awad E, Sabirova TM, Tretyakova NA et al (2021) A mini-review of enhancing ultrafiltration membranes (UF) for wastewater treatment: performance and stability. ChemEngineering. https://doi.org/10.3390/chemengineering

Shehata N, Egirani D, Olabi AG et al (2023) Membrane-based water and wastewater treatment technologies: Issues, current trends, challenges, and role in achieving sustainable development goals, and circular economy. Chemosphere 320. https://doi.org/10.1016/j.chemosphere.2023.137993

Shen X, Liu Q, Li H, Kuang X (2022) Membrane-free electrodeionization using graphene composite electrode to purify copper-containing wastewater. Water Sci Technol 86:1733–1744. https://doi.org/10.2166/wst.2022.298

Shayo GM, Elimbinzi E, Shao GN, Fabian C (2023) Severity of waterborne diseases in developing countries and the effectiveness of ceramic filters for improving water quality. Bull Natl Res Centre 47(1). https://doi.org/10.1186/s42269-023-01088-9

Shoshaa R, Ashfaq MY, Al-Ghouti MA (2023) Recent developments in ultrafiltration membrane technology for the removal of potentially toxic elements, and enhanced antifouling performance: a review. Environ Technol Innov 31. https://doi.org/10.1016/j.eti.2023.103162

Shahu VT, Thombre SB (2019) Air gap membrane distillation: a review. J Renew Sustain Energy 11(4). American Institute of Physics Inc. https://doi.org/10.1063/1.5063766

Singh A, Srivastava A, Saidulu D, Gupta AK (2022) Advancements of sequencing batch reactor for industrial wastewater treatment: major focus on modifications, critical operational parameters, and future perspectives. J Environ Manag 317. Academic Press. https://doi.org/10.1016/j.jenvman.2022.115305

Singh NB, Nagpal G, Agrawal S, Rachna (2018) Water purification by using adsorbents: a review. Environ Technol Innov 11:187–240. https://doi.org/10.1016/j.eti.2018.05.006

Singh V, Ahmed G, Vedika S, Kumar P, Chaturvedi SK, Rai SN, Vamanu E, Kumar A (2024) Toxic heavy metal ions contamination in water and their sustainable reduction by eco-friendly methods: isotherms, thermodynamics and kinetics study. Sci Rep 14(1). https://doi.org/10.1038/s41598-024-58061-3

Suhan MBK, Mahtab SMT, Aziz W et al (2021) Sudan black B dye degradation in aqueous solution by Fenton oxidation process: kinetics and cost analysis. In: Case studies in chemical and environmental engineering, vol 4. https://doi.org/10.1016/j.cscee.2021.100126

Sun AC, Kosar W, Zhang Y et al (2014) Vacuum membrane distillation for desalination of water using hollow fiber membranes. J Membr Sci 455:131–142. https://doi.org/10.1016/j.memsci.2013.12.055

Suyambulingam I, Gangadhar L, Sana SS et al (2023) Chitosan biopolymer and its nanocomposites: emerging material as adsorbent in wastewater treatment. Adv Mater Sci Eng 2023. https://doi.org/10.1155/2023/9387016

Ugarte P, Renda S, Cano M, Pérez J et al (2024) Air-gap membrane distillation of industrial brine: effect of brine concentration and temperature. Ind Eng Chem Res 63(3):1546–1553. https://doi.org/10.1021/acs.iecr.3c03415

Venkatesh G, Brattebø H (2011) Environmental impact analysis of chemicals and energy consumption in wastewater treatment plants: case study of Oslo, Norway. Water Sci Technol 63(5):1018–1031. https://doi.org/10.2166/wst.2011.284

Wang C, Park MJ, Gonzales RR et al (2023) Graphene oxide-based layer-by-layer nanofiltration membrane using inkjet printing for desalination. Desalination 549. https://doi.org/10.1016/j.desal.2022.116357

Wang C, Wang Y, Qin H et al (2020) Application of microfiltration membrane technology in water treatment. IOP Conf Ser Earth Environ Sci 571(1). https://doi.org/10.1088/1755-1315/571/1/012158

Wang JL, Liu X (2021) Forward osmosis technology for water treatment: recent advances and future perspectives. J Clean Prod 280:124354. https://doi.org/10.1016/j.jclepro.2020.124354

Wang Y, Li D, Li J et al (2022) Metal organic framework UiO-66 incorporated ultrafiltration membranes for simultaneous natural organic matter and heavy metal ions removal. Environ Res 208. https://doi.org/10.1016/j.envres.2021.112651

Waqas S, Harun NY, Sambudi NS et al (2023) A review of rotating biological contactors for wastewater treatment. Water (Switzerland) 15(10). https://doi.org/10.3390/w15101913

Wei Y, Wang D, Li G et al (2024) Research and application progress of electrochemical water softening technology in China. Desalin Water Treat 319. https://doi.org/10.1016/j.dwt.2024.100425

Worch E (2012) Adsorption technology in water treatment: fundamentals, processes, and modeling. De Gruyter, Berlin, Boston. https://doi.org/10.1515/9783110240238

Xiao J, Guo S, Wang D, An Q (2024) Fenton-like reactions recent advances and new trends. Chem A Eur J 30(24). https://doi.org/10.1002/chem.202304337

Yan Z, Jiang Y, Liu L et al (2021) Membrane distillation for wastewater treatment: a mini review. Water (Switzerland) 13(24). https://doi.org/10.3390/w13243480

Zainurin SN, Wan Ismail WZ, Mahamud SNI et al (2022) Advancements in monitoring water quality based on various sensing methods: a systematic review. Int J Environ Res Publ Health 19(21). https://doi.org/10.3390/ijerph192114080

Zhang J, Mirza NR, Huang Z et al (2021) Evaluation of direct contact membrane distillation coupled with fractionation and ozonation for the treatment of textile effluent. J Water Process Eng 40. https://doi.org/10.1016/j.jwpe.2020.101789

Zhai Y, Liu G, van der Meer WGJ (2022) One-step reverse osmosis based on riverbank filtration for future drinking water purification. Engineering 9:27–34. https://doi.org/10.1016/j.eng.2021.02.015

Zhao Z, He Q, Fan J, Sun W, Ray MB (2023) Effect of granular activated carbon adsorption on mitigating microfiltration membrane fouling by algal organic matter. Aqua Water Infrastruct Ecosyst Soc 72:1881–1893. https://doi.org/10.2166/aqua.2023.127

Ziemann E, Das AK, Manna P et al (2022) Crosslinked polyethersulfone membranes for organic solvent nanofiltration in polar aprotic and halogenated solvents. J Membr Sci 663. https://doi.org/10.1016/j.memsci.2022.120963

Contamination of Surface Water and Groundwater Quality

Prashant Shukla, Pankaj Sharma, Anuradha Sahu, Shivam Chaubey, and Navneet Sharma

Abstract The degradation of surface and groundwater quality significantly impacts environmental sustainability and public health. Water quality is vital for human use and achieving Sustainable Development Goals. However, groundwater contamination often hinders progress due to resource limitations for adaptation. This chapter examines natural and human-induced contamination sources, processes, and impacts on surface water and underground aquifers. Key contamination sources include industrial effluents, agricultural runoff, urban wastewater, and natural substances, introducing pollutants includes heavy metals, nutrients, pathogens, and organic chemicals into water systems. The chapter outlines contamination pathways, noting that point-source discharges typically pollute surface waters, while groundwater pollution occurs through leaching and infiltration from diffuse sources. Large-scale groundwater quality assessments are scarce compared to surface water and often rely on inadequate data. Contaminated water leads to biodiversity loss, ecosystem damage, and health risks. The chapter underscores the importance of comprehensive water quality management, encompassing advanced detection and remediation technologies, robust regulatory frameworks, and sustainable land use practices. It emphasises the need for multidisciplinary collaboration to effectively tackle the complexities of water contamination and protect water resources.

Keywords Effluents · Pollution · Contamination · Groundwater

P. Shukla (✉) · S. Chaubey
Division of Agricultural Engineering, ICAR-Indian Agricultural Research Institute, New Delhi, India
e-mail: shuklaprashant310@gmail.com

P. Sharma
Department of Soil and Water Engineering, Punjab Agricultural University, Ludhiana, Punjab, India

A. Sahu
Department of Agricultural and Food Engineering, Indian Institute of Technology, Kharagpur, India

N. Sharma
International Water Management Institute, New Delhi, India

Dimple et al. (eds.), *Surface Water and Groundwater Quality for Sustainable Utilization*, Springer Hydrogeology, https://doi.org/10.1007/978-981-96-8189-1_2

1 Introduction

According to UNESCO (2022), "introducing any substance into water in an undesirable concentration not normally present in water, e.g., microorganisms, chemicals, waste or sewage, which renders the water unfit for its intended use is water pollution". Pollutants can be originated from natural and human activities. Pollution sources can be classified into two categories according to where they enter water bodies: non-point sources, which are diffuse and release pollutants over a large area, such as runoff from agricultural and urban areas, and construction sites; and point sources, which are specific locations like pipes and sewers and include factories, sewage treatment plants, mines. Water quality is vital for both surface and groundwater due to its essential role in health, ecosystems, and the economy. Surface waters like rivers, lakes, and reservoirs are often used for drinking water and recreation, and contaminants in these waters can lead to severe health problems and disrupt various activities. Clean surface water also supports aquatic biodiversity and is critical in agriculture and industry, where poor water quality can damage crops, increase production costs, and threaten food security.

Groundwater, especially important in rural areas, is another key drinking water source. However, groundwater contamination has become a significant global concern in recent years (Mukherjee and Singh 2022). The use of pesticides, fertilisers, and excessive groundwater extraction for irrigation are significant causes of groundwater contamination in rural areas of developed nations. Groundwater pollution in urban areas is mostly caused by hazardous microorganisms and chemical pollutants from urban development, along with waste from industry and dwelling sources. Hazards to human health, ecological services, and sustainable socio-economic development include toxic metals, hydrocarbons, nanoparticles, microplastics, pesticides and other newly discovered pollutants (Li 2020). Contamination of groundwater is often harder to detect and remediate, leading to prolonged exposure to harmful substances. Contaminated groundwater can negatively impact crop health, soil quality, and industrial processes. Ensuring high water quality for both surface and groundwater protects human health, supports ecosystems, and maintains economic activities. This requires effective management practices, stringent regulations, and continuous monitoring to safeguard water resources and ensure their sustainable use for future generations.

This chapter aims to comprehensively understand contamination sources, including agricultural runoff, industrial discharge, urban development, and natural processes. It delves into various contaminants including heavy metals, pesticides, pathogens, and pharmaceuticals, and examines their effects on water quality, human health, and ecosystems. The breakdown of mineral deposits under the Earth's crust is the natural source of various pollutants detected in groundwater (Pandey et al. 2016; Subba Rao et al. 2021; He et al. 2020). However, the global population is growing at an accelerating rate, and pollution of human origin is becoming a bigger problem as cities become more industrialised and prosperous. The nations that are rapidly developing economically—many of which are in the eastern hemisphere—are those most affected by these global shifts (Lam et al. 2015). The chapter investigates the transport and fate of contaminants through processes like leaching, adsorption, advection and dispersion. Key objectives include assessing water quality monitoring techniques, such as sampling methods, analytical tools, and water quality indicators. It also explores remediation and management strategies, presenting physical, chemical, and biological treatment methods and best management practices (BMPs) for prevention. The chapter evaluates existing regulatory frameworks and standards, discussing their effectiveness in controlling contamination and highlighting areas for improvement.

2 Sources of Water Contamination

Surface water contamination primarily arises from agricultural runoff, industrial discharge, urban runoff, mining activities, and waste disposal. Agricultural runoff carries fertilisers, pesticides, and animal waste into rivers and lakes, introducing nutrients and toxins that harm aquatic ecosystems and water quality. Industrial discharge involves chemicals and heavy metals released from factories, often causing toxicity and thermal pollution, which disrupts aquatic life. Urban runoff, especially during heavy rains, washes pollutants from roads and urban areas into water bodies, with combined sewer overflows releasing untreated sewage. Mining activities contribute to acid mine drainage and sediment, while improper waste disposal from landfills and illegal dumping add pollutants to surface water.

Groundwater contamination stems from agricultural practices, industrial activities, residential sources, mining, and natural processes. Agricultural activities lead to nitrate leaching and pesticide infiltration, posing risks to drinking water. Industrial sources include chemical spills and leaks from underground storage tanks, introducing hazardous substances into aquifers. Residential sources, such as leaking septic systems and improper disposal of household chemicals, further contaminate groundwater. Mining activities cause contamination through excessive groundwater pumping and waste disposal.

Natural sources, like geological formations, release harmful substances like arsenic, fluoride, and radon, while over-pumping in the coastal areas causes seawater intrusion, also over-pumping in coastal areas can lead to affect water quality, salting freshwater aquifers. Effective management and coordinated efforts are essential to mitigate these contamination sources (Fig. 1).

Groundwater contamination arises from a wide range of sources. The main sources of groundwater contamination include agricultural chemicals, industrial waste discharges, heavy metals from mining activities, interactions between aquifers and polluted streams, and the disposal of household, municipal, and laboratory waste. There are various ways that contaminants can enter groundwater: seepage from the top layer or shallow subsoil; direct infiltration through wells; cross-contamination of wells that access different aquifers; and injection of contaminated water into freshwater aquifers. Furthermore, natural contaminants found in geological strata, such

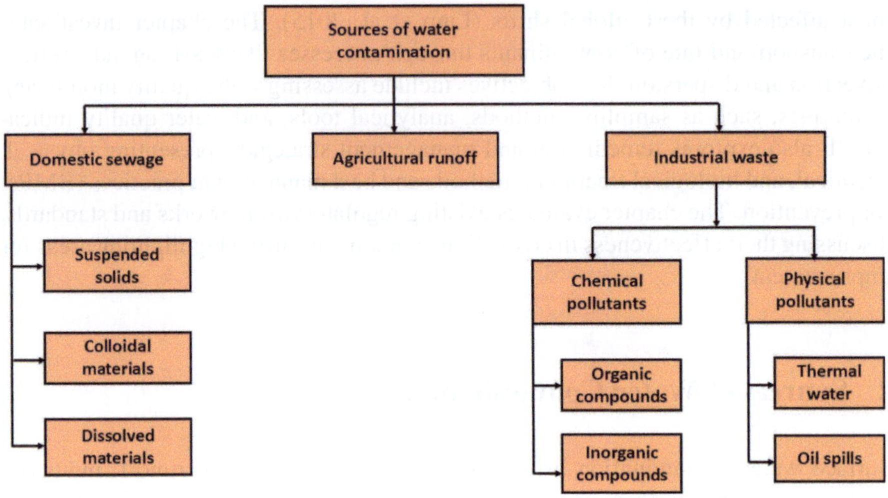

Fig. 1 Water contamination sources

as radon and arsenic, can interact with groundwater to further affect its quality. The solubility of compounds in the soil is affected by several elements, including temperature, hydrogeochemical processes, and the types of rocks and soils (De Giglio et al. 2015).

There are several sources of contamination listed below:

1. **Agricultural Sources**: Agriculture is responsible for 70% of global water usage and significantly contributes to water pollution by discharging substantial quantities of agrochemicals, organic matter, and sediments into water bodies. Nitrate, a prevalent chemical contaminant globally, particularly affects groundwater aquifers due to agricultural practices. The rapidly expanding livestock sector, outpacing crop production in many regions, generates substantial waste, notably manure, which poses serious threats to water quality. Key agricultural pollutants affecting water quality include nutrients, pesticides, salt composition, sediments, available organic carbon, pathogens, and metals, highlighting the urgent need to manage pollution measures. Poor agricultural practices like improper tillage increase soil erosion and sediment runoff into water bodies, where sediments play a crucial role in transporting various pollutants.
2. **Industrial Sources**: Industrial processes including manufacturing, cleaning, and other commercial activities produce effluent that is both suspended and dissolved in the water. The particular contaminants present in this effluent vary according to the industry. The production of iron and steels, power plants, mining, fracking for oil, food and beverage manufacturing, and industrial laundries are among the common businesses that generate large amounts of wastewater. Chemicals, oils, heavy metals, pesticides, silt, and other industrial by products are generally found as contaminants in industrial wastewater.

3. **Urban Sources**: India's increasing population, urbanization, rapid industrialization, and agricultural expansion have contributed to a decline in water quality. Large-scale industries discharge significant amounts of polluted liquid waste into river systems through canals. The primary water pollution source in India is the disposal of untreated sewage into both surfaces as well as in groundwater. Of the approximately 38,000 million litres per day (MLD) of sewage produced, only about 12,000 MLD can be treated, leaving a significant gap between sewage generation and treatment. Additionally, existing treatment facilities are often underutilized due to operational and maintenance issues (CPCB 2009).

Contaminants in Surface Water and Groundwater: Groundwater contamination occurs when harmful substances are introduced into groundwater as a result of human activities (Government of Canada 2017). In contrast to contamination of surface waters, groundwater pollution is imperceptible and challenging to remediate using existing technologies. Since groundwater pollutants frequently have no colour or smell, it might be difficult to identify their long-term negative impacts on human health (Chakraborti et al. 2015). Because groundwater is held in deep geological formations with lengthy residence durations, cleaning it up once contamination occurs is difficult and expensive (Wang et al. 2020). The natural purification process can take decades or even centuries, regardless of when the source of pollution is eliminated (Tatti et al. 2019).

Although the quantity of pollutants found in groundwater is steadily rising, they may be broadly classified into three groups, as Table 1 illustrates: chemical, biological, and radioactive contaminants. Both natural and man-made processes can produce these contaminants (Elumalai et al. 2020). Mineral deposits, brackish water, low-quality surface water, and saltwater are examples of natural sources of groundwater contamination. When human activities upset the natural environmental balance, these natural sources can become serious threats. Examples of such disturbances include acid mine drainage from mineral extraction, aquifer depletion leading to saltwater intrusion, and the leaching of hazardous chemicals from excessive irrigation (Wu et al. 2015; Li et al. 2018).

Common inorganic pollutants that contain nitrogen include nitrate, nitrite, and ammonia nitrogen. Human activity is the main source of nitrate, particularly industrial wastewater and agricultural runoff (Serio et al. 2018; Zhang et al. 2018; He et al. 2019; He and Wu 2019; Karunanidhi et al. 2019; Li et al. 2019). Groundwater nitrate contamination has been extensively reported worldwide (Fig. 2).

Inorganic Contaminants: Inorganic contamination involves the presence of harmful inorganic substances in water, soil, or air, typically resulting from natural processes or human activities. Certain chemical components and compounds, such as heavy metals like arsenic, lead, and mercury, as well as salts (like nitrates, chlorides, and sulphates) and other minerals, can be found in groundwater. They might originate from man-made activities including mining, industrial processes, agriculture, and inappropriate waste disposal, or they can come from natural sources such as the breakdown of minerals in the Earth's crust. Some of these compounds are hazardous, thus there can be serious concerns to both the environment and human health. Addressing

Table 1 Surface and groundwater Contaminants

Sr. no	Contaminant	Sources of contaminant
1	Aluminium (Al)	It can occur naturally in certain rocks and may also result from drainage from mining operations
2	Arsenic (As)	Natural processes, industrial operations, pesticide use, industrial waste, and the smelting of copper, lead, and zinc ores are some of the ways they can disseminate the environment
3	Chloride	Increased levels of sodium in potable water could be linked to mineral dissolution, industrial and residential waste, and saltwater intrusion
4	Copper (Cu)	Mineral leaching, mining operations, household and industrial waste, and metal plating are some of the ways it gets into the environment
5	Dissolved solids	It occurs naturally but can also be introduced into the environment from human activities like sewage, feedlots, and landfill leachate
6	Fluoride	It can be obtained to municipal water sources or found naturally. It is also widely employed in many different industrial applications
7	Hardness	The amount of calcium carbonate in the water indicates the quantity of dissolved metallic ions
8	Iron (Fe)	Along with being present naturally in rocks and sediments, it can also be brought in by mining, industrial waste, and corroding metal
9	Lead (Pb)	Industrial processes, mining, plumbing, coal, petrol and water treatment additives are some of the ways it gets into the environment
10	Manganese (Mn)	In addition to being present naturally as a mineral in rocks and sediments, it can also be introduced via mining and industrial waste
11	Zinc (Zn)	It is naturally found in water, especially in regions where it is mined
12	Sulfate	Waste from homes or businesses, mineral dissolution, and seawater intrusion can all result as a cause of elevated amounts
13	Sodium (Na)	It comes from the geological leaching of both subterranean and surface salt deposits, as well as the breakdown of other minerals
14	Mercury (Hg)	It is found as an organic mercury compound as well as an inorganic salt
15	Nickel (Ni)	It naturally occurs in soils, surface water and groundwater
16	Nitrate	Mineral deposits, soils, freshwater and saltwater systems, the atmosphere, and living things all naturally contain it. Additionally, sewage, feedlots, and fertilisers are some of the places where it enters the environment
17	Silver (Ag)	It is released into the environment during the mining, processing, and disposal of ore
18	Pesticides	As fungicides, insecticides, herbicides, and algicides, they are released into the environment
19	Volatile organic compounds	The production of plastics, dyes, rubbers, solvents, crude oil, pesticides, inks, medicines, preservatives, and other goods releases them into the environment
20	Coliform microorganisms	They are found in the environment naturally in the soils, plants, and intestines of warm-blooded living beings

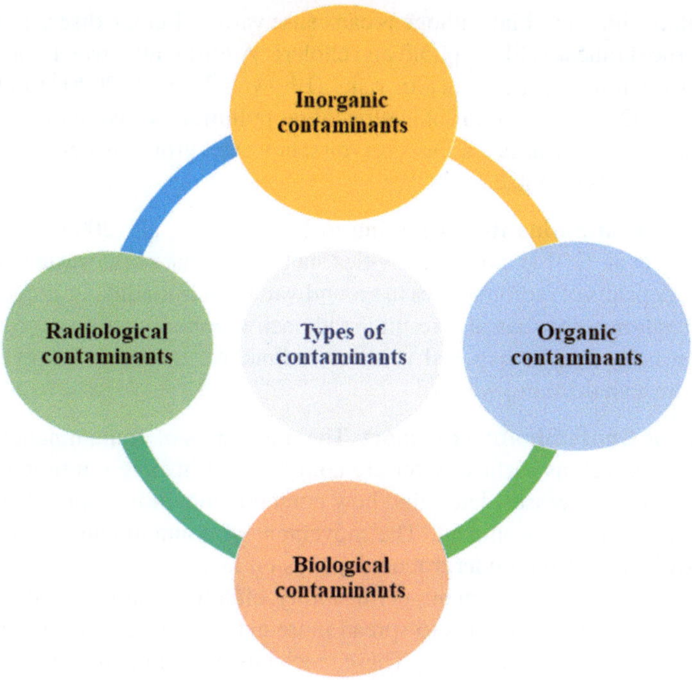

Fig. 2 Types of contaminants

inorganic contamination is often complex and costly because these contaminants do not break down over time and can remain in the environment, necessitating advanced treatment methods to remove them from polluted water or soil.

Organic Contaminants: Many of the organic pollutants that are frequently found in drinking water are categorised as endocrine disruptors or human carcinogens. The primary sources of biodegradable organic pollutants are industrial wastewater and home sewerage. Numerous of these chemicals can be converted by bacteria into stable inorganic substances. They are naturally derived from carbohydrates, proteins, lipids, and oils. While they don't directly harm living things, they can lower groundwater's dissolved oxygen content. Hydrocarbons, plasticisers, pesticides, medications, personal hygiene items, and natural oestrogens are examples of common organic pollutants (Meffe and Bustamante 2014; Lapworth et al. 2012). Owing to their gradual or non-existent breakdown, these substances may present a persistent risk to groundwater's quality as a source of drinking water (Schulze et al. 2019).

Biological Contaminants: These include algae and microorganisms such as bacteria, viruses etc. Naturally occurring algal species interact with some of these microbes, competing with them for limited resources, while other microorganisms are derived from natural sources (Lam et al. 2018; Flemming and Wuertz 2019). Drinking water contaminated with microbial pathogens can cause various human diseases, including severe diarrheal illnesses like typhoid and cholera. Additionally, water contaminated with viruses can also present health risks (Bhowmick et al. 2020; Lokhandwala and Gautam 2020). Due to eutrophication, algal pollution is frequently observed in surface waterways such as lakes and reservoirs; however, groundwater rarely contains significant amounts of algae.

Radiological Contaminants: According to Dahlgaard et al. (2004), Huang et al. (2012), Lytle et al. (2014), human activities such as nuclear plant waste can produce radioactive deposits of radionuclides in groundwater. Additionally, improper disposal of medical radioisotopes can also result in radioactive contamination in groundwater. The human body can be exposed to these radioactive materials in several ways, including through drinking water.

Mechanics of Contaminant Transport: The mechanics of contaminant transfer in groundwater as well as surface water are complex and involve multiple interacting processes. These processes determine how contaminants move, spread, and potentially change in the environment. The movement of contaminants, both chemical and bacteriological, from underground depends on factors such as the groundwater table's position, hydraulic gradient, permeability, effective grain size, soil types, and more. Generally, pollution tends to spread more extensively under saturated conditions compared to unsaturated ones. These mechanisms of contaminants of transport generally occur:

Advection: The main method via which pollutants are carried through groundwater and surface water, moving along with the flowing water. In groundwater, this occurs as water travels through the pores of soil, sand, or rock, with the flow's speed and direction determined by factors like the hydraulic gradient, aquifer permeability, and the medium's porosity. In rivers and streams, advection happens as water flows downstream, transporting contaminants. Variations in water flow velocity, influenced by topography and channel shape, directly affect the rate of contaminant movement.

Dispersion: Dispersion refers to the spreading of contaminants as they move through water, resulting in the dilution of the contaminant plume. In groundwater, dispersion is driven by the heterogeneity of the aquifer material, such as differences in pore sizes and flow paths, which cause contaminants to spread out and mix with uncontaminated water. In surface water, factors like turbulence, flow velocity variations, and channel irregularities contribute to dispersion. Over time and space, this process leads to a wider, less concentrated contaminant plume.

Diffusion: Due to concentration gradients, pollutants diffuse when they go from regions of higher concentration to areas of lower concentration. Unlike advection, which depends on the bulk movement of water, diffusion happens at the molecular level and is independent of water flow. Although diffusion is typically slower than advection and dispersion, it plays a significant role over long periods and in low-flow environments.

Biological Uptake: Biological uptake refers to the absorption of contaminants by living organisms, such as plants, animals, and microorganisms. This process can lead to the accumulation of pollutants in biological tissues, known as bioaccumulation. Biological uptake can reduce the concentration of contaminants in water, but it also poses risks to ecosystems and human health, especially if the contaminants are persistent like heavy metals or certain organic compounds.

Assessment and Monitoring of Water Quality: The water quality assessment and monitoring entails analysing a range of physical, chemical, and biological factors to ascertain the water's suitability for various applications and state of health. Under the Global Environmental Monitoring System (GEMS) Water Programme, the Central Pollution Control Board (CPCB) started conducting nationwide water quality monitoring in 1978. With the establishment of the National Programme for Monitoring Indian National Aquatic Resources (MINARS) in 1984, 113 sites in 10 river basins were covered. Today, the network has expanded to 870 stations monitoring rivers, lakes, and groundwater. Important factors that are measured are pathogens, heavy metals, pH, turbidity, temperature, and nutrients like phosphate and nitrogen. These assessments are crucial for ensuring water safety for drinking, supporting aquatic life, agriculture, and recreational use. Regular data collection from these sites helps define current conditions and track changes over time.

Sampling: The initial stage of water quality testing involves sampling, where using correct techniques is crucial to obtaining accurate results. Water is sampled directly from its source, including rivers, lakes, wells, or taps, with sampling points chosen based on contaminant locations and water usage intentions. Samples are gathered in sterile containers according to precise protocols to prevent contamination.

Analytical Methods: Various analytical methods have been developed and utilized to detect different types of water pollutants. Sample preparation is crucial in this process, often requiring cleanup and concentration steps essential for subsequent analyses. Two primary techniques used for sample preparation are Solid-phase extraction (SPE) and liquid–liquid extraction (LLE). SPE is commonly favoured for its efficient, selective preparation using cartridges containing suitable sorbents. Despite its advantages, SPE can experience issues like channelling and cartridge blockage in complex samples, prompting ongoing research to improve its effectiveness.

Water Quality Indicators: Water quality parameters cover a range of chemical, physical, and biological characteristics. Six key indicators include:

1. **Dissolved Oxygen (DO)**: Dissolved oxygen levels are crucial to sustain humans, plants as well as animals, as they indicate the amount of oxygen available to aquatic life. Both high and low dissolved oxygen (DO) levels can greatly impact aquatic ecosystems. Factors such as bacterial activity and water temperature influence DO levels, making it an essential parameter for assessing water quality in streams and lakes.
2. **Turbidity and Total Suspended Solids**: Turbidity reflects water clarity, which is determined by the concentration of TSS. TSS includes particles bigger than two

microns, including clay, silt, gravel, sand, and algae. Suspended solids, including decaying organic matter and pollutants like phosphorus, pesticides, or heavy metals, can cloud water and affect its quality. High concentrations result in less clear water, impacting aquatic life and overall water quality.

3. **Bioindicators**: Bioindicators are organisms used to assess ecosystem health, such as microalgae. These natural indicators, including plants, plankton, animals, and microbes, provide valuable information for evaluating water quality and detecting pollution. The natural indicators are plants, plankton, animals, and microbes which provide insights into water quality and pollution levels. They are influenced by factors such as light, water temperature, and suspended solids. Changes in bioindicator populations can reflect the environmental impact of human activities.
4. **Nitrates**: While nitrogen is a vital nutrient, high nitrate concentrations—often from human and animal waste, industrial pollutants, and agricultural runoff—can lead to excessive algae growth. This can deplete dissolved oxygen levels, harming aquatic life. Elevated nitrate levels are also a concern for human health, making monitoring essential for both public health and aquatic ecosystem protection.
5. **pH Scale**: The pH scale of water measures acidity or alkalinity, ranging from 0 to 14, with 7 being neutral. Low pH values indicate acidity, while higher values suggest alkalinity Several factors can alter pH levels in water, including acid rain, vehicle emissions, agricultural runoff, spills, sewer overflows, and other pollutants. Significant pH changes can harm aquatic life and affect overall water quality. pH values for freshwater uses are presented in Table 2.
6. **Water Temperature**: Water temperature affects the health of aquatic organisms and influences other water quality parameters, such as dissolved oxygen. It also impacts the susceptibility of organisms to parasites, pollution, and disease. Water temperature is a crucial indicator of water quality because it can fluctuate due to seasonal differences. The CPCB categorises water resources based on their intended uses to establish water quality standards for different water bodies.

Effects of Water Pollution: The World Health Organisation (WHO) estimates that 400,000 people die in India each year as a result of poor access to water, sanitation, and hygiene. Waterborne diseases cause a loss of 90 million workdays each year, with a loss of production treatment costs amounting to 6 billion (McKenzie and Ray 2004). Additionally, high arsenic levels pose a cancer risk to 10 million people, while over 65 million are threatened by fluorosis, which has become widespread across 17 to 20 states.

Environmental and Health Impacts: Water pollution poses a significant environmental and health threat globally, impacting ecosystems and human well-being alike. It severely damages ecosystems such as wetlands and rivers, leading to biodiversity loss and habitat degradation. Aquatic life suffers as pollutants alter pH levels, introduce toxic chemicals, and reduce oxygen levels, impairing their ability to breathe and survive. These contaminants can accumulate in their tissues, resulting in health issues and mortality. Furthermore, water pollution adversely affects human health by causing waterborne diseases and chemical toxicity, with long-term effects including

Table 2 Key water quality standards for different freshwater uses

Intended use	Group	Standards
Source of drinking water that undergoes disinfection but not conventional treatment	A	• Total coliform organisms should be 50 MPN/100 mL or fewer • pH: 6.5–8.5 • Dissolved oxygen (DO) $\geq$ 6 mg/L • Biochemical oxygen demand (BOD) $\leq$ 2 mg/L
Outdoor bathing (organised)	B	• Total coliform organisms should be MPN/100 ml shall be 500 or less • pH: 6.5–8.5 • DO $\geq$ 5 mg/l • BOD $\leq$ 3 mg/l or Less
Drinking water source that is first treated conventionally and then disinfected	C	• Total coliform organisms should be MPN/ 100 ml shall be $\leq$ 5000 • pH: 6–9 • DO should be 4 mg/l or more • BOD $\leq$ 3 mg/l
Wildlife, fisheries propagation	D	• pH: 6.5–8.5 • DO $\geq$ 4 mg/l or more • Free ammonia (as N) $\leq$ 1.2 mg/l
Irrigation, industrial cooling, controlled waste disposal	E	• pH 6.0–8.5 • EC $\leq$ 2250 micromhos/cm • Sodium absorption ratio (SAR) < 26 • Boron $\leq$ 2 mg/l

increased cancer risks and reproductive disorders. Thus, addressing water pollution is crucial for safeguarding both natural environments and public health.

3 Mitigation and Remediation Strategies

Pollution Prevention: Some important pollution prevention methods are discussed below:

Best Management Practices (BMPs): These are proactive strategies designed to decrease the amount of pollution entering water bodies. These practices are customized for specific environments, such as agricultural lands, urban areas, and industrial sites. For instance, in agriculture, BMPs may involve using vegetative buffers—areas of grass, shrubs, or trees planted along field edges to capture and absorb runoff before it reaches streams or rivers. BMPs for managing stormwater and lowering surface runoff in urban environments could include green infrastructure options like rain gardens, green roofs, and permeable pavements. Additionally, erosion control techniques such as silt fences and terracing can help prevent soil erosion and avoid sediment contamination of water sources.

Sustainable Agricultural Practices: Sustainable farming methods are essential for reducing farming's negative environmental effects, particularly when it comes to contaminated water. Crop rotation, cover crops, and decreased tillage are a few examples of practices that are crucial for preserving soil health, lowering the need for chemical inputs, and minimising runoff. Crop rotation, which helps avoid soil depletion and lowers the accumulation of pests and illnesses, entails switching up the crops planted on a specific plot of land each season. To prevent soil erosion and boost organic matter levels, cover crops such as rye or clover are planted in the off-season. Reduced tillage, or no-till farming, limits soil disturbance, preserves soil structure, and decreases erosion.

4 Treatment Technologies

Physical Treatment Methods: Physical treatment methods are the beginning steps in the water treatment way, focusing on the removal of solid contaminants through mechanical means. These methods include sedimentation, filtration, and adsorption. In sedimentation, water is held in a tank or basin, allowing heavier particles to settle to the bottom. This step is followed by filtration, where water passes through materials like sand, gravel, or membranes that trap and remove smaller particles. Adsorption involves the use of materials like activated carbon, which have large surface areas to attract and hold dissolved substances, removing them from the water.

Chemical Treatment Methods: Chemical treatment methods involve adding chemicals to water to disinfect, coagulate, or oxidize contaminants. Chlorination is a widely used method where chlorine is added to water to eliminate pathogens, ensuring it is safe for drinking. Another method, ozonation, involves introducing ozone gas into the water to oxidize and decompose both organic and inorganic substances. Coagulation and flocculation are processes where chemicals like alum are added to cause fine particles to clump together into larger flocs, which can then be easily removed by sedimentation or filtration. These methods are often used in combination with physical treatments to ensure thorough purification.

Biological Treatment Methods: Biological treatment methods utilize natural processes and organisms to degrade pollutants in water. One example is bioremediation, where bacteria, fungi, or plants are used to break down hazardous substances into less harmful ones. Constructed wetlands are another biological method, where natural or engineered wetlands are used to treat wastewater. In these systems, plants, soil, and microorganisms work together to filter out pollutants as water flows through. These methods are especially effective at addressing organic contaminants and are frequently used alongside other treatment processes to enhance overall water quality.

Regulatory Frameworks: Regulatory frameworks are legal structures established by governments to protect water quality and manage water resources. These frameworks set permissible levels for various contaminants, require regular monitoring of water bodies, and mandate corrective actions if standards are not met. For instance, the Act of Clean Water in the United States (US) provides comprehensive guidelines to maintain the integrity of surface waters, including limits on pollutant discharges and requirements for pollution control technologies. The European Union's (EU) Directive Water Framework similarly sets high standards for water quality across member states, promoting sustainable water management practices.

Integrated Water Resource Management (IWRM): It is a thorough method of managing water resources, that acknowledges the connections between ecosystems, land, and water. IWRM seeks to maintain the long-term sustainability of water resources while balancing the demands of multiple industries, including agriculture, urban development, and industry.

This approach requires coordinated efforts among diverse stakeholders, such as government agencies, local communities, and private organizations. By integrating policies and practices across different sectors, IWRM seeks to promote equitable and sustainable water use, reduce conflicts over water resources, and increase the water systems resilience to environmental changes.

5 Future Challenges and Research Directions

Climate change significantly threatens global water quality. Surface and groundwater contamination is made worse by rising temperatures, changing precipitation patterns, and a propensity for extreme weather events like floods and droughts. Higher temperatures accelerate evaporation, concentrating pollutants in reduced water volumes. Altered precipitation leads to more intense runoff, carrying pollutants from urban, agricultural, and industrial areas into water bodies. Additionally, climate change promotes harmful algal blooms that produce toxins, contaminating drinking water and damaging aquatic ecosystems. These combined effects make maintaining water quality increasingly difficult, necessitating adaptive management strategies and enhanced monitoring. Technological innovations are revolutionizing the way water quality is monitored and managed. Advanced sensors, remote sensing technologies, and data analytics are providing real-time information on water quality parameters, enabling more accurate and timely responses to contamination events. For example, smart sensors deployed in rivers, lakes, and groundwater wells can continuously monitor parameters like pH, temperature, dissolved oxygen, and contaminant levels, transmitting data in real-time to central monitoring stations. Remote sensing technologies, including satellite imagery and drones, allow for the large-scale monitoring of water bodies, identifying changes in water quality over time and detecting pollution sources. Additionally, the integration of these technologies with machine learning and artificial intelligence (AI) enables predictive modelling, helping to forecast potential contamination events and optimize water management strategies.

Public awareness and education are essential elements of effective water quality management. Educating communities about the sources and impacts of water contamination, as well as the importance of water conservation and protection, can lead to more responsible behaviours and greater support for water management initiatives. Public awareness campaigns can be targeted at various audiences, including farmers, industrial operators, urban residents, and policymakers, encouraging actions that reduce pollution and conserve water resources. For instance, educational programs can teach farmers about sustainable practices that minimize runoff, while urban residents can learn about the importance of reducing household pollutants. Additionally, public participation in water quality monitoring programs, such as citizen science initiatives, can enhance community engagement and provide valuable data for researchers and policymakers. Addressing the complex challenges of water contamination requires ongoing research in several key areas. Future research should concentrate on exploring the long-term effects of unfolding contaminants as pharmaceuticals, microplastics, and nanomaterials—on water quality and ecosystem health. Additionally, there is a need to investigate the development of highly effective and sustainable water treatment techniques, especially those capable of addressing a broad spectrum of contaminants at once.

6 Conclusion

This covers the contamination of surface as well as groundwater, emphasizing environmental and health impacts, mitigation and remediation strategies, and future challenges. Sources of contamination, including runoff from agricultural, and industrial discharges, and inappropriate waste management, have a substantial effect on aquatic ecosystems and also on human health. The degradation of habitats and decline in biodiversity highlights the urgency for effective mitigation strategies. Human health risks include waterborne diseases, chemical toxicity, and prolonged exposure to pollutants. Mitigation and remediation strategies discussed include pollution prevention through Best Management Practices (BMPs) and sustainable agricultural methods, as well as treatment technologies like physical, chemical, and biological processes. To effectively address water contamination, policymakers and practitioners should focus on strengthening regulatory frameworks is crucial. Governments need to enhance water quality regulations and enforce them strictly, especially to address emerging contaminants like microplastics and pharmaceuticals. Encouraging sustainable agricultural practices, such as minimizing pesticide use and implementing crop rotation, can greatly decrease agricultural runoff, which is the main source of contamination. Investing in advanced monitoring technologies, including real-time sensors, will improve early detection and response. Public awareness campaigns should be prioritized to increase understanding of water conservation and pollution prevention while engaging communities in citizen science can foster responsibility for water quality.

References

Bhowmick GD, Dhar D, Nath D, Ghangrekar MM, Banerjee R, Das S, Chatterjee J (2020) Coronavirus disease 2019 (COVID-19) outbreak: some serious consequences with urban and rural water cycle. NPJ Clean Water 3(1):32

Central Pollution Control Board (2009) Ministry of Environment and Forests

Chakraborti D, Rahman MM, Mukherjee A, Alauddin M, Hassan M, Dutta RN, Hossain MM et al (2015) Groundwater arsenic contamination in Bangladesh—21 years of research. J Trace Elements Med Biol 31:237–248

Dahlgaard H, Eriksson M, Nielsen SP, Joensen HP (2004) Levels and trends of radioactive contaminants in the Greenland environment. Sci Total Environ 331(1–3):53–67

De Giglio O, Quaranta A, Barbuti G, Napoli C, Caggiano G, Montagna MT (2015) Factors influencing groundwater quality. Towards an integrated management approach. Ann Igiene Med Prevent Comun 27(1):52–57

Elumalai V, Nethononda VG, Manivannan V, Rajmohan N, Li P, Elango L (2020) Groundwater quality assessment and application of multivariate statistical analysis in Luvuvhu catchment, Limpopo, South Africa. J Afr Earth Sc 171:103967

Flemming HC, Wuertz S (2019) Bacteria and archaea on Earth and their abundance in biofilms. Nat Rev Microbiol 17(4):247–260

Government of Canada (2017) Groundwater contamination. https://www.canada.ca/en/environment-climate-change/services/water-overview/pollution-causes-effects/groundwater-contamination.html

He S, Wu J (2019) Hydrogeochemical characteristics, groundwater quality, and health risks from hexavalent chromium and nitrate in groundwater of Huanhe Formation in Wuqi county, northwest China. Exposure and Health 11:125–137

He X, Wu J, He S (2019) Hydrochemical characteristics and quality evaluation of groundwater in terms of health risks in Luohe aquifer in Wuqi County of the Chinese Loess Plateau, northwest China. Hum Ecol Risk Assess Int J 25(1–2):32–51

He X, Li P, Ji Y, Wang Y, Su Z, Elumalai V (2020) Groundwater arsenic and fluoride and associated arsenicosis and fluorosis in China: occurrence, distribution and management. Expo Health 12(3):355–368

Huang CP, Lin TY, Chiao LH, Chen HB (2012) Characterization of radioactive contaminants and water treatment trials for the Taiwan research reactor's spent fuel pool. J Hazard Mater 233:140–147

Karunanidhi D, Aravinthasamy P, Subramani T, Wu J, Srinivasamoorthy K (2019) Potential health risk assessment for fluoride and nitrate contamination in hard rock aquifers of Shanmuganadhi River basin, South India. Hum Ecol Risk Assess Int J 25(1–2):250–270

Lam S, Nguyen-Viet H, Tuyet-Hanh TT, Nguyen-Mai H, Harper S (2015) Evidence for public health risks of wastewater and excreta management practices in Southeast Asia: a scoping review. Int J Environ Res Public Health 12(10):12863–12885

Lam TP, Lee TM, Chen CY, Chang JS (2018) Strategies to control biological contaminants during microalgal cultivation in open ponds. Biores Technol 252:180–187

Lapworth DJ, Baran N, Stuart ME, Ward RS (2012) Emerging organic contaminants in groundwater: a review of sources, fate and occurrence. Environ Pollut 163:287–303

Li P, Wu J, Tian R, He S, He X, Xue C, Zhang K (2018) Geochemistry, hydraulic connectivity and quality appraisal of multilayered groundwater in the Hongdunzi Coal Mine, Northwest China. Mine Water Environ 37(2):222–237

Li P, He X, Guo W (2019) Spatial groundwater quality and potential health risks due to nitrate ingestion through drinking water: a case study in Yan'an City on the Loess Plateau of northwest China. Hum Ecol Risk Assess Int J 25(1–2):11–31

Li P (2020) To make the water safer. Expo Health 12(3):337–342

Lokhandwala S, Gautam P (2020) Indirect impact of COVID-19 on environment: a brief study in Indian context. Environ Res 188:109807

Lytle DA, Sorg T, Wang L, Chen A (2014) The accumulation of radioactive contaminants in drinking water distribution systems. Water Res 50:396–407

McKenzie D, Ray I (2004) Household water delivery options in urban and rural India. In: Working Paper No. 224, Stanford Centre for International Development, Stanford University, USA

Meffe R, de Bustamante I (2014) Emerging organic contaminants in surface water and groundwater: a first overview of the situation in Italy. Sci Total Environ 481:280–295

Mukherjee I, Singh UK (2022) Environmental fate and health exposures of the geogenic and anthropogenic contaminants in potable groundwater of lower Ganga Basin, India. Geosci Front 13(3):101365

Pandey HK, Duggal SK, Jamatia A (2016) Fluoride contamination of groundwater and it's hydrogeological evolution in District Sonbhadra (UP) India. Proc Natl Acad Sci, India, Sect A 86:81–93

Schulze S, Zahn D, Montes R, Rodil R, Quintana JB, Knepper TP, Berger U et al (2019) Occurrence of emerging persistent and mobile organic contaminants in European water samples. Water Res 153:80-90

Serio F, Miglietta PP, Lamastra L, Ficocelli S, Intini F, De Leo F, De Donno A (2018) Groundwater nitrate contamination and agricultural land use: a grey water footprint perspective in Southern Apulia Region (Italy). Sci Total Environ 645:1425–1431

Subba Rao N, Dinakar A, Kumari BK, Karunanidhi D, Kamalesh T (2021) Seasonal and spatial variation of groundwater quality vulnerable zones of Yellareddygudem watershed, Nalgonda district, Telangana State, India. Arch Environ Contam Toxicol 80:11–30

Tatti F, Papini MP, Torretta V, Mancini G, Boni MR, Viotti P (2019) Experimental and numerical evaluation of groundwater circulation wells as a remediation technology for persistent, low permeability contaminant source zones. J Contam Hydrol 222:89–100

UNESCO (2022) United Nations World Water Development Report 2022: groundwater: making the invisible visible

Wang D, Wu J, Wang Y, Ji Y (2020) Finding high-quality groundwater resources to reduce the hydatidosis incidence in the Shiqu County of Sichuan Province, China: analysis, assessment, and management. Expo Health 12(2):307–322

Wu J, Li P, Qian H (2015) Hydrochemical characterization of drinking groundwater with special reference to fluoride in an arid area of China and the control of aquifer leakage on its concentrations. Environ Earth Sci 73:8575–8588

Zhang Y, Wu J, Xu B (2018) Human health risk assessment of groundwater nitrogen pollution in Jinghui canal irrigation area of the loess region, northwest China. Environ Earth Sci 77:1–12

Physical, Chemical, and Biological Parameters of Surface Water and Groundwater Quality

T. Raghavendra, V. Himabindu, and D. Bhagawan

Abstract During the late 1800s Industrial Revolution, countries around the world identified novel forms of pollution. Pollution is ubiquitous, affecting both the atmosphere and bodies of water. Additional information on the variability of pollutant levels should be provided. Water-borne diseases serve as a dependable measure of contamination in the natural environment. This chapter specifies a comprehensive impression of water quality measures from an ecological standpoint, encompassing the impact on people and other organisms. The three water quality categories are analysed in depth, focussing on their notable common attributes, which comprise physical, chemical, and biological elements, water quality assessed based on these parameters and their description, resources, implications, impacts, as well as of measurement.

T. Raghavendra
Environmental Science Division, H&S Department, CVR College of Engineering, Hyderabad, Telangana, India

V. Himabindu
University College of Engineering, Science and Technology, Jawaharlal Nehru Technological University Hyderabad, Hyderabad, India
e-mail: drvhimabindu@jntuh.ac.in

D. Bhagawan (✉)
Department of Environmental Science, School of Earth Sciences, Central University of Rajasthan, Ajmer, India
e-mail: bhagawan@curaj.ac.in

Dimple et al. (eds.), *Surface Water and Groundwater Quality for Sustainable Utilization*, Springer Hydrogeology, https://doi.org/10.1007/978-981-96-8189-1_3

Graphical Abstract

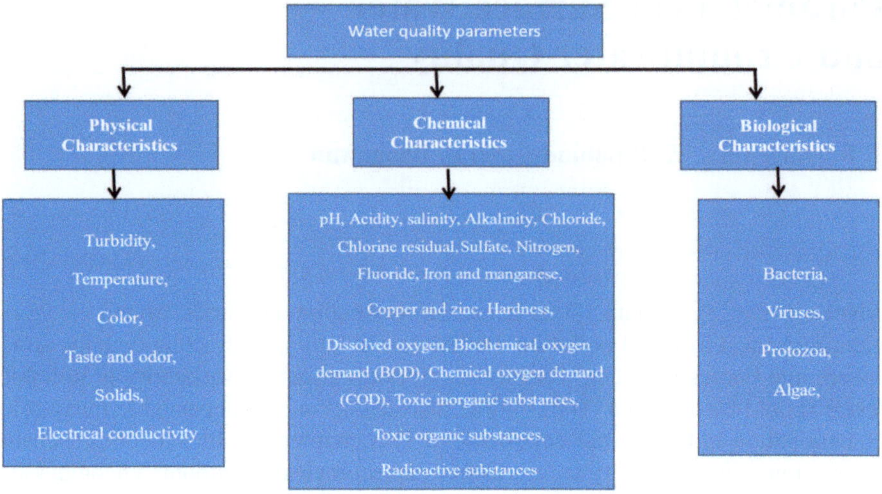

Keywords Surface water · Ground water · Water quality · Physical · Chemical · Biological parameters

1 Introduction

Water is the one element more vital to life than oxygen. As a result, in-depth research in the scientific literature has focused on water quality. The "quality" of water determines its physical, chemical, and biological characteristics. Shah et al. (2017) state that the assessment compares the condition of the water to that of living things and human activity. On Earth, life cannot exist without water, and the biosphere can only exist in water. Water is the only chemical substance that can exist in all three stages of life: solid, liquid, and gas, sleet, ice, snow, and hail), a liquid (water droplets and rain), or a gas (water vapor). This event occurs within the typical range of Earth's known air pressure and temperature. Moreover, water can quickly transition between different states.

The gaseous form of water controls the atmosphere's transpiration and moisture characteristics. The environment's water vapour molecules often remain in the state of gas for a short period, typically a few days, before they condense and form droplets of liquid or ice crystals. This process leads to the formation of clouds or fog. Likewise, fog and clouds exhibit a finite duration and frequently dissipate or release a significant amount of their moisture shortly after forming.

For most water sources on Earth, precipitation is the primary source. The hydrological cycle results from continuous moisture transfer between the Earth and the atmosphere. Evapotranspiration, precipitation, and runoff are all components of this cycle that control renewable freshwater's temporal and spatial distribution. Climate change and climatic conditions further complicate this distribution.

Earth's combined ocean and freshwater volume have remained constant across geological time. However, the proportion of ocean water to fresh water constantly fluctuates due to climatic variations primarily caused by human activity. Moreover, the water's movement throughout the hydrological cycle significantly impacts rainfall distribution and temperature adjustment. The cycle is essential for the formation of a liveable climate and the regulation of temperature.

The Earth's surface is predominantly water, with over 70% covered by this vital resource. The ocean alone accounts for around 97% of the total water on the planet. Nevertheless, the elevated salt concentration in ocean water renders it is non potable and unfit for other uses. Two percent of the left 3% is located exclusively within the glaciers and polar ice caps. By contrast, a mere one percent of water is available in streams, rivers, reservoirs, lakes, and groundwater suitable for the usage of mankind. Potable water must always be available in large amounts for all living things to survive and be healthy.

Even though it lacks calories and organic nutrients, potable water is essential for all living things, including humans. From the last few decades, global improvement is there in the accessibility of potable water. Approximately one billion individuals require accessibility to potable water, while an additional 2.5 billion require appropriate sanitary amenities. There is a direct association between the availability of pure water and the amount of money spent on goods and services, including employee remuneration, per person in a country's gross domestic product (GDP).

According to estimates by Kulshreshtha (1998), more than 50% of the global population will face the threat of water scarcity by 2025. McKinsey's 2009 estimate predicts that by 2030, the requirement for water in specific emerging nations will exceed 50% of their available resources. In the global economy, water plays a vital role as a solvent, serving in various chemical processes and being used in cooling infrastructure, transport, and businesses. According to Baroni et al. (2007), agricultural activities utilise around 70% of the Earth's freshwater resources.

The quality of water is assessed by analysing water's physical, chemical, and biological parameters with a particular set of criteria. This assessment is conducted to verify if the water is potable or if it poses any risks to the environment. Diversified water usage, including domestic, agricultural, industrial, mining, energy production, and forestry activities, contributes to water pollution, which can also affect the aquatic ecosystem as well as the availability of potable water.

Hydrological monitoring sites to measure water and keep track of its level, flow, and speed. Analyse water samples collected at frequent intervals to monitor water quality, and the results are crucial for identifying geographical and temporal patterns in groundwater and surface water.

Worldwide, regional, and local evaluations of water quality data provide insights into important characteristics of aquatic habitats while elucidating human activities' beneficial and detrimental effects. A thorough understanding of water quality might help conduct various water evaluations.

2 Classification of Water

Depending on the resources, water categorized into surface water and groundwater. Agricultural, industrial, and household pollutants can contaminate both potable (safe for drinking) and non-potable water. These pollutants include heavy metals, pesticides, fertilisers, toxic compounds, and oils (Grey 2017).

The water quality categories and their scientific descriptions:

1. Potable water: good in taste, safe and pure for human consumption, and household use.
2. Palatable water: considered to have components that are not unsafe to human health and have a pleasing appearance.
3. Polluted (Contaminated) water: includes undesirable contaminants and unsuitable for consumption or household purposes.
4. Infected water: hazardous chemicals and microorganisms contaminated the water.

3 Water Quality

Natural processes and human activities, either separately or in combination, influence surface or groundwater quality. These processes include the breakdown of minerals in bedrock, the movement of water into the atmosphere through evapotranspiration, The accumulation of dust and particulate matter from the wind, organic matter and minerals from the soil, run-off water, and biological pollutants entering aquatic habitats can cause changes in water's physical and chemical characteristics.

Water in nature can contain both non-dissolved and dissolved substances. Minerals and dissolved salts must be present in high-quality water, as these elements are critical to the health and strength of the species that depend on ecological resource. In both rural and urban areas, specific and diffuse pollution sources impact water quality. This problem is caused by runoff from agriculture, industrial discharge, and sewage discharge are some of the factors contributing to this problem. Droughts, floods, and end users' ignorance can all affect the quality of water. It's critical to take environmental cleanliness, storage, disposal, and hygiene into account while managing water resources (Stark et al. 2000).

The quality water is a critical concern for humanity as it closely correlates with human well-being. The water quality attributes of aquatic ecosystems result from various interactions involving physical, chemical, and biological factors. Due to age and characteristics, rivers, lakes, and estuaries constantly undergo geological and geochemical changes. Human activity can quickly upset the aquatic system's delicate balance, leading to pollution that most visibly manifests as fish death, bad taste, and stench.

4 Properties of Water

Two hydrogens and one oxygen atom comprise a water molecule, or H_2O. Every water molecule has an oxygen atom that is more attracted to electrons than is reasonable. When the hydrogen end has a positive charge and the oxygen end has a negative charge, a single polar covalent bond forms. The presence of this link imparts water with its polar characteristics. Water is electrically neutral because it has equal protons and electrons. The following dimensions broadly recognise and characterise the fundamental configuration of a water molecule:

O–H bond energy = 450 kJ/mol
H–O–H angle = 104.5°
O–H bond length = 95.7 picometers
Dipole moment = 1.83 debyes.

Water's distinctive characteristics are the result of intermolecular interactions among its polar molecules. Adjacent molecules form hydrogen bonds as the regions of slightly negative charge in one molecule are attracted to the areas of considerably positive charge in surrounding atoms. Essential characteristics of water include its heat capacity, dipole moment, dielectric constant, and ability to give and receive protons. It can also dissociate and combine with ions and other dipoles to form coordination complexes, as well as act as a conduit for energy transfer and storage.

To guarantee purity, the potable water or use in public or industrial activities must meet stringent criteria of the bureau of Indian standards (BIS 10500:2012), which established specific regulatory limits for over ninety potential pollutants in water (Table 2). These restrictions are essential to maintaining the purity of drinking water by preventing harmful substances that may lead to health problems or the spread of waterborne infections. In industrial facilities, water treatment is necessary to guarantee that its quality meets the required standards for certain crucial operations. Water characteristics are evaluated via physical, chemical, and biological elements.

Water quality measurements are important because different applications may have varied requirements. Dissolved oxygen (DO) is an essential factor in assessing a surface water quality. The DO levels in the water determines the degree of water pollution. Minimal levels of dissolved oxygen suggest significant water pollution and organic pollutants' use of dissolved oxygen. This page comprehensively describes the three categories of water quality parameters.

All living things require clean, safe, and plentiful freshwater to survive, as well as for ecosystems, regions, and economies to function well.

5 Parameters of Water Quality

There are three types of water quality parameters like, physical, chemical, and biological some of those were given in Table 1.

Table 1 List of water quality parameters

Physical characteristics	Chemical characteristics	Biological characteristics
Taste and odour, colour, temperature, turbidity, solids, electrical conductivity	pH, salinity, acidity, alkalinity, hardness, chloride, dissolved oxygen, biochemical oxygen demand (BOD), chemical oxygen demand (COD), chlorine residual, fluoride, sulphate, nitrogen, iron and manganese, copper and zinc, toxic inorganic and organic substances, radioactive substances	Bacteria, viruses, protozoa, algae, counts of specific organisms and groups of organisms

6 Detailed Characteristics

6.1 Physical Characteristics

It is possible to identify the water physical characteristics using the sight, senses of touch, taste, and smell (Table 1). For example, one can use touch to determine temperature, visual sense to discern colour, suspended particles, turbidity, and floating debris, and smell to detect taste and odour.

6.1.1 Turbidity

Turbidity refers to the degree of opacity or cloudiness in water. It quantifies the capacity of light to transmit through water. Suspended materials, including silt, organic debris, clay, plankton, and other particles, cause turbidity to the water (APHA 2023).

Turbidity refers to the amount of light-transmitting colloid and floating particles in water. Both aesthetic and health factors make it significant. Algal blooms, suspended sediments, plant nutrients, decomposing plant debris, and human activity all impact the transparency of natural water bodies. An affordable way to assess the levels of total suspended solids (TSS) is through turbidity. Except for relatively clear waters, it could be more sensible. However, it helps determine drinking water quality when treating water.

The turbidity in drinking water could be more pleasing, giving the water a disagreeable appearance. Turbidity can significantly affect the expenses associated with water treatment for various purposes. The presence of particles can serve as protective shelters for pathogenic microbes, effectively shielding them from the disinfection

Table 2 Permissible limits for water quality parameters as per BIS 10500:2012, along with their standard methods

S. no	Parameter	Unit	Method	Requirement (as per IS: 10500:2012)	
				Acceptable limit	Permissible limits in the absence of alternative sources
Organoleptic and physical parameters					
1	pH	–	IS 3025 (Part 11)	6.5–8.5	No relaxation
2	Color	Hazen	IS 3025 (Part 4)	5	15 Max
3	Odour	–	IS 3025 (Part 5)	Agreeable	Agreeable
4	Taste	–	IS 3025 (Part 8)	Agreeable	Agreeable
5	Turbidity	NTU	IS 3025 (Part 10)	1	5 Max
6	Total dissolved solids	mg/l	IS 3025 (Part 16)	500	2000 Max
Chemical parameters					
7	Total Alkalinity as $CaCO_3$	mg/l	IS 3025 (Part 23)	200	600 Max
8	Total Hardness as $CaCO_3$		IS 3025 (Part 21)	200	600 Max
9	Sulphates as SO_4		IS 3025 (Part 24)	200	400 Max
10	Chlorides as Cl		IS 3025 (Part 32)	250	1000 Max
11	Nitrate as NO_3		APHA 4500-NO_3	45	No relaxation
12	Aluminium as Al		APHA 3120B	0.03	0.2 Max
13	Calcium as Ca		IS 3025 (Part 40)	75	200 Max
14	Magnesium as Mg		IS 3025 (Part 46)	30	100 Max
15	Ammonia (as NH_3 N)		APHA 4500-NH_3	0.5 Max	No relaxation
16	Anionic detergents as MBAS		IS: 13428-2005	0.2	1.0 Max
17	Barium as Ba		APHA 3120B	0.7 Max	No relaxation

(continued)

Table 2 (continued)

S. no	Parameter	Unit	Method	Requirement (as per IS: 10500:2012)	
				Acceptable limit	Permissible limits in the absence of alternative sources
18	Fluoride as F		IS 3025 (Part 60)	1	1.5 Max
19	Total iron as Fe		IS 3025 (Part 53)	1.0 Max	No relaxation
20	Boron as B		APHA 3120B	0.5	2.4 Max
21	Chloramines as Cl_2		IS 3025 (Part 26)	4	No relaxation
22	Selenium as Se		IS 3025 (Part 56)	0.01	No relaxation
23	Silver as Ag	mg/l	APHA 3120B	0.1 Max	No relaxation
24	Phenolic compounds		IS 3025 (Part 43)	0.001	0.002 Max
25	Sulphide as H_2S		IS 3025 (Part 29)	0.05 Max	No relaxation
26	Copper as Cu		IS 3025 (Part 42)	0.05	1.5 Max
27	Manganese as Mn		IS 3025 (Part 59)	0.1	0.3 Max
28	Mineral oil		IS 3025 (Part 39)	0.5 Max	No relaxation
29	Zinc as Zn		IS 3025 (Part 49)	5	15 Max
30	*Pesticides residue parameters*				
	Aldrin/Dieldrin	μg/l	EPA method 8081b	0.03 Max	No relaxation
	Alachlor			20 Max	
	Atrazine			2 Max	
	Alpha HCH			0.01 Max	
	Beta-HCH			0.04 Max	
	Butachlor			0.04 Max	
	Chlorpyriphos			30 Max	
	Delta HCH			0.04 Max	

(continued)

Table 2 (continued)

S. no	Parameter	Unit	Method	Requirement (as per IS: 10500:2012)	
				Acceptable limit	Permissible limits in the absence of alternative sources
	2,4 Dichloro phenoxyacetiacid			30 Max	
	DDT			01 Max	
	Endosulfan (Alpha, Beta and Sulpahte)			0.4 Max	
	Ethion			03 Max	
	Gamma HCH (Lindane)			02 Max	
	Isoproturon			09 Max	
	Malathion, Malaoxon			190 Max	
	Methyl parathion, Methyl peraxon			0.3 Max	
	Monocrotophos			01 Max	
	Phorate, phorate sulphaxide, phorate sulphone			02 Max	
Trihalomethanes					
31	Bromoform	mg/l	APHA 6232	0.1 Max	No relaxation
	Dibromochloromethane		APHA 6232	0.1 Max	
	Bromodichloromethane		APHA 6232	0.06 Max	
	Chloroform		APHA 6232	0.2 Max	
Toxic substances					
32	Cadmium as Cd	mg/l	IS 3025 (Part 41)	0.003 Max	No relaxation
33	Cyanide as CN		IS 3025 (Part 27)	0.05 Max	
34	Lead as Pb		IS 3025 (Part 47)	0.01 Max	
35	Chromium as Cr		IS 3025 (Part 52)	0.05 Max	
36	Mercury as Hg		IS 3025 (Part 48)	0.001	
37	Molybdenum as Mo		APHA 3120B	0.07 Max	

(continued)

Table 2 (continued)

S. no	Parameter	Unit	Method	Requirement (as per IS: 10500:2012)	
				Acceptable limit	Permissible limits in the absence of alternative sources
38	Nickel as Ni		APHA 3120B	0.02 Max	
39	Arsenic as As		IS 3025 (Part 37)	0.01 Max	
40	Polychlorinated biphenyls		APHA	0.0005 Max	
41	Polynuclear aromatic hydrocarbons as PAH		APHA	0.0001 Max	
Biological parameters					
Total coli forms		MPN per 100 ml	IS 15185:2016	Shall not be detectable in any 100 ml of water	No relaxation
E. coli			IS 15185:2016	Shall not be detectable in any 100 ml of water	

process. Particulate matter can obstruct or harm the respiratory organs of fish, diminishing their immunity to illnesses, impeding their growth rates, impacting the development of eggs and larvae, and compromising the effectiveness of fishing techniques (Kiprono 2017).

A lot of harmful organic contaminants, like PCBs (polychlorinated biphenyls), PAHs (polycyclic aromatic hydrocarbons), mercury, chromium, lead, cadmium, and different pesticides, stick to particles in the air (Cole et al. 1999). As suspended particles have a higher potential to absorb solar heat than non-suspended particles, a rise in turbidity leads to higher water temperature, reducing food availability. The dissolved oxygen concentration decreases due to warm water's lower oxygen solubility than cold water (Cole et al. 1999).

A nephelometric turbidimeter is a tool for quantifying turbidity. Turbidity can be quantified using either NTU (nephelometric turbidity units) or TU (turbidity units). 1 mg/L of suspended silica is equivalent to one turbidity unit (TU). An ordinary person can perceive turbidity levels exceeding 5 NTU, but in wastewater, turbidity can reach up to 100 NTU (APHA 2023). Due to the process of natural filtering as the water percolates through the soil, groundwater often has relatively low turbidity (Viessman and Hammer 2004).

Although less frequently employed compared to other water quality criteria in this compilation, turbidity denotes the degree of cloudiness in water. Turbidity sensors are intended to quantify water's light transmission capacity. Increases in silt, clay, and organic component concentrations can cause turbidity. The primary problem is how turbidity in water appears visually. Everyone finds turbid water repulsive (Davis 2010).

6.1.2 Temperature

Temperature affects a substance's taste, thickness, capacity to dissolve, smell, and chemical reactions (APHA 2023). It also impacts the process of chlorination and sedimentation, as well as BOD (Davis 2010).

The ability to dissolve oxygen exhibits a decrease in warm water as opposed to cold water. The water temperature has a notable impact on aquatic life. Trout and salmon flourish and procreate at colder temperatures, whereas bass and sunfish prefer warmer temperatures for their health. The temperature in aquatic environments generally corresponds to the average daily air temperature. In aquatic environments, it affects reproduction, migration, and assessment time. It also affects the oxygen solubility in water, the photosynthetic activity of aquatic plants and algae. Temperature affects the solubility, viscosity, chemical reactions, odours, and water taste. The temperature of the water has an impact on sedimentation, chlorination, and biological oxygen requirements. The optimal temperature range for water is between 50 and 60 °F.

6.1.3 Colour

The colour in water is primarily an objection with the aesthetic aspect of water quality. The colour in water can create the perception that it is unsuitable for consumption, even though it may be entirely potable. The coloration of water can signify the organic components, such as chemicals and algae. In latest studies, scientists have employed colour as a quantitative measure to evaluate the existence of potentially dangerous or poisonous chemical substances in water. The clear water is crucial for most users, whether they use it for domestic or industrial purposes. Colour analysis can assist in evaluating the expenses associated with water discoloration. Colour can be eradicated from water by filtration methods, sedimentation, and coagulation. The degradation of organic matter, mostly from plants, can change the colour of water. Soil, rocks, and other inorganic materials can also affect water colour. Changes in watercolour may result in aesthetic issues, but they do not affect how the water tastes. Colour can be measured precisely by contrasting a water sample with reference solutions or coloured glass discs.

When trying to determine water's colour, it is crucial to distinguish between its apparent and actual colour. The apparent colour is composed of particulate matter and dissolved pigments. The actual hue of water can be determined once all particulate matter has been removed from the water by filtration. Note that colour can be quantified on a scale from 0 to 70 colour units. Due to its inherent lack of pigmentation, pure water does not possess any chromatic components. The presence of decomposed organic matter, specifically flora, and inorganic substances like soil, stones, and boulders contribute to the discoloration of water. This discoloration is considered undesirable from an aesthetic standpoint rather than due to any health concerns (APHA 2023). It is possible to quantify the colour of a water sample by comparing it to glass discs or colour-regulated solutions. At a concentration of 1 mg/L, a potassium chloroplatinate (K_2PtCl_6) solution produces a colour unit.

One can express the hue of a water sample in the following way: The apparent colour of a water sample encompasses both the dissolved and suspended components of its colour. We measure the actual colour by filtering the water sample to exclude suspended matter. One can assess the colour on a transparent scale ranging from 0 to 70 colour units. According to the APHA 2023 standard, water in its pure form lacks any colour, indicating a measurement of 0 colour units.

6.1.4 Taste and Odor

Introducing foreign compounds into water can alter its flavour and create Odors. These substances encompass organic compounds, suspended gases, and inorganic chemicals. Most of these pollutants come from natural, agriculture and residential sources. Water can acquire taste and odour from foreign substances such as dissolved gasses, inorganic chemicals, and organic molecules. These components may come from domestic, agricultural, or natural source (Hammer 2011).

Taste and odour are subjective human experiences that reflect the perceived quality of water. The human experience of taste encompasses the sensations of sourness (caused by hydrochloric acid), saltiness (caused by sodium chloride), sweetness (caused by sucrose), and bitterness (caused by caffeine). Simple chemicals elicit sour and salty flavours. Nevertheless, the sensations of sweetness and bitterness are generated by intricate chemical molecules. Odour is generated by releasing gases resulting from the breakdown of organic materials or the presence of chemicals introduced into the wastewater. Specialized instruments, like the Portable H_2S meter, built specifically for this purpose, quantify hydrogen sulphide odour.

6.1.5 Solids

Total solids (TDS) are inorganic materials with trace amounts of organic components dissolved in water. The main ingredients usually consist of anions such as nitrate, carbonate, hydrogen carbonate, chloride, sulphate, and cations of calcium, magnesium, sodium, and potassium. Evaporating the water and then drying the residue until it reaches a constant weight between 103 and 105 °C are the steps involved in calculating the total solids content of water.

There are three categories of solids: settleable, suspended, and filterable. Settleable materials, such as fine particles of silt and dense organic substances, descend naturally due to gravitational force. The classification of suspended particles and filterable solids is based on the particles' size and ordinary glass-fibre filters' capacity to hold onto suspended solids.

The importance of suspended particles in water is substantial for various reasons. The solids may comprise algal blooms, which can cause significant eutrophication in any aquatic environment. They will decrease the amount of light that can pass through the surface waters and disrupt the growth of aquatic plants. The deposition of these substances in rivers and lakes can result in septic conditions and unpleasant surroundings. Additionally, their presence may suggest the discharge of unacceptable sewage effluent.

Solid substances can exist in either a suspended or dissolved state when introduced into water. When a glass fibre filter filters a water sample, it traps solid particles on its surface. It is typical to measure the total dissolved solids when assessing the number of solid particles in water. Measuring the total dissolved solids enables the determination of the quantity of organic substances in the water.

Total dissolved solids (TDS) levels classify water into three basic categories:

The concentration of TDS in freshwater is typically below 1,500 mg/L.

The TDS concentration in brackish water usually falls within the range of 1,500–5,000 mg/L.

The TDS concentration of saline water exceeds 5,000 mg/L.

Solid matter in water manifests as either dissolved or suspended particles. A glass fibre filter can identify these two particles by filtering the water sample (Tchobanoglous et al. 2003). The filter effectively collects the particulate matter floating in the water while allowing the dissolved substances to flow through. Place the filtrate of the sample in a small container and expose it to evaporation, leaving behind the solid particles as a residue (APHA 2023).

The term "fixed solids" refers to the combined number of suspended solids (TSS) and dissolved solids (TDS) that remain after heating to dryness at a specific temperature. As per the APHA 2023 standard, these volatile solids disperse when exposed to a temperature of 550 °C.

6.1.6 Electrical Conductivity (EC)

EC of water refers to its capacity to transmit or conduct an electric current (Tchobanoglous et al. 2003). Since ions transmit electrical current, conductivity directly correlates with the concentration of ions in a solution (APHA 2023). Consequently, water can utilize it as a fundamental criterion to evaluate the application for agriculture. The units used in the United States for electrical conductivity are micromhos per centimetre (μmho/cm). In contrast, the International System of Units (SI) uses milli Siemens per meter (mS/m) or Deci Siemens per meter (dS/m). The value of (mS/m) equals ten μmho/cm, equivalent to 1000 μS/cm or one dS/m. Water in its pristine state does not conduct electricity efficiently. The ultra-pure water will have the electrical conductivity of 5.5×10^{-6} S/m. Drinking water typically exhibits an electrical conductivity that falls within the range of 0.005–0.05 S/m. Seawater has 5 S/m.

According to the APHA 2023 guidelines, EC can be used to approximate the Total Dissolved Solids measurement of water.

$$\text{TDS (mg/L)} \cong \text{EC (dS/m or } \mu\text{mho/cm)} \times (0.55 - 0.7)$$

When replenishing groundwater with treated wastewater, total dissolved solids can be employed to assess the concentration of ions in the water. According to APHA 2023, the electrometric approach is the standard measurement method.

Another fundamental physical characteristic that you should be familiar with pertains to electrical conductivity, which quantifies the capability of a water sample or a comparable solution to transmit or pass electrical currents. The conductivity of water will rise in proportion to the quantity of ions present.

Conductivity is a crucial metric in water quality assessment due to its high sensitivity in detecting water contamination levels. High conductivity indicates a significant presence of impurities in the water. However, drinkable water and ultra-pure water are essentially non-conductive to electricity.

Water conducts electricity due to the dissolution of positively and negatively charged ions. Sodium (Na^+), calcium (Ca^{+2}), potassium (K^+), and magnesium (Mg^{+2}) are the primary positively charged ions. Chloride (Cl^-), sulphate ($SO_4{}^{-2}$), carbonate ($CO_3{}^{-2}$), and bicarbonate (HCO_{3-}) are the primary anions present in water. Phosphates ($PO_4{}^{-3}$) and nitrates ($NO_3{}^{-2}$) have negligible effects on conductivity. However, they possess substantial biological significance. While not intrinsically captivating, conductivity is a crucial indicator of water's hardness, pH, and concentration of dissolved particles. Water's electrical conductivity varies depending on its origin, including groundwater, agricultural runoff, municipal wastewater, or rainfall. Hence, conductivity can serve as an indicator of either groundwater seepage or a sewage leak.

6.2 Chemical Characteristics

The primary cause of exposure to drinking water's chemical components is the potential for long-term exposure to harmful health effects. Some of chemical elements in present in water can cause severe health impacts, even with just one instance of exposure. The chemical poisoning of drinking water can lead to a significant number of severe health issues. The following are the water's primary chemical features (Table 2).

6.2.1 pH

The pH level is one of the most critical factors in evaluating water quality. The negative logarithm of the concentration of hydrogen ions explains pH (Edzwald 2010). Hammer (2011) employs a dimensionless numerical value to denote a solution's acidity or alkalinity level. The pH scale uses numbers to indicate how acidic or alkaline water is. According to Tchobanoglous et al. (2003), an abundance of hydrogen ions (H^+) indicates acidic water, whereas an abundance of hydroxyl (OH) ions indicates alkaline water.

The pH range is 0 to 14, with seven being neutral. A pH below 7 means the solution is acidic, while a pH above 7 means the solution is alkaline. Clean water is considered neutral when the temperature is 25 °C and the pH is about 7.0.

The pH of ordinary rainfall, approximately 5.6, signifies a somewhat acidic characteristic resulting from atmospheric carbon dioxide gas (APHA 2023). The WHO recommends a pH permissible limit is 6.5–8.5 for drinking water. This range is essential for living organisms' optimal health and survival (WHO 2011). The APHA 2023 recommendations provide two distinct procedures for detecting the pH level: the electrometric and the colorimetric methods.

Extreme pH levels, whether too high or too low, can negatively affect water's usability. An elevated pH level results in a bitter taste and reduces chlorine disinfection efficiency, necessitating extra chlorine use (Campanella et al. 2016). As the pH level increases, the oxygen concentration in water also increases. Water with a high acidity level, as indicated by a low pH, can corrode and degrade metals and other materials. Water pH can be altered by pollution which leads to detrimental effects on aquatic organisms and flora (APHA 2023).

The impact of pH on animals and plants:

To thrive, a particular pH level is necessary for most aquatic species, including flora and fauna.

Even a slight change in pH can harm their overall health.

Acidic water with a relatively low pH can hinder fish egg hatching, damage membranes and injures fish and aquatic insects' (Kiprono 2017).

Water with excessively high or low pH values can be lethal.

Most fish cannot survive in water with a pH level lower than four or higher than 10, and only a few species can withstand water with a pH below three or over 11.

Because of their skin's exceptional sensitivity to pollutants, amphibians are highly susceptible to low pH levels, putting them at significant risk of endangerment.

According to Kiprono (2017), acid rain's reduction in pH levels could be the cause of the global decline in amphibian populations.

The impact of pH on chemical compounds in water:

Heavy metals, like chromium lead, and cadmium, are mostly dissolve in water with high acidity (low pH).

pH variations cause changes in the chemical structures of some compounds in water.

This is crucial because the solubility of some heavy metals significantly rises when dissolved in water, increasing their potential for harm.

Consequently, it can potentially impact aquatic plants and animals (Edzwald 2010).

In water with a neutral or acidic pH, Nevertheless, when the water's alkalinity increases (resulting in a higher pH level), the toxicity of ammonia to these species intensifies.

6.2.2 Salinity

Salinity is a quantitative evaluation of the amount of salt present in water. The presence of dissolved ions such as sodium, chloride, carbonate, and sulphate, however, increases the salinity of different saline fluids. Salts and other contaminants have an impact on the water quality that is used for agriculture or consumption. In addition, they impact aquatic organisms, with each type of creature having a specific range of salinity that it can withstand. High salt water can be unfit for residential, agricultural, or industrial purposes. Furthermore, the water's ionic content can be crucial. Cladocerans, sometimes known as water fleas, have a higher sensitivity to potassium chloride than sodium chloride when both are present at equal concentrations.

6.2.3 Acidity

Acidity is a property that measures the quantity of acid in a solution. The ability of water to completely neutralize a strong base up to a particular pH level indicates how acidic the water is. Carbon dioxide, mineral acids, and hydrolysed salts such as aluminium and ferric sulphate are all common sources of water acidity. Acids can affect several activities, such as biological activity, chemical reactions, and corrosion (APHA 2023).

Carbon dioxide, whether derived from the environment or produced by aquatic organisms through respiration, elevates the acidity of water upon dissolution, resulting in carbonic acid (H_2CO_3) formation. We evaluate the acidity level by titrating with a standard sodium hydroxide solution (0.02 N) and using an indicator called phenolphthalein (APHA 2023). This concerns quantifying the acidity level in a exact solution. Mineral acids, hydrolysed salts, and carbon dioxide primarily contribute to acidity. When added to water, acids can affect various processes, including biological activity, chemical reactions, and corrosion. A pH sensor determines the acidity of water.

6.2.4 Alkalinity

The natural water alkalinity mainly arises from the formation of bicarbonates during processes that occur in the soils through which the water filters. It quantifies the water's ability to counteract acidic effects and indicates its buffering capacity. Furthermore, we can attribute this to the existence of hydroxides and carbonates.

Alkalinity plays a crucial role in safeguarding fish and aquatic organisms by acting as a protective barrier against sudden fluctuations in pH levels. Organisms, particularly those that live in water, have the most favourable growth conditions when the pH level is between 6.0 and 9.0. Surface waters with high alkalinity can mitigate the acid rain effects and other acidic pollutants. This strategy is implemented to prevent detrimental fluctuations in pH levels and protect aquatic creatures.

The composition of rocks and soils, the presence of salts, specific plant activity, and the discharge of industrial effluent all influence the alkalinity levels in streams. Oligotrophic lakes, which have low nutrient levels, typically exhibit lesser alkalinity. Conversely, eutrophic lakes with high nutrition levels tend to display more alkalinity. Alkalinity measures water's ability to neutralise acids. The main goal of evaluating the alkalinity is to verify the accurate amounts of lime and soda needed for water softening, which is particularly helpful in lowering boiler corrosion. Bicarbonate ions, carbonate ions, and hydroxide ions enhance water's alkalinity; when its pH is more than 7.0, the water is considered alkaline.

6.2.5 Chloride

Groundwater, streams, and lakes naturally contain chloride. However, if the chloride concentration in freshwater is relatively high, around 250 mg/L or more, it could be a sign of wastewater pollution (Chatterjee 2001). Several resources, such as high-chloride rocks, agricultural runoff, and wastewater, can introduce chlorides into surface water.

Chloride ions Consuming water that contains chloride ions (Cl) does not present any health hazards to the general population. However, due to elevated levels of chloride ions, most individuals may experience an unpleasant, salty taste. Although most chlorides are considered safe for human consumption, table salt, due to its high sodium content, has been associated with cardiovascular and kidney issues (WHO 2011). In animal and plant organisms, minimal amounts of chlorides are necessary for fundamental cellular processes.

One can taste sodium chloride at a concentration of 250 mg/L, but you typically can't detect magnesium or calcium chloride by taste until they reach values of 1000 mg/L. The regulations for potable water stipulate that chloride values should not be more than 250 mg/L. Several methods are available for measuring the chloride in water, the most used technique is titration using silver nitrate (APHA 2023).

6.2.6 Residual Chlorine

Natural water should not contain chlorine (Cl_2); however, humans intentionally introduce it to kill microorganisms and make it safe for use. Although chlorine is inherently dangerous as a gas, it poses no risk to human health when present in a diluted aqueous solution. The recommended amount of residual chlorine in potable water is

0.2 mg/L. According to Husain et al. (2010), the water distribution system's residual concentration ensures the water's superior hygienic quality.

Chlorine can undergo a chemical reaction with organic substances present in water, creating harmful molecules known as trihalomethanes (THMs). Carcinogens classify these compounds, including chloroform ($CHCl_3$). We often conduct chlorine residual measurement using either a colour comparator test kit or a spectrophotometer (APHA 2023).

6.2.7 Hardness

Water hardness is an inherent quality that can improve its taste and make it more appealing to drink. The natural occurrence of calcium and magnesium minerals in the water is responsible for its hardness. Indicators of a hard water supply include inadequate soap foaming and scum. Carbonates cause transient hardness, while non-carbonates cause permanent hardness. Boiling water effectively eliminates the transient hardness that water minerals cause.

The following is a quantitative assessment of the level of hardness, measured in mg/L or ppm as calcium carbonate ($CaCO_3$).

The softness level is measured in mg/L as $CaCO_3$ and ranges from 0 to 100 mg/L.

The recommended range for moderate water hardness is 100–200 mg/l as $CaCO_3$.

The hardness level should be between 200 and 300 mg/L as $CaCO_3$.

Extremely challenging: 300–500 mg/L as $CaCO_3$.

Exceptionally high: 500–1,000 mg/l $CaCO_3$.

Hardness is caused by high mineral concentrations in water. If you neglect to properly maintain your water supply, the presence of dissolved ions may lead to the formation of scale deposits on your hot water pipes. Showering with water with a high mineral content can impede the formation of a lather when using soap. Magnesium and calcium ions can be present in water due to their infiltration from rocks and soil. In general, groundwater typically exhibits higher hardness than surface water. Hard water commonly refers to water that contains a significant amount of minerals (APHA 2023).

The leading causes of water hardness in natural water sources are calcium (Ca^{2+}) and magnesium (Mg^{2+}) ions. These ions come into direct contact with rocks and soil, particularly deposits of limestone, and end up in the water. The ions also include bicarbonates, sulphates, and occasionally chlorides and nitrates (Davis and David 2008).

Carbonates and bicarbonates generate temporary hardness, which boiling can eliminate, while sulphates and chlorides primarily cause permanent hardness, which persists even after boiling. Water containing a hardness level exceeding 300 mg/L is generally classified as complex, while a hardness level exceeding 150 mg/L is often noticeable to most individuals. Conversely, water with a hardness level below 75 mg/L is considered soft (APHA 2023; Hammer 2011).

From a health perspective, water hardness levels up to 500 mg/L are considered healthy. However, exceeding this threshold may lead to a laxative effect. We often assess hardness through titration using EDTA (ethylene diamine tetraacetic acid) and indicators like Eriochrome Black and Blue. We typically measure the concentration in mg/L of calcium carbonate (APHA 2023).

6.2.8 Major Ions in Water

Water supplies contain a variety of trace ions that impact the chemical properties and make up most of the mineral content in natural water.

Sodium can have important implications for people's health. Water frequently washes sodium salts out of the land environment into both groundwater and surface water due to their high solubility.

Potassium is a vital dietary component found in drinking water. However, it can have a laxative effect when present in excessive amounts.

Calcium is essential for human nutrition and is crucial in developing teeth and bones. People commonly refer to it as limestone, and it is the major component causing hardness to water.

Magnesium is a highly profuse element in the earth surface. High quantities of magnesium sulphates can induce a laxative effect in certain individuals. Furthermore, when present in large concentrations, it has a disagreeable flavour. The elevated concentration could potentially be dangerous to individuals with heart and kidney conditions. Taste preferences establish the limits on chloride concentrations in drinking water.

Water containing excessive levels of sulphate may have a harsh taste. It also contributes to the occurrence of unpleasant smells. An excessive amount of bicarbonate increases water's salinity and total solid content. Boiling quickly eliminates the carbonate content, making it a transient form of water hardness Even small amounts of nitrates can negatively affect the blood's ability to carry oxygen, causing health problems in infants under six months of age and pregnant women.

6.2.9 Sulphate

Natural water and wastewater contain sulphate ions (SO_4^{2-}). The increased amounts of sulphate in natural water are commonly a result of the discharge of natural sodium sulphate deposits (Glauber's salt) or magnesium sulphate (Epson salt). While the excessive use of chemicals in drinking water may result in disagreeable flavours or unwanted laxative effects, it does not cause a significant health concern to the people (Davis and David 2008).

6.2.10 Nitrogen

The four nitrogen components observed in water and wastewater are in the forms of organic nitrogen, nitrite, nitrate, and ammonia pollution results from breaking organic molecules and ammonia into nitrogen in sewage-contaminated water. Nitrites and nitrates are the products of microorganisms' transformation of these substances. Nitrate, or nitrogen, is a mineral necessary for plant progression but has the potential to stunt the development of certain plants (APHA 2023).

Elevated levels of nitrates in surface water can increase the growth of algae, which leads to a deterioration of water quality. Agricultural areas frequently use chemical fertilizers, which can lead to groundwater contamination with nitrates. Drinking water with a nitrate concentration exceeding 10 mg/L poses an immediate and severe health risk to newborns. Nitrate ions in the bloodstream produce a chemical reaction with haemoglobin, which diminishes the blood's capacity to carry oxygen. This medical disorder is referred to as "blue baby" or methemoglobinemia (APHA 2023; Hammer 2011).

6.2.11 Fluoride

Drinking water with an ideal concentration of fluoride ions (F^-) is beneficial for maintaining optimal oral health. 1.0 mg/L fluoride is sufficient to efficiently inhibit tooth decay, especially in children. An overabundance of fluoride causes dental fluorosis, a disorder characterised by discoloured teeth (Davis 2010; WHO 2011). The WHO recommends the maximum permissible concentration of fluoride in drinking water at 1.4 mg/L in hot areas and 2.4 mg/L in cold areas of the country (APHA 2023).

6.2.12 Iron and Manganese

Even in trace concentrations, iron (Fe) and manganese (Mn) give drinking water a unique bitter taste. However, they may not have any negative health impacts (APHA 2023). Ferrous (Fe^{2+}) and manganous (Mn^{2+}) ions are the common forms of these metals in groundwater. When exposed to air, these ions change chemically into the insoluble ferric (Fe^{3+}) and manganic (Mn^{3+}) forms. Because of this, the water is murky and unfit for drinking. Additionally, these ions may stain plumbing fixtures and textiles in shades of brown or black. Use various instrumental methods to measure them, such as inductively coupled plasma, cold vapor, flame, electrothermal, and atomic absorption spectrometry (APHA 2023).

6.2.13 Copper and Zinc

In low levels, copper (Cu) and zinc (Zn) are non-toxic. Both are important and advantageous for human well-being and development, as well as plant and animal growth. However, they have the potential to induce unpleasant flavours in potable water. When zinc levels are in the water, it gives it a milky appearance. To measure copper (Cu) and zinc (Zn) use the same techniques as those specified in APHA (2023).

6.2.14 Dissolved Oxygen

Dissolved oxygen (DO) is the quantity of oxygen gas in a liquid. Water is infiltrated through diffusion from the encompassing atmosphere, through exposure to air, and as a byproduct of photosynthesis. Most aquatic organisms need dissolved oxygen for their survival and growth. Trout and stoneflies depend on a substantial level of dissolved oxygen (DO), whereas catfish, worms, and dragonflies can tolerate a relatively lower level of DO.

Inadequate oxygen content in water can result in the mortality of both adult and juvenile organisms, hindered growth, unsuccessful survival of eggs and larvae, and alterations in the species composition within a specific aquatic environment. The presence of low oxygen levels (DO < 3 mg/L) in a water body results in impaired cellular activity and disturbs the balance of circulatory fluids in the aquatic system, ultimately resulting in mortality. The test is an essential assessment of water contamination, as defined by the APHA in 2023.

As dissolved oxygen concentration rises, water quality improves. The solubility of oxygen in water is moderate, and it reacts quickly to temperature changes. For example, at 20 °C, a chemical's solubility limit in water is about 9 mg/L. The concentration is 14.6 mg/L at 0 °C (Abbas et al. 2014).

Salinity, temperature, and water pressure influence the amount of dissolved oxygen in the solution. However, levels of dissolved oxygen do not directly correlate with public health. Many individuals find drinking water with minimal or no oxygen content disagreeable. There are three primary methods for detecting dissolved oxygen levels: the electrometric technique, noted for its speed and cost-effectiveness; the Winkler titration process, a more traditional method; and the colorimetric technique (APHA 2023).

6.2.15 Biochemical Oxygen Demand (BOD)

For survival, bacteria and other microbes consume organic materials. Metabolism produces oxygen. This method lowers the amount of dissolved oxygen in a sample of water. A high enough concentration of organic matter in the water will use a lot of dissolved oxygen to help the organic matter break down. Nevertheless, this proposes difficulties as underwater flora and fauna necessitate dissolved oxygen for survival.

The dilution method can measure the biological oxygen demand, or the amount of dissolved oxygen needed by aerobic organisms to decompose organic molecules in water within a given time frame and temperature range. People frequently utilize BOD levels to measure the organic purity of water, indicating the extent of contamination. Increased BOD levels are indicative of water pollution.

A widely used method for quantifying it involves calculating the quantity of oxygen consumed per litre of the specimen throughout five days (BOD5) at 20 °C. During organic matter decomposition, microbes, including bacteria and fungi, consume the organic matter, leading to its final oxidation. Microorganisms will use more oxygen as they exert more effort, leading to an elevated level of biochemical oxygen demand (BOD), which in turn reduces the oxygen available for other aquatic species. Bacteria and other microbes utilise organic compounds as sources of nourishment. They use oxygen to break down organic matter (APHA 2023; Tchobanoglous et al. 2003).

The organic substances are decomposed into less complex chemicals, such as carbon dioxide and water, and the microorganisms utilise the energy produced from this process to support their development and reproduction (Abbas et al. 2014). In the absence of constant natural or artificial replenishment, the levels of DO in water will decline as microorganisms degrade organic matter. The bacteria use more BOD as water organic matter increases. BOD, or biochemical oxygen demand, is an indicator of pollution load in wastewater. A high BOD indicates more pollution, while a low BOD indicates less pollution (Abbas et al. 2014).

Microorganisms require a significant amount of time to fully break down organic matter, typically 20 days or longer in normal conditions. BOD is a variable that is time dependent. There will be no oxygen consumption at the initial time (time = 0), and the biochemical oxygen demand (BOD) will equal zero. With time, microorganisms consume oxygen, leading to a rise in biochemical oxygen demand (BOD). After reaching the biological oxygen demand limit (BODL), the organic components completely decompose (Abbas et al. 2014).

6.2.16 Chemical Oxygen Demand

Chemical Oxygen Demand (COD) calculates how much oxygen, under certain conditions (oxidising agent, temperature, and duration), is needed to oxidize the organic materials in the water. COD, which reflects the result of wastewater discharge on the ecosystem, is an essential metric for assessing water quality. Higher COD levels suggest that the sample contains more organic material that is susceptible to oxidation. The breakdown of this material results in a decrease in oxygen levels within the water.

The BOD to COD ratio quantifies the amount of organic matter in water that naturally occurring microorganisms in the surrounding environment can decompose. COD is a quantitative measure of both biodegradable and non-biodegradable organic compounds (Tchobanoglous et al. 2003). The test involves using potent oxidising agents such as potassium dichromate, sulphuric acid, and temperature. You can obtain the outcome within a mere two hours. The COD values consistently go beyond the BOD values for the identical sample (APHA 2023).

6.3 Heavy Metals

“Heavy metal” refers to highly dense metallic chemical components that can be harmful or toxic even in trace amounts. Distinguished examples of heavy metals are lead (Pb), chromium (Cr), nickel (Ni), copper (Cu), cobalt (Co), mercury (Hg), cadmium (Cd), arsenic (As), and nickel (Ni). These are the geological milieu’s fundamental constituents. They enter the body via food, water, and, to a lesser extent, air inhalation.

Specific micronutrients, like copper, selenium, and zinc are maintaining the human body’s metabolic activities. However, they can become poisonous and lead to several serious diseases when present in high concentrations. There are many different types of inorganic hazardous chemicals in water, although their concentrations are negligible. Even in small quantities, they can pose a risk to public health. Industrial activity and improper management of hazardous waste produce some poisonous substances, while others occur naturally. These entities can be classified into two clearly defined categories.

6.3.1 Metallic Compounds

In this group are some hazardous heavy metals, such as arsenic (As), barium (Ba), thallium (Tl), selenium (Se), cadmium (Cd), chromium (Cr), lead (Pb), mercury (Hg), and silver (Ag). Each metal is dangerous in its own way. Some chemicals, like arsenic (As) and hexavalent chromium ($Cr^{6+)}$, are hazardous and can hurt you immediately. Some, like thallium (Tl), lead (Pb), mercury (Hg), and cadmium (Cd), might not hurt you right away, but they can cause health problems in the long run. Smith (2016) and WHO (2003a, b): Scientists can use spectrophotometers, inductively coupled plasma (ICP), or atomic absorption photometers to find low levels of heavy metals.

6.3.2 Non-metallic Compounds

This category comprises nitrates (NO_3) and cyanides (CN_3). We have already discussed nitrate in the previous section, specifically its relationship to nitrogen. Cyanide induces oxygen deprivation by binding to the haemoglobin sites, thereby inhibiting oxygen transport by red blood cells. This leads to a condition known as cyanosis, characterised by a blue skin colour (Hammer 2011; WHO 2011). Additionally, it induces chronic impacts on the central nervous system and thyroid. People often conduct the quantification of cyanide using colorimetric, titrimetric, or electrometric techniques (APHA 2023).

6.4 Toxic Organic Substances

The literature has documented the presence of over 100 hazardous chemical compounds in water. These pollutants are typically artificial and not naturally present in water. The substances mentioned encompass insecticides, pesticides, solvents, detergents, and disinfectants (Davis 2010; Tchobanoglous et al. 2003). Advanced instrumental methods for conducting tests include mass spectrometry, gas chromatography (GC), and high-performance liquid chromatography (HPLC) (APHA 2023).

6.5 Radioactive Substances

Possible origins of radioactive compounds in water encompass the disposal of nuclear power plant byproducts, industrial waste, or the use of radioactive chemicals in medical research, as well as the extraction of radioactive elements through mining. Radioactive compounds undergo decay, resulting in the emission of beta, alpha, and gamma radiation.

In humans and other animals exposure to the radiation can harm living tissues genetically and physically (Cothern 2014; Skeppstrom et al. 2007). The natural occurrence of radon gas in groundwater and its high volatility, which allows inhalation during showering, pose a significant health risk. When it comes to drinking water, there are widely accepted standards that are regularly applied to measure alpha particles, beta particles, photon emitters, radium-226 and -228, and uranium (Cothern 2014). In water quality applications, the unit of radioactivity is the picocurie per litre (pCi/L). One picocurie is approximately equal to two atoms undergoing disintegration each minute. It can be determined using a range of advanced instrumental methods (Skeppstrom et al. 2007).

6.6 Biological Parameters of Water Quality

The presence or absence of organisms can serve as a dependable indicator of water quality. Ecologists can analyse fish and bug populations in natural aquatic environments and assess the overall water quality. Consequently, an aquatic environment exhibiting abundant, varied, and harmonious organisms is regarded as a thriving ecosystem. Some species possess a well-established ability to tolerate specific pollutants, which enables them to serve as indicators of the presence of toxins (APHA 2023; Davis 2010; Tchobanoglous et al. 2003).

Microorganisms are there widely in environment. The human body sustains a typical community of microorganisms in the digestive system, with a notable fraction composed of coliform bacteria (Wiesmann et al. 2007). Although wastewater has a high concentration of germs per millilitre, most of these bacteria are not pathogenic. According to Wiesmann et al. (2007), dangerous bacteria are only likely present in wastewater if they contain waste from individuals with disorders.

6.6.1 Bacteria

Due to their cellular form and method of obtaining nutrients, we classify bacteria as unicellular organisms. The three main cell shapes of bacteria are the rod-shaped bacillus, the sphere-shaped cocci, and the spiral-shaped spirillum (Wiesmann et al. 2007; APHA 2023).

In about 30 min, a solitary bacterial cell can undergo maturation and division, creating two new cells. Bacteria reproduce quickly in the right environment, including enough food, a comfortable temperature, and a suitable pH. In just 24 h, bacterial culture can grow to a density of 20 million cells per millilitre.

The exponential proliferation of observable bacterial colonies on a suitable, nourishing substrate enables the identification and quantification of bacteria in water (Mara and Horan 2003). Multiple differentiations exist among the diverse types of bacteria. One differentiation is based on their method of food metabolism. Aerobic bacteria depend on oxygen for their metabolic processes, whereas anaerobic bacteria can only survive in an oxygen-deprived environment. Facultative bacteria are a type of bacterium that may survive in environments with or without oxygen (Wiesmann et al. 2007).

Bacteria exhibit reduced growth and reproduction rates under cold temperatures. With each incremental 10°C temperature rise, the growth rate and reproduction rate double until it reaches the optimal temperature. Most bacterial species thrive at an optimal temperature of around 35 °C (Mara and Horan 2003). Bacteria cause some dangerous aquatic illnesses, including cholera, leptospirosis, tularemia, shigellosis, and typhoid and paratyphoid fever. Sometimes, gastroenteritis outbreaks occur due to inadequate application of proper sanitary practices. These outbreaks might involve one or more diseases (Tchobanoglous 2010).

Bacteria are unicellular bacteria that can quickly metabolize nutrients and reproduce promptly under optimal water pH, food availability, and temperature conditions. Measuring the number of bacteria in a water sample is difficult because they overgrow. Generally, bacteria have a slow rate of reproduction in colder water. Water with high bacterial contamination can cause several waterborne illnesses, including cholera, tularaemia, and typhoid.

6.6.2 Algae

Algae, small organisms classified as plants, contain photosynthetic pigments like chlorophyll. Autotrophic organisms need sunlight to convert inorganic elements into organic matter. During this process, they emit oxygen and take in carbon dioxide. Additionally, they are essential to the stabilization ponds' wastewater treatment process. Algae cause taste and odour difficulties. Some algae species can harm the environment and present health impacts to the public. For example, if livestock and other domestic animals ingest water contaminated with blue-green algal species, they might die (Wiesmann et al. 2007; Mara and Horan 2003).

Algae consisting of pigments that can carry out photosynthesis. These plants survive because solar radiation effectively transforms inorganic materials into organic ones. As this process continues, the algae take in carbon dioxide and expel oxygen. Stabilization ponds are a necessary component of the wastewater treatment process, which depends on algae. Algae cause two primary problems: an offensive stench and a decline in flavour. Certain types of algae pose significant hazards to human health, and blue-green algae have the potential to be lethal to cattle.

6.6.3 Viruses

Viruses are the smallest biological entities, containing all the genetic information needed for reproduction. Only a high-resolution electron microscope can observe them. Viruses are obligatory intracellular parasites that depend on their host organism for survival and replication.

They can pass through filters that are resistant to microorganisms. Waterborne viral infections are acknowledged as the primary causes of infectious hepatitis and poliomyelitis. Water treatment plants employ a disinfection method that is highly efficient in eliminating most waterborne viruses (Tchobanoglous 2010; Wiesmann et al. 2007; Mara and Horan 2003).

Viruses are minuscule biological entities that can harm an individual's well-being. Only powerful electron microscopes can observe them. All viruses are dependent on parasites for their survival. Due to their small size, viruses can easily penetrate most filters. Specific aquatic viruses have the potential to induce hepatitis and other related health complications. Although treating viruses can be challenging, most water treatment resources should be able to eradicate bacteria effectively through the disinfection process.

6.6.4 Protozoa

Protozoa are single-celled microbes that consume solid organic particles, bacteria, and algae. Larger, multicellular organisms then consume them as nourishment. Aquatic protozoa are microorganisms found suspended in water and commonly known as zooplankton. The cysts they generate resist disinfection (Tchobanoglous et al. 2003; Wiesmann et al. 2007; Davis 2010).

6.6.5 Indicator Organisms

Coliforms, a bacterium, are an essential biological indicator for assessing water quality and detecting contamination. Bodily waste consistently expels large quantities of pathogenic coliforms from the human digestive system (Wiesmann et al. 2007). Therefore, sewage-polluted water invariably contains coliforms.

Escherichia coli is a coliform bacterium commonly found in home sewage. The probability of discovering them is excellent, even in modest water pollution. The untreated sewage sample, with a volume of 100 mL, has an estimated quantity of 3 million Escherichia coli bacteria (APHA 2023). Coliform bacteria are highly durable microorganisms with a longer lifespan in water than other pathogens that cause diseases. The two primary techniques for screening for coliform bacteria are the membrane filter and multiple-tube fermentation processes. The following part will describe the fundamental approach because coliform bacteria testing is essential for protecting public health (APHA 2023; Cothern 2014).

7 Summary

Nature's water is never completely free of impurities. Precipitation provides most the Earth's water sources. Precipitation involves water movement over the surface (runoff) and through the ground (infiltration), picking up various dissolved contaminants that significantly affect its usability. Because of its distinct physical, chemical, and biological properties, water is essential to the survival of both plants and animals. These attributes directly influence the features and function of aquatic life. Water quality features establish the rules for discharging waste into water bodies. Furthermore, these attributes intricately link to enhancing water quality and establishing policy measures.

8 Conclusion

The definition, origins, impacts, effects, and measurement techniques of surface and groundwater quality's physical, chemical, and biological characteristics are reviewed. We also address the categorization of water based on quality, providing a distinct definition for each category.

References

Abbas SH, Ismail IM, Mostafa TM, Sulaymon AH (2014) Biosorption of heavy metals: a review. J Chem Sci Technol 3:74–102

APHA. Lipps WC, Braun-Howland EB, Baxter TE (eds) (2023) Standard methods for the examination of water and wastewater, 24th edn. APHA Press, Washington DC

Baroni L, Cenci L, Tettamanti M, Berati M (2007) Evaluating the environmental impact of various dietary patterns combined with different food production systems. Eur J Clin Nutr 61(2):279–286

Campanella B, Onor M, D'Ulivo A, Giannecchini R, D'Orazio M, Petrini R et al (2016) Human exposure to thallium through tap water: a study from Valdicastello Carducci and Pietrasanta (northern Tuscany, Italy). Sci Total Environ 548:33–42

Chatterjee A (2001) Water supply waste disposal and environmental pollution engineering (including Odour, noise and air pollution and its control), 7th edn. Khanna Publishers, Delhi

Cole S, Codling I, Parr W, Zabel T, Nature E, Heritage SN (1999) Guidelines for managing water quality impacts within UK European Marine Sites

Cothern CR (2014) Radon, radium, and uranium in drinking water. CRC Press, Boca Raton

Davis ML, David A (2008) Introduction to environmental engineering, 4th edn. McGraw-Hill Companies, New York

Davis ML (2010) Water and wastewater engineering—design principles and practice. McGraw-Hill, New York

Edzwald JK (2010) Water quality and treatment a handbook on drinking water. McGraw-Hill, New York

Gray N (2017) Water technology, 3rd edn. CRC Press, London

Hammer MJ (2011) Water and wastewater technology, 7th edn. Pearson Education, Upper Saddle River

Husain A, Bushara, Rao GP, Damak S, Ali MM (2010) UNESCO – EOLSS sample chapters physical, chemical and biological aspects of water – general characteristics of water. Encyclopedia of Desalination and Water Resources (DESWARE)

Kiprono SW (2017) Fish parasites and fisheries productivity in relation to exreme flooding of Lake Baringo, Kenya [PhD]. Kenyatta University, Nairobi

Kulshreshtha (1998) A global outlook for water resources to the year 2025. Water Resourc Manag 12:167–184

Mara D, Horan NJ (2003) Handbook of water and wastewater microbiology. Elsevier, London

Shah C (2017) Which physical, chemical and biological parameters of water determine its quality

Skeppstrom O (2007) Uranium and radon in groundwater an overview of the problem. Eur Water EWRA 17/18:51

Smith DL, Cooper MJ, Kosiara JM, Lamberti GA (2016) Body burdens of heavy metals in Lake Michigan wetland turtles. Environ Monit Assess 188:128

Stark JR, Hanson PE, Goldstein RM, Fallon JD, Fong AL, Kroening SE, Andrews WJ (2000) Water quality in the Upper Mississippi River Basin, Minnesota, Wisconsin, South Dakota. United States Ceol Surv Circ 1211:1995–1998

Tchobanoglous (2010) The evolution of wastewater treatment. H2O New York City Water Summit, 9 Apr 2010, p 30, New York, NY

Tchobanoglous G, Burton FL, Stensel HD, Metcalf & Eddy, Inc. (2003) Wastewater engineering: treatment and reuse, 4th edn. Tata McGraw-Hill Limited, New Delhi

Viessman W, Hammer MJ (2004) Water supply and pollution control, 7th edn. New Jersey Pearson Prentice Hall, Upper Saddle River

Wiesmann U, Choi IS, Dombrowski E-M (2007) Fundamentals of biological wastewater treatment. Wiley, Darmstadt

World Health Organization (2011) Guidelines for drinking-water quality, 4th edn. WHO, Geneva
World Health Organization (2003a) Overview of the world health report 2003: shaping the future Geneva. http://www.who.int/whr/2003/en/overview_en.pdf
World Health Organization (2003b) Chromium in Drinking-Water. World Health Organization (WHO), Geneva

Soft Computing Methods for Surface Water and Groundwater Quality Analysis

Salim Heddam

Abstract Since several decades, monitoring and control of water quality continue to attract a great deal of interest and, indeed, there have been a number of developments in this subject. However, the application of machine learning algorithms for water quality has significantly increased which have explored in number in the early of last years. Hereafter, we explore the capability of the various machine learning models, i.e., adaptive boosting (AdaBoost), Categorical Boosting (CatBoost), Extreme Gradient Boosting (XGBoost), and Light Gradient-Boosting Machine (LightGBM), for predicting critical water quality variables namely, the Phycocyanin fluorescence water in situ (PC). The PC variable was modelled using various water quality variables, i.e., water pH, water specific conductance (SC), water temperature (T_w), dissolved oxygen concentration (DO), and discharge (Q). The performances of the four machine learning models were compared based on numerical performances, i.e., the root mean squared error (RMSE), the mean absolute error (MAE), the coefficient of correlation (R), and the Nash-Sutcliffe efficiency (NSE), and graphical visualization. In depth, analysis and comparison between the models were done by highlighting some important conclusions. Results obtained revealed that, PC can be predicted with high precision and accuracies using all models and the XGBoost exhibited the high performances with R $\approx$ 0.922, NSE $\approx$ 0.849, RMSE $\approx$ 0.271 mg/L, and MAE $\approx$ 0.191 mg/L.

Keywords Modelling · Predicting · Water quality · Phycocyanin · DO · T_w · SC · pH · DO

1 Introduction

For decades, the control of water quality has been considered as a priority for ensuring public health (Sarafaraz et al. 2024). The number of monitoring stations has increased dramatically and quickly through the effects of the water quality deterioration (Sajib

S. Heddam (✉)
Faculty of Science, Agronomy Department, University 20 Août 1955 Skikda, Skikda, Algeria
e-mail: s.heddam@univ-skikda.dz

Dimple et al. (eds.), *Surface Water and Groundwater Quality for Sustainable Utilization*, Springer Hydrogeology, https://doi.org/10.1007/978-981-96-8189-1_4

et al. 2024), which was developed into an ecological and environmental issue for which a robust control strategy is necessary (Essamlali et al. 2024). Thus, the evaluation of various water quality variables, i.e., dissolved oxygen (*DO*), water turbidity (*TU*), temperature (*Tw*), pH, electrical conductivity (EC), become of great importance (Jayaraman et al. 2024). In response to the rapid and great industrialization and the expansion of human activities, an increasing pollution rate of water bodies has been recorded worldwide and it has been subject to considerable debate (Huang et al. 2024). Modelling, monitoring and simulating the fluctuation of water quality variables over times and space is of great importance, as it helps in the protection of the aquatic species, better understanding of the nature of water pollution, and for better use of the available dataset across various regions (Rahaman et al. 2024).

The use of in situ measured data has proven beneficial for controlling the fluctuation and variability of waters in temporal and spatial scales and field measurements have been used since several decades. In addition to in situ measurement, the use of numerical models has shown excellent performances. Presently, the numerical models employed for simulating water quality under different scenarios, are based on the successive use of various water quality variables, leading to a complicated model. Actually, there is a need for introducing new alternatives methods that can help in better understanding the interaction between ensembles of water quality variables. However, some particular water quality variables have attracted the attention of researchers worldwide. Among them, we can cite dissolved oxygen concentration in river (DO), water turbidity (TU), chlorophyll a (Chla) concentration, and cyanobacteria harmful algal blooms. Accurate estimation of phytoplankton chlorophyll-a (Chla) and phycocyanin pigment concentration (PC) is of particular importance as it is an indicator of the harmful algal bloom, water turbidity as an indicator of water clarity and dissolved oxygen concentration as an indicator of water organic pollution and nutrient enrichment. During the last few years, a variety of models has been developed for estimating various water quality variables and the major part of these models are based on machine learning algorithms (ML), and it has demonstrated that the predictive accuracies of the ML models is mainly governed by the variability of water quality variables over time and space. Consequently, there has been an increasing number of published papers related to the application of ML for modelling water quality variables.

Busari et al. (2024) compared between ANN, SVR, RFR, LSTM and GRU for modelling chlorophyll-a (Chl-a) concentration as an indicator of harmful algal blooms, measured in the Boyd Millpond, Reedy river watershed located in South Carolina, USA. They used water pH, TU, DO, SC, and T_w as potential features for modelling Chl-a concentration. Obtained results showed that, RFR model having an R^2, MAE, and RMSE of approximately $\approx$0.87, $\approx$0.97, and $\approx$3.53, outperformed all other models. Qu et al. (2023) use various ML models namely, XGBoost, RFR, SVR and GBR, for modelling Chl-a concentration in the Taihu lake ecosystem research station, China. Compared to all proposed models, the XGBoost exhibited the best performances with the highest R^2 ($\approx$0.950) and lowest RMSE ($\approx$3.690) values, and it outperformed the other models. In another study, Rodríguez-López et al. (2024) use in situ, meteorological, and remote sensing data for modelling Chl-a concentration measured in the lake Maihue, Chile. They compared between seasonal autoregressive integrated moving averages with exogenous regressors (SARIMAX), dynamic generalized linear model (DGLM), and the LSTM models. Obtained results revealed that: (i) using in situ variables as input for the models, moderate to weak performances were obtained with R^2 less than ≈0.65, (ii) using meteorological variables as input for the models, excellent performances were obtained by the DGLM with R^2 ≈0.98, and (iii) using satellite data, weak to moderate performances were obtained with R^2 less than ≈0.56. Kruk (2023) conducted a study in the Polish part of the Vistula lagoon (VL), located in the southern part of the Baltic Sea coast. They used a combination of physico-chemical, nutritional and phytoplankton data for developing the XGBoost model. The XGBoost model shows high fluctuation among the physicochemical, nutritional, and phytoplankton for which the accuracies for the testing dataset were found to be ≈0.7872, ≈0.6724, and ≈0.6979, respectively.

Gao et al. (2024) compared between SVR, MLR, ANN, and partial least squares regression (PLSR) for modelling soybean Chl-a concentration. For developing the models, they used the landsat-8 spectral response function for extracting eight vegetation indices incorporated as features for predicting Chl-a concentration. The superior performances were obtained using the ANN model with $R^2 \approx 0.846$ and RMSE ≈ 0.384 for the validation dataset. Mamun et al. (2024) applied four ML models namely, RFR, ANN, K-nearest neighbors (KNN), and adaptive boosting (AdaBoost), for modelling Chl-a concentration using sentinel-2 multi-spectral imager (MSI) and landsat-8 operational land imager (OLI) satellite imagery surface reflectance (SR) products. Among all proposed models, the RFR model provided superior numerical performances with R^2 and RMSE of approximately ≈0.610 and ≈12.51, respectively. Chen et al. (2024) compared between Bayesian model averaging (BMA), LSTM, SVR and RFR for forecasting Chl-a concentration from 1 to 7 days. The study was conducted at lake Taihu, China, by combining meteorological, hydrological and water quality. Obtained results revealed that, models performances have significantly decreased from 1 to 7 days ahead forecasting, for which the mean R^2 was decreased from ≈0.950 to ≈0.680. Recently, phycocyanin pigment concentration (PC) has received a great deal of attention and some published paper can be found in the literature. Aowtrakool et al. (2024) applied a dynamic model for modelling PC concentration, growth, biomass concentration, and remaining ammonium concentrations. The dynamic model provided excellent performance for PC concentration with R^2 and RMSE of 0.88 and 0.0016, respectively. Maity et al. (2024) developed an empirical growth model based on the combination of principal component analysis (PCA) and multiple linear regression (MLR) for modelling cyanobacterium *Leptolyngbya valderiana* showing high R^2 value of 0.99 and mean absolute percentage error (MAPE) of 6.2%.

By comprehensively reviewing the published papers we can conclude that, a wide range of ML have been proposed and successfully applied for modelling water quality variables and PC concentration. Inspired by this idea, in this chapter, we aim to present an application of various machine learning for modelling PC in river freshwater. We apply four boosting regression models namely: (i) categorical boosting (CatBoost), (ii) light gradient-boosting machine (LightGBM), (iii) extreme gradient boosting

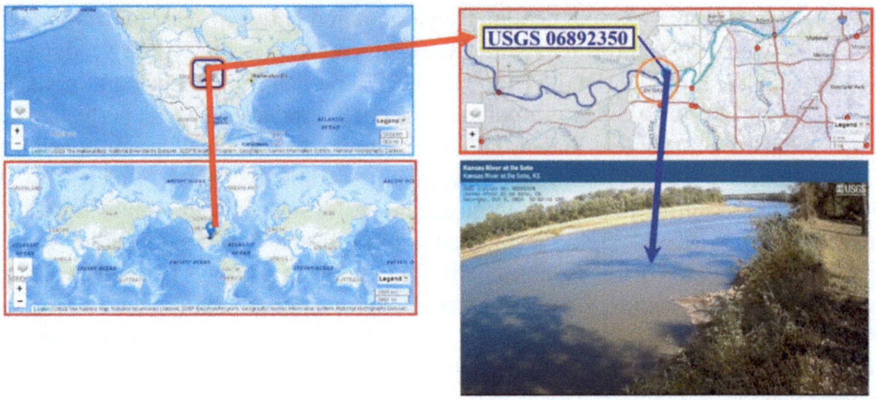

Fig. 1 Location map for the USGS water quality station

(XGBoost), (iv) and adaptive boosting (AdaBoost). The four models were developed using data collected at one USGS (United States Geological Survey) station.

2 Case Study and Dataset

In the present chapter, we use data collected form the United State Geological Survey (USGS) available at https://waterdata.usgs.gov/nwis/. We selected three stations namely: (i) USGS 06892350 Kansas R at Desoto, Leavenworth County, Kansas, (USA) (Latitude 38°59′00″, Longitude 94°57′52″ NAD27). Map showing the location of the water quality station is depicted in Fig. 1 with the calculated statistic descriptive in Table 1. Water quality variables samples measurements used for models' development were collected during the time period described as follow. Data are used for modelling phycocyanin pigment concentration (PC) based on five water quality variables namely, water temperature (T_w), specific conductance (*SC*), water *pH*, dissolved oxygen concentration (DO), and discharge (*Q*), collected during the period ranging from 26 June 2017 to 06 November 2023 (2015 data). For each station, data are randomly divided into 70% for training and 30% for validation. The correlation matrix for all data is depicted in Fig. 2.

3 Machine Learning Models

In the present chapter, we compare various machine-learning models (ML) to predict three water quality variables. All of the methods we develop belong to the general form of "*function approximation in machine learning*". The algorithms consist in

Table 1 Summary statistics of variables

Variables	Subset	Unit	X_{mean}	X_{max}	X_{min}	S_x	C_v	R
USGS 06892350 Kansas R at Desoto, Leavenworth County, Kansas, (USA)								
Q	Training	Ft^3/s	8002.54	70100.00	788.00	10351.94	1.294	−0.495
	Validation	Ft^3/s	7364.26	65900.00	798.00	10323.35	1.402	−0.498
	All data	Ft^3/s	7811.21	70100.00	788.000	10344.955	1.324	−0.496
T_w	Training	°C	15.955	32.800	0.000	9.492	0.595	0.183
	Validation	°C	16.889	32.500	0.000	9.558	0.566	0.244
	All data	°C	16.235	32.800	0.000	9.519	0.586	0.204
SC	Training	uS/cm	821.66	1640.000	211.000	244.382	0.297	0.438
	Validation	uS/cm	846.00	1590.000	315.000	235.844	0.279	0.402
	All data	uS/cm	828.96	1640.000	211.000	242.053	0.292	0.428
pH	Training	/	8.381	9.000	7.500	0.293	0.035	0.657
	Validation	/	8.415	9.000	7.600	0.297	0.035	0.686
	All data	/	8.391	9.000	7.500	0.295	0.035	0.667
DO	Training	mg/L	10.682	17.000	5.800	2.320	0.217	0.082
	Validation	mg/L	10.601	15.800	6.300	2.246	0.212	0.044
	All data	mg/L	10.658	17.000	5.800	2.298	0.216	0.069
PC	Training	RFU	1.191	3.920	0.110	0.672	0.564	1.000
	Validation	RFU	1.263	3.960	0.150	0.698	0.552	1.000
	All data	RFU	1.213	3.960	0.110	0.680	0.561	1.000

directly linking a given set of input variables "*features*" to one modelled variable. The methods are briefly described below.

3.1 *Adaptive Boosting (AdaBoost)*

The adaptive boosting algorithm (AdaBoost) is a machine learning models belonging to the family of boosting tree and it is based on the idea of combining an ensemble of weak learner to construct a solid learner, it was introduced by Freund and Schapire (1996, 1997). By linearly combining the weak learner, the weight of each sample is included during the learning process of the model, this leads to an improvement in the training process of the algorithm making the calculated error significantly decreased in an exponentially manner (Li et al. 2024). The final response of the AdaBoost model is obtained by combining through weighted aggregation the weak models previously developed, which helped in providing a strong regressor (Yingjie et al. 2024).

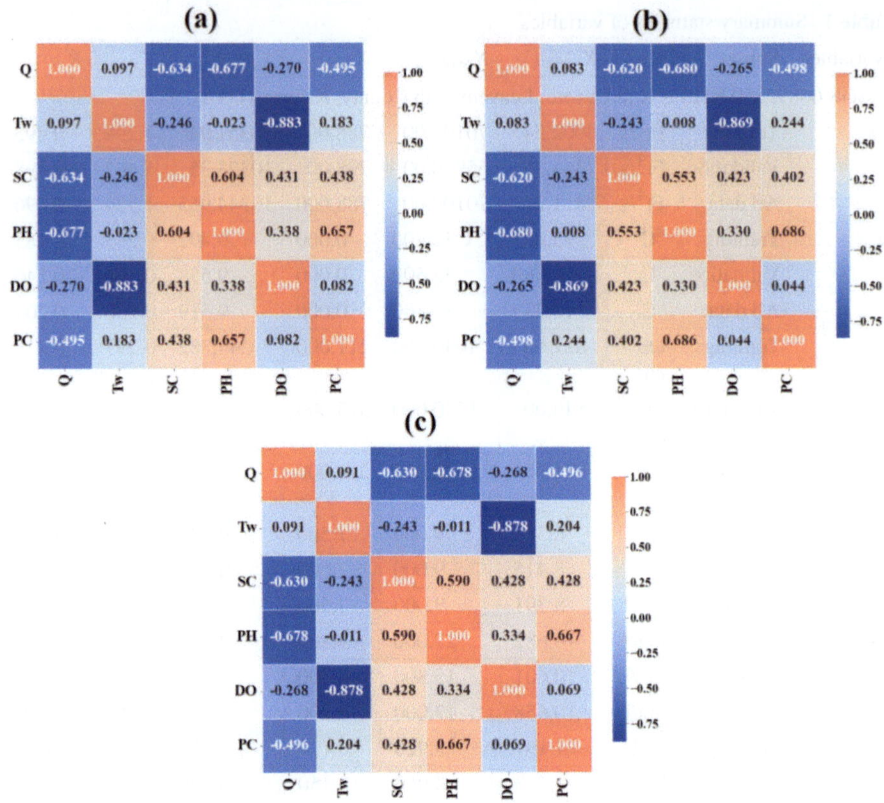

Fig. 2 The correlation heatmaps between all variables: **a** training, **b** validation, and **c** all data

3.2 *Categorical Boosting (CatBoost)*

The categorical boosting (CatBoost) is one of the boosting algorithms proposed by Dorogush et al. (2018), Prokhorenkova et al. (2018). The CatBoost start by randomly sorting the available dataset and it use the same idea as the gradient boosting decision trees (GBDT) having the capability to handles categorical features. Similar to the major decision tree algorithms, the CatBoost receives the responses of all weak trees and provides the final response. The mains idea behind the development of the CatBoost algorithm is based on improving the GBDT by introducing the so-called ordered boosting. The training dataset are ordered in a specific order (β), then, the CatBoost builds an ensemble of regressors for which the number of instances could be increased one-by-one in the same specific order (Omer and Shareef 2022; Xu et al. 2023).

3.3 Light Gradient Boosting Machine (LightGBM)

The light gradient boosting machine (LightGBM) was developed by Ke et al. (2017). The LightGBM is based on the concept of GBDT and they combine two paradigms: the exclusive features bundling and the gradient-based one-side sampling (Jas et al. 2024), which have contributed significantly in improving the handling process of large dataset and high features dimension spaces (Zhang et al. 2023). The objective of the LightGBM training algorithm is to develops a regression function having the capability to decrease the loss function calculated between the actual, i.e., measured variable and the predicted variable, which itself is the combination of an ensemble of basic weak decision trees functions (Pan et al. 2023).

3.4 Extreme Gradient Boosting (XGBoost)

The extreme gradient boosting (XGBoost) was developed by Chen and Guestrin (2016). The XGBoost works using an iterative strategy by calculating an average value obtained from an ensemble of weak regressors. This can be achieved by splitting the data continuously, which lead to the construction of an ensemble of regression trees models (Zhou et al. 2023; Long et al. 2023). From a computational point of view, the XGBoost adopts a forward method based on the concept of addition, for which, the calculated value at the iteration t can be obtained by the summation of the pre step value and the new added value calculated the model (Long et al. 2023).

4 Different Result and Numerical Analysis

In this section, the performances of the four ML models, i.e., CatBoost, LightGBM, XGBoost, and AdaBoost were evaluated and compared. The models were applied for modelling PC concentration. Furthermore, we have compared between various input combinations as depicted in Table 2. For each combination, we have different water quality variables as input. Two different scenarios were evaluated in the present study, namely: (i) modelling PC using only water quality variables, i.e., without periodicity, and (ii) modelling PC using water quality variables and the periodicity, i.e. the year, month and days numbers. The models were evaluated using four numerical indices namely, RMSE, MAE, R and NSE, calculated as follow:

$$RMSE = \sqrt{\frac{1}{N}\sum\nolimits_{i=1}^{N}[(O)_i - (P)_i]^2}, \quad (0 \leq RMSE < +\infty) \tag{1}$$

$$MAE = \frac{1}{N}\sum\nolimits_{i=1}^{N}|(O)_i - (P)_i|, \quad (0 \leq MAE < +\infty) \tag{2}$$

Table 2 The input combinations of different models

USGS 06892350 Kansas R at Desoto, Leavenworth County, Kansas, (USA)					
Models				Input variables	Output
CatBoost1	LightGBM1	XGBoost1	AdaBoost1	Q, T_w, SC, pH, DO	PC
CatBoost2	LightGBM2	XGBoost2	AdaBoost2	Q, T_w, SC, pH	PC
CatBoost3	LightGBM3	XGBoost3	AdaBoost3	Q, SC, pH, DO	PC
CatBoost4	LightGBM4	XGBoost4	AdaBoost4	Q, SC, pH	PC
CatBoost5	LightGBM5	XGBoost5	AdaBoost5	Q, T_w, SC	PC
CatBoost6	LightGBM6	XGBoost6	AdaBoost6	Q, SC	PC

$$R = \left[\frac{\sum_{i=1}^{N} \left((O)_i - \overline{O}\right)\left((P)_i - \overline{P}\right)}{\sqrt{\sum_{i=1}^{N} \left((O)_i - \overline{O}\right)^2 \sum_{i=1}^{N} \left((P)_i - \overline{P}\right)^2}} \right] \tag{3}$$

$$NSE = 1 - \left[\frac{\sum_{i=1}^{N} [(O)_i - (P)_i]^2}{\sum_{i=1}^{N} [(O)_i - \overline{O}]^2} \right], \quad (-\infty < NSE \leq 1) \tag{4}$$

where $\overline{O}$ and $\overline{P}$ are the mean measured and mean predicted variables (i.e., PC), respectively, and O_i and P_i specify the observed and predicted variable for ith observations, and N shows the number of data points. The models were calibrated using 70% of the data and validated using the remaining 30% of the data. Further comparison between the models' performances was done using graphical comparison, i.e., the scatterplot, boxplot, violinplot and Taylor diagram. Obtained results are discussed hereafter for each station separately.

4.1 *Modelling PC Concentration Without Periodicity*

Figure 3 displays the Scatterplot between measured and predicted PC concentration obtained by the best models without the inclusion of the periodicity. Numerical performances are depicted in Table 3. From the obtained results, we can clearly see that, during the training stage, all models performed very well for all input combinations, exhibiting excellent numerical performances. We clearly see that, the high predictive accuracy was achieved with R, NSE, RMSE and MAE values of approximately, ≈0.983, ≈0.966, ≈0.124, and ≈0.091, obtained by the XGBoost1, and the lowest predictive accuracy was obtained by the AdaBoost6 having an R, NSE, RMSE and MAE values of approximately, ≈0.681, ≈0.379, ≈0.529, and ≈0.446. Always during the training stage, if we compare the models tacking into account the mean values of the four numerical indices, we can see that there are a significant and noticeable differences between the models. High predictive accuracy tends to have a high mean R and NSE, and very low mean RMSE and MAE values. The mean R and NSE values were ranged from ≈0.761 to ≈0.959, and from ≈0.480 to ≈0.917, respectively. Obtained results showed that for all cases, the XGBoost had distinct numerical values both for the R and NSE and for the RMSE and MAE, where excellent mean values of ≈0.959, ≈0.917, ≈0.186 and ≈0.137 were obtained. Yet, the R and NSE which are usually assumed to have near 1.00 values in a perfect prediction case, have an approximately perfect values, showing the high capabilities of the XGBoost modes in correctly predicting DO concentration. During the validation stage as seen in Table 3, the performances of the models were slightly decreased and the models based on the sixth input combination having the *Q* and *SC* as input variables were the poorest among all models. The CatBoost1 was found to be the best model having the high R and NSE values of ≈0.883, and ≈0.778, respectively, and the lowest RMSE and MAE values of 0.328 and 0.236, respectively. The CatBoost1 was followed by the LightGBM1 and XGBoost1, while the AdaBoost1 was ranked in the last place. All models were further compared using the boxplot and violinplot (Fig. 4) and the Taylor diagram (Fig. 5). We can clearly see from the figures that, the AdaBoost is the lowest models exhibiting the high prediction errors.

4.2 Modelling PC Concentration with Periodicity.

By analyzing the results reported in Table 4 for the validation stage, we can see that the performances of the models were significantly increased by the inclusion of periodicity, i.e., the year, month and days number. Generally, there was a noticeable increase in the predictive accuracy and the correlation between measured and calculated PC tends to be higher than for the training dataset. The AdaBoost had the poorest performances with mean R, NSE, RMSE and MAE values of approximately ≈0.806, ≈0.546, ≈0.469 and ≈0.397, respectively, whereas the best performances were guaranteed by the LightGBM models, with mean R, NSE, RMSE and MAE values of approximately ≈0.907, ≈0.823, ≈0.293 and ≈0.205, respectively, slightly superior to the XGBoost models. The CatBoost exhibited a good numerical performance, implying that these algorithms have excellent capability in handling nonlinear relationship between water quality variables. Furthermore, analyzing the models' performances at different input combination provides further insight into the performances of the developed models. We can see that, a strong linear relationship is observed with acceptable R and NSE values for all models, except for AdaBoost6, for which the models' performances were decreased. We can see that, using only *Q*, SC as input variables, the models provided relatively acceptable performances with R and NSE values ranging from ≈0.759 to ≈0.895 and from ≈0.481 to ≈0.801, and the AdBoost6 was found to be the poorest one with R and NSE values of approximately, ≈0.759 and ≈0.481, respectively. Further comparison between the models with and without comparison revealed that: (i) the R, NSE, RMSE and MAE values of the AdaBoost1 were improved by 3.767%, 4.754%, 3.579%, and 0% respectively, (ii) the R, NSE, RMSE and MAE values of the CatBoost1 were improved by 3.851%, 8.098%, 15.244%, and 17.373% respectively, (iii) the R, NSE,

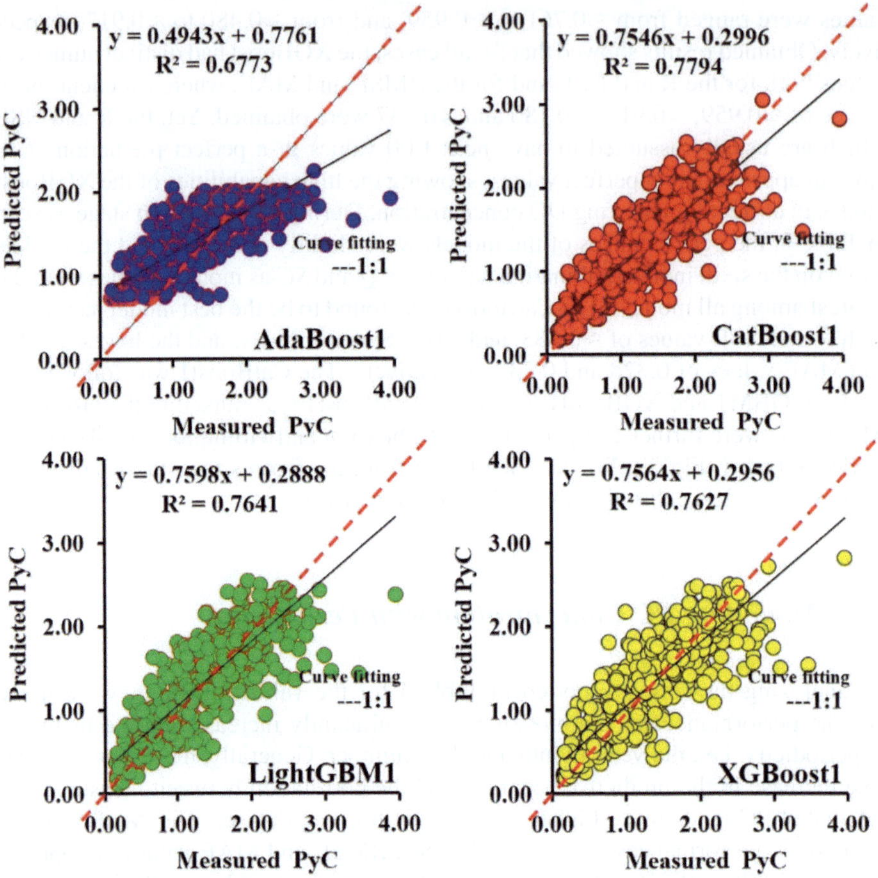

Fig. 3 Scatterplot of measured and predicted PC concentration using different ML models for the validation dataset without periodicity

RMSE and MAE values of the LightGBM1 were improved by 4.805%, 9.817%, 17.699%, and 16.667%, respectively, and (iv) the R, NSE, RMSE and MAE values of the XGBoost1 were improved by 5.613%, 11.417%, 20.294%, and 20.084%, respectively. It can be concluded that, the high improvement rate was gained using the XGBoost approaches. The performances of the models were further compared using graphical plotting of measured and predicted PC concentration. Scatterplot of measured and predicted PC concentration using the best model with periodicity are depicted in Fig. 6 for which we can see that, the data points are less scattered using the XGBoost model. The boxplot and violinplot of measured and predicted data for the validation stage are depicted in Fig. 7. Finally, the models were further compared using Taylor diagram, for which the weakness of the AdaBoost model is more obvious.

Table 3 Results of Phycocyanin concentration modelling without periodicity

Models	Training				Validation			
	R	NSE	RMSE	MAE	R	NSE	RMSE	MAE
AdaBoost1	**0.834**	**0.594**	**0.428**	**0.366**	**0.823**	**0.589**	**0.447**	**0.367**
AdaBoost2	0.747	0.397	0.521	0.443	0.745	0.424	0.529	0.445
AdaBoost3	0.788	0.507	0.471	0.402	0.780	0.506	0.490	0.404
AdaBoost4	0.767	0.514	0.468	0.395	0.762	0.509	0.488	0.408
AdaBoost5	0.751	0.491	0.479	0.402	0.714	0.458	0.513	0.420
AdaBoost6	0.681	0.379	0.529	0.446	0.643	0.355	0.560	0.466
CatBoost1	**0.950**	**0.900**	**0.212**	**0.157**	**0.883**	**0.778**	**0.328**	**0.236**
CatBoost2	0.937	0.876	0.237	0.176	0.861	0.742	0.354	0.254
CatBoost3	0.926	0.854	0.256	0.188	0.848	0.719	0.369	0.266
CatBoost4	0.881	0.772	0.320	0.247	0.793	0.629	0.425	0.321
CatBoost5	0.903	0.810	0.293	0.218	0.796	0.633	0.422	0.310
CatBoost6	0.914	0.832	0.275	0.209	0.643	0.385	0.547	0.405
LightGBM1	**0.981**	**0.962**	**0.131**	**0.091**	**0.874**	**0.764**	**0.339**	**0.234**
LightGBM2	0.972	0.942	0.161	0.112	0.861	0.742	0.354	0.252
LightGBM3	0.966	0.931	0.176	0.125	0.846	0.715	0.372	0.266
LightGBM4	0.930	0.861	0.250	0.187	0.785	0.614	0.433	0.325
LightGBM5	0.954	0.907	0.204	0.143	0.767	0.583	0.450	0.320
LightGBM6	0.876	0.762	0.328	0.247	0.656	0.418	0.532	0.401
XGBoost1	**0.983**	**0.966**	**0.124**	**0.091**	**0.873**	**0.762**	**0.340**	**0.239**
XGBoost2	0.978	0.955	0.143	0.104	0.859	0.738	0.357	0.253
XGBoost3	0.974	0.947	0.155	0.112	0.831	0.688	0.389	0.271
XGBoost4	0.948	0.895	0.218	0.162	0.783	0.610	0.435	0.326
XGBoost5	0.964	0.925	0.183	0.134	0.773	0.592	0.445	0.313
XGBoost6	0.906	0.813	0.290	0.216	0.661	0.423	0.529	0.399

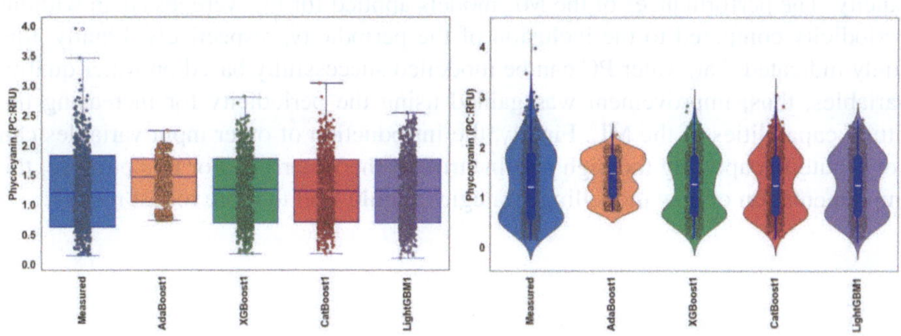

Fig. 4 Comparison between measured and predicted PC concentration using different ML models for the validation dataset without periodicity: boxplot (left panel) and violinplot (right panel)

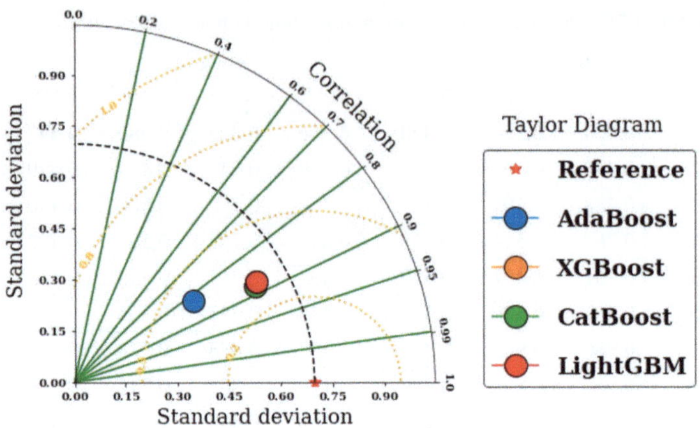

Fig. 5 Taylor diagram for models' comparison between measured and predicted PC concentration for the validation dataset without periodicity

5 Conclusions

In various previous published papers, researchers worldwide have showed that machine learning (ML) algorithms can be used for modelling water quality variables and they demonstrated that the availability of in situ measured data with high precision and continuity in measurement have helped in developing solid predictive models. In the present study, we propose the application of several ML algorithms for modelling a critical water quality variable namely, phycocyanin pigment concentration (PC). The PC variable was modeled using data collected from one USGS station, and with different input variables combination. Thus, we have compared between, AdaBoost, CatBoost, LightGBM, and XGBoost. From the obtained results we can draw various conclusions. High predictive accuracies were obtained using all models and the AdaBoost was found to be the less accurate model, without or without periodicity. The performances of the ML models applied for PC were less than without periodicity compared to the inclusion of the periodicity, respectively. Finally, this study indicated that, water PC can be modelled successfully based on water quality variables, thus, improvement was gained using the periodicity for increasing the fitting capabilities of the ML. Finally, the introduction of other input variables can contribute in capturing the high nonlinearity of the water PC. For this purpose, the implementation of new modelling strategies should lead to more robust results.

Table 4 Results of Phycocyanin concentration modelling with periodicity

Models	Training				Validation			
	R	NSE	RMSE	MAE	R	NSE	RMSE	MAE
AdaBoost1	**0.865**	**0.627**	**0.410**	**0.355**	**0.854**	**0.617**	**0.431**	**0.367**
AdaBoost2	0.833	0.586	0.432	0.368	0.830	0.584	0.450	0.382
AdaBoost3	0.802	0.517	0.466	0.395	0.803	0.527	0.479	0.404
AdaBoost4	0.821	0.593	0.428	0.364	0.811	0.578	0.453	0.384
AdaBoost5	0.807	0.515	0.467	0.402	0.777	0.490	0.498	0.418
AdaBoost6	0.787	0.511	0.469	0.400	0.759	0.481	0.502	0.425
CatBoost1	**0.972**	**0.944**	**0.159**	**0.117**	**0.917**	**0.841**	**0.278**	**0.195**
CatBoost2	0.965	0.930	0.178	0.130	0.909	0.826	0.291	0.207
CatBoost3	0.926	0.854	0.256	0.188	0.848	0.719	0.369	0.266
CatBoost4	0.961	0.922	0.187	0.135	0.908	0.824	0.292	0.207
CatBoost5	0.903	0.810	0.293	0.218	0.796	0.633	0.422	0.310
CatBoost6	0.952	0.901	0.211	0.156	0.885	0.781	0.326	0.237
LightGBM1	**0.991**	**0.981**	**0.093**	**0.064**	**0.916**	**0.839**	**0.279**	**0.195**
LightGBM2	0.989	0.976	0.103	0.071	0.918	0.842	0.277	0.194
LightGBM3	0.987	0.974	0.108	0.074	0.910	0.828	0.289	0.201
LightGBM4	0.984	0.968	0.120	0.084	0.918	0.842	0.277	0.198
LightGBM5	0.984	0.966	0.123	0.084	0.887	0.787	0.322	0.224
LightGBM6	0.979	0.957	0.140	0.099	0.895	0.801	0.311	0.216
XGBoost1	**0.992**	**0.984**	**0.085**	**0.062**	**0.922**	**0.849**	**0.271**	**0.191**
XGBoost2	0.990	0.980	0.095	0.070	0.919	0.844	0.275	0.193
XGBoost3	0.991	0.982	0.091	0.066	0.913	0.833	0.285	0.200
XGBoost4	0.988	0.975	0.107	0.077	0.913	0.834	0.284	0.199
XGBoost5	0.988	0.974	0.108	0.078	0.876	0.768	0.336	0.233
XGBoost6	0.983	0.965	0.126	0.091	0.883	0.779	0.327	0.228

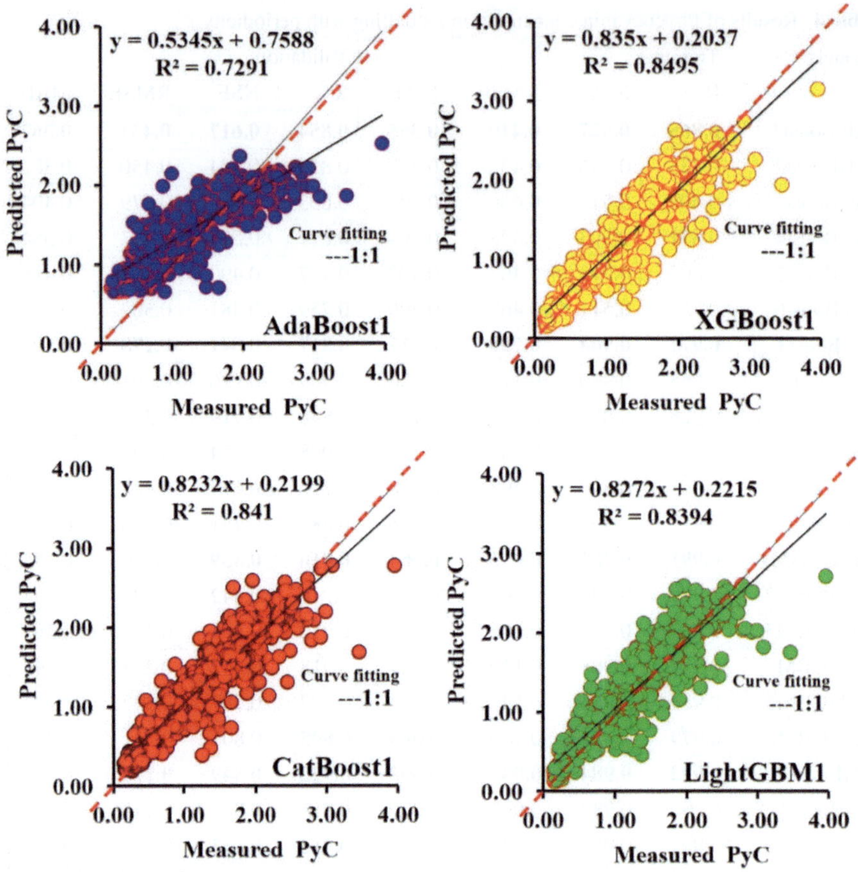

Fig. 6 Scatterplot of measured and predicted PC concentration using different ML models for the validation dataset with periodicity

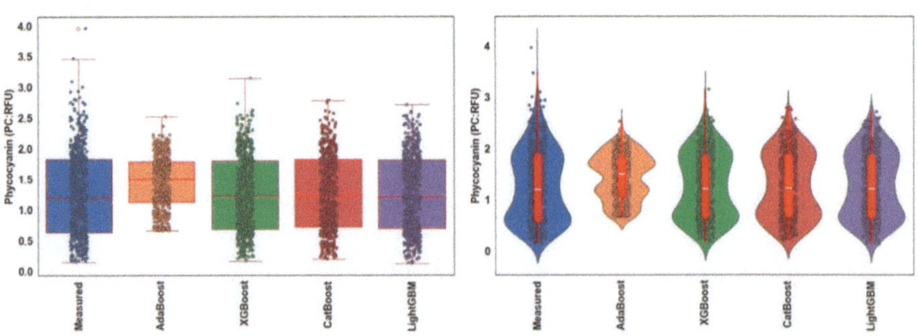

Fig. 7 Comparison between measured and predicted PC concentration using different ML models for the validation dataset with periodicity: boxplot (left panel) and violinplot (right panel)

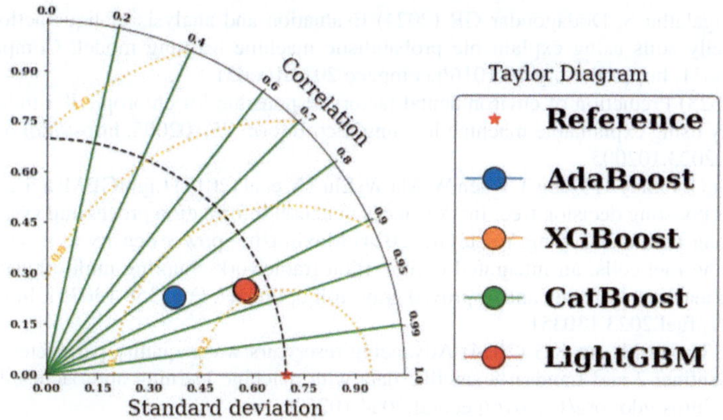

Fig. 8 Taylor diagram for models comparison between measured and predicted PC concentration for the validation dataset with periodicity

References

Aowtrakool N, Sopitthummakhun A, Laomettachit T, Ruengjitchatchawalya M (2024) A dynamic model for predicting biomass and phycocyanin yields in Arthrospira (Spirulina) platensis: a guidance for effective batch cultivation. Algal Res 103709. https://doi.org/10.1016/j.algal.2024.103709

Busari I, Sahoo D, Harmel RD, Haggard BE (2024) Prediction of chlorophyll-a as an index of harmful algal blooms using machine learning models. J Nat Resour Agric Ecosyst https://doi.org/10.13031/jnrae.15812

Chen C, Chen Q, Yao S, He M, Zhang J, Li G, Lin Y (2024) Combining physical-based model and machine learning to forecast chlorophyll-a concentration in freshwater lakes. Sci Total Environ 907:168097. https://doi.org/10.1016/j.scitotenv.2023.168097

Chen T, Guestrin C (2016) XGBoost: a scalable tree boosting system. In: Proceedings of the 22nd ACM Sigkdd international conference on knowledge discovery and data mining, pp 785–794. https://doi.org/10.48550/arXiv.1603.02754

Dorogush AV, Ershov V, Gulin A (2018) CatBoost: gradient boosting with categorical features support. arXiv:1810.11363

Essamlali I, Nhaila H, El Khaili M (2024) Advances in machine learning and IoT for water quality monitoring: a comprehensive review. Heliyon. https://doi.org/10.1016/j.heliyon.2024.e27920

Freund Y, Schapire RE (1996) Experiments with a new boosting algorithm. In: ICML, vol 96, pp 148–156

Freund Y, Schapire RE (1997) A decision-theoretic generalization of on-line learning and an application to boosting. J Comput Syst Sci 55(1):119–139

Gao C, Li H, Wang J, Zhang X, Huang K, Song X, Li F (2024) Combined use of spectral resampling and machine learning algorithms to estimate soybean leaf chlorophyll. Comput Electron Agric 218:108675. https://doi.org/10.1016/j.compag.2024.108675

Huang S, Xia J, Wang Y, Lei J, Wang G (2024) Water quality prediction based on sparse dataset using enhanced machine learning. Environ Sci Ecotechnol 100402. https://doi.org/10.1016/j.ese.2024.100402

Jayaraman P, Nagarajan KK, Partheeban P, Krishnamurthy V (2024) Critical review on water quality analysis using IoT and machine learning models. Int J Inf Manag Data Insights 4(1):100210. https://doi.org/10.1016/j.jjimei.2023.100210

Jas K, Mangalathu S, Dodagoudar GR (2024) Evaluation and analysis of liquefaction potential of gravelly soils using explainable probabilistic machine learning model. Comput Geotech 167:106051. https://doi.org/10.1016/j.compgeo.2023.106051

Kruk M (2023) Prediction of environmental factors responsible for chlorophyll a-induced hypereutrophy using explainable machine learning. Eco Inform 75:102005. https://doi.org/10.1016/j.ecoinf.2023.102005

Ke G, Meng Q, Finley T, Wang T, Chen W, Ma W, Liu TY et al (2017) LightGBM: a highly efficient gradient boosting decision tree. In: Advances in neural information processing systems, vol 30

Li HW, Wang L, Liu JN, Yang Y, Lu GL (2024) Maximizing power density in proton exchange membrane fuel cells: an integrated optimization framework coupling multi-physics structure models, machine learning, and improved gray wolf optimizer. Fuel 358:130351. https://doi.org/10.1016/j.fuel.2023.130351

Mamun M, Hasan M, An KG (2024) Advancing reservoirs water quality parameters estimation using Sentinel-2 and Landsat-8 satellite data with machine learning approaches. Eco Inform 102608. https://doi.org/10.1016/j.ecoinf.2024.102608

Maity S, Tiwari R, Mallick N (2024) Application of empirical modelling to predict the growth of a marine cyanobacterium Leptolyngbya valderiana BDU 41001 for production of bioethanol and C-phycocyanin under semi-outdoor cultivation. J Clean Prod 473:143507. https://doi.org/10.1016/j.jclepro.2024.143507

Long X, Gu X, Lu C, Li Z, Ma Y, Jian Z (2023) Prediction of the jump height of transmission lines after ice-shedding based on XGBoost and Bayesian optimization. Cold Reg Sci Technol 213:103928. https://doi.org/10.1016/j.coldregions.2023.103928

Omer ZM, Shareef H (2022) Comparison of decision tree-based ensemble methods for prediction of photovoltaic maximum current. Energy Convers Manag X 16:100333. https://doi.org/10.1016/j.ecmx.2022.100333

Prokhorenkova L, Gusev G, Vorobev A, Dorogush AV, Gulin A (2018) CatBoost: unbiased boosting with categorical features. In: Advances in neural information processing systems, vol 31

Pan Z, Lu W, Bai Y (2023) Groundwater contaminated source estimation based on adaptive correction iterative ensemble smoother with an auto LightGBM surrogate. J Hydrol 620:129502. https://doi.org/10.1016/j.jhydrol.2023.129502

Qu S, Wang JJ, Kumar A, Yu ZG, Zhao WQ (2023) Identification of driving factors for chlorophyll-a in multi-stable shallow lakes of China employing machine-learning methods. Ecohydrology 16(8):e2590. https://doi.org/10.1002/eco.2590

Rahaman MH, Sajjad H, Hussain S, Masroor M, Sharma A (2024) Surface water quality prediction in the lower Thoubal river watershed, India: a hyper-tuned machine learning approach and DNN-based sensitivity analysis. J Environ Chem Eng 112915. https://doi.org/10.1016/j.jece.2024.112915

Rodríguez-López L, Alvarez D, Bustos Usta D, Duran-Llacer I, Bravo Alvarez L, Fagel N, Urrutia R et al (2024) Chlorophyll-a detection algorithms at different depths using in situ, meteorological, and remote sensing data in a Chilean Lake. Remote Sens 16(4):647. https://doi.org/10.3390/rs16040647

Sarafaraz J, Kaleybar FA, Karamjavan JM, Habibzadeh N (2024) Predicting river water quality: an imposing engagement between machine learning and the QUAL2Kw models (case study: Aji-Chai, river, Iran). Results Eng 21:101921. https://doi.org/10.1016/j.rineng.2024.101921

Sajib AM, Diganta MTM, Moniruzzaman M, Rahman A, Dabrowski T, Uddin MG, Olbert AI (2024) Assessing water quality of an ecologically critical urban canal incorporating machine-learning approaches. Eco Inform 102514. https://doi.org/10.1016/j.ecoinf.2024.102514

Xu Y, Li J, Yu X, Xiao L, Luo T, Wei C, Li L (2023) Composition prediction of pore solution in hardened concrete materials based on machine learning. Dev Built Environ 16:100285. https://doi.org/10.1016/j.dibe.2023.100285

Yingjie Z, Yibo A, Weidong Z (2024) Physics informed ensemble learning used for interval prediction of fracture toughness of pipeline steels in hydrogen environments. Theoret Appl Fract Mech 130:104302. https://doi.org/10.1016/j.tafmec.2024.104302

Zhang W, Wu C, Tang L, Gu X, Wang L (2023) Efficient time-variant reliability analysis of Bazimen landslide in the three gorges reservoir area using XGBoost and LightGBM algorithms. Gondwana Res 123:41–53. https://doi.org/10.1016/j.gr.2022.10.004

Zhou X, Zhao C, Bian X (2023) Prediction of maximum ground surface settlement induced by shield tunneling using XGBoost algorithm with golden-sine seagull optimization. Comput Geotech 154:105156. https://doi.org/10.1016/j.compgeo.2022.105156

Chemical Analysis Method for Surface and Ground Water Quality Assessment

Malkhan Singh Jatav, Venkatesh Gaddikeri, Akshay V. Dahiwale, Sourabh Nema, Dilip Barman, Sudesh Singh Choudhary, Jitendra Rajput, and Nandlal Kushwaha

Abstract Water quality assessment is a fundamental aspect of environmental monitoring, directly influencing public health, agricultural sustainability, and industrial processes. This chapter presents a comprehensive review of chemical analytical techniques for surface and groundwater quality evaluation, categorized into classical wet chemical methods and instrumental approaches. Classical techniques, including gravimetric, titrimetric, and colorimetric methods, provide cost-effective and robust analytical frameworks. In contrast, instrumental techniques, such as electrochemical (potentiometry, voltammetry, coulometry), spectroscopic (UV-Vis spectroscopy, atomic absorption spectroscopy, Fourier-transform infrared spectroscopy), chromatographic (gas and ion chromatography), and mass spectrometric (ICP-MS) methods, offer enhanced precision, sensitivity, and specificity for detecting trace contaminants. This critical evaluation underscores the strengths and limitations of these methodologies, particularly regarding detection limits, selectivity, and matrix effects.

1 Introduction

Water is a fundamental natural resource, critical to human survival and sourced primarily from surface and groundwater. The quality and quantity of water impact various sectors, including agriculture, industry, domestic use, and ecosystem health. The quality of water, defined by its chemical, physical, and biological characteristics, is essential in determining its suitability for human consumption and ecological stability. Among these characteristics, the chemical properties of water are particularly important, as they directly influence water's safety, usability, and impact on

M. S. Jatav (✉) · A. V. Dahiwale · S. Nema · D. Barman · S. S. Choudhary
National Institute of Hydrology, North Western Regional Centre, Jodhpur, Rajasthan, India
e-mail: msjatav.nihr@gov.in

V. Gaddikeri · J. Rajput · N. Kushwaha
Indian Council of Agricultural Research, Indian Agricultural Research Institute, New Delhi, India

Dimple et al. (eds.), *Surface Water and Groundwater Quality for Sustainable Utilization*, Springer Hydrogeology, https://doi.org/10.1007/978-981-96-8189-1_5

both human and environmental health. Assessment of these chemical properties is, therefore, important for effective water quality management.

However, Chemical methods of water quality assessment involve identification and quantification of the chemical components within water samples (Kenkel 2002), which includes, dissolved gases, metals, nutrients, pesticides, and organic compounds. These constituents significantly impact water's suitability and functionality for drinking, irrigation, industrial applications, and aquatic ecosystems.

In general, chemical methods used for water quality analysis are divided into two categories, based on the measurement techniques and how the tests are performed. i.e. classical analysis (wet chemical) method and instrumental analysis methods (Harvey 2000).

2 Classical Analysis (Wet Chemical) Method

These methods are fundamentally based on basic chemical reactions between the analyte and an added reagent. They use only a simple balance, without reliance on additional mechanical or electronic instruments (Harvey 2000). Additionally, these methods generally produce easily detectable products, such as colored compounds or solid precipitates. Common classical analysis methods for water quality analysis of surface and groundwater include gravimetric analysis, titrimetric (volumetric) analysis, colorimetric analysis, and conductometric analysis.

2.1 Gravimetric Method

Gravimetric analysis, also known as gravimetry, is a quantitative chemical analysis method that relies on the precise measurement of mass or changes in mass. This technique is commonly utilized to quantify specific substances or elements in water samples. The main principle of gravimetry involves converting the analyte of interest into a compound with a well-defined and stable composition, which can then be isolated and accurately weighed.

Any gravimetric analysis process typically includes the following general steps: (a) preparing a solution containing a known quantity of the sample, (b) isolating the target constituent, (c) precisely weighing the isolated compound, and (d) calculating the amount of the constituent in the original sample based on the measured mass.

Additionally, these methods are further subdivided as follows:

- Precipitation Gravimetry
- Volatilization Gravimetry
- Particulate Gravimetry

- **Precipitation Gravimetry**

This method relies on the principle of converting the analyte, or substance being measured, into an insoluble precipitate by adding a specific reagent known as the precipitant. The precipitate, which has a known and stable chemical composition, is isolated from the solution, usually by filtration. After isolation, the precipitate is dried or ignited to eliminate moisture or volatile components, and ensures the accurate mass measurement. The mass of the precipitate is then used to calculate the concentration of the analyte in the original sample based on stoichiometric relationships (Egbuna et al. 2018). This method is commonly used to assess water quality parameters such as sulfate (SO_4^{2-}), chloride (Cl^-), and certain metal ions like silver (Ag^+) and lead (Pb^{2+}). However, Specific reagents are used to precipitate these ions: for example, barium chloride ($BaCl_2$) is used to turn sulfate into barium sulfate ($BaSO_4$), while silver nitrate ($AgNO_3$) is used to convert both chloride and silver into silver chloride ($AgCl$) (Zhang et al. 2023). Additionally, sodium sulfate (Na_2SO_4) or sulfuric acid (H_2SO_4) can be employed to precipitate lead as lead sulfate ($PbSO_4$).

- **Volatilization Gravimetry**

The basic principle of this method is same as precipitation method only difference is that it converts the analyte into a volatile compound through heating or chemical reaction. The volatile compound is then captured or condensed, and its mass is used to determine the analyte concentration (Bharty n.d.). For example, carbonate (CO_3^{2-}) is converted into carbon dioxide (CO_2) by treating with an acid (commonly hydrochloric acid, HCl), and the CO_2 is absorbed by a solid alkaline reagent to measure the carbonate content.

- **Particulate Gravimetry**

This method involves measuring particulate matter by filtering a sample to capture suspended particles, then drying and weighing the filter. The mass difference gives the concentration of particulate matter (Garland et al. 2018). Commonly used to assess Total Suspended Solids (TSS) in water samples.

2.2 *Titrimetric (Volumetric) Method*

The titrimetric method, also called volumetric analysis, measures the volume of a reagent that reacts in a specific way with the substance being analysed, i.e. analyte. However, it is very accurate method for determining the concentration of the analyte because it measures how much of a solution with a known concentration is needed to completely react with the analyte (Kozak and Townshend 2019).

The general titrimetric method of analysis is based on the following chemical reaction;

$$\mathbf{aA + bB \rightarrow Product}$$

In this reaction, the molecules of the analyte **A** react with **b** molecules of the reagent **B** to form the product. The reagent **B** is known as the titrant, which is a solution with a known concentration. During the titration, the titrant is gradually added from a burette until the amount of **B** is chemically equivalent to the amount of **A** (Grinberg and Rodriguez 2018).

The general titrimetric method of analysis is based on the following chemical reactions, which can be categorized into following four different types of titrimetric analysis:

- **Titrations Based on Acid–Base Reactions (Acidimetry and Alkalimetry)**

This type of titrimetric method includes titrations in which a base reacts with a standard acid, known as acidimetry, and those in which an acid reacts with a standard base, referred to as alkalimetry. In these reactions, hydrogen ions (H^+) from the acid combine with hydroxide ions (OH^-) from the base to produce water (H_2O). The basic chemical reaction can be represented as follows (Mosher and Kèlter 2023);

$$HA + OH^- \rightarrow A^- + H_2O$$

where,

HA	An acid
OH^-	A hydroxide ion
A^-	A conjugate base of the acid
H_2O	Water.

- **Titrations Based on Redox (Oxidation–Reduction) Reactions**

Redox titration is a titration method that involves a change in oxidation number or the transfer of electrons between reacting substances known as a redox titration (Mosher and Kelter 2023) his type of titration typically involves the reaction between an oxidizing agent and a reducing agent, often facilitated by a redox indicator. The standard solution used in redox titrations can be either an oxidizing agent or a reducing agent. And common redox reaction generally represented as follows;

$$Ce^{4+} + Fe^{2+} \rightleftharpoons Ce^{3+} + Fe^{3+}$$

In this reaction; Cerium (IV) ions (Ce^{4+}) are reduced to Cerium (III) ions (Ce^{3+}), Iron (II) ions (Fe^{2+}) are oxidized to Iron (III) ions (Fe^{3+}).

This reaction is commonly used in redox titrations, especially in determining the concentration of one of the elements, either Ce^{4+} or Fe^{2+}. While transfer of electrons between these elements defines the redox nature of the reaction.

- **Complexometric Titration**

The titration between metal ion solution and the complexing agent (chelating agent) in presence of a suitable indicator is called complexometric titration. In this titrimetric analysis formation of the colored complex is used to indicate the endpoint. This method involves the quantitative estimation of metal ions in solution by titration through complexation with a strong chelating ligand. Chelating ligands provide enhanced stability to the metal complex due to the chelate effect (Kozak and Townshend 2019). Ethylenediaminetetraacetic acid (EDTA) is a widely used and highly effective complexing agent for metal ions. For the determination of metal ion concentration using EDTA, the general reaction represented as follows:

$$M^{n+} + EDTA^{2-} \rightarrow M\text{-}EDTA$$

where,

Mn^{+} Represents a metal ion such as Fe^{3+}, Zn^{2+}, etc.
$EDTA^{2-}$ Chelating agent.

This reaction indicates the formation of a 1:1 complex between the metal ion and EDTA, which is used as the basis for determining the metal ion concentration in solution (Kozak and Townshend 2019).

- **Precipitation Titration**

This method involves the formation of a precipitate and is characteristic of precipitation titrations. In this type of titrations, the formation of a solid precipitate indicates the completion of the reaction (Waldhoff and Spilker 2004). For example, silver nitrate ($AgNO_3$) is commonly used to precipitate halide ions such as chloride (Cl^-), bromide (Br^-), and iodide (I^-) and thiocyanate ions (SCN^-) from their respective solutions. These precipitation reactions are used to measure the number of ions present in the solution, and allows the accurate determination based on the amount of solid formed (Bouchonnet 2013).

The reaction can be represented as:

$$AgNO_3 + X^- \rightarrow AgX \downarrow + NO_3^-$$

where,

X can be Cl^-, Br^-, I^-, or SCN^-.

This balanced equation describes the reaction in which silver ions (Ag^+) react with halide or thiocyanate ions (X^-) to form a solid precipitate (AgX), marking the endpoint of the titration.

3 Instrumental Analysis

Besides classical methods, this approach uses specialized instruments other than just balances to carry out chemical analysis. Most of chemical analysis falls into this category. The instrumental methods either be used to observe and study the chemical reactions between the analyte and a reagent, or they may be used to measure a specific property of the analyte. A wide variety of instrumental techniques is available for this type of analysis. Among the most common methods for assessing the quality of surface and groundwater are electrochemical methods, spectroscopic methods, chromatographic methods, and mass spectrometry.

3.1 Electrochemical Methods

Electrochemical methods are key techniques that are widely used to assess the quality of surface and groundwater. Since the 1970s, these techniques have advanced from basic techniques for measuring salinity to more sophisticated sensors, such as voltametric and potentiometric sensors to more advanced voltametric sensors that can measure multiple chemicals simultaneously (Saputra 2023). These methods can be adopted in both laboratory and field applications to provides a better understanding of water pollution and contamination. Additionally, the development of small and precise sensors (microsensors) has made it possible to measure even minor chemical changes, that helps to assess how trace elements and contaminants behaviour in natural water sources.

At the first level, interfacial electrochemical methods are divided into static methods and dynamic methods. In static methods, no current passes between the electrodes, and the concentrations of species in the electrochemical cell remain unchanged, or static e.g. potentiometry. On other hand dynamic methods uses current flow and concentration change as the result of a redox reaction e.g. voltametry. Moreover, they are further subdivided by whether one choose to control the current or the potential (Faulkner 1983). In addition to this another common method i.e. Coulometric methods.

- **Potentiometric Method**

Potentiometric analysis is a simple yet powerful technique used to measure the concentration of ions in a solution by determining the electrical potential difference between two electrodes. These two electrodes are remains immersed in the solution and one acting as the reference and the other as the indicator (Mosher and Kelter 2023).

The principle of this method is based on measuring the potential difference between two electrodes. This difference between two electrodes changes with the ionic concentration of the sample as a titrant is added. The indicator electrode

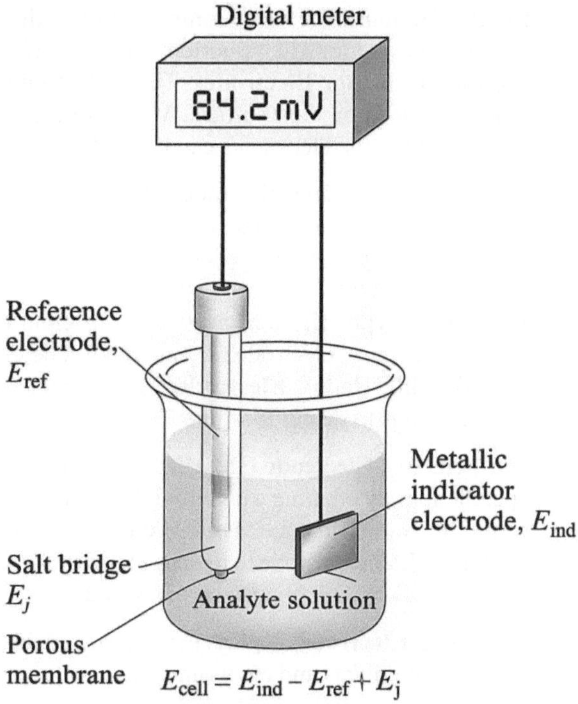

Fig. 1 A cell for potentiometric determinations (Ghazizadeh and Asadollahzadeh 2021)

responds to this change in potential, while the reference electrode maintains a stable potential throughout the process (Zdrachek and Bakker 2021).

In potentiometric titration, a salt bridge is used to keep the solution (called the analyte) apart from the reference solution as shown in Fig. 1. This stops them from mixing and causing incorrect results. However, the overall potential of the cell can be calculated using the following equation:

$$E_{cell} = E_{ind} - E_{ref} + E_j$$

where,

E_{cell} Total cell potential
E_{ind} Indicator electrode potential
E_{ref} Reference electrode potential
E_j Junction potential.

There are many types of electrodes used in potentiometry, each with specific utility, and they should be used carefully. Some of the commonly used electrodes in water quality analysis are given below:

(a) **Indicator Electrodes**: These electrodes respond to changes in ion concentration in a solution. An ideal indicator electrode should react quickly and be selective for specific ions. The most common type is the ion-selective electrode (ISE), used to measure ions like fluoride, nitrate, or pH.

(b) **Metallic Indicator Electrodes**: These electrodes develop a potential based on redox reactions on their surface. Metals like gold, platinum, silver, and copper are used. Metallic electrodes can be further classified into four types:

   i. **First Type**: Metal electrodes placed in solutions containing their ions, like a copper electrode in a copper ion solution.
   ii. **Second Type**: Used to determine ions like chloride, using electrodes like silver.
   iii. **Third Type**: These measure ions different from the electrode metal, such as a mercury electrode for calcium ion determination.
   iv. **Metallic Redox Electrodes**: Inert metals (e.g., Pt, Au) used in redox reactions to detect ions like Ce^+.

(c) **Ion-Selective Electrode (ISE)**: These electrodes measure the concentration of specific ions by creating a potential difference. They have a selective membrane to minimize interference. The pH electrode is the most common ISE, using a glass membrane sensitive to hydrogen ions (H^+). Other ions like fluoride, nitrate, and gases like CO_2 can also be measured.

   i. **Glass Electrode**: A glass membrane sensitive to ions like H^+ is used in pH measurements and environmental tests, such as detecting dissolved CO_2 in water.
   ii. **Crystalline Electrode**: Made of solid inorganic salt crystals, these electrodes are highly selective and operate through ion-exchange mechanisms. They are used to measure ions that can enter the crystal structure.
   iii. **Reference Electrodes**: These electrodes provide a stable and known potential during measurements. A common reference electrode is the Silver/ Silver Chloride Electrode (Ag/AgCl), which consists of a silver wire (Ag) coated with silver chloride (AgCl) placed in a solution saturated with potassium chloride (KCl). The half-cell reaction for this electrode is:

$$AgCl_{(s)} + e^- \rightarrow Ag_{(s)} + Cl^-_{(aq)}$$

This reaction ensures a stable reference potential, commonly used in electrochemical experiments and water quality analysis (Grinberg and Rodriguez 2016).

- **Voltametric Method**

Voltametry is a dynamic electrochemical technique developed by Professor Jaroslav Heyrovský in 1922 (Sahtani 2022). This method analyses how current and voltage change during redox reactions in an electrochemical cell. By applying a potential to an electrode, voltametry induces oxidation or reduction of electroactive species at the electrode surface. The resulting current response provides understandings into the reaction kinetics, mechanisms, and concentration of analytes based on their redox characteristics.

Typically, the electrochemical cell in voltametry consists of three electrodes i.e. a working electrode, reference electrode, and a counter (auxiliary) electrode as shown in Fig. 2 (Marsili 2020). These electrodes are connected to a device called potentiostat, which controls the potential at the working electrode relative to the reference electrode. When potential is applied to the working electrode, it oscillates around a set value, allows current to flow. This current then induces either oxidation or reduction of the analyte in solution, with the current magnitude directly proportional to the analyte concentration (Feldman et al. 1990). By monitoring current as function of the applied voltage, analyte concentrations in dilute solutions can be determined accurately (Scholz 2015).

Typically, Modern voltametric instruments consist of three main components:

i. **Potentiostat**

   The Potentiostat is the core of the instrumentation. Its primary role is to precisely control the applied potential and monitor the resulting current in the system. Modern potentiostats are equipped with advanced components such as converters, electrometers, amplifiers, and microprocessors. These allow for precise modulation of potential, often based on pre-defined waveforms, which can be linear, triangular, square, or differential pulse waveforms.

ii. **Personal computer**

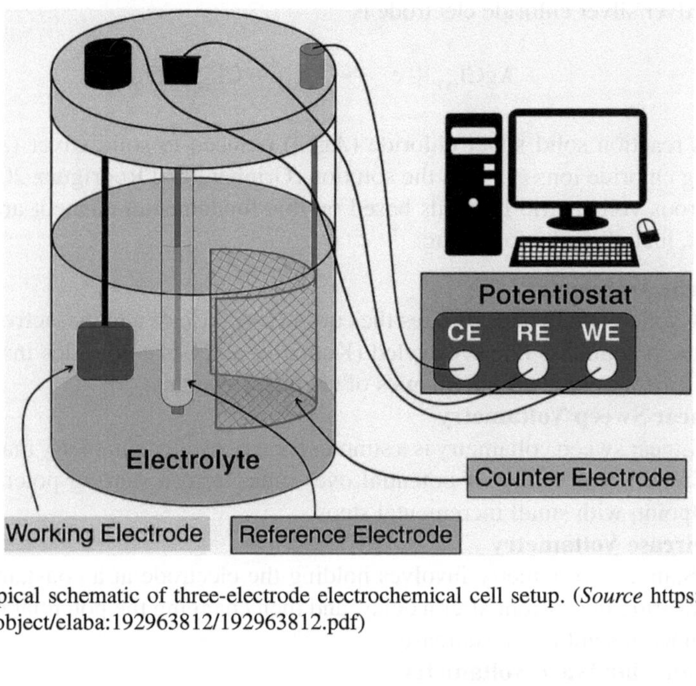

Fig. 2 Typical schematic of three-electrode electrochemical cell setup. (*Source* https://epublications.vu.lt/object/elaba:192963812/192963812.pdf)

Modern voltametric experiments are commonly performed using computer-controlled instruments. The software allows for accurate control of the modulation and waveform properties which increases sensitivity and speed of the experiments.

iii. **Electrochemical cell**

The electrochemical cell used to hold the analyte dissolved in a suitable solvent and ionic electrolyte. Typically, these cells are made from materials like glass, polyethylene, or Teflon. Generally, inside these electrochemical cells, the following components are presents (Grinberg and Rodriguez 2016).

(a) **Working Electrode**

These are the electrodes where the redox reactions of interest occur. The material of the working electrode is chosen based on its stability within the potential range of interest. Common working electrode materials include mercury, gold, platinum, and carbon. Mercury electrodes offer a wide negative potential range, making them ideal for hydrogen ion reduction, while platinum is durable and stable in organic solvents. Carbon electrodes are widely used due to their broad potential window, high conductivity, and structural stability.

(b) **Reference Electrode**

The reference electrode provides a stable and known potential against which the working electrode potential is measured (Sagiroglu et al. 2024). Common reference electrodes include the saturated calomel electrode (SCE) and the silver/silver chloride (Ag/AgCl) electrode. For example, the half-reaction for the silver/silver chloride electrode is

$$AgCl_{(s)} + e^- \rightarrow Ag_{(s)} + Cl^-_{(aq)}Cl^-_{(aq)}$$

In this reaction solid silver chloride (AgCl) reduced to solid silver (Ag) while generating chloride ions (Cl^-) in the solution (Grinberg and Rodriguez 2016).

Numerous voltametric methods based on this fundamental concept are popular nowadays, including the following:

i. **Cyclic Voltametry**

Cyclic voltametry measures the current response of a redox-active solution as the potential is linearly cycled (Kumar et al. 2024), provides insights into the kinetics and thermodynamics of chemical systems.

ii. **Linear Sweep Voltametry**

Linear sweep voltametry is a simpler form of cyclic voltametry that involves a single linear sweep of potential over time, from a starting potential to an endpoint, with small incremental steps.

iii. **Staircase Voltametry**

Staircase voltametry involves holding the electrode at a constant voltage, measuring the current after a delay, and then changing the potential in a linear manner, resembling a staircase.

iv. **Triangular Wave Voltametry**

In triangular wave voltametry, the potential is linearly scanned to a value beyond the reaction point and then returned, producing a voltammogram with distinct current peaks for the forward and reverse reactions.

v. **Reverse Pulse Voltametry**

Reverse pulse voltametry measures currents associated with reduction or oxidation reactions, with the pulse designed to maximize current flow while minimizing electron transfer outside the pulse.

vi. **Differential Pulse Voltametry**

Differential pulse voltametry applies a series of short pulses with rising baseline voltage, measuring the current just before each pulse, allowing for precise detection of substances like catecholamines.

vii. **Square Wave Voltametry**

Square wave voltametry uses a symmetrical square wave superimposed on a staircase scan to enhance signal detection while reducing background noise.

viii. **Rotated Electrode Voltametry**

Rotated electrode voltametry speeds up mass transfer by spinning the electrode, quickly achieving a steady state and resulting in higher current responses.

ix. **Stripping Voltametry**

Stripping voltametry detects trace elements by preconcentrating the analyte onto the working electrode, followed by a potential scan that strips the analyte back into solution, with current proportional to concentration.

- **Coulometric Method**

Coulometry is an electrochemical technique used to measure the total charge, in coulombs, consumed during the redox reaction of an analyte at an electrode. It should not be confused with colorimetry, which is a spectroscopic method. Unlike voltametric and amperometry techniques, coulometry does not rely on mass transport current control to create a concentration-dependent signal (Worsfold et al. 2019). Coulometry regarded as an absolute method because it typically does not require calibration, as the electrical charge can be measured with high precision.

The main principle involved in the coulometry is the measurement of the quantity of the electricity which is directly proportional to the chemical reaction at the electrode. This is given by the Faraday's first law (Worsfold et al. 2019):

$$W = \frac{M_r \times Q}{\mathrm{F\,n}}$$

where,

Q Consumed current
Mr Relative molecular weight
N Number of moles
F Faraday's constant (96487C mol^{-1}).

Additionally, in coulometry, the analyte is fully consumed during the reaction, making it an absolute method, where calibration is not required since the charge is measured with high precision.

Modern colorimetric instruments typically consist of three main components:

i. **Light Source**

The light source emits light across the ultraviolet (UV), visible, or near-infrared (NIR) spectrum. Common types of light source are

- **Tungsten Lamps**: Ideal for visible light applications due to their broad emission spectrum.
- **Deuterium Lamps**: Used for UV measurements, providing a consistent output in the UV range.

ii. **Optical System**

The optical system comprises elements that direct and manipulate the emitted light. They generally comprise

- **Monochromator**: Utilizes prisms or diffraction gratings to isolate specific wavelengths. It allows for precise selection of the wavelength that corresponds to the analyte's absorbance peak.
- **Filters**: Selectively transmit certain wavelengths, typically used in simpler instruments for specific analytes.
- **Cuvette Holder**: Holds the sample cuvette in the light path. It ensures stability during measurements, minimizes stray light, and maintains proper alignment of the light path through the sample.

iii. **Detector**

It basically measures the intensity of light transmitted through the sample and compares it to the light emitted by the source. The commonly used detectors are

- **Photodiodes**: Commonly used for detecting visible light, providing quick response times.
- **Photomultiplier Tubes (PMTs)**: Highly sensitive detectors suitable for low-concentration samples in both UV and visible ranges, capable of detecting minute changes in light intensity.
- **Charge-Coupled Devices (CCDs)**: Advanced detectors that allow simultaneous measurement of multiple wavelengths, enhancing the instrument's analytical capabilities.

The colorimetric method has a wide range of applications in water quality assessment, including the determination of inorganic and organic compounds, carbonate concentration, and even uranium and plutonium due to minimal interference with the sample (Nollet and Gelder 2013). It has also been successfully used to determine over 50 elements.

3.2 Spectroscopic Methods

Spectroscopy is another chemical analysis technique used to analyse the interaction between matter and electromagnetic radiation. Its origins can be traced back to the early nineteenth century, with significant contributions by Joseph von Fraunhofer, a German physicist known for his pioneering work in this field (Shah et al. 2020). Over time, the field of spectroscopy has grown to include various methods and technologies, each designed to analyse different types of interactions between matter and electromagnetic radiation.

At its core, spectroscopy is based on how matter interacts with electromagnetic radiation. This interaction can occur through several mechanisms, including absorption, emission, scattering, diffraction, and resonance. The specific interaction mechanism depends on the nature of the material being analysed and the characteristics of the electromagnetic radiation involved (Vogel et al., 1989).

3.2.1 Classification of Spectroscopic Method:

There are several ways to classify spectroscopic methods, and each classification shows how these techniques help us to understand the matter and how it interacts with electromagnetic radiation.

i. **Based on Nature of the Interaction Mechanism**

A fundamental way to classify spectroscopic techniques is by the mechanism of interaction between matter and radiation. Based on this, the methods are classified as follows:

(a) **Absorption Spectroscopy**

In absorption-based techniques, matter absorbs electromagnetic radiation, causing a transition between energy levels within atoms or molecules. This results in a measurable reduction in the intensity of the transmitted radiation, which provides insights into the structure or composition of the material. Commonly used techniques that utilize absorption spectroscopy include UV-Visible (UV-Vis) spectroscopy, which assesses concentrations of substances in the ultraviolet and visible light ranges, and Atomic Absorption Spectroscopy (AAS), which detects trace metals in water samples through atomic absorption (Cooper and Klymkowsky 2020).

(b) **Emission Spectroscopy**

In emission spectroscopy, matter emits radiation after being excited by an external energy source such as heat or light. The emitted radiation is characteristic of the material's composition and energy structure, offering valuable information about the substance (Pérez-Juste and Faza, n.d.). Moreover, the emission process can be observed by various methods such as flame photometry, where heat induces substances to emit light, and fluorescence spectroscopy, in which materials absorb high-energy radiation and subsequently emit light at lower energy levels. This method is particularly valuable for identifying the composition and behavior of materials across different states (solid, liquid, or gas), and provide insights into their properties and interactions under varying conditions.

(c) **Scattering, Diffraction, and Reflection Spectroscopy**

These spectroscopic methods rely on the interaction of radiation with matter through processes such as scattering and deflection. Scattering occurs when radiation interacts with matter and is deflected or scattered. By analysing the scattered radiation, we can gain insights into molecular structures and dynamics. In diffraction-based methods, electromagnetic radiation is diffracted by the regular atomic arrangement of crystalline materials. A common example of these methods is Raman spectroscopy which is used to study vibrational modes in molecules, providing detailed structural information and X-ray Diffraction (XRD) which widely employed in the field materials science and crystallography for determining crystal structures.

(d) **Resonance or Coherence Spectroscopy**

In resonance spectroscopy, matter absorbs radiation at specific resonant frequencies, typically within the radiofrequency or microwave range. These techniques are highly sensitive to the molecular environment and can reveal intricate details about atomic or nuclear structures (Bergström 2015). Most common method based on this principle is Nuclear Magnetic Resonance (NMR) Spectroscopy which extensively used in chemistry and biochemistry for structural elucidation (Franca and Nollet 2017).

ii. **Based on Scale of Application**

Spectroscopic methods can be further categorized based on the scale or level at which they are applied, specifically whether the interaction of electromagnetic radiation (EMR) is studied at the atomic or molecular level. Based on this distinction, these methods are classified as follows:

(a) **Atomic Spectroscopy**

These method focuses on the interaction of EMR with individual atoms, and examine the energy levels within an atom and the transitions between these levels.

(b) **Molecular Spectroscopy**

In this examines how EMR interacts with molecules, and explores the energy levels associated with both individual atoms and the bonds between them, as well as the overall structure of the molecule. It specifically focuses on how molecules interact, including their vibrations and rotation.

For example, ASS method is used to detect and quantify metals in samples at the atomic level, whereas IR spectroscopy is used for molecular structure analysis by studying molecular vibrations.

iii. **Based on Specific Technology Involved**

In spectroscopy, different parts of the electromagnetic spectrum like UV light, visible light, infrared, and microwaves are used for different purposes. Each focus on specific properties of materials, However, the choice of technology depends on the sample type and the specific information needed. The common methods are

(a) **UV-Visible (UV-Vis) Spectroscopy**
This technique operates in the UV and visible ranges of the electromagnetic spectrum, analysing electronic transitions in molecules. It is commonly used to determine the concentration of substances in solution through absorbance measurements based on the Beer-Lambert law.

(b) **Infrared (IR) Spectroscopy**
This technique is used to study molecular vibrations and provides detailed insights into functional groups and molecular structures by utilizing the infrared range of the electromagnetic spectrum.

(c) **Raman Spectroscopy**
This technique complements infrared methods by probing molecular vibrations, particularly in symmetric molecules where infrared signals may be weak or absent.

In this section, the focus is on the most commonly used spectroscopic methods for the assessment of surface and groundwater quality.

A. **Absorption Spectroscopy**

Absorption spectroscopy used to measure the concentration of a substance in a sample by analysing the light absorbed by that substance. It relies on the interaction of light with matter, where specific wavelengths of light are absorbed by the molecules of the sample, leading to transitions between different energy levels. Some of common absorption based spectroscopic method described below.

- **UV-Visible (UV-Vis) Spectroscopy**

UV-Visible spectroscopy is based on the principle of measuring the absorption of light by various molecules in a sample within the ultraviolet (200–400 nm) and visible (400–800 nm) ranges of the electromagnetic spectrum. When UV or visible light is absorbed by these molecules, their energy content increases, resulting in electronic transitions. This absorption generates a distinct spectrum that can be used to identify and quantify the compounds present (Rizwan and Gwenin 2021).

The relationship between the amount of light absorbed and the concentration of absorbing molecules in a solution is described by Beer-Lambert Law (Rizwan and Gwenin 2021). This law states that when a beam of monochromatic light passes through a solution containing an absorbing substance, the decrease in radiation intensity is proportional to both the concentration of the absorbing substance and the path length of the light through the solution.

Mathematically,

$$A = \log(I_0/I) = \varepsilon Ci$$

where,

A Absorbance (no units)
ε Molar absorptivity (L mol^{-1} cm^{-1})
C Concentration of the solution (mol L^{-1})
I_0 Intensity of light upon a sample cell
I Intensity of light departing the sample cell
i Path length of the sample cell (cm).

According to this law, higher concentration of molecules capable of absorbing light at a specific wavelength result in greater light absorption.

Additionally, this technique effectively measures nitrate (NO_3^-) concentrations ranging from 0.1 to 100 mg/L and widely adopted for analysing nitrates, phosphates (PO_4^{3-}), and organic carbon in different water sources. Also, UV-Vis spectroscopy is effective for detecting heavy metals in water samples. The atomic absorption spectrophotometers operate on the principle of UV-Visible absorption spectroscopy, is commonly employed for this purpose.

- **Atomic Absorption Spectroscopy (AAS)**

AAS is a well-established chemical analytical technique first introduced by Alan Walsh in year 1953 (Ali and Aboul-Enein 2006). Since then, AAS has become widely used method due to its high reliability in analysing trace elements such as lead (Pb), cadmium (Cd), and mercury (Hg), which are the critical for potable water quality analysis according to regulatory standards (Sardans et al. 2010). In addition to these elements, AAS is capable of detecting approximately 70 different elements, making it a versatile tool for various applications in different fields. One of the primary advantages of AAS is its ability to measure the concentration of a specific element in a sample without interference from other elements, thus eliminating the need for prior separation. Additionally, this method's sensitivity enables it to quantify metals at levels as low as parts per billion (ppb).

The principle of AAS is operates on the principle of absorption of radiation by unexcited (ground state) atoms. When a sample containing metal salts or metallic compounds is introduced into a flame, it undergoes vaporization, resulting in the production of gaseous atoms. While a fraction of these atoms may become excited due to thermal energy, the majority remain in their ground state. These ground state atoms possess the unique capability to absorb light at specific wavelengths, that are unique to the element. As the light passes through the flame containing these atoms, a portion of it is absorbed. The amount of light absorbed is directly proportional to the number of ground state atoms present, which correlates to the concentration of the element in the sample (Müllertz et al. 2016).

The total amount of radiation absorbed by the atoms can be expressed mathematically as:

$$\upsilon = \frac{\pi e2}{\mathrm{mC}} * \mathrm{Nf}$$

where

- υ Frequency absorbed by the atoms
- e Charge on electron
- m Mass of electron
- C Velocity of light
- N Total number of atoms
- f Oscillator strength.

The process of AAS involves several steps for accurate analysis, some of most cruicial steps are as follows:

- **Sample Introduction**: The sample is prepared and introduced into a nebulizer, where it is converted into an aerosol.
- **Atomization**: The aerosol is transported into a flame (or furnace), where it is atomized, producing free gaseous atoms.
- **Radiation Source**: A hollow cathode lamp emits light tuned to the target element, which then passes through the atomized sample.
- **Detection**: The transmitted light is measured by a photodetector, which records the intensity of the light before and after it penetrates the sample.
- **Data Analysis**: The absorbed light intensity is analysed to determine the concentration of the target element in the sample.

However, these steps are performed using an instrument known as an atomic absorption spectrometer. Figure 3 illustrates a typical atomic absorption spectrometer along with its fundamental components.

B. **Infrared (IR) Spectroscopy**

It is also known as vibrational spectroscopy, which utilizes the vibrational transitions occurring within molecules. Commonly used to measure inorganic and organic compounds in water samples, as it provides valuable information for identifying and determining the structures of compounds. The technique is performed using an instrument called an infrared spectrometer (or spectrophotometer), which generates an infrared spectrum for analysis.

The functioning of IR spectroscopy based on the concept of absorption of infrared lights, typically in the range of 4000–400 cm^{-1}, which induces vibrational transitions in chemical bonds within molecules. Each functional group (e.g., O–H, C=O, C–H) has a unique vibrational frequency, resulting in a characteristic absorption band that can be used to identify and quantify specific compounds in a water sample.

However, one of significant advancement in IR spectroscopy is Fourier Transform Infrared Spectroscopy (FTIR), which enhanced the capabilities of this technique and broadens its application in the field of water quality analysis and detection of various elements.

- **Atomic Absorption Spectroscopy (AAS)**

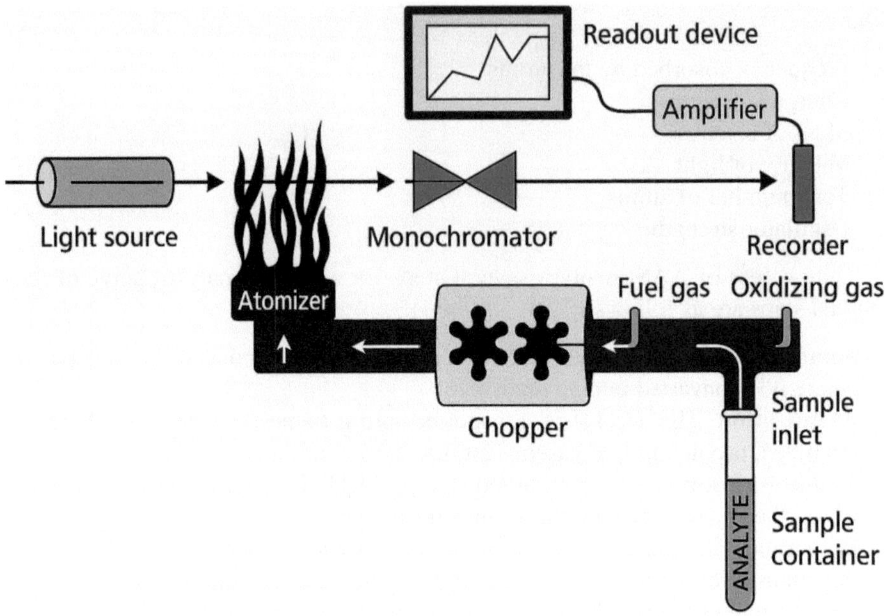

Fig. 3 Typical atomic absorption spectrometer with fundamental components. (*Source* https://scienceinfo.com/atomic-absorption-spectrophotometry/)

FTIR is used for assessing the quality of surface and groundwater, although it is not typically a primary tool for direct elemental analysis, such as for heavy metals. Instead, FTIR is effective for detecting and analysing the organic pollutants, functional groups of dissolved compounds, and certain inorganic compounds in water. The increasing application of this technique in evaluating surface and groundwater quality is linked to its capability to identify and quantify various organic and inorganic compounds (e.g., pesticides, hydrocarbons).

The primary difference between conventional spectroscopy and FTIR lies in their measurement approaches. In conventional spectroscopy, measurements are conducted in the frequency domain, where radiant power expressed as absorbance or transmittance, is recorded as a function of frequency. In contrast, time domain spectroscopy records radiant power based on time. Despite these different approaches, both types of measurements hold the same information. The conversion between time and frequency domains is achieved through a mathematical technique known as Fourier Transform

In FTIR, Michelson interferometer is commonly used for time domain frequency measurement. An interferometer typically consists of two mirrors: one stationary and one movable. The movable mirror moves at a steady speed. Radiation from the source passes through a beam splitter, splitting the beam so that one part reaches the movable mirror while the other reflects off the fixed mirror. The two beams then converge again at the beam splitter, producing constructive or destructive interference

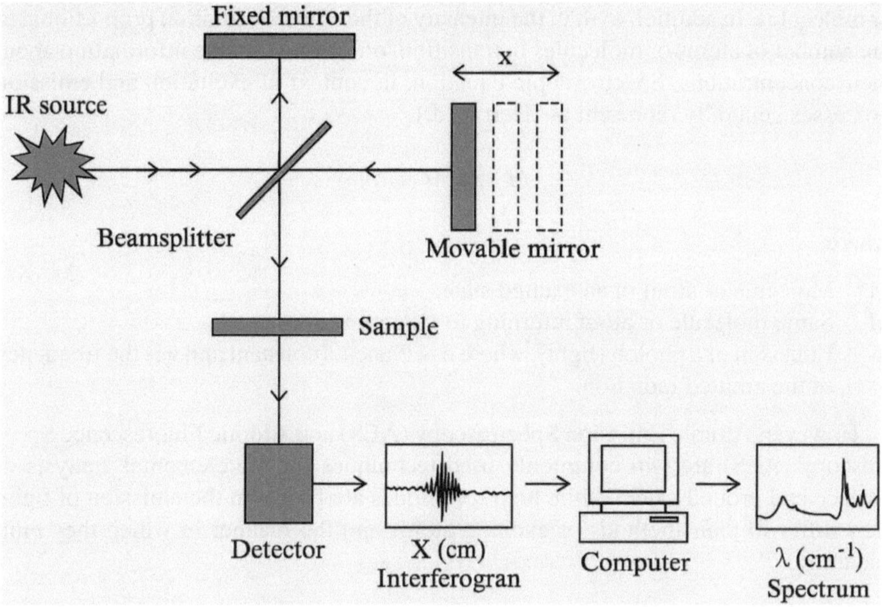

Fig. 4 Schematic representation of instrumentation of FTIR (Ojeda and Dittrich 2012)

based on the difference in path length. For monochromatic light, this interference produces a sine wave signal at the detector, converting the frequency domain signal into a time-domain signal, known as an interferogram. The presence of multiple frequencies causes the interferogram to become more intricate. Fourier transform is applied to the interferogram to convert it back into the frequency domain spectrum. If a sample is placed in the path, certain frequencies are absorbed, and the Fourier transform yields the IR spectrum of the sample.

The general instrumentation for FTIR, which is crucial for sample analysis in water quality assessment, is illustrated in Fig. 4.

C. **Emission Spectroscopy**

Emission spectroscopy is a significant analytical technique utilized for the identification and measurement of various substances. This technique works on the principle that when a molecule M, is excited to a higher energy state, M* and subsequently returns to a lower energy state, it emits some amount of the absorbed energy as radiation the excitation energy required to raise the molecule to a higher energy state can be supplied by external sources such as flames, electrical discharges, or plasma, which rises the sample to high temperatures.

The net energy emitted during this process corresponds to the energy difference between the initial and final states, resulting in the generation of radiation at a specific wavelength. This wavelength provides information about the type of atom or molecule involved in the transition, allows for qualitative identification of atoms or molecules. In addition to that, the intensity of the emitted radiation proportional to the number of atoms or molecules in transition, offers quantitative information about their concentrations. Spectroscopic equation, in context of excitation and emission processes generally represent as (Hertz n.d.)

$$M^* \rightarrow M + h\nu$$

where,

- M* Molecule or atom in an excited state.
- M Same molecule or atom returning to its ground state.
- hν Emission of a photon (light), where h is Planck's constant and ν is the frequency of the emitted radiation.

However, Atomic Emission Spectroscopy (AES) and Atomic Fluorescence Spectroscopy (AFS) are two commonly used techniques for the elemental analysis of surface and groundwater. While both techniques are based on the emission of light, they differ in their methods of exciting atoms and the manner in which they emit radiation.

- **Atomic Emission Spectroscopy (AES)**

In AES, atoms present in a sample are excited by an external energy source, such as a flame, plasma, or electrical discharge (Omanwa et al. 2020), as illustrated in Fig. 5. This energy increases atoms to higher energy levels, and when they return to their ground state, they emit light at characteristic wavelengths. Then this emitted light is used to analyse, identify and quantify the elements present in the sample (Patnaik 2017).

The excitation sources used in AES, including flames, plasma, electrical arcs, and sparks, are particularly effective for multi-element analysis (Grinberg and Rodriguez 2018), making this technique valuable for water quality assessment and environmental monitoring.

- **Atomic Fluorescence Spectroscopy (AFS)**

AFS is another common type of emission spectroscopy that differs slightly from AES in its method of excitation and emission. In AFS, atoms are excited by absorbing the electromagnetic radiation (light). Once the atoms absorb this energy, they transition to

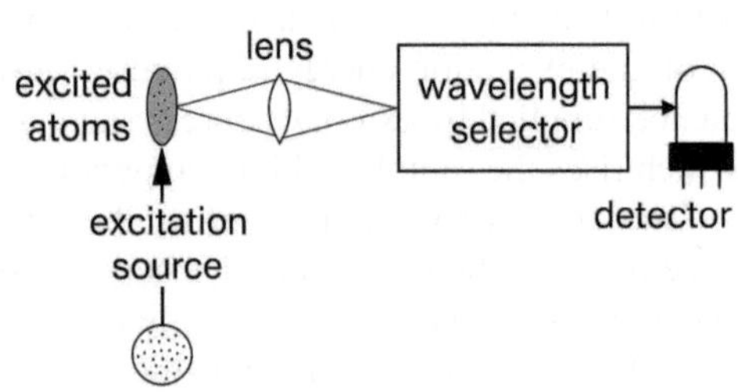

Fig. 5 Typical instrumentation of atomic emission spectrometer. (*Source* https://www.chemicool.com/definition/atomic_emission_spectroscopy_aes_oes.html)

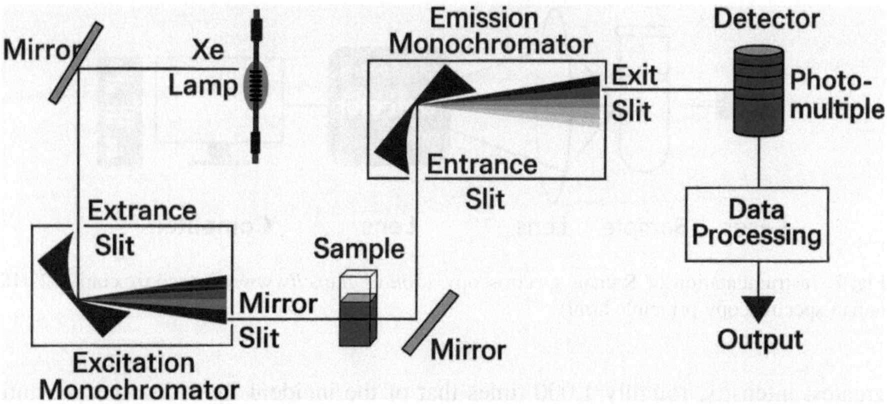

Fig. 6 Typical instrumentation of atomic emission spectrometer (Gomes et al. 2019)

a higher energy state. As the excited atoms return to their lower energy state (ground state), they emit light in the form of fluorescence. By measuring the intensity of this emitted light using the suitable detectors, it is possible to determine the concentration of specific elements in water samples (John Kenkel 2014).

The excitation sources commonly used in AFS include lasers or lamps, which provide the necessary electromagnetic radiation to excite the atoms. The typical instrumentation of an atomic fluorescence spectrometer is presented in Fig. 6. This method is particularly effective for detecting trace elements such as mercury (Hg), zinc (Zn), and selenium (Se) in surface and groundwater, as well as in various environmental, biological, and agricultural samples.

D. **Raman Spectroscopy**

Raman spectroscopy is a common type of molecular spectroscopy. This method focuses on measuring the inelastic scattering of monochromatic light, usually emitted by a laser, as it interacts with the molecular vibrations within a sample. This interaction provides information about the vibrational energy levels of molecules, making raman spectroscopy valuable for characterizing molecular structures, identifying substances, and analysing chemical compositions in various fields, including chemistry, biology, and material science (Rehman et al. 2012).

When light interacts with sample materials, it generally leads to both elastic and inelastic scattering. In elastic scattering, such as Rayleigh scattering, there is no energy transfer between the incident light and the molecules, resulting in no change to the wavelength. Conversely, inelastic scattering, (e.g. Raman scattering), involves an energy transfer between the light and the molecules, this energy transfer alters the wavelength of the scattered light, which is called the Raman effect.

The interaction of light with a sample's vibrational modes, whether the sample is in solid, gas, or liquid form, results in energy losses and gains known as Stokes and anti-Stokes scattering. Rayleigh scattering, as an elastic process, exhibits the

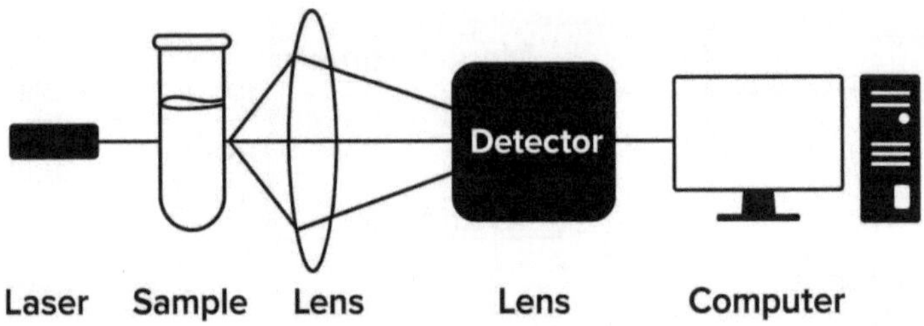

Fig. 7 Instrumentation of Raman spectroscopy. (*Source* https://www.sciencedoze.com/2022/10/raman-spectroscopy-principle.html)

greatest intensity, roughly 1,000 times that of the incident light. In contrast, anti-Stokes scattering, which involves energy gain, is less prominent due to its weaker intensity compared to Stokes scattering, which involves energy loss. Most photons are scattered elastically in Rayleigh scattering, resulting in no energy change. Additionally, Raman spectrum, which plotted with wavelength versus scattering intensity and Raman shift ($\Delta v\,cm^{-1}$), reveals unique spectral lines corresponding to the vibrational states of each material. This intensity of the Raman peaks then used to determine the concentration of substances. Ultimately, makes Raman spectroscopy as an effective method for quantitatively detecting contaminants in water (Zhen Li and Li 2016).

A Raman spectrometer consists of several key components, including an excitation source (often a laser), a light collection system, a monochromator to filter the scattered light, a detector to measure the intensity, and a data processing system to analyse the results. Figure 7 illustrates the fundamental components and outlines the general operational principle.

Raman spectroscopy covers a range of advanced techniques, such as standard Raman spectroscopy, resonance Raman scattering (RRS), surface-enhanced Raman scattering (SERS), and FT-Raman spectroscopy, which are extensively utilized for identifying pollutants in water (Zhen Li and Li 2016). These techniques can detect various contaminants, including chlorates (ClO_3^-, ClO^-, ClO_4^-), heavy metals {mercury (Hg), lead (Pb), chromium (Cr), arsenic (As)}, anions{cyanide (CN^-)}, ferricyanide [{$Fe(CN)_6$}$^{3-}$), nitrate (NO_3^-), nitrite (NO_2^-), asbestos {$Mg_3(Si_2O_5)(OH)_4$}, and sulphate (SO_4^{2-}), as well as bacteria, viruses, and other microbial pathogens, making them essential for effective environmental monitoring and water quality analysis.

E. **Mass Spectrometry**

Mass spectrometry (MS) is common and widely used chemical-analytical technique used to identify the composition of a sample by measuring the mass of its atoms or molecules. This technique involves transforming a small amount of the sample into ions, which are then analysed in a high-vacuum environment using electric and/or magnetic fields.

Mass spectrometer is an instrument that generally used to perform the mass spectroscopy. It works on the basis of the concept that a tiny quantity of sample is

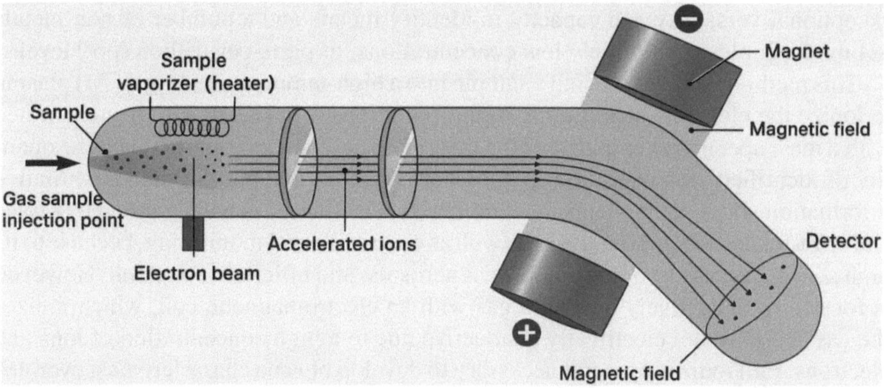

Fig. 8 Instrumentation of mass spectroscopy (MS). (*Source* https://boisestate.pressbooks.pub/chemistry/chapter/2-3-atomic-structure-and-symbolism/)

converted into a gas phase or another suitable form and then ionized, discharging the resultant charged particles into a high-vacuum, enclosed environment. Depending on the type of device, they are exposed to either electric or magnetic fields The device identifies these ions by calculating their mass-to-charge ratio through an analysis of the forces acting on them. This method offers excellent sensitivity and usually destroys the material, requiring just a small amount for examination. Ion abundances are arranged according to increasing mass-to-charge ratios on a graph known as a mass spectrum, which presents the results (Vadakedath et al. 2022).

Since, mass spectrometers work in a high vacuum environment because they produce and work with gas-phase ions. They are composed of three fundamental components as shown in Fig. 8, and the sample has undergone the subsequent procedures:

- **Ionization**: Samples in the form of gas or vapour are ionized at the instrument source.
- **Mass-Selective Analyzer**: It extracts and accelerates ions from earlier processes using electronic lenses, increasing their kinetic energy. The ions are then 'filtered' in the analyser based on their mass to charge ratio, which consists of a number of instruments that connect numerous mass analysers in series.
- **Ion Detector**: After separation, ions strike the detectors sensor, measuring electrical charge and amplifying weak ionic current

Finally, the mass spectrometer's display device displays the mass spectrum obtained by processing the detector signal.

- **Inductively coupled plasma mass spectrometry (ICP-MS)**

An inductively coupled plasma is used to ionize and produce atomic ions in liquid samples in inductively coupled plasma mass spectrometric (ICP-MS) method. This method is a sophisticated hybrid mass spectrometry technique and renowned for its exceptional versatility and capacity to identify metals and a number of non-metals in liquid samples at extremely low concentrations, or parts-per-trillion (ppt) levels.

This method includes placing a sample into a high-temperature Argon (Ar) plasma to ionize the elements, resulting in monovalent ions. These ions are then examined with a mass spectrometer to determine their mass-to-charge ratio (m/z) and the quantity of identified ions over time. ICP-MS provides both qualitative and quantitative information about components in a sample. This technique is widely utilized in trace and ultra-trace elemental analysis, as well as environmental monitoring, because to its high sensitivity, use of liquid samples as aerosols, and efficient ionization. However, is formed by inductively heating a gas with an electromagnetic coil, which ionizes the gas and makes it electrically conductive due to a high concentration of ions and electrons. Full ionization is not necessary to develop plasma characteristics; even 1% ionization can produce plasma (Bogaerts et al. 2002). In spectrochemical analysis, plasmas are normally electrically neutral, with each positive ion charge balanced by a free electron. These plasmas are mostly composed of singly charged positive ions, with few negative ions, resulting in about equal amounts of ions and electrons per volume.

ICPs typically operate in two modes: capacitive (E) with low plasma density and inductive (H) with high plasma density. External inputs trigger the switch from E to H heating mode (Lee and Chung 2006). However, ICP-MS commonly uses the H mode. Figure 9 illustrates the general instrumentation layout for ICP-MS.

Typically, for the detection and measurement of trace metals and pollutants, including lead (Pb), mercury (Hg), arsenic (As), and cadmium (Cd), at very low concentrations (parts-per-trillion levels), ICP-MS is frequently utilized in surface and groundwater quality monitoring (Reddy et al. 2012). It ensures water safety by

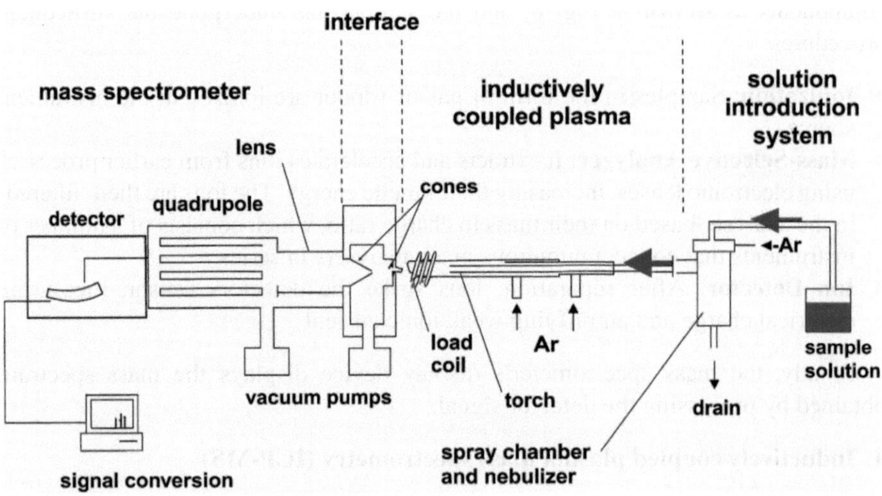

Fig. 9 Instrumentation of inductively coupled plasma mass spectrometry (ICP-MS) (Rezaei et al. 2019)

conforming to regulatory criteria, assists in heavy metal pollution monitoring, and tracks contamination origins via isotope analysis.

3.3 Chromatographic Methods

Chromatography is another popular laboratory technique used to separate and analyse the components of a mixture in both gaseous and liquid states. The terms "chromatography" and "chromatogram" were first introduced by the Russian botanist Mikhail Tsvet at the beginning of the twentieth century (Giddings et al. 2024). Since its inception, chromatography has significantly evolved. Modern chromatography systems incorporate advanced technology, including computer software, to control high-efficiency miniature columns, allowing for the separation of even trace amounts of complex samples with high precision and high timesaving.

In essence, it operates on the partitioning principle, in which the mixture is divided into two phases: a mobile phase that passes through the stationary phase and a stationary phase that remains fixed in situ, frequently on a surface or inside a column. Depending on the type of chromatography, the mobile phase may be a liquid or a gas, while the stationary phase may be a solid or a viscous liquid (Rood 1998).Because of their varied affinities and solubilities, the mixture's components interact with each phase in distinct ways, causing them to migrate and separate differentially. Nevertheless, separation makes it possible to precisely identify and measure each component in a mixture, which is why chromatography is a crucial technique in chemical analysis.

Numerous factors, such as the types of phases involved, the separation method employed, and the physicochemical phenomena listed in Table 1, can be used to categorize chromatographic procedures.

Chromatographic analysis methods are frequently used to evaluate the quality of water. For instance, identifying contaminants including pesticides, herbicides, and industrial effluents is one way to assess the quality of surface water (ISO 15680 2003). Standard chromatographic procedures for the analysis of volatile organic compounds and pesticides in ground water samples, as outlined in EPA Method 524.2 and EPA Method 537 by the U.S. Environmental Protection Agency, are widely adopted and recognized.

However, materials must be properly prepared before chromatographic analysis, which frequently entails filtering, extraction, and concentration procedures. The kind of pollutants and the particular chromatographic technology to be employed determine the preparation process to be utilized. Readers can refer to the EPA Guidance Document for comprehensive information.

This article covers basic chromatographic procedures that serve as the foundation for more sophisticated approaches.

A. **Column Chromatography**

Table 1 .

Criteria	Method name	Mobile phase	Stationary phase
Nature of phase	Liquid phase chromatography (LC)	Liquid (e.g., water, methanol, or acetonitrile)	Solid or liquid-coated solid (e.g., silica gel or polymer)
	Gas chromatography (GC)	Gas (e.g., helium or nitrogen)	Liquid-coated solid inside a column (e.g., polysiloxane-coated capillary column)
Separation process involved	Liquid/solid chromatography (adsorption chromatography)	Liquid (e.g., ethanol or water)	Solid adsorbent (e.g., alumina or silica gel)
	Ion chromatography (IC)	Buffered solution (e.g., sodium chloride solution)	Solid with ionic sites (e.g., ion-exchange resin)
Physicochemical phenomena involved	Partition chromatography	Liquid solvent (e.g., a mixture of water and alcohol)	Solid support (e.g., paper) with a liquid phase adsorbed on it
	Affinity chromatography	Liquid (e.g., buffer solution)	Solid with specific ligands (e.g., agarose beads with antibody ligands)

Column chromatography is a fundamental type of chromatography and serves as the basis for many later developments in the field. It operates as a form of adsorption chromatography, where the separation of components depends on their degree of adsorption to the stationary phase. This technique uses a polar solid substance packed into a vertical glass or metal column as the stationary phase. Over time, it has led to the development of several advanced chromatographic methods, such as High-Performance Liquid Chromatography (HPLC), Ultra-Performance Liquid Chromatography (UPLC), and Gas Chromatography (GC) (Badri et al. 2023).

The basic idea behind column chromatography is that when a mixture of the sample in the mobile phase is added at the top of the column, the various components of the mixture move at different rates. Greater adsorption affinity components move more slowly and elute last, whereas lower affinity components move more quickly and elute first. In a reversible process, the solute molecules adhere to the column.

However, The equation below describes the rate of movement of the components (Grinberg and Rodriguez 2018):

$$R = ds/dm$$

where,

R Rate of movement of a component,
ds Distance moved by the solute,

dm Distance moved by the mobile phase.

- **A crucial step and element in column chromatography**

The component/processes listed below are necessary to carry out column chromatography.

- **Vertical column**: a vertical column with a knob at the bottom, preferably made of glass, is required for column chromatography. This column should have no marks or graduations and should be cylindrical in form, like a burette. Also following are the essential component
- **Stationary phase**: particles in the stationary phase, which is frequently made up of fine sand or another suggested adsorbent, should be consistent in size and shape and devoid of impurities.
- **Mobile phase**: chromatography-grade solvents, which can be a single solvent or a combination of solvents designed to meet the particular separation needs, make up the mobile phase.
- **Column Plugging Mechanism**: to retain the stationary phase while allowing the escape of solvent and sample, a cotton wool or asbestos pad is used to plug the bottom exit of the column.
- **Adsorbents**: two main adsorbents used in column chromatography are silica gel (SiO_2) and alumina (Al_2O_3). They come in a variety of mesh sizes and are labeled on the box as "silica gel 60" or "silica gel 230–400." The mesh size of the sieve used to categorize the particles is indicated by these values. For gravity chromatography, larger particles with lower mesh values work better, while flash chromatography prefers smaller particles with higher mesh values.
- **Solvent characteristics**: polarity of the solvent has a considerable impact on the transport of chemicals through the column. Polar solvents compete with polar molecules in the mixture for adsorption sites, increasing the solubility and flow of polar substances. Even polar molecules move swiftly in a highly polar solvent; but extremely polar solvents can generate rapid flow, resulting in poor separation. In contrast, if the solvent is not sufficiently polar, molecules may fail to elute efficiently (Kumari et al. 2022). As a result, choosing a suitable eluting solvent is critical, as polar interactions dominate chemical interactions with silica or alumina in column chromatography.

However, Fig. 10 illustrates the basic concepts and crucial elements of column chromatography.

- **General step by step procedure of column chromatography**

- **Column Preparation**: The process starts with the column's preparation, which involves packing the stationary phase into a vertical column and moistening it with the mobile phase. The stationary phase is secured and kept from escaping by placing a cotton or asbestos pad at the bottom.

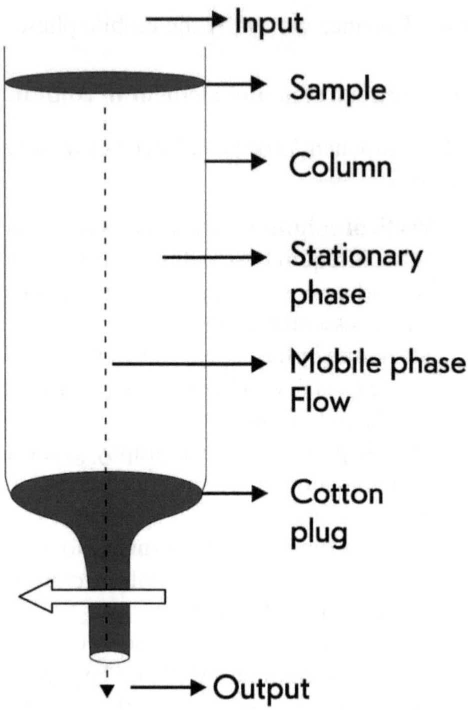

Fig. 10 Typical instrumentation of column chromatography. (*Source* https://www.studyread.com/column-chromatography/)

- **Sample Application**: To ensure uniform distribution, the sample to be separated is put on top of the packed stationary phase. A second cotton or asbestos pad is positioned between the sample and the stationary phase to prevent disruption during elution.
- **Mobile Phase Addition**: The mobile phase is then introduced over the sample, saturating it and facilitating its movement through the stationary phase. A collecting beaker is positioned at the bottom of the column to capture the eluent as it flows out
- **Elution Process Separation**: The components of the sample are transported lower by the mobile phase as it percolates through the stationary phase. Depending on their affinities for the stationary and mobile phases, different components migrate at different speeds; those that prefer the mobile phase elute first.
- **Components**: The migration of sample components results in their separation, with each component traveling at its own rate. This separation occurs as a function of the adsorption affinities, allowing for distinct elution patterns.
- **Formation of Colored Bands**: As components separate, colored bands form within the column, each representing a specific group of compounds in the mixture. This visualization is crucial for assessing the success of the separation.
- **Fraction Elution**: To selectively elute particular components, the mobile phase can be changed. This targeted elution might take hours or days, depending on the sample size, column length, and mobile phase and packing material parameters.

This method is particularly beneficial for concentrating and evaluating low-concentration contaminants, allowing for exact identification and quantification that is required for water quality monitoring and environmental protection initiatives.

B. **Gas Chromatography**

Gas chromatography (GC) is a sophisticated chromatographic method that analyzes and detects components in a sample based on their specific affinity for the adsorbent material in a chromatographic column. In this method, sample is delivered in either liquid or gaseous form using a gas chromatography syringe and then injected into the column through an intake port, where it is vaporized. The vaporized sample is moved through the column by a constantly flowing carrier gas, known as the mobile phase. Hydrogen is a common carrier gas and is commonly used in thermal conductivity detectors (TCD). The sample components are separated and detected at the detection port under strictly regulated temperature settings, and the findings are displayed on a computer as chromatographic peaks. As, carrier media in chromatography can be liquid or gas, depending on the kind of chromatography used. Gas chromatography separates and detects volatile substances using a gaseous carrier medium, whereas liquid chromatography (such as High-Performance Liquid Chromatography - HPLC) uses a liquid carrier.

The components within the sample travel the column at varying speeds, which are determined by three major factors:

- The physical properties of the components,
- Their chemical properties,
- Their interaction with the stationary phase within the column.

Every separated component is electronically detected and recognized as it leaves the column. The components of the sample are separated by the stationary phase in the column, and each component leaves the column at a different retention time. The column temperature and the flow rate of the carrier gas are other variables that affect retention time.

A typical gas chromatograph includes the following important components:

- Inlet port for sample injection
- Adsorption column for separation
- Detector port for component identification
- Flow controller to regulate carrier gas flow.

Furthermore, perfect temperature management is essential for successful molecular separation since temperature influences how quickly components flow along the column. Lower temperatures enhance separation but require more elution time beside this hydrogen, has a high separation efficiency, is widely employed as a carrier gas, while helium is favoured due to its safety and compatibility with various detectors.

In gas chromatography, the fundamental principle is the partitioning of sample components between two phases: a stationary phase and a mobile gas phase. Additionally, based on the nature of the stationary phase, gas chromatography is further classified as Liquid Chromatography (GLC) where stationary phase is liquid and Gas-Solid Chromatography (GSC) where stationary phase is solid.

- **Gas–Liquid Chromatography (GLC)**

It is the most widely used gas chromatographic method for water quality analysis. In this method, the carrier gas vaporizes the sample and transports it through the columns. The components are separated according to their vapor pressures and affinity for the stationary phase. The affinity of a material for the stationary phase is described by an equilibrium constant known as the partition coefficient (K_c), which may be represented by the following equation (Nallakukkala and Lal 2022):

$$K_c = [A]_S/[A]_m$$

where,

$[A]_S$ Concentration of compound A in the stationary phase,
$[A]_m$ Concentration of compound A in the mobile phase.

The distribution constant K_c drives compound transport in the column, and K_c differences facilitate effective compound separation. Furthermore, K_c is temperature dependent and affected by the chemical characteristics of the stationary phase. The separation of distinct sample components can be improved by increasing the temperature or changing the stationary phase.

The successful functioning of the gas chromatographic technique depends on a number of crucial elements. The apparatus is precisely built in order to ensure the precise and effective detection of analytes, as seen in Fig. 11. This is done by combining several interrelated components, which usually consist of the following elements:

- **Gas Inlets**: Gas is supplied from cylinders through filtered piping to ensure high purity and regulated pressure.
- **Pneumatic Control**: These controls regulate gas flow and pressure to various components of the instrument, including carrier gas and detector flows.
- **Injector**: The injector volatilizes the sample for introduction into the carrier stream. Common types of injectors include, Split/Splitless, Programmed Thermal Vaporizing (PTV) and Cool-on-column (COC) for liquid samples.
- **Column**: The retention of analytes in gas chromatography occurs through interactions with the stationary phase, which can be classified into three main types: dispersive, dipole, and hydrogen bonding interactions. Columns used in gas chromatography can be either packed or capillary.
- **Column Oven**: The column oven controls the temperature for separation, supporting the column. It can operate in isothermal or temperature-programmed modes.
- **Detector**: The detector amplifies the analyte's response, generating signals for chromatogram production. Several types of detectors are commonly employed, each with specific advantages depending on the analysis. These include the Flame

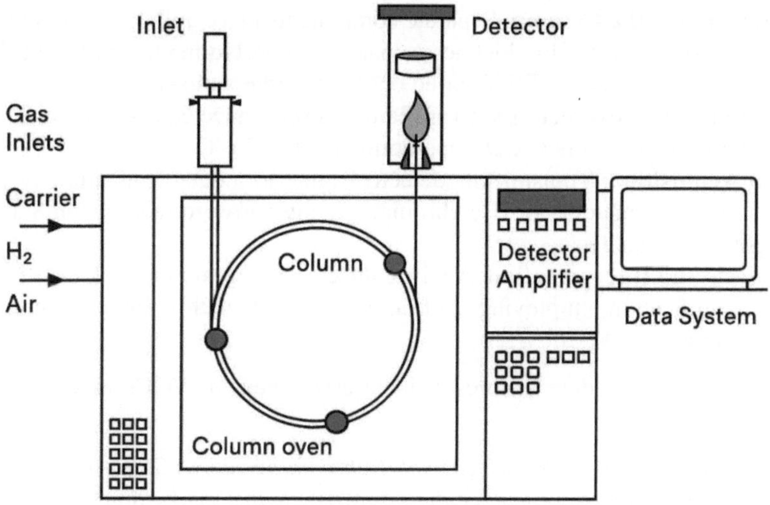

Fig. 11 General instrumentation of gas–liquid chromatography (Wang 2020)

Ionization Detector (FID), Electron Capture Detector (ECD), Flame Photometric Detector (FPD), Nitrogen Phosphorus Detector (NPD), Thermal Conductivity Detector (TCD), and Mass Spectrometer (MS).

- **Data System**: The data system digitizes the detector's signal to create the chromatogram, facilitating quantitative and qualitative analysis for sample identification and quantification.

- **General step by step procedure of Gas–Liquid Chromatography**

The step-by-step procedure for identifying and quantifying volatile organic compounds (VOCs) in water quality assessments is outlined as follows:

- **Sample Preparation**: Collect a sample volume of 1–2 mL and filter it using a 0.45 μm or 0.22 μm filter (Nylon, PTFE, or PVDF) to remove particulates. Familiarize yourself with the chemical formulas of target compounds like Benzene (C_6H_6), Toluene (C_7H_8), and Chloroform ($CHCl_3$).
- **Sample Injection**: Inject 1–2 μL of the prepared sample into the gas chromatograph using an appropriate injector. A Split/Splitless injector is suitable for routine analyses, while a Cool-on-Column (COC) injector is recommended for thermally sensitive samples.
- **Vaporization**: Rapidly vaporize the sample in the injector at approximately 250–300 °C
- **Carrier Gas Flow**: Use commonly recommended helium (He) gas as the carrier at flow rates of 1–5 mL/min. Hydrogen (H_2) may be used for enhanced sensitivity.

- **Separation in the Column**: Separate components in a capillary column (30 m × 0.32 mm ID, 0.1 μm film thickness) coated with Polydimethylsiloxane (PDMS) or Polyethylene Glycol (PEG) based on various interactions.
- **Detection**: Utilize detectors such as Flame Ionization Detectors (FID) or Thermal Conductivity Detectors (TCD), operating at 200–300 °C.
- **Data Acquisition**: Transmit the detector signal to a data acquisition system to generate a chromatogram. Use chromatography software (e.g., ChemStation or Empower) for analysis.
- **Analysis and Interpretation**: Analyze the chromatogram for peak identification and quantification, employing calibration standards with known concentrations for accuracy.

This concise procedure ensures reliable assessment of VOCs in water quality evaluations.

(This chapter does not address gas-solid chromatography (GSC), which is more commonly used in material science than in water quality evaluation).

- **Ion Chromatography**

Ion exchange chromatography, often known as ion chromatography, is a method used to separate ions and polar compounds based on their affinity for ion exchangers. This approach employs the reversible exchange of ions, which allows target ions in the sample solution to be replaced with ions connected to the ion exchanger. This process is governed by the following principles:

- **Retention Mechanism**: The retention of solute ions occurs through their attraction to charged sites on the stationary phase. This interaction forms the basis for the separation of different ionic species.
- **Types of Ion Exchangers**: Ion exchangers are classified into two types: anion exchangers, which include positively charged groups that attract and bind anions, and cation exchangers, which contain negatively charged groups that bind cations. Anion exchangers separate acidic samples, whereas cation exchangers separate positively charged ions like protonated bases.
- **Selectivity Factors**: The charge, hydrated radius, and polarizability of the ions all have an impact on ion exchange selectivity. Higher-charged ions bind more firmly to the exchanger, whilst smaller hydrated radii interact more efficiently with the stationary phase. Furthermore, more polarizable ions interact more strongly with the stationary phase.

A typical IC apparatus consists of numerous basic components as shown in Fig. 12, which usually contain the following elements.

- **Pump**: Essential for providing a continuous flow of eluent through the injector, column, and detector, ensuring optimal separation conditions at pressures up to 4000 psi.

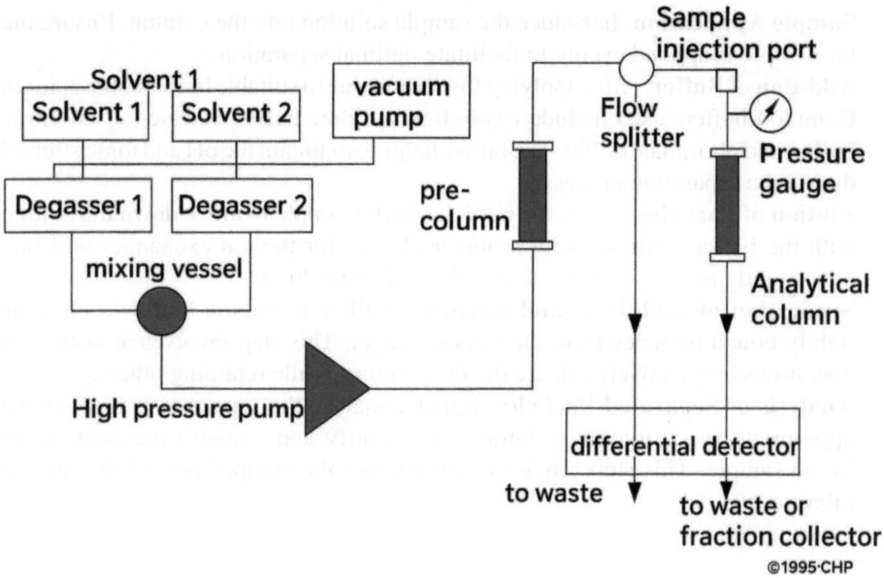

Fig. 12 Typical instrumentation of ion chromatography. (*Source* https://www.chemicool.com/definition/high_performance_liquid_chromatography_hplc.html)

- **Injector**: Facilitates sample introduction using an injection valve, allowing for the direct injection of liquid samples and the dissolution of solid samples. It accommodates volumes ranging from 0.1 to 100 μL with high precision.
- **Columns**: Composed of materials such as stainless steel, titanium, glass, or inert plastics (e.g., PEEK), these components vary in diameter (2 mm to 5 cm) and length (3 cm to 50 cm) depending on the specific application. A guard column is also utilized to protect the primary separation column from contaminants.
- **Suppressor**: Reduces the background conductivity of the eluent, enhancing the sensitivity of ion detection by converting ionic eluents into water.
- **Detectors**: Measure conductivity changes as ions elute from the column, commonly employing electrical conductivity detectors for this purpose.
- **Data System**: Manages data acquisition and analysis; a pre-programmed computing integrator suffices for routine analyses, while more complex tasks may necessitate a data station or minicomputer for enhanced functionality.

- **General step by step procedure of Ion Chromatography**

- **Preparation of the Column**: Begin by filling a chromatography column with the chosen ion exchanger, which is typically made of styrene and divinyl benzene. Select the appropriate type of exchanger based on the charge of the ions to be separated use an anionic exchanger for anions and a cationic exchanger for cations.

- **Sample Application**: Introduce the sample solution into the column. Ensure that the sample is applied evenly to facilitate optimal separation.
- **Addition of Buffer**: After applying the sample, add a suitable buffer to the column. Common buffers used include tris-buffer, pyridine buffer, acetate buffer, citrate buffer, and phosphate buffer. The buffer helps to maintain the pH and ionic strength during the separation process.
- **Elution of Particles**: Allow the particles in the sample to move down the column with the buffer. Particles with a higher affinity for the ion exchanger will bind more tightly and travel more slowly through the column.
- **Separation of Tightly Bound Particles**: Utilize a specific buffer to elute the tightly bound particles from the ion exchanger. This step involves adjusting the conditions to selectively release the desired ions while retaining others.
- **Analysis of Separated Particles**: Finally, analyse the separated particles using appropriate spectroscopic techniques to identify and quantify the components in the sample. This step provides insights into the composition of the original mixture.

4 Conclusion

Chemical analysis methods are critical for determining the quality of surface and groundwater, which has a direct influence on human health, agriculture, industry, and ecosystems. These approaches measure essential chemical elements such as metals, nutrients, and organic molecules, providing critical data for assessing water appropriateness. The evaluation approaches are classified as classical (wet chemical) and instrumental procedures, with each providing significant benefits in terms of accuracy and efficiency. As environmental pressures raise concerns about water quality, these analytical approaches will remain critical for successful water management, providing safe and sustainable water supplies for a variety of uses.

References

Franca AS, Nollet LML (2017) Spectroscopic methods in food analysis Franca, Adriana S. morawa.at. https://www.morawa.at/detail/ISBN-2244036060098/Franca-Adriana-S./Spectroscopic-Methods-in-Food-Analysis

Ali I, Aboul-Enein HY (2006) Instrumental methods in metal ion speciation. CRC Press. https://doi.org/10.1201/9781420019407

Badri S, Charan MS, Kumar DL (2023) A review on columns used in chromatography. J Multidiscipl Res 17–23. https://doi.org/10.37022/tjmdr.v3i3.502

Bergström J (2015) 2—Experimental characterization techniques. In: Bergström J (ed) Mechanics of solid polymers. William Andrew Publishing, pp 19–114. https://doi.org/10.1016/B978-0-323-31150-2.00002-9

Bharty DMK (nd) Paper: FS-201 (Forensic Chemistry)

Bogaerts A, Neyts E, Gijbels R, van der Mullen J (2002) Gas discharge plasmas and their applications. Spectrochim Acta, Part B 57(4):609–658. https://doi.org/10.1016/S0584-8547(01)00406-2
Bouchonnet S (2013) Introduction to GC-MS coupling. CRC Press. https://doi.org/10.1201/b13910
Cooper MM, Klymkowsky MW (2020) Chapter 2: Spectroscopy: how we know what we know about the structure of matter. https://openbooks.lib.msu.edu/oclue/chapter/chapter-2-spectroscopy-how-we-know-what-we-know-about-the-structure-of-matter/
Egbuna C, Ifemeje JC, Udedi SC, Kumar S (eds) (2018) Phytochemistry: volume 1: fundamentals, modern techniques, and applications. Apple Academic Press. https://doi.org/10.1201/9780429426223
Faulkner LR (1983) Understanding electrochemistry: some distinctive concepts. J Chem Educ 60(4):262. https://doi.org/10.1021/ed060p262
Feldman BJ, Gheller SF, Bailey GF, Newton WE, Schultz FA (1990) Electrochemical cells for voltammetry, coulometry, and protein activity assays of small-volume biological samples. Anal Biochem 185(1):170–175. https://doi.org/10.1016/0003-2697(90)90275-e
Garland C, Delapena S, Pennise D (2018) An alternative technique for determining gravimetric particle mass deposition on filter substrate: the particle extraction method. Open J Air Pollut 7(4), Article 4. https://doi.org/10.4236/ojap.2018.74016
Ghazizadeh M, Asadollahzadeh H (2021) A rapid cadmium determination based on ion selective membrane potentiometric sensor by bis (salicylaldehydo)ethylenediimine as carrier. Anal Methods Environ Chem J 4(2):25–33. https://doi.org/10.24200/amecj.v4.i02.136
Giddings JC, Keller RA (2024) Chromatography | Definition, Types, & Facts | Britannica. https://www.britannica.com/science/chromatography
Gomes AJ, Lunardi CN, Rocha FS, Patience GS (2019) Experimental methods in chemical engineering: fluorescence emission spectroscopy. https://doi.org/10.1002/cjce.23506
Grinberg N, Rodriguez S (2016) Ewing's analytical instrumentation handbook, 4th edn. Taylor & Francis. https://books.google.co.in/books?id=ks3zrQEACAAJ
Grinberg N, Rodriguez S (eds) (2018) Ewing's analytical instrumentation handbook, 4th ed. CRC Press. https://doi.org/10.1201/9781315118024
Harvey D (2000) Modern analytical chemistry. McGraw-Hill
Hertz H (nd) The units (sec)-1 are called Hertz (Hz)!
ISO 15680:2003 (2003) ISO. https://www.iso.org/standard/27983.html
Kenkel J (2014) Analytical chemistry for technicians. Routledge & CRC Press. https://www.routledge.com/Analytical-Chemistry-for-Technicians/Kenkel/p/book/9781439881057
Kenkel J (2002) Analytical chemistry for technicians. CRC Press. https://www.taylorfrancis.com/books/mono/https://doi.org/10.1201/9781420056709/analytical-chemistry-technicians-john-kenkel
Kozak J, Townshend A (2019) Titrimetry | Overview*. In: Worsfold P, Poole C, Townshend A, Miró M (eds) Encyclopedia of analytical science, 3rd ed. Academic Press, pp 111–120. https://doi.org/10.1016/B978-0-12-409547-2.14419-1
Kumar A, Pandit S, Sharma K, Agrawal S, Kuhad RC, Mathuriya AS, Dikshit PK, Mishra SK, Seth CS, Prasad R (2024) Microbial degradation of cellulose extracted from wheat bran for bioelectricity production using microbial fuel cell. Process Saf Environ Prot 190:574–585. https://doi.org/10.1016/j.psep.2024.06.076
Kumari VBC, Patil SM, Ramu R, Shirahatti PS, Kumar N, Sowmya BP, Egbuna C, Uche CZ, Patrick-Iwuanyanwu KC (2022) Chapter 5—Chromatographic techniques: types, principles, and applications. In: Egbuna C, Patrick-Iwuanyanwu KC, Shah MA, Ifemeje JC, Rasul A (eds) Analytical techniques in biosciences. Academic Press, pp 73–101. https://doi.org/10.1016/B978-0-12-822654-4.00013-0
Lee M-H, Chung C-W (2006) On the E to H and H to E transition mechanisms in inductively coupled plasma. Phys Plasmas 13(6):063510. https://doi.org/10.1063/1.2212387
Marsili R (ed) (2020) Techniques for analyzing food Aroma. CRC Press. https://doi.org/10.1201/9781003067573

Mosher M, Kelter P (2023) An introduction to chemistry. Springer International Publishing. https://doi.org/10.1007/978-3-030-90267-4
Müllertz A, Perrie Y, Rades T (eds) (2016) Analytical techniques in the pharmaceutical sciences. Springer. https://doi.org/10.1007/978-1-4939-4029-5
Nallakukkala S, Lal B (2022) Standard analytical techniques for analysis of wastewater. In: Gas hydrate in water treatment. Wiley, pp 155–205. https://doi.org/10.1002/9781119866145.ch8
Nollet LML, Gelder LSPD (eds) (2013) Handbook of water analysis, 3rd ed. CRC Press. https://doi.org/10.1201/b15314
Ojeda JJ, Dittrich M (2012) Fourier transform infrared spectroscopy for molecular analysis of microbial cells. In: Navid A (ed) Microbial systems biology: methods and protocols. Humana Press, pp 187–211. https://doi.org/10.1007/978-1-61779-827-6_8
Omanwa SK, Sonekar RP, Bajaj NS (2020) Borate phosphors: processing to applications. Routledge & CRC Press. https://www.routledge.com/Borate-Phosphors-Processing-to-Applications/Omanwar-Sonekar-Bajaj/p/book/9781032075754
Patnaik P (2017) Handbook of environmental analysis: chemical pollutants in air, water, soil, and solid wastes, 3rd ed. CRC Press. https://doi.org/10.1201/9781315151946
Pérez-Juste I, Faza ON (nd) 1 Interaction of radiation with matter.
Reddy VH, Prasad PMN, Reddy AVR, Reddy YVR (2012) Determination of heavy metals in surface and groundwater in and around Tirupati, Chittoor (Di). Andhra Pradesh, India
Rehman I, Movasaghi Z, Rehman S (2012) Vibrational spectroscopy for tissue analysis. CRC Press. https://doi.org/10.1201/b12949
Rezaei F, Vanraes P, Nikiforov A, Morent R, De Geyter N (2019) Applications of plasma-liquid systems: a review. Materials 12(17), Article 17. https://doi.org/10.3390/ma12172751
Rizwan M, Gwenin C (2021) Chapter 5—Nanomaterials in renewable energy: UV-visible spectroscopy characterization and applications. In: Devasahayam S, Hussain CM (eds) Nano tools and devices for enhanced renewable energy. Elsevier, pp 103–120. https://doi.org/10.1016/B978-0-12-821709-2.00017-7
Rood D (1998) Gas chromatography problem solving and troubleshooting. J Chromatogr Sci 36(1):55–56. https://doi.org/10.1093/chromsci/36.1.55
Sagiroglu MZ, Demirel ED, Mutlu S (2024) Accurate ion type and concentration detection using two bare electrodes by machine learning of non-faradaic electrochemical impedance measurements of an automated fluidic system. J Electroanal Chem 961:118256. https://doi.org/10.1016/j.jelechem.2024.118256
Sahtani K (2022) Production of activated carbon nanofibers using natural catalysts and their use in miRNA biosensors. http://hdl.handle.net/11452/28943
Saputra HA (2023) Electrochemical sensors: basic principles, engineering, and state of the art. Monatshefte Für Chemie - Chemical Monthly 154(10):1083–1100. https://doi.org/10.1007/s00706-023-03113-z
Sardans J, Montes F, Peñuelas J (2010) Determination of As, Cd, Cu, Hg and Pb in biological samples by modern electrothermal atomic absorption spectrometry. Spectrochim Acta, Part B 65(2):97–112. https://doi.org/10.1016/j.sab.2009.11.009
Scholz F (2015) Voltammetric techniques of analysis: the essentials. ChemTexts 1(4):17. https://doi.org/10.1007/s40828-015-0016-y
Shah N, Arain MB, Soylak M (2020) Chapter 2—Historical background: milestones in the field of development of analytical instrumentation. In: Soylak M, Yilmaz E (eds) New generation green solvents for separation and preconcentration of organic and inorganic species. Elsevier, pp 45–73. https://doi.org/10.1016/B978-0-12-818569-8.00002-4
Vadakedath S, Kandi V, Godishala V, Pinnelli VBK, Alkafaas SS, Elkafas SS (2022) The principle, types, and applications of mass spectrometry: a comprehensive review. Biomed Biotechnol 7(1), Article 1. https://doi.org/10.12691/bb-7-1-2
Vogel AI, Jeffery GH, Vogel AI (1989) Vogel's textbook of quantitative chemical analysis, 5th ed. Longman Scientific & Technical, Wiley

Waldhoff H, Spilker R (eds) (2004) Handbook of detergents, Part C: analysis. CRC Press. https://doi.org/10.1201/9781420030334

Wang X (2020) Historical perspective on identifying and controlling spontaneous combustion. In: Wang X (ed) Spontaneous combustion of coal: characteristics, evaluation and risk assessment. Springer International Publishing, pp 29–72. https://doi.org/10.1007/978-3-030-33691-2_2

Worsfold P, Townshend A, Poole CF, Miró M (2019) Encyclopedia of analytical science. Elsevier Science. https://books.google.co.in/books?id=Nu2SDwAAQBAJ

Zdrachek E, Bakker E (2021) Potentiometric sensing. Anal Chem 93(1):72–102. https://doi.org/10.1021/acs.analchem.0c04249

Zhang P, Yang M, Lan J, Huang Y, Zhang J, Huang S, Yang Y, Ru J (2023) Water quality degradation due to heavy metal contamination: health impacts and eco-friendly approaches for heavy metal remediation. Toxics 11:828. https://doi.org/10.3390/toxics11100828

Zhen Li JW, Li D (2016) Applications of Raman spectroscopy in detection of water quality. Appl Spectrosc Rev 51(4):333–357. https://doi.org/10.1080/05704928.2015.1131711

Physical Analysis Methods for Surface Water and Ground Water Quality Quantification

Barnali Saha, P. Shabana, Ramlakhan Singh Thakur, Soubhagya Sekhar Nayak, and S. C. Haokip

Abstract The assessment of water quality is critical for environmental monitoring, public health, and resource management. This chapter explores the various physical analysis methods employed for the quantification of surface water and groundwater quality. This study focuses on the fundamental physical parameters such as temperature, pH, turbidity, electrical conductivity, Total Dissolved Solids (TDS), Dissolved Oxygen (DO), salinity, flow rate, water level, suspended solids, colorimetry, and radiation absorption and scattering. Each method's principles, instrumentation, and application in both field and laboratory settings are meticulously reviewed. Key instruments include thermometers, digital pH meters, turbidimeters, Secchi discs, conductivity meters, TDS meters, DO meters, refractometers, flow meters, water level loggers, and colorimeters. The paper emphasizes the importance of regular calibration and maintenance of these instruments to ensure accuracy and reliability. Furthermore, it discusses the advantages and limitations of field-based versus laboratory-based analyses, highlighting the need for precise data collection and interpretation through statistical and geospatial methods. By providing a comprehensive overview of physical analysis techniques, this chapter aims to enhance the understanding and application of these methods in monitoring and managing water quality effectively.

Keywords Water quality · Groundwater quality · Electrical conductivity

1 Introduction

Water quality refers to the chemical, physical, and biological characteristics of water that determine whether or not it is suitable for many purposes, including drinking, agriculture, recreation, and the preservation of aquatic life. The assessment and

B. Saha (✉) · P. Shabana · R. S. Thakur · S. C. Haokip
ICAR-Central Institute of Agricultural Engineering, Bhopal, India
e-mail: saha26barnali@gmail.com

S. S. Nayak
Water Resources, Department of Agriculture and Farmers Empowerment, Bhubaneswar, Odisha, India

Dimple et al. (eds.), *Surface Water and Groundwater Quality for Sustainable Utilization*, Springer Hydrogeology, https://doi.org/10.1007/978-981-96-8189-1_6

preservation of water quality are vital in environmental science, as they impact ecosystem health, human well-being, and the sustainability of natural resources. Regular monitoring of water quality plays a key role in identifying harmful contaminants, reducing the risk of waterborne diseases, and protecting biodiversity. It also supports the sustainable management of water resources, ensuring compliance with environmental regulations and fostering resilience to climate change (UN Water 2021). Additionally, water quality data informs policy and decision-making, helping to ensure the safe use of water, promote sustainable agriculture, and maintain ecological balance (WHO 2020).

Groundwater and Surface water are the two major supplies of freshwater, essential for human consumption, agriculture, industry, and maintaining biodiversity. Surface water includes lakes, rivers, and reservoirs, which are readily accessible and widely used for drinking water supply, irrigation, and industrial processes. Groundwater, found in aquifers underneath the Earth's surface, provides a more reliable supplies of water, especially in arid regions, supporting agriculture, domestic use, and ecological functions. Both sources are interconnected; surface water can recharge groundwater through infiltration, and groundwater can discharge into rivers and lakes. Effective management of both resources is crucial to prevent overexploitation, pollution, and depletion, as contamination in one source can affect the other. Protecting these water supplies is vital for sustaining ecosystems, supporting human development, and adapting to increasing water demand and climate change.

Among the various approaches available, physical analysis methods play a crucial role in assessing the physical characteristics of water, which are vital indicators of its overall quality. Physical analysis methods focus on quantifying parameters such as temperature, electrical conductivity (EC), turbidity, and total suspended solids (TSS). These metrics offer valuable information on the physical condition of the water, which can have an impact on its chemical makeup and biological well-being. For instance, turbidity, a quantitative measure of the purity of water, can indicate the presence of suspended particles that may carry pollutants or reduce light penetration, thereby affecting aquatic life (APHA 2017). Similarly, electrical conductivity is a key parameter for assessing the concentration of dissolved salts, which can influence water suitability for drinking and irrigation (WHO 2011). The importance of physical analysis methods in water quality assessment is underscored by their ability to provide rapid and cost-effective measurements. These methods are often employed as part of routine monitoring programs to identify alterations in the characteristics of water that may result from natural processes or human activities, such as agricultural runoff, industrial discharges, or urban development (US EPA 2020). By regularly monitoring these physical parameters, it is possible to identify trends and potential threats to water quality, enabling timely interventions to protect and manage water resources effectively.

Within this framework, the subsequent discourse will examine diverse techniques of physical analysis employed to measure the quality of surface-level water and groundwater. The focus will be on their application, significance, and the insights they provide into the health and sustainability of water resources.

2 Key Physical Parameters for Surface Water and Groundwater Quality Quantification

Evaluation of water quality is based on assessing numerous physical parameters, which constitute key markers of both groundwater and surface water quality. These parameters not only reflect the physical state of the water but also influence its chemical composition and biological interactions. The primary physical parameters commonly assessed include temperature, electrical conductivity (EC), turbidity and total suspended solids. These measurements are crucial in understanding water's ecological health and suitability for human consumption and agricultural use.

2.1 *Temperature*

Temperature is a key physical parameter in assessing both surface water and groundwater quality, as it profoundly affects chemical, biological, and physical processes. In surface waters like rivers, lakes, and streams, temperature impacts the solubility of gases, particularly oxygen. Warmer water holds less dissolved oxygen, which can lead to hypoxia, stressing or killing aquatic life and altering the metabolic rates of organisms. Higher temperatures can also increase the toxicity of pollutants and are influenced by features resembling air temperature, solar radiation and human activities including industrial discharges and urban runoff (Wetzel 2001; APHA 2017). For instance, thermal pollution from industrial sources can raise water temperatures, causing thermal stress on ecosystems and fostering harmful algal blooms (Caissie 2006).

Groundwater temperature tends to be more stable due to its insulation from direct sunlight and atmospheric conditions but varies with aquifer depth and geological characteristics. Temperature affects the solubility and movement of minerals and gases in groundwater. Higher temperatures can enhance the dissolution of minerals, increasing the concentrations of magnesium (Mn), calcium (Ca) ions and influence microbial activity (Freeze and Cherry 1979; Hem 1985). Furthermore, the temperature of groundwater regulates the procedures of water treatment and the rate of pipe corrosion in water delivery systems (WHO 2011). Therefore, regular monitoring of temperature is crucial for effective water quality management, helping to understand and address variation in both surface-level water and groundwater environments.

2.2 *pH*

pH is a crucial physical parameter in evaluating both the quality of surface-level water and groundwater, reflecting the hydrogen ion concentration and indicating acidity or alkalinity on a scale from 0 to 14, with 7 being neutral. In surface water sources such as lakes, rivers, and streams, pH levels can naturally fluctuate due to factors like rainfall, biological activity, and organic material. As an example, photosynthesis process of aquatic plants boosts pH during the day by lowering carbon di-oxide (CO_2) level, whereas respiration process lowers pH level at night (Wetzel 2001). Human activities, including industrial discharges, agricultural runoff, and acid rain, can drastically alter pH, often making water more acidic. Acidification can harm aquatic life, especially organisms with calcium-based shells or exoskeletons, by disrupting their physiological processes (APHA 2017).

Groundwater pH tends to be more stable due to the buffering effects of surrounding soil and rock. Nonetheless, the aquifer's geological makeup has an impact on it. For instance, the breakdown of carbonate minerals in groundwater flowing through limestone formations typically results in a higher pH (Hem 1985). On the other hand, regions with high levels of organic matter or acid rain can cause a decrease in the pH of groundwater, which can cause pipe corrosion, heavy metal leakage, and unfit water for agricultural or drinking use (WHO 2011).

2.3 *Turbidity*

A key physical indicator for evaluating the quality of groundwater and surface water is turbidity. It indicates water cloudiness caused by suspended particles like silt, clay, organic matter, algae, and microorganisms. Measuring turbidity provides insights into water clarity, essential for evaluating potential water quality issues affecting ecological health, drinking water, and industrial use.

In surface water, turbidity can fluctuate due to natural processes such as erosion, storm runoff, and seasonal changes, as well as human activities like construction and agriculture. High turbidity levels in rivers and lakes reduce light penetration, impairing photosynthesis in aquatic plants and disrupting food chains. Suspended particles can also transport pollutants such as heavy metals, pesticides, and pathogens, increasing health risks to aquatic life and humans (Davies-Colley and Smith 2001). Elevated turbidity may even clog fish gills, leading to stress or death in sensitive species (APHA 2017).

Groundwater generally has lower turbidity due to natural filtration through soil and rock. However, conditions such as fine particles, colloidal matter, or microbial contamination from septic systems or runoff can increase turbidity (Freeze and Cherry 1979). While turbidity itself may not directly harm groundwater quality, it can indicate the presence of harmful substances or hydrological changes. Elevated turbidity often requires additional water treatment to ensure safety for drinking and industrial purposes (WHO 2011).

2.4 Electrical Conductivity (EC) and Salinity

Electrical conductivity (EC) and salinity are essential physical factors that are utilised to evaluate the quality of both surface-level water and groundwater. The electrical conductivity of water, or EC, is an indicator of its dissolving ion concentration that is mostly related to the amount of salt present. Salinity is the aggregate amount of the salts dissolved in water. It is frequently employed in conjunction with EC for assessing the quality of water, particularly for ecological, related to agriculture, and drinkable usages.

In **surface water**, EC and salinity levels can vary due to natural processes such as rainfall, evaporation, and the geology of the surrounding area. Human activities, such as agricultural runoff, industrial discharges, and urban development, can significantly increase EC and salinity levels. High salinity, often detected through increased EC, can harm aquatic ecosystems by reducing species diversity, limiting freshwater habitats, and disrupting biological processes (APHA 2017). Elevated salinity also affects the water's suitability for irrigation, potentially leading to soil salinization and reduced crop yields (WHO 2011).

In **groundwater**, EC and salinity are influenced by the mineral composition of aquifers, with higher levels often found in areas with naturally saline formations or where seawater intrusion occurs. Over-extraction of groundwater can exacerbate salinity issues, making the water unsuitable for drinking or agricultural use (Freeze and Cherry 1979). Monitoring EC helps identify salinity problems in groundwater, which are critical for managing water resources, especially in arid or coastal regions (Hem 1985).

2.5 Total Suspended Solids and Total Dissolved Solids

Total Suspended Solids (TSS) and Total Dissolved Solids (TDS) are crucial physical parameters for assessing both the quality of surface-level water and groundwater. While TDS comprises the dissolved salts, mineral substances, and molecules of organic matter, TSS is the dry amount of particulates that are suspended in water, such as silt, organic debris, and microbes.

In **surface water**, TSS levels fluctuate due to erosion, storm runoff, and human activities like agriculture and construction. Elevated TSS reduces light penetration, impairs photosynthesis, and can carry pollutants, such as pesticides and heavy metals, affecting aquatic ecosystems (Davies-Colley and Smith 2001). High TSS can also clog fish gills, reducing oxygen levels and increasing water temperatures (APHA 2017). TDS, on the other hand, produced by natural processes like as the weathering of rocks and human activities, including wastewater discharge and agricultural runoff. High TDS levels can degrade water quality by impacting taste, corrosion of pipes, and harming aquatic life due to the presence of excessive salts (WHO 2011).

In **groundwater**, TSS is generally low due to natural filtration through soil and rock. However, contamination from industrial waste or septic systems can introduce suspended particles, signaling potential pollution (Freeze and Cherry 1979). TDS in groundwater is largely influenced by the geological composition of aquifers and can rise due to factors like over-extraction and seawater intrusion, reducing water suitability for drinking and irrigation (Hem 1985).

2.6 *Dissolved Oxygen (DO)*

Dissolved oxygen (DO) is a significant physical indicator that is utilised to evaluate the quality of both surface-level water and groundwater. It refers to the quantity of oxygen gas dissolved in water, critical for aquatic ecosystems and water chemistry. In surface water, dissolved oxygen is a important source for the survival of fish, invertebrates, and aerobic microorganisms. High DO levels generally indicate good water quality, while low levels can result in hypoxia, negatively impacting aquatic life (Wetzel 2001). Groundwater, conversely, often has lower DO levels due to limited atmospheric exchange and organic matter decomposition.

Temperature, water flow, and biological processes influence DO levels. Cold water holds more oxygen than warm water, and flowing waters, like rivers, tend to have higher DO due to greater mixing with the atmosphere. In contrast, stagnant water bodies and groundwater have lower oxygen levels, often approaching anoxic conditions in deep aquifers (Ward et al. 2003). DO measurements, commonly in mg/lt, provide a direct indicator of water quality. Waters with DO levels more than 5 mg/lt are generally considered healthy for aquatic organisms, while levels below 3 mg/L signal potential stress (USEPA 2006).

2.7 *Water Level and Flow Rate*

Water level and flow rate are fundamental physical parameters for the quality assessment of both surface-level water and groundwater. They provide essential information about the hydrodynamics, quantity, and health of aquatic systems.

Water level is an essential measure in both surface and groundwater systems. In surface water, fluctuations in water level, driven by seasonal variations, rainfall, and anthropogenic activities, can indicate hydrological health. For instance, declining water levels may signal over-extraction, drought, or poor recharge rates, which affect water quality by concentrating pollutants and reducing the habitat for aquatic species (Krause et al. 2007). In groundwater, water levels indicate aquifer health and recharge rates. A decline in groundwater levels often indicates overuse or reduced recharge, which can cause the decreasing of water availability, and increased the amount of presence pollutants, and the potential for land subsidence (Sophocleous 2002).

The amount of water flowing across a cross-sectional area in a given amount of time is measured by flow rate, which is commonly expressed in cubic meters per second. In surface water systems, flow rate affects water quality by influencing dilution, dispersion of pollutants, and oxygenation. Higher flow rates in rivers and streams promote greater oxygen exchange between water and the atmosphere, which can mitigate pollution impacts and improve water quality (Allan and Castillo 2007). Furthermore, high flow rates help transport sediments and nutrients, sustaining ecosystems and preventing stagnation (Poff et al. 1997). In groundwater systems, flow rate affects the transport of contaminants, as slower rates increase the potential for chemical reactions between water and geological materials. Contaminant migration in groundwater is highly dependent on flow rates and aquifer characteristics. In areas with low flow rates, pollutants can accumulate, making groundwater more susceptible to contamination (Fetter 1999).

2.8 *Colorimetry*

The shade of colour of water samples is assessed employing a technique termed the colorimetry to identify the level of purity of the water. This method involves assessing the color intensity of a sample, which can indicate the existence of various substances and pollutants. In surface-level water, Colorimetry helps detect the concentration of dissolved and suspended substances, including organic matter, algae, and sediments. For instance, color measurements can be used to estimate the levels of chlorophyll-a, a proxy for algal biomass, which is crucial for assessing eutrophication and overall water quality (APHA 2017). Discoloration in surface water can also be linked to industrial discharges or runoff, affecting the aesthetic and ecological value of water bodies (Hoffmann et al. 2005).

In groundwater, Colorimetry is less commonly used but can still provide valuable information. Color changes in groundwater can indicate the existence of contaminants which contains iron, manganese, or organic compounds. For example, a yellow or brown hue might suggest high concentrations of iron or organic matter (Hem 1985). While less visible than in surface water, changes in groundwater color can still offer insights into contamination and changes in aquifer conditions.

2.9 *Radiation Absorption and Scattering*

Radiation absorption and scattering are critical physical processes used in the quality and purity quantification of surface-level water and groundwater. These processes are essential for understanding the optical properties of water, which can reveal insights into water quality and the presence of various substances.

Radiation absorption refers to the process by which water absorbs specific wavelengths of light, leading to changes in the water's spectral properties. This process is influenced by the amount of dissolved and suspended substances in water. For surface water, high absorption of certain wavelengths can indicate the presence of organic matter, chlorophyll, or pollutants such as metals and nutrients (Coble 1996). Absorption measurements are commonly used in remote sensing and water quality monitoring to assess algal blooms and the overall health of aquatic ecosystems (Gordon et al. 1983). In groundwater, absorption can give the idea about the presence of dissolved organic carbon and other contaminants. However, due to limited access and the typically darker environment of groundwater, absorption studies are less common but still useful in certain contexts, such as when assessing pollution plumes or natural organic matter (Coble 2007).

Scattering refers to the redirection of light by particles suspended in water. In surface water, scattering is influenced by the concentration and size of suspended particles such as sediments, phytoplankton, and organic matter. Increased scattering can indicate higher levels of suspended sediments or algal blooms, affecting water clarity and quality (Horne and Goldman 1994). Scattering measurements, often obtained using in situ sensors or remote sensing technology, are crucial for assessing turbidity and the ecological status of aquatic systems (Smith and Baker 1981). In groundwater, scattering is less frequently measured due to the typically lower particle concentrations compared to surface waters. However, in cases of high sediment load or during groundwater-surface water interactions, scattering measurements can provide valuable information about sediment transport and water quality.

3 Key Instruments for Physical Analysis of Water Quality

Measuring physical parameters provides a comprehensive understanding of water quality, helping to identify potential pollutants, assess ecological health, and ensure the safety of water for various applications. Physical analysis, which involves the measurement of different physical parameters, plays a crucial role in understanding water quality. Regular monitoring helps in detecting and addressing water quality issues promptly.

3.1 Temperature Measurement

Temperature measurement in surface water and groundwater is crucial for understanding environmental conditions and water quality dynamics. Accurate temperature measurements using appropriate instruments are essential for monitoring and managing water quality effectively. The thermometers used for water quality assessment are field thermometers and data-logging thermometers.

3.1.1 Field Thermometers

Field thermometers measure temperature based on the physical properties of materials that change with temperature, such as the expansion of mercury or changes in electrical resistance. These can be mercury/alcohol thermometers, digital thermometers, thermocouple sensors and infrared thermometers. Digital Thermometers are widely used for their precision and ease of use, digital thermometers provide quick and accurate temperature readings, making them ideal for field measurements. Thermocouple sensors offer high accuracy and can measure a wide range of temperatures. They are often used in more advanced water quality studies, including continuous monitoring in rivers, lakes, and groundwater. Infrared Thermometers are useful for non-contact measurements, infrared thermometers can quickly assess the temperature of water bodies, particularly at the surface level.

- Field Settings:
  - Direct Measurement: Used for on-site temperature measurements in natural environments, such as water bodies, soil, or air.
  - Data Collection: Field thermometers are employed to gather temperature data for environmental studies, field research, and assessments of physical conditions.
  - Ease of Use: Portable and easy to use, suitable for immediate and simple temperature measurements.
- Laboratory Settings:
  - Pre-Testing: Used for calibrating and validating temperature-sensitive instruments before or after laboratory experiments.
  - Quality Control: Ensures that laboratory conditions remain within desired temperature ranges to maintain accuracy and consistency in experimental results.

3.1.2 Data Logging Thermometers

Data Logging Thermometers are deployed in the field to continuously record temperature over time. They are valuable for tracking long-term trends and detecting gradual changes in water quality. These can be standalone data loggers, remote data loggers and multi-channel data loggers. Standalone Data Loggers are portable devices with built-in sensors and memory. Often include an LCD screen for real-time display of measurements and buttons for configuration. Remote Data Loggers are equipped with wireless or cellular communication capabilities to transmit data to a remote server or cloud-based system for real-time monitoring. Multi-Channel Data Loggers measure and record temperature from multiple sensors or locations simultaneously. Useful for complex experiments or monitoring large systems.

- Field Settings:

  - Environmental Monitoring: Used to track temperature variations in natural settings, such as water bodies, soils, and atmospheric conditions. Provides continuous data for assessing environmental changes over time.
  - Remote Locations: Ideal for monitoring temperature in remote or difficult-to-access locations where continuous monitoring is needed without regular manual checks.

- Laboratory Settings:

  - Experiment Monitoring: Used to record temperature variations during experiments or process monitoring to ensure that conditions remain consistent.
  - Data Analysis: Provides detailed temperature data for analyzing trends, validating experimental conditions, and ensuring reproducibility of results.

3.2 pH Measurement

pH measurement is a critical aspect of water quality analysis, as pH affects many chemical and biological processes. Two common methods for measuring pH are using pH meters and pH indicator strips. Each method has its own principles, instrumentation, and applications in different settings.

3.2.1 pH Meter

The pH meter is a key instrument in assessing the quality of water which measures the acidity or alkalinity of the water on a scale of 0–14. It consists of a reference electrode and a pH electrode with a glass membrane which generates a voltage proportional to hydrogen ion concentration. Whereas the reference electrode provides a stable voltage reference. It works based on the principle of electrochemical potential. When the water sample comes in contact with the glass electrode electrochemical potential is generated and based on that there is a voltage generation which is converted into the pH value by the calibration curve.

- Field Settings:

  - Portable Units: Handheld pH meters are used for on-site measurements in various environments, including rivers, lakes, and industrial sites. They are designed to be rugged and easy to use in field conditions.
  - Real-Time Monitoring: Some portable pH meters offer data logging and connectivity features for remote monitoring and continuous data collection.

- Laboratory Settings:

  - Bench-Top Units: Laboratory pH meters are typically more accurate and offer more advanced features, such as multiple calibration points and data logging.

  - Routine Testing: Used in water quality testing, research, and quality control in various industries. Regular calibration with standard solutions is crucial for maintaining accuracy.

3.2.2 pH Strips

pH indicator strips use color-changing reagents that respond to hydrogen ion concentration. The strip changes color based on the pH level of the solution. Each strip is typically designed for a specific pH range and provides a color chart for comparison to determine the pH. The strip is usually made of paper or plastic coated with pH-sensitive dyes that change color according to pH. An accompanying chart or scale is used to compare the color of the strip after immersion to determine the pH value.

- Field Settings:
  - Simple and Portable: pH indicator strips are ideal for quick, on-site measurements. They are easy to use and do not require calibration or complex equipment.
  - Cost-Effective: Suitable for informal testing, such as in educational settings, gardening, or simple water quality checks.
- Laboratory Settings:
  - Limited Use: While less common in advanced laboratories due to lower precision, pH indicator strips can be used for preliminary or qualitative assessments.
  - Educational Purposes: Useful for teaching basic concepts of pH and acid–base chemistry.

3.2.3 Refractometers

Refractometers are useful for measuring the refractive index of water, offering insights into its salinity, dissolved solids, and other key parameters. It measures the bending of light as it passes through a water sample. The extent of light bending depends on the concentration of dissolved substances, such as salts. A handheld analog refractometer consists of a prism and a scale for the measurement. A few drops of sample are placed on the prism and secured with a cover plate for uniform distribution of the sample over the prism. Place the prism end of the refractometer towards a light source and adjust the focus until the scale becomes clearly visible. As the refractometer readings are influenced by the temperature changes, calibration of the refractometer with distilled water is necessary before use. Some refractometers have the Automatic Temperature Compensation (ATC) feature that automatically adjusts the readings to account for temperature variations.

- Field Settings:

- Portable Units: Handheld refractometers are widely used in the field for quick salinity measurements. They are useful for marine research, aquaculture, and environmental monitoring.
- Simple and Durable: Refractometers are valued for their simplicity, portability, and durability in field conditions.

- Laboratory Settings:
  - Precision: In laboratories, more sophisticated refractometers with precise calibration and digital readouts are used for detailed analysis. They are employed in water quality testing, chemical analysis, and research.
  - Calibration and Maintenance: Laboratory refractometers are regularly calibrated with standard solutions to ensure accuracy.

3.3 Total Dissolved Solids and Total Suspended Solids Measurement

3.3.1 TDS Meter

It is a hand-held device that measures the Total Dissolved Solids in a water sample. The TDS meter works by detecting the electrical conductivity of the water, which is then converted into a TDS value, typically expressed in parts per million (ppm). TDS meters only provide a general measure of the total concentration of dissolved ions, such as salts and minerals, but they do not identify or quantify individual components. TDS meters do not indicate whether the dissolved solids in the water are harmful or not.

3.3.2 Gravimetric Method

It quantifies the suspended solids by filtering the water sample, drying the filter, and weighing the residue.

- Field settings:

  Less practical due to equipment and time constraints.
- Laboratory settings:

  Common method for precise measurements. The sample is filtered, and the filter with the residue is weighed to determine the concentration of suspended solids.

3.4 *Dissolved Oxygen Measurement*

Dissolved oxygen (DO) is a critical parameter in water quality assessment, affecting aquatic life and water chemistry. The measurement of DO can be performed using various methods, including the Winkler method and oxygen electrodes. Both methods are essential for different aspects of water quality monitoring. The Winkler method, while precise, is more labor-intensive and traditionally used for verification and reference. Oxygen electrodes, on the other hand, offer convenience, speed, and real-time data, making them suitable for a wide range of applications in environmental monitoring and research.

3.4.1 Dissolved Oxygen (DO) Meter

Dissolved Oxygen (DO) meters are specialized instruments used to measure the concentration of oxygen dissolved in water. DO meters can be equipped with electrochemical sensors (polarographic or galvanic cells) or optical sensors. The former detects the amount of oxygen in the water by measuring the electrical current generated by the reaction of oxygen with specific chemicals in the sensor. Whereas the latter rely on luminescence or fluorescence. These sensors measure the light emitted or absorbed by a luminescent dye that reacts with oxygen, providing a reading of the DO concentration.

- Field Settings:
  - Portable DO Meters: Handheld or portable oxygen meters equipped with oxygen electrodes are used for on-site measurements. They are designed to be rugged and easy to use in field conditions.
  - Real-Time Monitoring: Some portable meters offer real-time data logging and telemetry for continuous monitoring of DO levels in various environments.
- Laboratory Settings:
  - Bench-Top Meters: Laboratory DO meters are typically more precise and feature advanced calibration and measurement options. These meters are used for detailed analysis in controlled settings.
  - High Sensitivity: Suitable for research, quality control, and detailed water quality studies, providing accurate and reproducible results.

3.4.2 Winkler Method

The Winkler method is a classic chemical approach for determining dissolved oxygen concentration in water. The principle involves fixing the oxygen in a water sample by chemical reagents, and then measuring the amount of iodine liberated, which is proportional to the amount of dissolved oxygen present. The water sample is first treated with manganese (II) sulfate, which reacts with dissolved oxygen to form manganese (III) hydroxide. And then an alkaline iodide solution is added, which reacts with the manganese (III) hydroxide to liberate iodine. The liberated iodine is then titrated with a thiosulfate solution, and the amount of thiosulfate used is proportional to the dissolved oxygen concentration.

- Field Settings:
  - Field Test Kits: Portable Winkler method test kits are available for field use. These kits include reagents and titration equipment suitable for on-site measurements.
  - Procedure: Field kits require careful handling and timing but are useful for quick assessments of dissolved oxygen in natural water bodies.
- Laboratory Settings:
  - Laboratory Kits: More sophisticated Winkler method setups in laboratories allow for precise measurement and analysis. They include burettes, pipettes, and titration equipment.
  - Accuracy and Precision: The method is considered accurate and is often used as a reference method in laboratories for validating other DO measurement techniques.

3.5 *Flow Rate and Water Level Analysis*

Instruments for flow rate and water level analysis are essential for quantifying surface water and groundwater quality, with different tools suited for field and laboratory settings.

3.5.1 Flow Meters

These are used to measure the rate of water flow in open channels or pipes. Common types include electromagnetic, ultrasonic, and mechanical flow meters. Electromagnetic flow meters, such as the Endress + Hauser Promag series, are non-invasive and effective for a wide range of applications (Endress + Hauser 2023). Ultrasonic flow meters, such as those by Siemens, use sound waves to measure flow velocity and are particularly useful for measuring flow in rivers and streams (Siemens 2023).

3.5.2 Current Meters

Instruments like the Marsh-McBirney Flo-Mate 2000 are used to measure water velocity in rivers and streams. These meters are essential for calculating flow rate using the velocity-area method (Marsh-McBirney 2023).

3.5.3 Pressure Transducers

These are used to measure the water level by detecting the hydrostatic pressure exerted by the water column. Instruments such as the Keller Series 21 pressure transducer are widely used for groundwater level monitoring (Keller 2023).

3.5.4 Manual Dipmeters

These are used for measuring static water levels in wells or tanks in a laboratory setting. They provide a simple and direct measurement method (Boyle et al. 2014).

3.5.5 Electronic Water Level Meters

For more precise laboratory measurements, electronic water level meters, such as the Geosense range, are employed. They offer high accuracy and data logging capabilities (Geosense 2023).

Accurate measurement of flow rate and water level is crucial for effective water quality management. Field instruments like flow meters and pressure transducers provide real-time data essential for environmental monitoring, while laboratory instruments such as permeameters and water level loggers offer detailed analysis under controlled conditions.

3.6 Colorimeters

A colorimeter is an instrument used to detect and measure the concentration of substances in the water. They work by measuring the intensity of color produced when a specific reagent reacts with a target chemical in the water. This color intensity correlates with the concentration of the substance being tested. Colorimeters can be used to screen for a wide range of chemicals, including fluoride, chlorine, cyanide, dissolved oxygen, iron, zinc, phosphorus, calcium, silica, bromine, ammonia, and more. By providing accurate and reliable measurements, colorimeters are valuable tools in ensuring that water meets safety and regulatory standards.

- Field settings:

  Portable colorimeters are used for on-site water quality testing. They are designed for ease of use and quick results, making them suitable for field surveys and environmental monitoring.
- Laboratory settings:

  Used for precise and accurate measurements of water quality parameters. Laboratory colorimeters are typically more sensitive and may offer a broader range of wavelengths.

4 Merits and Demerits of Field-Based Versus Laboratory-Based Analyses Methods

The assessment of surface and groundwater quality is essential for preserving healthy ecosystems and providing safe water for human use. Various analytical methods are used to quantify water quality parameters, each having its unique benefits and drawbacks. This review focuses on the strengths and limitations of field-based and laboratory-based analyses, both of which play integral roles in water quality monitoring and management.

Field-based analyses offer significant advantages, particularly in terms of providing immediate and real-time data. These analyses involve using portable instruments that allow for the rapid evaluation of water quality parameters directly at the sampling site. Commonly measured parameters include temperature, pH, conductivity, turbidity, and dissolved oxygen (D.O). The ability to collect data in situ is a major benefit, as it minimizes the risks associated with sample transport, such as contamination and changes in sample composition (Sawyer et al. 2003). Moreover, field-based analyses enable the assessment of spatial variability across large areas, facilitating comprehensive environmental monitoring.

One of the key strengths of field-based analyses is their cost-effectiveness and efficiency. They enable the collection of a number of observations within a short timeframe, which is particularly valuable in extensive surveys and routine monitoring programs. Additionally, field-based methods are often less expensive than laboratory analyses, making them accessible for use in regions with limited resources.

Despite these advantages, field-based analyses have certain limitations. Portable instruments used in the field are often less sensitive and less precise than laboratory equipment. This can result in data with higher variability and potentially lower accuracy (APHA 2017). Some environmental factors like temperature fluctuations, humidity, and interference from other substances can influence the reliability of field measurements. Furthermore, field-based analyses may be limited in their ability to detect low concentrations of pollutants or perform complex chemical analyses, which are critical for understanding the full scope of water quality issues.

Laboratory-based analyses, on the other hand, are conducted in controlled environments using sophisticated instrumentation, which offers higher precision and accuracy. Laboratories are equipped to perform comprehensive a of water samples, including the detection of trace contaminants and detailed chemical, biological, and physical characterizations. Instruments such as gas chromatographs, mass spectrometers, and atomic absorption spectrometers are capable of detecting low concentrations of pollutants that may not be detectable in the field. Laboratory analyses can also include microbiological assessments, which are essential for determining the safety of water for human consumption (Bartram and Ballance 1996).

The controlled environment of a laboratory reduces the influence of external factors, ensuring the reliability and accuracy of results. This level of control allows for the execution of complex analytical procedures that can provide a more comprehensive understanding of water quality dynamics. Laboratory analyses are particularly valuable when high precision is required, such as in legal and regulatory contexts where stringent quality standards must be met (Rice et al. 2017).

However, laboratory-based analyses are not without challenges. They require significant time and resources, making them more expensive and less accessible than field-based methods. The time lag between sample collection and analysis can also pose a challenge, especially when immediate results are necessary for decision-making. Furthermore, the process of transporting samples from the field to the laboratory introduces the risk of contamination or changes in sample integrity (Meybeck and Helmer 1989).

The choice between field-based and laboratory-based analyses is based on the objectives of the study, the parameters of interest, and the resources available. Field-based analyses are particularly useful for preliminary assessments, routine monitoring, and when immediate results are required. They are also advantageous in situations where cost and time efficiency are priorities. In contrast, laboratory-based analyses are more suited for detailed studies that require high precision and accuracy, particularly when the detection of trace elements or complex analyses is necessary (Sillanpaa 2014).

The integration of field-based and laboratory-based methods can provide a comprehensive approach to water quality assessment. Field analyses can be used for initial screenings and to identify areas of concern, while laboratory analyses can be employed for in-depth investigations and confirmation of findings. This complementary approach allows for a more comprehensive understanding of problems related to water quality, which ultimately supports the development of effective management strategies.

5 Need for Precise Data Collection

Surface and groundwater systems are inherently dynamic, with numerous interacting physico-chemical processes influencing the quality of water. These systems are also subject to natural variability due to factors such as seasonal changes, weather events, and geological influences (Hirsch et al. 1991). To accurately assess the state of these water bodies, data collection must be precise enough to capture the variability and trends in water quality parameters over time and across different spatial scales. For instance, monitoring programs need to ease the problems by detecting both short-term fluctuations and long-term trends in parameters like nutrient concentrations, pH, temperature, dissolved oxygen, and turbidity.

The accuracy of data regarding water quality is critically based on the methodologies used for sampling and analysis. Precise data collection involves using standardized sampling protocols and high-quality analytical techniques to minimize errors and ensure the reliability of results (Bartram and Ballance 1996). The selection of appropriate sampling locations, frequencies, and methods is vital to representatively capture the variability in water bodies, both spatially and temporally.

In the context of groundwater, precision is particularly important due to the complex subsurface environments and the ability for contamination from multiple sources, which includes agricultural runoff, industrial discharges, and wastewater effluents. Groundwater quality assessments require accurate data on parameters such as total dissolved solids, heavy metals, nitrate levels, and microbial contamination. These assessments often involve the use of monitoring wells, which must be strategically placed and properly maintained to obtain reliable data.

The integration of advanced technologies has significantly enhanced the precision of data collection in water quality monitoring. Remote sensing technologies, such as satellite imagery and aerial drones, provide spatially extensive data that can supplement traditional ground-based sampling methods. These technologies offer the advantage of covering large areas quickly and can be used to detect changes in parameters like chlorophyll concentration and surface temperature. Additionally, the use of automated sensors and data loggers allows for continuous monitoring, providing high-frequency data that can capture transient events such as stormwater runoff or pollution spills.

Quality assurance and quality control (QA/QC) measures are crucial for components of precise data collection. QA/QC protocols involve routine calibration of instruments, validation of analytical methods, and regular checks for consistency and accuracy (American Public Health Association 2017). Proper training of personnel involved in data collection and analysis is also crucial to minimize human errors and ensure adherence to standardized procedures.

Precise data collection is particularly important for regulatory and compliance purposes. Many water quality standards and guidelines, such as those established by the U.S. Environmental Protection Agency (EPA) and the World Health Organization (WHO), require rigorous data to ensure that water bodies meet safety and quality criteria (EPA 2003). Non-compliance with these standards can have significant legal and financial implications, highlighting the importance of reliable data for decision-making and enforcement.

The need for precision in data collection is also driven by the potential risks posed by pollutants to human health and ecosystems. Contaminants like heavy metals, pesticides, and pathogens can significantly affect public health, especially in communities that depend on untreated or insufficiently treated water sources. Precise data collection enables the identification and quantification of these contaminants, supporting risk assessments and the creation of targeted mitigation strategies (Backhaus and Faust 2012).

In addition to human health concerns, precise data collection is essential for understanding the effects of pollutants on aquatic ecosystems. Changes in the quality of water can affect the distribution and abundance of aquatic life, alter food web dynamics, and disrupt ecosystem services like nutrient cycle and water purification. Accurate data on parameters like nutrient loading, sedimentation rates, and Pollutant concentrations are crucial for evaluating the health of aquatic ecosystems and for devising effective conservation and management strategies (Bunn and Arthington 2002).

Moreover, precise data collection supports the growth and validation of water quality models, which are used for predicting the impacts of various scenarios and management actions on water resources. Models rely on high-quality data to accurately simulate the behaviour of water bodies under different conditions, such as changes in land use, climate, and pollutant loads. By providing reliable inputs, precise data collection enhances the predictive capabilities of models, supporting evidence-based decision-making and planning.

6 Statistical and Geospatial Methods for Data Interpretation

The quantification of surface and groundwater quality involves more than just the collection of data; it requires sophisticated methods for interpretation to extract meaningful insights. Statistical and geospatial techniques are essential for analyzing water quality data, identifying trends and patterns, and providing actionable information for management and policy-making. These methods enable scientists and policymakers to better understand the complex interactions within aquatic ecosystems and to address challenges posed by pollution, climate change, and human activities.

Statistical methods play a crucial role in interpreting water quality data by offering tools to manage variability and uncertainty. Descriptive statistics, including means, medians, and standard deviations, are typically the initial step in summarizing water quality data. These statistics help in understanding the central tendencies and dispersion of various parameters, such as pH, temperature, dissolved oxygen (D.O), and nutrient concentrations. Inferential statistics, including hypothesis testing and confidence intervals, are used to draw conclusions about water quality trends and to determine the significance of observed changes over time (Rosenblad 2017).

Multivariate statistical methods, namely principal component analysis (PCA) and cluster analysis, are especially effective for analyzing complex water quality data sets with multiple interrelated variables. PCA simplifies data by pinpointing the key variables that account for the most variance, allowing researchers to focus on the most significant factors affecting water quality (Jolliffe and Cadima 2016). Cluster analysis groups similar observations, helping to identify distinct water quality patterns and classify water bodies based on their characteristics.

Time-series analysis is another powerful statistical tool used to interpret water quality data. By examining data collected over time, researchers can identify temporal patterns, such as seasonal fluctuations or long-term trends (Hipel and McLeod 1994). Techniques like autoregressive integrated moving average (ARIMA) models and spectral analysis are utilized to forecast the future quality of water and assess the impacts of interventions or regulatory measures (Box et al. 2015).

Geospatial methods complement statistical analyses by providing a spatial context to water quality data. Geographic Information Systems (GIS) are highly preferred in water quality studies to integrate, visualize, and analyze spatial data (Goodchild

2010). GIS allows researchers to create detailed maps that display the distribution of water quality parameters across different regions, facilitating the identification of pollution hotspots and areas at risk (Burrough et al. 2015). By overlaying data related to water quality with other spatial layers, such as land use, topography, and hydrology, GIS helps in understanding the relationships between human activities and water quality.

Spatial interpolation techniques, such as kriging and inverse distance weighting (IDW), are employed to estimate water quality parameters at unsampled locations based on observed data (Goovaerts 1997). Kriging is a geostatistical technique that offers both estimates and measures of uncertainty, enabling a more informed interpretation of spatial patterns. These techniques enhance the spatial resolution of water quality assessments, supporting more accurate and comprehensive evaluations.

Remote sensing is another valuable geospatial tool that complements ground-based data collection. Satellite imagery and aerial photography provide large-scale, continuous observations of water bodies, enabling the monitoring of parameters such as chlorophyll concentration, surface temperature, and turbidity. Remote sensing data can be combined with GIS to analyze temporal changes in water quality and evaluate the effects of land use changes, climate variability, and extreme weather events.

The integration of statistical and geospatial methods allows for a holistic interpretation of water quality data. By combining temporal and spatial analyses, researchers can identify both short-term events, such as pollution spills, and long-term trends, such as eutrophication or groundwater depletion. This integrated approach also supports the development of predictive models that simulate the impacts of different management scenarios on water quality. These models rely on high-quality data and advanced statistical techniques to provide reliable forecasts and inform decision-making.

Effective water quality management requires not only the identification of problems but also the evaluation of potential solutions. Statistical and geospatial methods facilitate this by allowing for the evaluation of the effectiveness of management interventions, such as adopting best management practices (BMPs) or building wastewater treatment facilities. By analyzing pre- and post-intervention data, researchers can quantify the improvements in water quality and identify areas where additional efforts are needed.

Moreover, these methods support the development of water quality standards and regulations by providing the scientific evidence needed to establish safe levels of pollutants and to prioritize monitoring and enforcement efforts. Statistical analyses help in setting threshold values that protect human health and aquatic life, while geospatial analyses ensure that these standards are applied consistently across different regions.

7 Conclusions

In this section, we explored the important role of physical analysis methods in assessing surface water and groundwater quality. These methods, which include techniques such as turbidity measurement, temperature profiling, conductivity analysis, and sediment analysis, provide foundational data that help us understand the physical characteristics of water bodies. This data is crucial for assessing pollution levels, ecosystem health, and the effectiveness of water management practices.

The strength of physical analysis lies in its ability to offer direct, observable metrics that serve as indicators of potential contamination and water quality changes. When combined with chemical and biological analyses, these methods enable a comprehensive assessment, ensuring that water resources are monitored holistically. Moreover, technological advancements such as remote sensing and the use of sensors have enhanced the accuracy, efficiency, and scope of physical water quality measurements. However, these methods come with limitations, including site-specific variability, the need for periodic recalibration, and sensitivity to environmental factors. Despite these challenges, the integration of physical analysis with other assessment tools offers a robust approach to water quality management, ensuring that both surface and groundwater systems are safeguarded against pollution, over-extraction, and climate-induced changes.

The future of water quality monitoring will likely involve a greater reliance on automated systems and real-time data acquisition, providing more dynamic insights into water conditions. Nonetheless, the principles and methods discussed here will continue to be essential for ensuring that our water resources remain safe and sustainable for future generations.

References

Allan JD, Castillo MM (2007) Stream ecology: structure and function of running waters. Springer

American Public Health Association (APHA) (2017) Standard methods for the examination of water and wastewater. American Public Health Association

Backhaus T, Faust M (2012) Predictive environmental risk assessment of chemical mixtures: a conceptual framework. Environ Sci Technol 46(5):2564–2573

Bartram J, Ballance R (1996) Water quality monitoring: a practical guide to the design and implementation of freshwater quality studies and monitoring programmes. CRC Press

Box GE, Jenkins GM, Reinsel GC, Ljung GM (2015) Time series analysis: forecasting and control. Wiley

Boyle J, Siewert R, Helmer R (2014) Hydrological and water quality monitoring. Wiley

Bunn SE, Arthington AH (2002) Basic principles and ecological consequences of altered flow regimes for aquatic biodiversity. Environ Manag 30:492–507

Burrough PA, McDonnell RA, Lloyd CD (2015) Principles of geographical information systems. Oxford University Press, USA

Caissie D (2006) The thermal regime of rivers: a review. Freshw Biol 51(8):1389–1406

Coble PG (1996) Characterization of marine and terrestrial DOM in seawater using excitation-emission matrix spectroscopy. Mar Chem 51(4):325–346

Coble PG (2007) Marine optical biogeochemistry: the chemistry of ocean color. Chem Rev 107(2):402–418

Davies-Colley RJ, Smith DG (2001) Turbidity suspeni) ed sediment, and water clarity: a review 1. JAWRA J Am Water Resou Assoc 37(5):1085–1101

Endress and Hauser. (2023). Promag series electromagnetic flow meters. Retrieved from Endress and Hauser

Environmental Protection Agency (EPA) (2003) Guidelines establishing test procedures for the analysis of pollutants. Environmental Protection Agency

Fetter F (1999) Frank Fetter on population versus prosperity. Popul Dev Rev 25(3):577–590

Food & Agriculture Org. US EPA (2020) Water quality standards handbook. U.S. Environmental Protection Agency

Geosense (2023) Electronic water level meters. Retrieved from Geosense

Goodchild M (2010) Twenty years of progress: GIScience in 2010. J Spat Inf Sci 1:3–20

Goovaerts P (1997) Geostatistics for natural resources evaluation. Oxford University Press, USA

Gordon HR, Brown OB, Evans RH (1983) Exact Rayleigh scattering calculations for use with the Nimbus-7 coastal zone color scanner. Appl Opt 22(4):609–620

Hem JD (1985) Study and interpretation of the chemical characteristics of natural water. Department of the Interior, US Geological Survey, vol 2254

Hipel KW, McLeod AI (1994) Time series modelling of water resources and environmental systems. Elsevier

Hirsch RM, Alexander RB, Smith RA (1991) Selection of methods for the detection and estimation of trends in water quality. Water Resour Res 27(5):803–813

Hoffmann J, Müller B, McCoy E (2005) The role of colorimetry in water quality assessment. J Water Res 39(17):4053–4062

Horne AJ, Goldman CR (1994) Limnology. McGraw-Hill

Jolliffe IT, Cadima J (2016) Principal component analysis: a review and recent developments. Philos Trans R Soc A Math Phys Eng Sci 374(2065):20150202

Keller (2023) Series 21 pressure transducer. Retrieved from Keller

Krause P, Boyle DP, Bäse F (2007) Comparison of different efficiency criteria for hydrological model assessment. Adv Geosci 5:89–97

Marsh-McBirney (2023) Flo-Mate 2000 current meter. Retrieved from Marsh-McBirney

Meybeck M, Helmer R (1989) The quality of rivers: from pristine stage to global pollution. Palaeogeogr Palaeoclimatol Palaeoecol 75(4):283–309

Poff NL, Allan JD, Bain MB, Karr JR, Prestegaard KL, Richter BD, Stromberg JC (1997) The natural flow regime: a paradigm for river conservation and restoration. Bioscience 47(11):769–784

Rice EW, Baird RB, Eaton AD, Clesceri LS (2017) Standard methods for the examination of water and wastewater

Rosenblad A (2017) Multivariate statistical methods: a primer. J Stat Softw 78:1–4

Sawyer CN, McCarty PL, Parkin GF (2003) Chemistry for environmental engineering and science. McGraw-Hill

Siemens (2023) Ultrasonic flow meters. Retrieved from Siemens

Sillanpaa M (2014) Natural organic matter in water: characterization and treatment methods. Butterworth-Heinemann

Smith RC, Baker KS (1981) Optical properties of the clearest natural waters (200–800 nm). Appl Opt 20(2):177–184

Sophocleous M (2002) Interactions between groundwater and surface water: the state of the science. Hydrogeol J 10:52–67

United States. Environmental Protection Agency. Office of Water Regulations, United States. Environmental Protection Agency. Office of Water Planning (1996) National water quality inventory: report to congress. The Office. US EPA (2020) National water quality inventory report to congress. United States Environmental Protection Agency

USEPA (2006) National recommended water quality criteria. Environmental Protection Agency, U.S

Ward JV, Tockner K, Schiemer F (2003) Biodiversity of floodplain river ecosystems: ecotones and connectivity. Regul Rivers: Res Manage 15(1–3):125–139

Water UN (2021) Progress on the level of water stress: global status and acceleration needs for SDG indicator 6.4. 2, 2021

Wetzel RG (2001) Limnology: lake and river ecosystems. Gulf Professional Publishing

World Health Organization (2011) Guidelines for drinking-water quality, 4th ed. World Health Organization

World Health Organization (2020) Water, sanitation, hygiene and waste management for COVID-19: technical brief, 03 March 2020 (No. WHO/2019-NcOV/IPC_WASH/2020.1). World Health Organization

Remote Sensing and Google Earth Engine for Surface and Groundwater Quality Assessment

Nobin Chandra Paul, Rahul Banerjee, P. Navyasree, Bharti, Pankaj Das, and Tauqueer Ahmad

Abstract Remote sensing technology is a crucial asset in assessing and monitoring the condition of surface and groundwater, offering a cost-effective and time-efficient way to gather data over vast areas. This chapter investigates the application of remote sensing techniques in evaluating and tracking water quality over large regions, with a particular emphasis on surface and groundwater conditions, and water and salinity indicators. Essential, indices like the Normalized Difference Water Index (NDWI), Modified NDWI (MNDWI), and Normalized Difference Turbidity Index (NDTI) are vital for locating and outlining water bodies and identifying water quality characteristics such as turbidity. Additionally, the chapter explores indices designed for salinity assessment, including the Normalized Difference Salinity Index (NDSI) and other Salinity Indices (SIs), utilizing satellite data. Advanced tools like Google Earth Engine are employed for processing and analyzing spatio-temporal remote sensing datasets. A practical case study is presented to demonstrate the real-world utility of these indices, highlighting their effectiveness in evaluating water quality and salinity levels.

Keywords Google Earth Engine · NDWI · MNDWI · NDSI · Salinity indices · Surface water quality

N. C. Paul · P. Navyasree
School of Social Sciences & Policy Support, ICAR-National Institute of Abiotic Stress Management, Baramati, Maharashtra, India
e-mail: navyasree.ponnaganti@icar.gov.in

R. Banerjee (✉) · Bharti · P. Das · T. Ahmad
Division of Sample Surveys, ICAR-Indian Agricultural Statistics Research Institute, New Delhi, India
e-mail: rahuliasri@gmail.com

Dimple et al. (eds.), *Surface Water and Groundwater Quality for Sustainable Utilization*, Springer Hydrogeology, https://doi.org/10.1007/978-981-96-8189-1_7

1 Introduction

Water quality is crucial for environmental health, human well-being, and sustainable development. Traditional approaches to assessing water quality, which involve on-site sampling and laboratory analysis, often encounter challenges such as limited spatial coverage, high costs, and time-consuming procedures. On the other hand, remote sensing offers an efficient, cost-effective, and comprehensive method for monitoring water quality across large areas and over different time scales (Gorelick et al. 2017). Remote sensing technology facilitates the acquisition of continuous, high-resolution data over extensive geographic areas, making it an essential tool for monitoring water quality (McFeeters 1996; Xu 2006; Paul et al. 2018). Remote sensing offers several advantages for water quality assessment, such as extensive spatial coverage, consistent temporal data, cost efficiency, and access to both multispectral and hyperspectral data (Gorelick et al. 2017). This provides a comprehensive overview of water bodies and their surrounding environments (Huete 1988). Another significant benefit is temporal consistency, as satellites can repeatedly image the same locations over time, facilitating the analysis of changes and trends in water quality. Remote sensing also provides a cost-effective solution by minimizing the need for extensive fieldwork and laboratory analysis, thus reducing the overall expenses associated with water quality monitoring. Moreover, remote sensing instruments can capture data across multiple wavelengths, enabling the identification and measurement of various water quality parameters. Remote sensing indices are mathematical constructs that combine various spectral bands to highlight specific earth surface characteristics. For water quality assessment, certain indices are especially beneficial. The Normalized Difference Water Index (NDWI) uses the green and near-infrared (NIR) spectral bands to improve the detection of water bodies, effectively separating them from vegetation and soil (McFeeters 1996). The Modified Normalized Difference Water Index (MNDWI) uses green and shortwave infrared (SWIR) bands to provide improved identification of water bodies, especially in urban areas where buildings can hinder the detection of water (Xu 2006). The Normalized Difference Turbidity Index (NDTI) uses red and green spectral bands to evaluate water turbidity, a crucial measure of water clarity and quality (Lacaux et al. 2007). The Normalized Difference Salinity Index (NDSI) uses near-infrared (NIR) and red bands to track salinity levels in water bodies, aiding in the identification of areas affected by saltwater intrusion or salinization (Dwivedi and Sreenivas 1998).

Google Earth Engine (GEE) is a robust cloud-based platform designed for analyzing environmental data on a large scale (Moore and Hansen 2011). GEE offers access to a vast collection of satellite imagery and geospatial data, as well as sophisticated tools for data processing and analysis. In the field of water quality assessment, GEE is valuable due to its scalability, easy data access, advanced analytical capabilities, and promotion of collaboration and reproducibility. With its cloud computing capabilities, GEE can handle large datasets, making it possible to perform analyses on regional, national, or global scales. It provides access to a vast archive of both historical and current satellite imagery from sources such as Landsat, Sentinel,

MODIS, and more, facilitating long-term monitoring and trend analysis. GEE's suite of tools includes features for image preprocessing, statistical analysis, and visualization, enabling complex analyses that might be difficult to perform with traditional desktop software. Additionally, its cloud-based nature fosters collaboration among researchers and ensures that analyses are reproducible and shareable, promoting transparency and scientific rigor.

Integrating remote sensing indices with the capabilities of Google Earth Engine allows researchers to effectively analyze both spatial and temporal variations in water quality, providing valuable insights for managing water resources. This chapter includes a practical case study that illustrates the real-world application of these indices, showcasing their effectiveness in assessing water quality and salinity levels. The study focuses on the semi-arid regions of western Maharashtra. This introduction lays the groundwork for a thorough exploration of remote sensing techniques in water quality assessment, highlighting the importance of key indices and advanced analytical tools such as Google Earth Engine.

2 Remote Sensing in Water Quality Assessment

Remote sensing is the process of gathering information about the Earth's surface and atmosphere using sensors mounted on satellites. These sensors capture and measure reflected or emitted electromagnetic radiation across different spectral bands, delivering essential information about the physical and chemical characteristics of surface water and groundwater. Remote sensing can be categorized into two main types: passive and active. Passive remote sensing depends on natural energy sources, such as sunlight, and includes instruments like multispectral and hyperspectral imagers. These sensors capture data across various wavelengths, enabling the identification of different water quality parameters, such as turbidity and suspended sediments (Liu et al. 2003). Active remote sensing, by contrast, utilizes artificial energy sources such to illuminate targets and measure the reflected signals. This technique is especially valuable for acquiring detailed topographic data (Kumar and Singh 2019). Remote sensing provides several key advantages for water quality assessment, including broad spatial coverage, consistent temporal data, and cost-effectiveness. These features enable ongoing and thorough monitoring of water bodies across large areas and over extended periods, which is essential for tracking changes in water quality, managing water resources, and addressing environmental issues effectively (Dekker and Peters 1993).

2.1 Key Indices for Surface Water Quality

Remote sensing indices are crucial for assessing surface water quality, as they provide valuable insights into various parameters. By, employing different spectral bands, these indices emphasize specific features of water bodies, facilitating the tracking of changes and the detection of potential problems. Here are some important indices used in assessing surface water quality:

2.1.1 Normalized Difference Water Index (NDWI)

The Normalized Difference Water Index (NDWI) is a widely used tool for identifying water bodies and assessing water quality. Proposed by McFeeters in 1996 and refined in 2013, the NDWI utilizes green and near-infrared (NIR) bands to enhance water visibility while minimizing the influence of land vegetation and soil. The NDWI is calculated using the following equation:

$$NDWI = \frac{GREEN - NIR}{GREEN + NIR} \tag{1}$$

This index is particularly effective in distinguishing water from land features and is used for mapping water bodies and monitoring changes over time (Shen and Li 2010). For water quality assessment, NDWI is instrumental in monitoring surface water, enabling to track seasonal changes, drought effects, and flood occurrences. Although, NDWI is primarily designed for water body detection, it can also serve as an indirect indicator of certain water quality parameters. Typically, clear water exhibits higher NDWI values compared to turbid or polluted water. Consequently, changes in NDWI values over time can signal alterations in water quality, such as increased sediment load or the occurrence of algal blooms. As a baseline layer, NDWI is valuable for more detailed water quality assessments when combined with other indices and parameters to evaluate aspects like turbidity, chlorophyll concentration etc.

2.1.2 Modified NDWI (MNDWI)

The MNDWI enhances the NDWI by replacing the near-infrared (NIR) band with a shortwave infrared (SWIR) band, improving the detection of water bodies in areas with high turbidity (Xu 2006). The formula for MNDWI is:

$$MNDWI = \frac{GREEN - SWIR}{GREEN + SWIR} \tag{2}$$

MNDWI is more accurate than NDWI in urban areas where buildings and other structures can interfere with water body detection. It helps in accurately mapping urban water bodies, which is crucial for urban planning and management. Like the NDWI, the MNDWI can also function as an indirect measure of water quality (Sodhi et al. 2024).

Clear water bodies typically exhibit higher MNDWI values, whereas turbid or polluted waters present lower values. Variations in MNDWI over time can reflect changes in water quality. The MNDWI is particularly effective for identifying and mapping wetlands, especially in areas where wetlands are mixed with urban development, as it provides a clearer distinction of wetland boundaries compared to the NDWI. During flood events, MNDWI can more accurately define the extent of flooding by differentiating water from built-up areas and vegetation, which aids in emergency response and flood risk management. This adjustment makes MNDWI more sensitive to water with high sediment or organic matter content, making it valuable for evaluating water quality in rivers, lakes, and coastal zones.

2.1.3 Normalized Difference Turbidity Index (NDTI)

The Normalized Difference Turbidity Index (NDTI) determines water clarity by analyzing how light reflects off water particles. The index relies on the idea that clean water appears greener than murky water. As turbidity increases, the reflectance in the red spectrum also rises (Lacaux et al. 2007; Elhag et al. 2019). Turbidity, a crucial indicator of water quality, represents the concentration of particles suspended in water. The NDTI is utilized to estimate turbidity levels and is computed using the red and green spectral bands from remote sensing imagery, with the formula:

$$NDTI = \frac{RED - GREEN}{RED + GREEN} \tag{3}$$

The NDTI is highly effective for detecting variations in turbidity levels and has several important applications. It aids in mapping and monitoring water body clarity, with higher turbidity levels, which signify increased suspended particles, resulting in elevated NDTI values. During flood events, the NDTI can identify areas with high sediment concentrations, providing valuable information for disaster response, recovery efforts, and long-term flood management strategies. Overall, the NDTI is a crucial tool for assessing and monitoring water turbidity, offering essential insights into water quality and environmental health.

2.2 Salinity Indices for Groundwater Assessment

Salinity is a key factor influencing the quality of both surface and groundwater. High salinity levels can impair water usability and negatively impact aquatic ecosystems. Several indices have been developed to assess salinity through remote sensing data, offering valuable insights for effective water resource management.

2.2.1 Normalized Difference Salinity Index (NDSI)

The NDSI evaluates salinity levels in water bodies by analyzing the NIR and red bands. Saline waters have unique spectral signatures that differ from those of freshwater. The NDSI is useful for identifying areas with high salinity and tracking changes in salinity over time, which is essential for effective groundwater resource management (Major et al. 1990; Khan et al. 2001; Khan et al. 2007).

$$NDSI = \frac{RED - NIR}{RED + NIR} \tag{4}$$

By applying these indices and leveraging advanced analytical tools such as GEE, combined with validation through ground truth observations, a thorough understanding of water quality can be obtained. This methodology not only enhances water resource management but also aids in the formulation of policies designed to protect and improve water quality.

2.2.2 Salinity Index 1

Salinity Index 1 (SI1) utilizes green and red spectral bands to assess water salinity by analyzing their reflectance characteristics (Khan et al. 2005; Douaoui et al. 2006).

$$SI1 = \sqrt{GREEN \times RED} \tag{5}$$

SI1 is useful for estimating salinity levels in water bodies. Higher values generally indicate increased salinity, which can affect water quality and ecosystem health.

3 A Case Study: Leveraging Remote Sensing for Water Quality Surveillance

During the development of water quality metrics, the Sentinel-2 dataset from the COPERNICUS/S2_SR_HARMONIZED collection is employed, focusing on data between January 1, 2000, and January 31, 2022. The COPERNICUS/S2_SR_HARMONIZED dataset, provided by the European Space Agency (ESA) through the Sentinel-2 mission, offers high-resolution, atmospherically corrected surface reflectance data from Sentinel-2 satellites (**Source**: Harmonized Sentinel-2 MSI: MultiSpectral Instrument, Level-2A | Earth Engine Data Catalog | Google for Developers). This dataset includes images captured in multiple spectral bands, including visible, NIR, and SWIR regions, which are crucial for various remote sensing applications. To begin with, the region of interest, in this case is five tehsils of the Pune district, Maharashtra which comprises of Velhe, Bhor, Purandhar, Baramati & Indapur (Fig. 1). The data processing involves a crucial step where cloud contamination is addressed through the "maskS2clouds" function of GEE. This function employs the QA60 band to detect and exclude clouds and cirrus clouds. By examining specific bits within the quality assurance band-bit 10 for clouds and bit 11 for cirrus-the function effectively identifies and masks these cloud-related elements. Only, pixels where both flags are zero are considered clear and thus suitable for analysis. The imagery is then scaled by dividing the values by 10,000 to correct for surface reflectance. The filtered dataset is then subjected to further refinement. It is restricted to images with less than 10% cloud cover to enhance the quality of the data. The GEE facilitates this entire process by allowing efficient filtering, processing, and analysis of voluminous satellite data. By mapping the cloud masking function over the dataset, GEE ensures that only clear and high-quality images are used for generating indices such as the MNDWI, NDTI, NDSI, and the salinity indices. These indices provide valuable insights into water quality and environmental conditions by leveraging the detailed and harmonized surface reflectance data from Sentinel-2, processed to offer a clear and accurate representation of the Earth's surface. This approach leverages GEE's computational capabilities to handle and analyze extensive satellite imagery, ensuring accurate and reliable water quality assessments.

In the maps depicted in Figs. 2 and 3, the spatial distribution of NDWI and MNDWI values across the selected tehsils of Pune district is clearly illustrated, highlighting

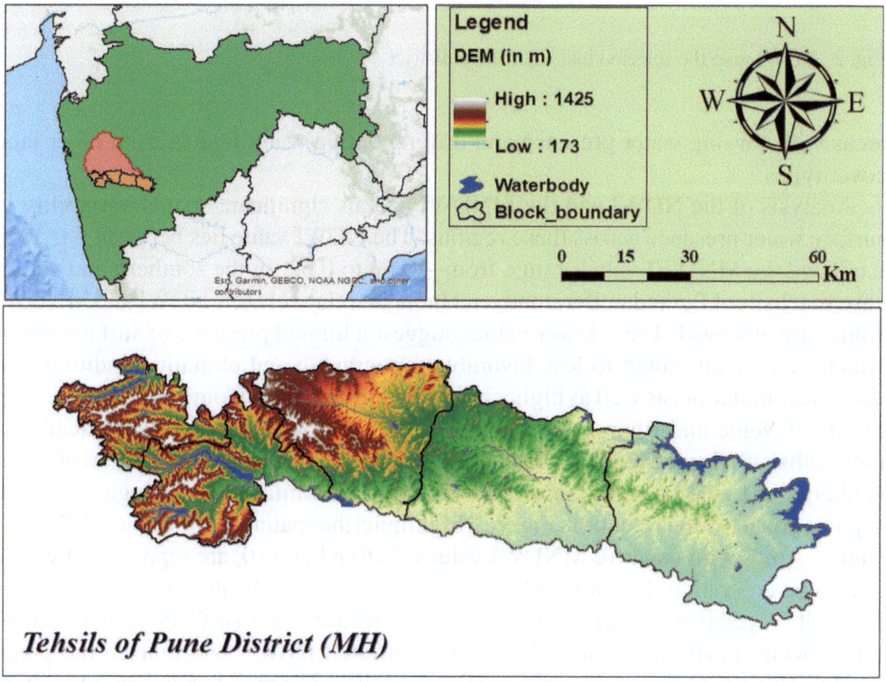

Fig. 1 Study area map of tehsils of the Pune district

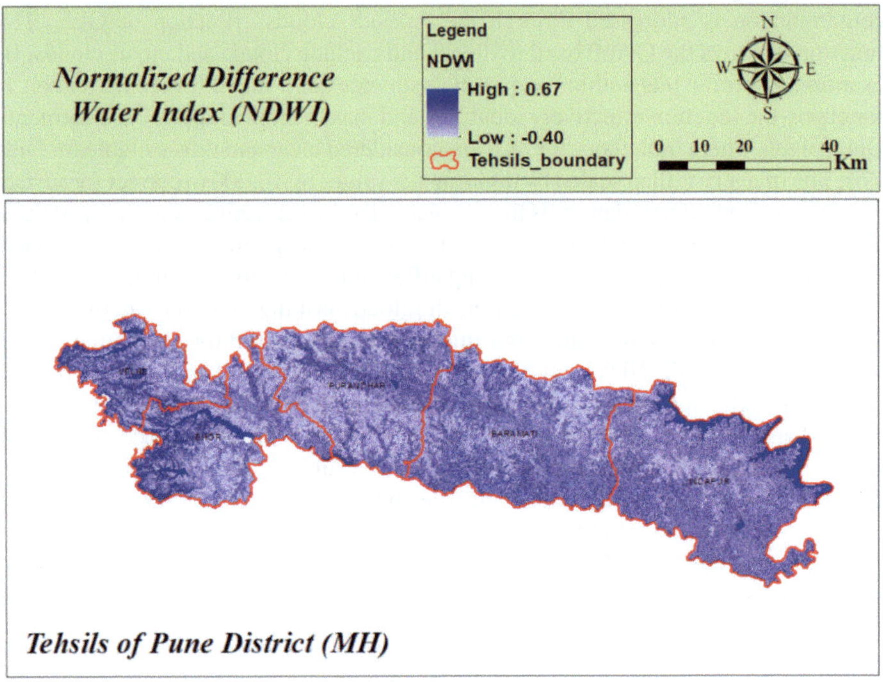

Fig. 2 NDWI map the selected tehsils of Pune district

areas with varying water presence and differentiating water bodies from other land cover types.

Analysis of the NDWI and the MNDWI reveals significant spatial variability in surface water presence across these regions. The NDWI value lies between −0.40 to 0.67, and the MNDWI values range from −0.63 to 0.62. In the southern and south-eastern tehsils of Purandar, Baramati, and Indapur, relatively low NDWI and MNDWI values are observed. These lower values suggest a limited presence of surface water, which may be attributed to less favorable topographic and climatic conditions for water accumulation, as well as higher built-up land coverage. Conversely, the western tehsils of Velhe and Bhor exhibit higher NDWI and MNDWI values, indicating a more substantial presence of surface water bodies. This higher concentration of water is likely due to the favorable topography and higher rainfall in these regions, which support greater water retention. Figure 3 highlights the spatial distribution of MNDWI values. Areas with negative MNDWI values, falling below 0, are represented with a light blue coloration. Negative NDWI values generally indicate non-water features like soil, vegetation, or urban areas. This is because these surfaces tend to reflect more electromagnetic radiation in the near-infrared (NIR) band than in the green band, resulting in a negative value when the NIR and green bands are subtracted. In contrast, areas with higher MNDWI values, exceeding 0.5, are visually represented with a dark blue color. These higher values are associated with water bodies,

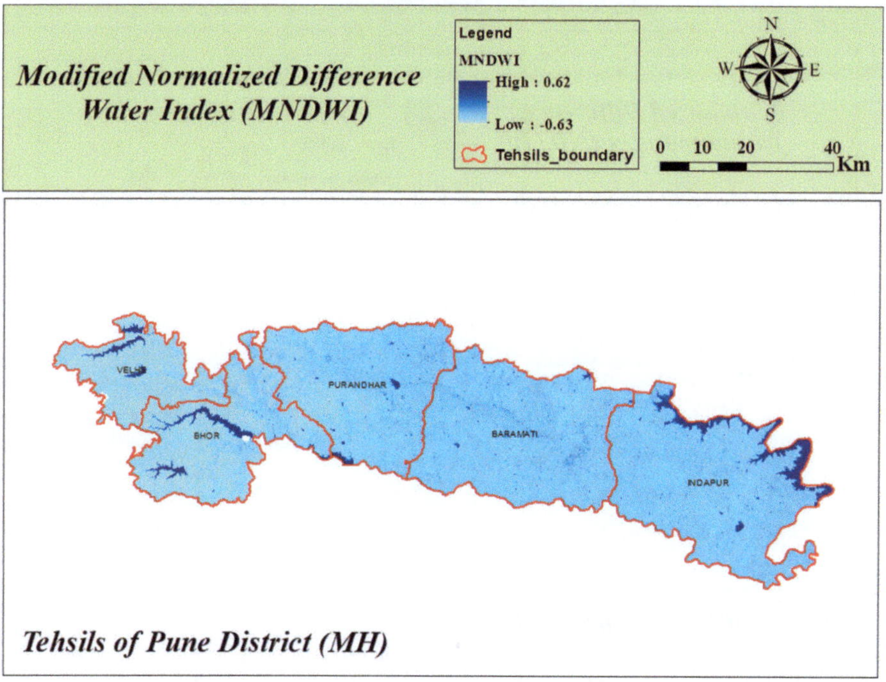

Fig. 3 MNDWI map of the selected tehsils of Pune district

mainly found in the western part of the district. The dark blue coloration effectively distinguishes these water bodies from other land cover types, providing a clear visual indication of their presence and extent within the study area. By analyzing NDWI and MNDWI indices, we obtain a detailed map of surface water locations in the selected tehsils of Pune district. This knowledge is valuable for water resource management, identifying regions with adequate water availability and those vulnerable to water scarcity.

Figure 4 illustrates the map of NDTI, which provides valuable insights into water turbidity levels by reflecting the concentration of suspended particles in water bodies. In the study area, NDTI values lies between −0.375 to 0.196, indicating varying levels of water turbidity across different regions. Negative NDTI values, ranging from −0.375 to 0, signify very low or negligible turbidity, indicating exceptionally clear water. These negative values are indicative of clean water conditions. In contrast, NDTI values from 0 to 0.196 represent low to moderate turbidity levels. As NDTI values approach 0.196, turbidity increases, suggesting a moderate presence of suspended particles. This increase could be attributed to natural sedimentation processes, moderate runoff, or some degree of pollution. Higher NDTI values near 0.196 imply a greater concentration of suspended particles, which may result from natural factors such as sedimentation and organic matter, or anthropogenic sources such as runoff from agricultural or urban areas.

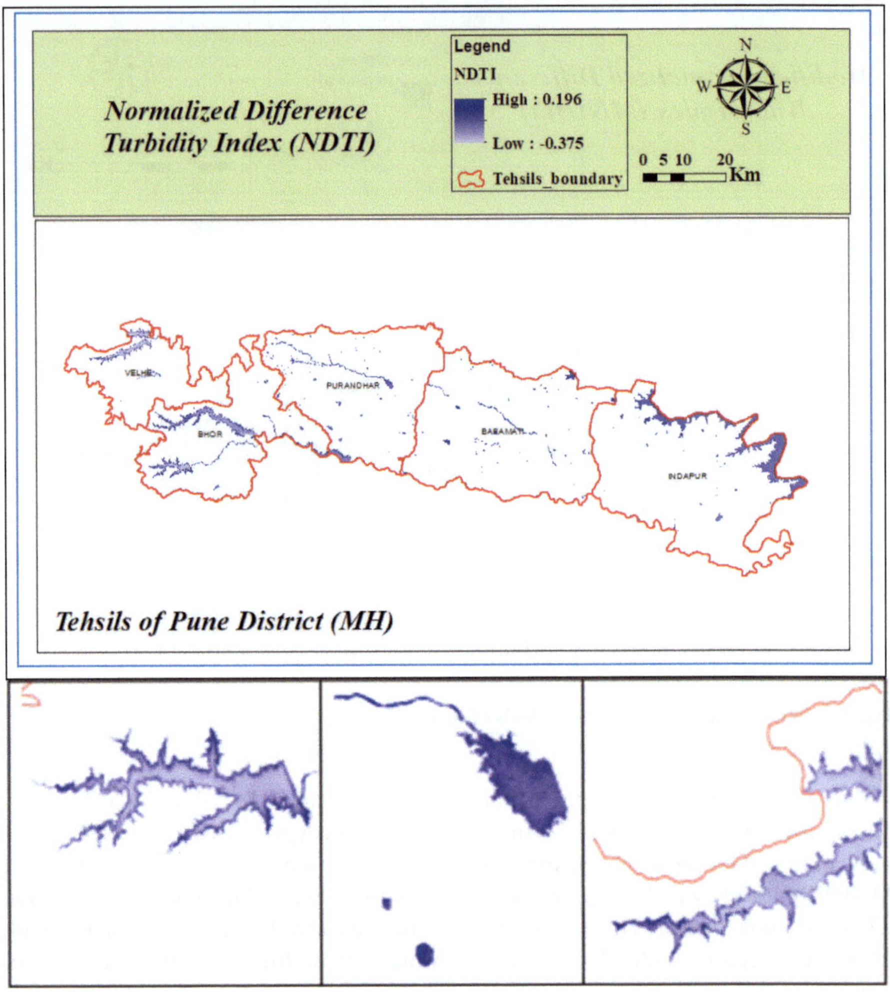

Fig. 4 NDTI map of the waterbody extracted from the study area

Figures 5 and 6 present the maps for NDSI and SI1, respectively, highlighting regions with varying salinity levels. The NDSI values range from −0.857 to 0.444, while the SI1 values range from 0.026 to 0.500. Higher values for both indices are mainly observed in the southeastern tehsils of Purandhar, Baramati, and Indapur. These elevated values indicate higher salinity levels in these regions, which could be attributed to factors such as irrigation practices, soil types, and climatic conditions conducive to salt accumulation. In contrast, the western tehsils of Bhor and Velhe exhibit lower values for both NDSI and SI1, suggesting lower salinity levels. These areas benefit from favourable topographical and hydrological conditions that promote natural leaching of salts, resulting in reduced soil salinity. The spatial

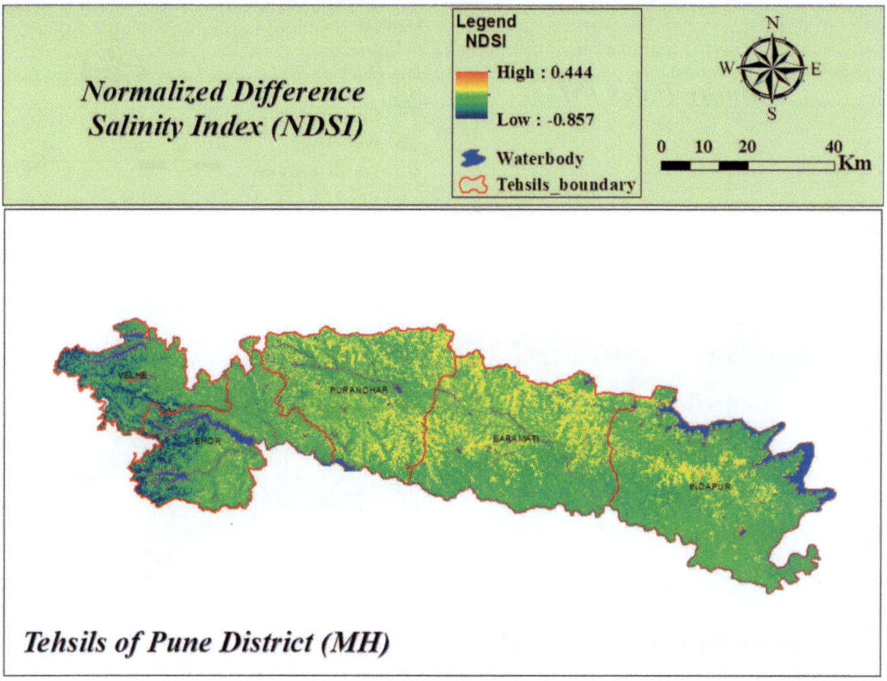

Fig. 5 NDSI map of selected tehsils of Pune district

patterns depicted in the maps of Figs. 5 and 6 provide a clear visual representation of the salinity distribution.

In summary, the analysis of NDSI and SI1 indices highlight significant spatial variability in soil salinity across the selected tehsils of Pune district. This information is vital for developing effective salinity management strategies and ensuring sustainable agricultural practices in the region. Nevertheless, it is crucial to validate the generated index maps against in-situ soil sample data to establish a reliable methodology. This validation, which is presently out of scope for the current study, would involve collecting soil samples and measuring Electrical Conductivity (EC) to validate the generated salinity indices.

4 Conclusions

Remote sensing has significantly transformed water quality assessment by offering a highly efficient, cost-effective, and comprehensive method for monitoring both surface and groundwater quality. The integration of satellite technology with advanced analytical tools like Google Earth Engine (GEE) has enabled researchers and policymakers to access and analyze extensive datasets, offering unparalleled

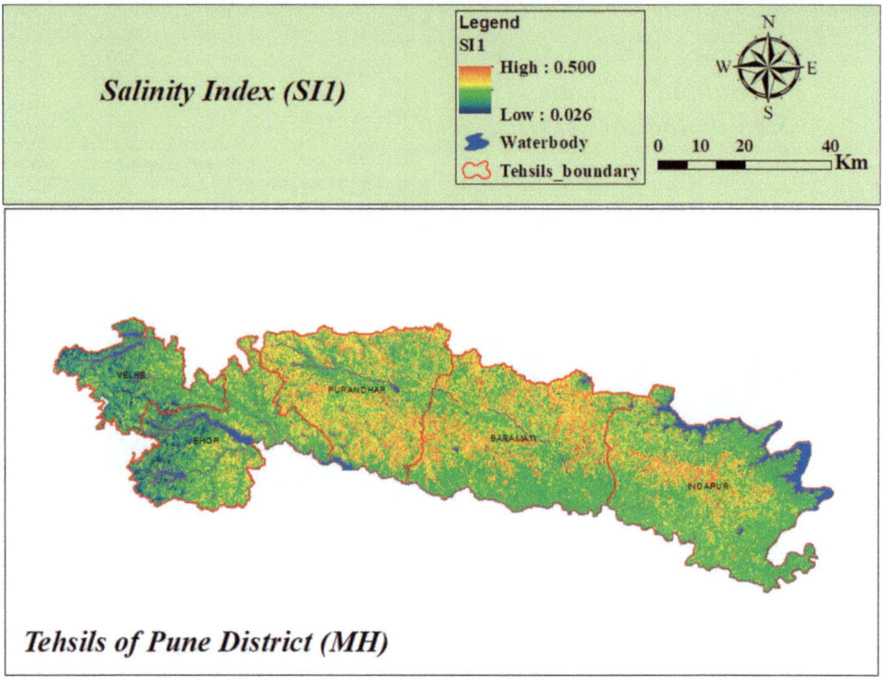

Fig. 6 Salinity indices 1 (SI1) of the study area

insights into water quality across vast areas. Remote sensing enhances monitoring capabilities by providing continuous observations across vast areas and extended timeframes. This is crucial for detecting and assessing changes in water quality due to pollution, sedimentation, and salinity fluctuations. The use of indices such as the NDWI, Modified NDWI, NDTI and NDSI provides valuable information on water body delineation, turbidity levels, and salinity concentrations. The clear visual representation of these indices further aids in identifying and delineating water bodies, supporting effective decision-making for sustainable water management in the region. Once validated with ground truth data, the generated indices maps will be more reliable and accurate, thereby supporting effective water management and policy-making. In conclusion, remote sensing has emerged as an invaluable asset, providing detailed, high-resolution data that supports informed decision-making and sustainable water resource management for the future.

References

Dekker AG, Peters SWM (1993) The use of the thematic mapper for the analysis of eutrophic lakes: a case study in the Netherlands. Int J Remote Sens 14(5):799–821. https://doi.org/10.1080/01431169308904379

Douaoui AEK, Nicolas H, Walter C (2006) Detecting salinity hazards within a semiarid context by means of combining soil and remote-sensing data. Geoderma 134(1–2):217–230. https://doi.org/10.1016/j.geoderma.2005.10.009

Dwivedi RS, Sreenivas K (1998) Delineation of salt-affected soils and waterlogged areas in the Indo-Gangetic plains using IRS-1C LISS-III data. Int J Remote Sens 19(14):2739–2751. https://doi.org/10.1080/014311698214488

Elhag M, Gitas I, Othman A, Bahrawi J, Gikas P (2019) Assessment of water quality parameters using temporal remote sensing spectral reflectance in arid environments. Saudi Arabia. Water 11(3):556. https://doi.org/10.3390/w11030556

Google Earth Engine. https://earthengine.google.com/platform/

Gorelick N, Hancher M, Dixon M, Ilyushchenko S, Thau D, Moore R (2017) Google earth engine: planetary-scale geospatial analysis for everyone. Remote Sens Environ 202:18–27. https://doi.org/10.1016/j.rse.2017.06.031

Huete AR (1988) A soil-adjusted vegetation index (SAVI). Remote Sens Environ 25:295–309. https://doi.org/10.1016/0034-4257(88)90106-X

Khan NM, Rastoskuev VV, Shalina EV, Sato Y (2001) Mapping salt-affected soils using remote sensing indicators—a simple approach with the use of GIS IDRISI. In: 22nd Asian conference on remote sensing, vol 5, no 9. Center for Remote Imaging, Sensing and Processing (CRISP), National University of Singapore, Singapore

Khan NM, Rastoskuev VV, Sato Y, Shiozawa S (2005) Assessment of hydrosaline land degradation by using a simple approach of remote sensing indicators. Agric Water Manag 77(1–3):96–109. https://doi.org/10.1016/j.agwat.2004.09.038

Kumar A, Singh SK (2019) Remote sensing applications in water quality assessment. In: Singh VP, Yadav S, Singh RN (eds) Water resources management. Springer, pp 215–236

Lacaux JP, Tourre YM, Vignolles C, Ndione JA, Lafaye M (2007) Classification of ponds from high-spatial resolution remote sensing: application to Rift Valley fever epidemics in Senegal. Remote Sens Environ 106(1):66–74. https://doi.org/10.1016/j.rse.2006.07.012

Liu Y, Islam MA, Gao J (2003) Quantification of shallow water quality parameters by means of remote sensing. Prog Phys Geogr 27(1):24–43. https://doi.org/10.1191/0309133303pp357ra

Major DJ, Baret F, Guyot G (1990) A ratio vegetation index adjusted for soil brightness. Int J Remote Sens 11(5):727–740. https://doi.org/10.1080/01431169008955053

McFeeters SK (1996) The use of the normalized difference water index (NDWI) in the delineation of open water features. Int J Remote Sens 17(7):1425–1432. https://doi.org/10.1080/01431169608948714

Moore RT, Hansen MC (2011) Google earth engine: a new cloud-computing platform for global-scale earth observation data and analysis. In: AGU fall meeting abstracts, vol 2011, pp IN43C-02

Paul NC, Sahoo PM, Ahmad T, Sahoo RN, Krishna G, Lal SB (2018) Acreage estimation of mango orchards using hyperspectral satellite data. Indian J Hortic 75(1):27–33. https://doi.org/10.5958/0974-0112.2018.00005.1

Shen L, Li C (2010) Water body extraction from landsat ETM+ imagery using Adaboost algorithm. In: Proceedings of the 2010 18th international conference on geoinformatics, Beijing, 18–20 June 2010, pp 1–4. https://doi.org/10.1109/GEOINFORMATICS.2010.5567762

Sodhi TS, Dahiphale P, Arora M (2024) Assessment of water dynamics and water quality parameters over Gobind Sagar Reservoir, India, based on high-resolution remote sensing approach. Water Conserv Sci Eng. https://doi.org/10.1007/s41101-024-00268-6

Xu H (2006) Modification of normalised difference water index (NDWI) to enhance open water features in remotely sensed imagery. Int J Remote Sens 27(14):3025–3033. https://doi.org/10.1080/01431160600589179

Integrating Remote Sensing and GIS for Surface and Groundwater Quality Assessment

Anuradha Sahu, Gaurav Mourya, and Prashant Shukla

Abstract Remote sensing (RS) techniques have transformed environmental monitoring by providing efficient and comprehensive methods for assessing surface and groundwater quality. These technologies provide a efficient and cost-effective approach to monitoring surface and groundwater quality across various spectral, spatial, and temporal resolutions. The major factors influencing surface and groundwater quality include industrial pollution, agricultural runoff, and improper waste disposal. A case study is presented in this chapter to enhance understanding of the relationship and role of Remote Sensing and Geographic Information Systems (GIS) in monitoring water quality. Different remote sensing indices, such as the Normalized Difference Vegetation Index (NDVI), Normalized Difference Turbidity Index (NDTI), Normalized Difference Water Index (NDWI) and Total Suspended Matter (TSM) were calculated using band ratios to monitor water quality and validated with the actual surface data. The groundwater quality was estimated using Weighted Arithmetic Water Quality Index method (WAWQI) and spatial variation map of groundwater quality index was generated by interpolation technique in GIS with the point data acquired from CGWB. The study result showed that the value of NDVI (0.59 to −0.21), NDWI (0.21 to −0.73), NDTI (0.416 to −0.24), and TSM (0-2) were observed and for pre monsoon it was found that only, 30% of groundwater have good water quality, 44% have very poor quality, and for post monsoon season only, 21% of groundwater have good water quality, 33% have very poor quality in the study region.

Keywords Surface water quality · GIS · Groundwater quality · Remote sensing

A. Sahu (✉) · G. Mourya
Department of Agricultural and Food Engineering (AgFE), Indian Institute of Technology, Kharagpur, India
e-mail: anuradhasahu12998@gmail.com

G. Mourya
e-mail: mouryagaurav@kgpian.iitkgp.ac.in

P. Shukla
Department of Soil and Water Conservation Engineering, Indian Agriculture Research Institute, New Delhi, India
e-mail: shuklaprashant310@gmail.com

Dimple et al. (eds.), *Surface Water and Groundwater Quality for Sustainable Utilization*, Springer Hydrogeology, https://doi.org/10.1007/978-981-96-8189-1_8

1 Introduction

Water is essential for life, and surface and groundwater are a highly valuable resource, especially in rural India, where it is a significant drinking water source. However, toxic elements from industries, landfills, and non-point sources like pesticides and fertilizers have increased groundwater pollution (Gholizadeh et al. 2016). Therefore, evaluating water quality is crucial for current use and its potential as a sustainable resource. According to Dandge and Patil (2022) and the CWMI by NITI Aayog, 70% of India's water supply is now polluted, reflecting a rapid decline in the quality of both groundwater and surface water. In the modern era, ensuring access to clean and safe water from natural aquifers has become a paramount concern. Monitoring water quality and tracking the evolution of contaminants across time and space is essential for maintaining the integrity of the water resources. Although traditional techniques used to monitor water are highly reliable, they have high costs, laborious procedures and require expensive instruments and laboratory tests (McCarthy et al. 2017; Adjovu et al. 2022). For the long-term sustainability of our aquifers, engaging stakeholders including the public in water quantity and quality monitoring is indispensable. To achieve this objective, there must be used of Remote sensing techniques, have transformed environmental monitoring of surface and groundwater quality as well as water resource management due to its ability to provide regular and consistent observations across large areas. With a consistent revisit rate, it enables well-structured time series analyses, offering a comprehensive view of changes over time. GIS provides comprehensive analyses and visualization tools for trends on different regions' quality of groundwater or surface fresh waters obtained from various field-based monitoring networks. The availability of numerous data products enhances its utility, complementing in situ sampling by providing additional context and coverage. Additionally, the mostly free and open access to remote sensing data makes it a cost-effective tool for widespread environmental monitoring and management. These technologies provide a non-invasive, efficient, and cost-effective approach to monitoring and managing water resources on a wide spatial and temporal scale. (Gholizadeh et al. 2016; Usali and Ismail 2010; Pereira et al. 2020; Abbas and Khan 2007). As we navigate through endeavours aimed at ensuring that water remains clean so that it can be consumed by everyone, it is important for us to welcome new approaches that give power to the people who live in these areas promoting sustainable management of water resources (Chen et al. 2021).

This chapter aims to provide an overview of various indices used for estimating surface and groundwater quality through remote sensing and GIS by analyzing satellite data at appropriate spatial and temporal scales. A study was conducted in Bhimgarh, a region within the Seoni district of Madhya Pradesh, to assess water quality. For surface water quality different indices was used and Groundwater quality was evaluated using the Weighted Arithmetic Water Quality Index technique (Patki et al. 2021; Islam et al. 2020). Continuous maps of spatially distributed groundwater quality were generated using remote sensing and GIS, identifying areas in the Seoni district where quality parameters exceed acceptable limits.

2 Materials and Methods

2.1 *Study Area*

The study examines the Bhimgarh Dam, also known as Sanjay Sarovar Dam, located in the Seoni district of Madhya Pradesh, India. Seoni district covers an area of 8,758 km^2, located between the coordinates 21°36′ to 22°57′ North in latitude and 79°19′ to 80°17′ East in longitude. The district has four distinct seasons, with the south-west monsoon season accounting for the majority of its annual average rainfall of 1,323.7 mm. The region's geology is characterized by Archean rocks and basaltic lava flows, with groundwater presence and movement influenced by secondary factors such as weathering, joints, and fractures. The topography is undulating, with the Satpura plateau hills extending from the south to the north and the Deccan plateau ranging in elevation from 325 to 740 m above mean sea level (MSL) (Fig. 1).

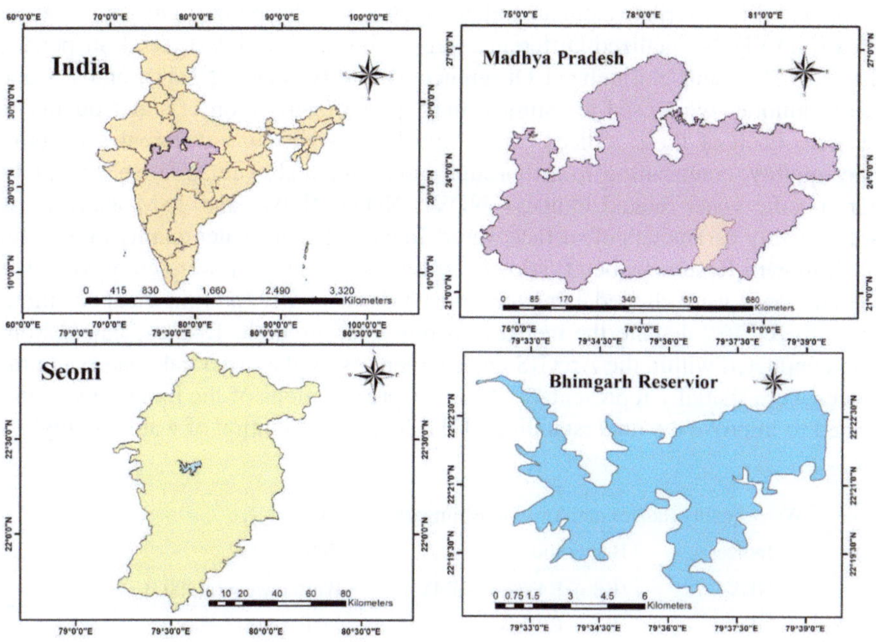

Fig. 1 Study area map of Bhimgarh reservoir

2.2 Data Used

The Bhimgarh reservoir's water quality parameters were analyzed using Landsat 9 data. It was introduced on September 27, 2021, and contains nine multispectral bands from the Operational Land Imager (OLI) and two from the Thermal Infrared Sensor. The spatial resolution of Landsat-9 data is 30 m, and the thermal infrared sensor resolution is 100 m. In this study band 3, band 4 and band 5 (visible green, red and near infrared) have been used to calculate the different water qualities indices. Groundwater quality data for 39 locations in the Seoni district were collected both the pre- and post-monsoon periods, from the Central Ground Water Board (CGWB). The analysis included parameters such as pH, nitrate, total hardness, calcium, magnesium, sulphate, fluoride, chloride, potassium, electrical conductivity, sodium, and total alkalinity. The outcomes were then compared with the Bureau of Indian requirements' drinking water quality requirements (2012).

In this case study, the shapefile for the Bhimgarh Reservoir was derived using Google Earth Engine. The satellite data extracted from the USGS Earth Explorer. Using visibility criteria to estimate indices such as the Normalized Difference Water Index (NDWI), Normalized Difference Vegetation Index (NDVI), Total Suspended Matter (TSM), and Normalized Difference Turbidity Index (NDTI), and RS and GIS techniques were used to estimate the "spatial distribution of water quality in the area. Each of these indices offers valuable insights into different aspects of water quality, contributing to a thorough environmental evaluation. In this study, four specific water-related indices—NDVI, NDTI, NDWI and TSM analyzed to assess quality parameters of surface water. To calculate the water quality indices for the Bhimgarh Reservoir, band ratios from Landsat 9 imagery were employed. The relevant bands were clipped using the reservoir's shapefile, and Empirical equations were utilized to calculate the indexes, as presented in Table 1. These calculations were conducted within the ArcGIS environment using the raster calculator, and the indices were spatially represented. "Visual representations of the indices were also created to improve the understanding of the spatial distribution of water quality".

Table 1 Water quality indices employed in Bhimgarh reservior

No	Indices	Band ratio	References
1	NDVI	(B8 − B4)/(B8 + B4)	Wardlow et al. (2007)
2	NDWI	(B3 − B8)/(B3 + B8)	McFeeters (1996)
3	NDTI	(B4 − B3)/(B4 + B3)	Elhag et al. (2019)
4	TSM	(B4/B3)	Gitelson and Merzlyak (1997)

2.3 Surface Water Quality Assessment Using RS Techniques

Water quality indices are specific measurements derived from satellite images that emphasize the detection and analysis of water bodies. These indices are crucial for extracting surface water using geospatial techniques, aiding in the detection, mapping, and ongoing monitoring of water resources. They are generally computed using various spectral bands from satellite sensors, including the green, red, visible, shortwave infrared (SWIR), and near-infrared (NIR) ranges.

Normalized Difference Water Index (NDVI) is used to evaluate vegetation density surrounding water bodies, which can be an indicator of water quality Dense, healthy vegetation near water sources often indicates good water quality, as it helps stabilize soil and reduce runoff. It is obtained by means of the red and near-infrared (NIR) spectral satellite bands (Nourani et al. 2024). The NDWI formula

$$\mathrm{NDVI} = \frac{\mathrm{NIR} - \mathrm{R}}{\mathrm{NIR} + \mathrm{R}} \tag{1}$$

It ranges from -1 to 1, where elevated values signify dense vegetation, while reduced values indicate sparse or no vegetation. By analyzing the surrounding vegetation, NDVI can provide insights into the ecological health of water bodies, as dense riparian vegetation often correlates with better water quality. This index is useful for environmental monitoring, conservation, and water resource management.

Normalized Difference Water Index (NDWI) is an indices were utilized in remote sensing to identify and monitor surface water bodies. It is determined applying near-infrared (NIR) and green satellite spectral bands (McFeeters 1996). The NDWI formula is:

$$\mathrm{NDWI} = \frac{(\mathrm{Green} - \mathrm{NIR})}{(\mathrm{Green} + \mathrm{NIR})} \tag{2}$$

It typically varies between -1 and $+1$, with values nearer $+1$ indicating the presence of water, while negative values suggest dry or non-water areas. This makes it especially useful for separating water features from soil and vegetation.

Normalized Difference Turbidity Index (NDTI) is a spectral indices used to evaluate the turbidity or cloudiness of water, often affected by suspended particles like pollutants or sediments. NDTI is derived from the red and green bands of satellite (Elhag et al. 2019). The formula for NDTI is:

$$\mathrm{NDTI} = \frac{(\mathrm{Red} - \mathrm{Green})}{(\mathrm{Red} + \mathrm{Green})} \tag{3}$$

NDTI is used to detect and monitor variations in water turbidity, which reflects changes in water quality. Higher NDTI values suggest increased turbidity, indicating cloudier or more opaque water, often due to sediments, algae, or pollutants. NDTI values (0 to +1) signifying greater turbidity and negative values (−1 to 0) indicating clearer water, while values near zero suggest moderate turbidity levels.

Total Suspended Matter (TSM): It indicates the concentration of suspended elements in the water, which encompasses sediments, plankton, and organic matter. Monitoring TSM is essential for evaluating surface water quality, as elevated levels can signal pollution or erosion. TSM is commonly estimated using remote sensing techniques, by using the ratio of red band and green band. Total Suspended Matter (TSM) varies widely by water body and is measured in milligrams per liter (mg/L) (Toming et al. 2017; Raag et al. 2014). The formula for TSM is

$$\text{TSM} = \frac{\text{Red}}{\text{Green}} \tag{4}$$

- **Higher TSM values** suggest a greater concentration of suspended matter, which may indicate poor water quality, high sediment loads, or pollution.
- **Lower TSM values** reflect clearer water with fewer suspended particles.

2.4 Groundwater Quality Estimation Using Geospatial Techniques

Groundwater quality was evaluated using the Weighted Arithmetic Water Quality Index (WAWQI) method. The concept of a WQI was initially introduced by Horton in (1965) and has been refined over time with various methodologies for comparing physicochemical parameters to established standards (Brown et al. 1972). WQIs function as simplified representations of intricate water quality situations, involving the selection of specific variables along with the establishment of suitable weighting and aggregation techniques (Patel et al. 2022). "According to Ramakrishnaiah et al. (2009), the WQI was assessed using weighted arithmetic index technique". In the initial phase, twelve key parameter of water quality were selected to measure the quality of drinking water. Each parameter's relative weight (W_i) was determined using the Table 2. Although the WAWQI method was originally designed for groundwater, it was widely applied in evaluating general water quality and successfully adapted for groundwater analysis.

This study employed the WAWQI method to assesed twelve frequently measured water quality parameters, including pH, magnesium, total hardness, sulphate, calcium, chloride, nitrate, fluoride, electrical conductivity, sodium, potassium, and total alkalinity. The suitability of these parameters for drinking criteria was estimated in accordance with the Bureau of Indian Standards (BIS 2012).

The WAWQI method was employed in four steps:

1. Identification of parameters which were utilize for water quality.
2. Scaling of quality ratings for water quality parameters.

Table 2 Showing parameters and relative weight calculated

Parameter	Standards as per BIS	Ideal value	K	W_i (relative weight)
pH	8.5	7	0.76111	0.08954
EC	300	0	0.76111	0.00254
HCO_3	200	0	0.76111	0.00381
Cl	250	0	0.76111	0.00304
SO_4	200	0	0.76111	0.00381
NO_3	45	0	0.76111	0.01691
F	1	0	0.76111	0.76111
TH	200	0	0.76111	0.00381
Ca	75	0	0.76111	0.01015
Mg	30	0	0.76111	0.02537
Na	200	0	0.76111	0.00381
K	10	0	0.76111	0.07611

3. Determination of unit weight (W_i) for each parameter, in opposition to the standard value (S_i).
4. Calculation of the overall WQI by Summing the subindex criteria values.

The Water Quality Index (WQI) were estimated using Eq. (1) which was used to compute the unit weight (W_i) for each parameter as follows:

$$W_i = \frac{K}{S_i} \tag{5}$$

where, K is the proportionality constant, S_i is each parameter standard value, and W_i is the weight of the ith parameter.

$$K = \frac{K}{\sum \frac{1}{S_i}} \tag{6}$$

Equation (3) were utilized for calculating the scale of quality rating (Q_i) for every parameter:

$$Q_i = \frac{V_i - V_0}{S_i - V_0} \tag{7}$$

Equation (4) were utilized for evaluating the scale of quality rating for pH.

$$Q_i = \frac{V_i - 7}{S_i - 7} \tag{8}$$

where V_0 is the ideal value of parameters and the concentration value is represented by V_i for the ith parameter for analysis.

The pH, which has an ideal value of 7 and all other parameters have an ideal value of zero.

Equation (5) may apply to show the final equation.

$$SI_i = \frac{\sum W_i Q_i}{\sum W_i} \tag{9}$$

There exist i parameters that are considered, and SI_i is the subindex of the ith parameter.

The standard values, ideal values and unit weights for the variables of water quality are given in Table 2.

$$WQI = \sum SI_i \tag{10}$$

Interpolation Techniques for spatial mapping of GWQI

The spatial mapping of groundwater quality parameters were interpolated by utilizing by spatial modelling, aided by the ArcGIS 10.8. The GWQI were utilized to mapped out in different geographical distributions using the IDW interpolation technique. This technique works well for assessing observations between measurement locations and for spatial interpolation. In IDW interpolation, each observation is calculated as a weighted average of the surrounding sample points, where the interpolated value is primarily affected by the nearest points. Researchers have used this method extensively, as shown by "Tabios and Salas (1985), and Mueller et al. (2004)".

3 Results and Discussion

3.1 Monitoring of Surface Water Quality

NDVI values analyzed in the Bhimgarh Reservoir range from 0.59 to −0.21. Low value of NDVI (NDVI < 0.2) suggest vegetation under moisture stress, while higher values (NDVI > 0.2) indicate healthier, more vigorous vegetation. The results indicate an enhancement in the water quality of the reservoir, promoting vegetation growth both within and around the surrounding area. Positive NDVI values suggest active vegetation growth, which can enhance sediment deposition under aerobic conditions, thereby improving water quality (Das et al. 2021; Wardlow et al. 2007). NDWI of the reservoir range from 0.21 to −0.73, suggesting that the water quality is not optimal (Guha et al. 2020). The total suspended matter (TSM) in the reservoir varies from 0 to 2, indicating a moderate concentration of suspended particles (Lacaux et al. 2007). The NDTI range from 0.41 to −0.24, where a decrease in turbidity levels and negative NDTI values indicate relatively clean water. In contrast, NDTI values above 0.4 suggest higher turbidity (Das et al. 2021). Figure 2 represent the different indices map of study area.

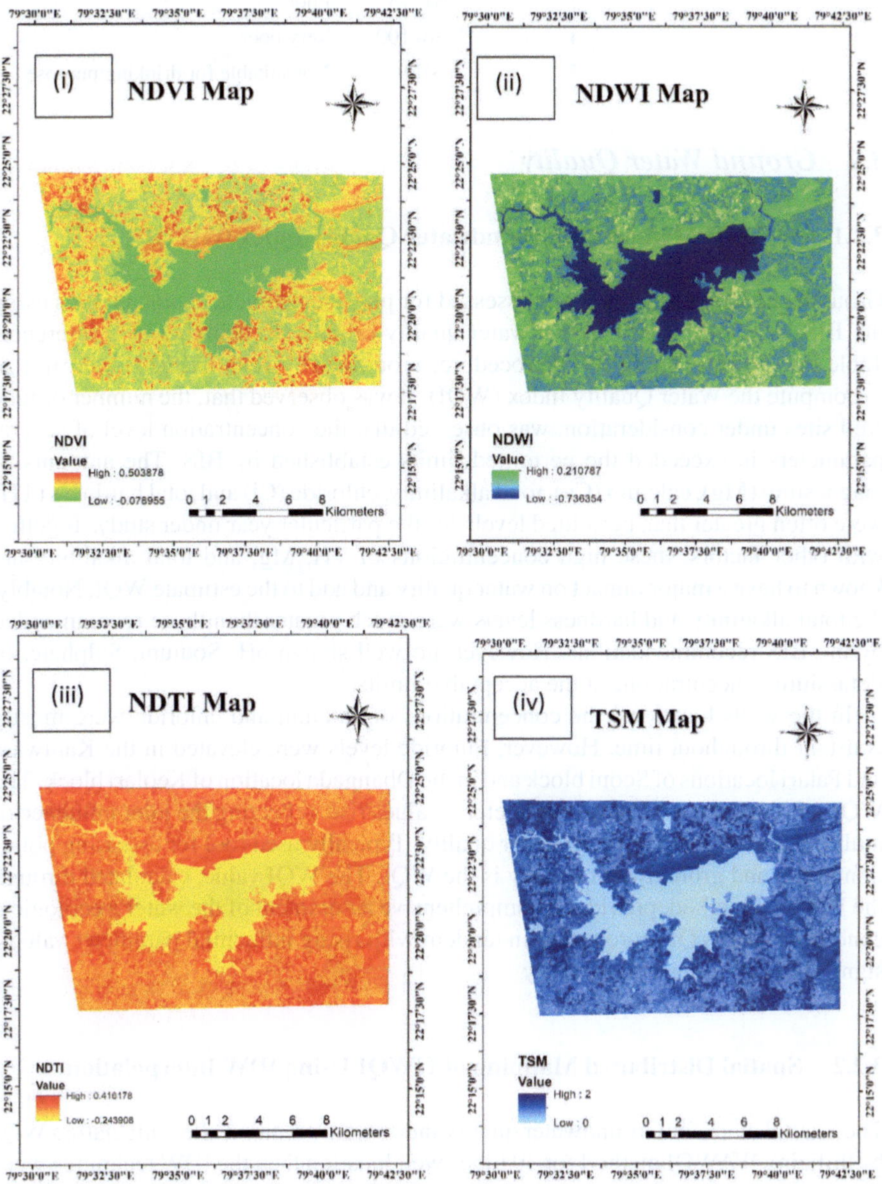

Fig. 2 (i) NDVI, (ii) NDWI, (iii) NDTI, (iv) TSM maps of the Bhimgarh Reservior

Table 3 Groundwater classification based on water quality index

S. no	GWQI	Groundwater quality
1	0–25	Excellent
2	26–50	Good
3	51–75	Poor
4	76–100	Very poor
5	>100	Not suitable for drinking purpose

3.2 *Ground Water Quality*

3.2.1 WAWQI Method for Groundwater Quality Index (GWQI)

Groundwater from tube well was assessed for physicochemical characteristics using the BIS standard (10500: 2012) water quality criteria for drinking as a reference. Table 3 is showing, the WAWQI procedure, as outlined by Tokatli (2020), were utilize to compute the Water Quality Index (WQI). It was observed that, the number of tube well sites under consideration, was observed that the concentration level of several parameters in exceeded the permitted limits established by BIS. The amounts of magnesium (Mg), calcium (Ca), total alkalinity, chloride (Cl) and total hardness (TH) were often greater than permitted levels for the particular year under study. Together with other factors, these high concentrations of TH, Mg, and total alkalinity are known to have a major impact on water quality and add to the estimate WQI. Notably, the total alkalinity and hardness levels were much greater than those recommended by the BIS recommendations. However, no well shown pH, Sodium, Sulphate, or Potassium concentrations at the acceptable limits.

In the wells locations, the concentrations of calcium and chloride were mostly constant throughout time. However, Fluoride levels were elevated in the Kaniwara and Palari locations of Seoni block and in the Dhangada location of Keolari block. The WQI score increases as these parameters' values rise, reflecting the joint influence of multiple factors on the overall water quality. Typically used tool for assessing pollution levels and groundwater quality is the WQI. The WQI value, calculated through the WAWQI method, provides a comprehensive assessment of the water's ecological condition. The WQI score grows in tandem with rising influential parameter values, signifying declining water quality.

3.2.2 Spatial Distributed Mapping of GWQI Using IDW Interpolation

The Spatial maps for Groundwater quality index were prepared after calculating WQI by utilizing WAWQI method for 39 tubes well location then the IDW technique were accomplished in order to generate spatial maps for pre and post monsoon. The IDW method uses adjacent weighted locations weights corresponding to the distance from sampled points to determine an average value for unsampled sites. With higher values assigning closer points greater weight, the IDW power coefficient modifies the effect of adjacent points.

Pre-monsoon

In the study region, during the pre-monsoon it was observed that the 3% of groundwater sources were classified as having excellent water quality, 44% were identified as having poor water quality, 30% had good water quality, 18% were deemed very poor, and 5% were considered unsuitable for human consumption prior to the monsoon season.

GWQI map of the bhimgarg reservior was created using IDW interpolation technique which was shown in Fig. 3. According to the spatial map well sites of Ugli, Rahiwara, Nagan Deori, Mehta, Masurbhanwari, Kudari, Khamaria, Kauria, Kalyanpur, Ghunai, Ghansor1, Gaurabibi, Chunatola, Chhapara, Borghat, Bamandehi, and Ari, the area come under "poor water" quality, Similarly at the Bamhodi, Banjal, Dhanaura, Dharamkuan, Kaniwara, Kauria, and Palari well sites very poor quality of water is found. Dhangada and Pandiachhappara well sites fall under the category of unfit for consumption in. The Amagarh well site is the only one with excellent water quality. Additionally, Banjari, Dargada, Dhuma, Gharghatia, Gorakhpur, Keolari, Kudopar, Madai, Makarjhir, Pipardahi, Sahasna, and Seoni1 are included in the region with good quality of groundwater in the pre monsoon period.

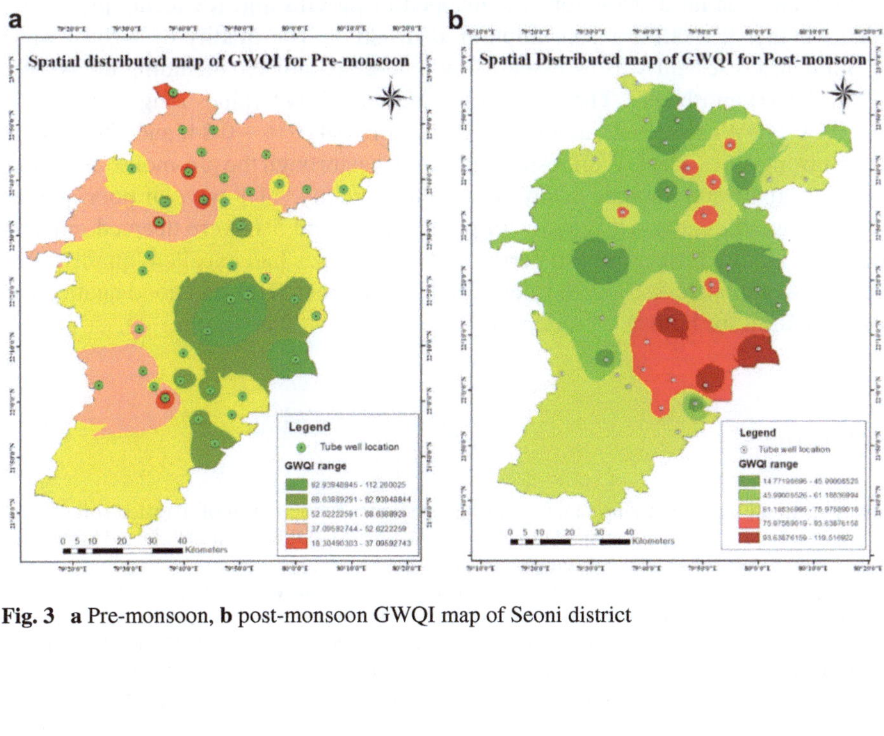

Fig. 3 **a** Pre-monsoon, **b** post-monsoon GWQI map of Seoni district

Post-monsoon

During the post-monsoon season, approximately 13% of groundwater sources exhibited excellent quality, while 33% had poor quality, 25% had very poor quality, and 21% were classified as good quality. Additionally, 8% of sources were deemed unsuitable for human consumption. As shown in Fig. 3, the spatial map indicates that the well sites at Amagarh, Bamandehi, Dargada, Dharamkuan, Masurbhanwari, Kudari, Kalyanpur, Rahiwara, Gaurabibi, Kudopar, Makarjhir, Nagan Deori, and Pipardahi are characterized by "poor water" quality. Very poor water quality was observed at the well sites of Ari, Bamhodi, Borghat, Chunatola, Dhanaura, Gharghatia, Mehta, Gorakhpur, Dhangada, and Madai. The well sites at Kauria, Kaniwara, and Pandiachhappara were categorized as having water unfit for consumption. The Banjari, Chhapara, Dhuma, Ghansor, and Keolari site is the only with excellent water quality. Moreover, the areas of Banjari, Banjal, Ghunai, Ugli, Khamaria, Lakhnadon, Palari, Sahasna, and Seoni were identified as having good groundwater quality at the time after the monsoon.

4 Conclusion

The groundwater quality index in Seoni district was evaluated using the WAWQI method, and spatial distribution was mapped using GIS and RS techniques determine whether it's safe for consumption as drinking. The quality of surface water of Bhimgarh reservoir was assessed using numerous indices, including the NDWI, NDVI, TSM) and NDTI. The results revealed that NDVI values ranged from 0.59 to −0.21, NDWI from 0.21 to −0.73, NDTI from 0.416 to −0.24, and TSM values ranged from 0 to 2. The GWQI results indicated that through the pre-monsoon period, 3% of groundwater sources exhibited excellent quality, 44% were of poor quality, 30% had good quality, 18% had very poor quality, and 5% were deemed unfit for drinking. During post-monsoon period, 13% of sources had excellent quality, 33% were of poor quality, 25% were categorized as very poor, 21% had good quality, and 8% were unfit for human consumption.

References

Abbas A, Khan S (2007) Using remote sensing techniques for appraisal of irrigated soil salinity. In: International congress on modelling and simulation (MODSIM). Modelling and Simulation Society of Australia and New Zealand, pp 2632–2638

Adjovu GE, Stephen H, Ahmad S (2022) Monitoring of total dissolved solids using remote sensing band reflectance and salinity indices: a case study of the imperial county section, AZ-CA, of the Colorado River. In: World environmental and water resources congress 2022, pp 1132–1145

Brown RM, McClelland NI, Deininger RA, O'Connor MF (1972) A water quality index—crashing the psychological barrier. In: Indicators of environmental quality: proceedings of a symposium held during the AAAS meeting in Philadelphia, Pennsylvania, December 26–31, 1971. Springer US, pp 173–182

Chen X, Liu L, Zhang X, Li J, Wang S, Liu D, Song K et al (2021) An assessment of water color for inland water in China using a Landsat 8-derived Forel–Ule index and the Google Earth Engine platform. IEEE J Select Top Appl Earth Observ Remote Sens 14:5773–5785

Dandge KP, Patil SS (2022) Spatial distribution of ground water quality index using remote sensing and GIS techniques. Appl Water Sci 12(1):7

Das S, Kaur S, Jutla A (2021) Earth observations based assessment of impact of COVID-19 lockdown on surface water quality of Buddha Nala, Punjab, India. Water 13(10):1363

Elhag M, Gitas I, Othman A, Bahrawi J, Gikas P (2019) Assessment of water quality parameters using temporal remote sensing spectral reflectance in arid environments, Saudi Arabia. Water 11(3):556

Gholizadeh MH, Melesse AM, Reddi L (2016) A comprehensive review on water quality parameters estimation using remote sensing techniques. Sensors 16(8):1298

Gitelson AA, Merzlyak MN (1997) Remote estimation of chlorophyll content in higher plant leaves. Int J Remote Sens 18(12):2691–2697

Guha S, Govil H, Gill N, Dey A (2020) Analytical study on the relationship between land surface temperature and land use/land cover indices. Ann GIS 26(2):201–216

Horton RK (1965) An index number system for rating water quality. J Water Pollut Control Fed 37(3):300–306

IS10500, BIS (2012) Indian standard drinking water–specification (second revision). Bureau of Indian Standards (BIS), New Delhi, pp 1–11

Islam MS, Azadi MA, Nasiruddin M, Islam MS (2020) Water quality index of Halda River, southeastern Bangladesh. Am J Environ Eng 10(3):59–68

Lacaux JP, Tourre YM, Vignolles C, Ndione JA, Lafaye M (2007) Classification of ponds from high-spatial resolution remote sensing: application to Rift Valley Fever epidemics in Senegal. Remote Sens Environ 106(1):66–74

McCarthy MJ, Colna KE, El-Mezayen MM, Laureano-Rosario AE, Méndez-Lázaro P, Otis DB, Muller-Karger FE et al (2017) Satellite remote sensing for coastal management: a review of successful applications. Environ Manag 60:323–339

McFeeters SK (1996) The use of the normalized difference water index (NDWI) in the delineation of open water features. Int J Remote Sens 17(7):1425–1432

Mueller TG, Pusuluri NB, Mathias KK, Cornelius PL, Barnhisel RI, Shearer SA (2004) Map quality for ordinary kriging and inverse distance weighted interpolation. Soil Sci Soc Am J 68(6):2042–2047

Nourani V, Ghaffari A, Behfar N, Foroumandi E, Zeinali A, Ke CQ, Sankaran A (2024) Spatiotemporal assessment of groundwater quality and quantity using geostatistical and ensemble artificial intelligence tools. J Environ Manag 355:120495

Patel P, Mehta D, Sharma N (2022) A review on the application of the DRASTIC method in the assessment of groundwater vulnerability. Water Supply 22(5):5190–5205

Patki VK, Jahagirdar S, Patil YM, Karale R, Nadagouda A (2021) Prediction of water quality in municipal distribution system. Mater Today Proc 13

Pereira OJ, Merino ER, Montes CR, Barbiero L, Rezende-Filho AT, Lucas Y, Melfi AJ (2020) Estimating water pH using cloud-based landsat images for a new classification of the Nhecolândia Lakes (Brazilian Pantanal). Remote Sens 12(7):1090

Raag L, Sipelgas L, Uiboupin R (2014) Analysis of natural background and dredging-induced changes in TSM concentration from MERIS images near commercial harbours in the Estonian coastal sea. Int J Remote Sens 35(18):6764–6780

Ramakrishnaiah CR, Sadashivaiah C, Ranganna G (2009) Assessment of water quality index for the groundwater in Tumkur Taluk, Karnataka State, India. J Chem 6(2):523–530

Tabios GQ, Salas JD (1985) A comparative analysis of techniques for spatial interpolation of precipitation. JAWRA J Am Water Resourc Assoc 365–380

Tokatli C (2020) Application of water quality index for drinking purposes in dam lakes: a case study of Thrace region. Sigma J Eng Natl Sci 38(1):393–402

Toming K, Kutser T, Uiboupin R, Arikas A, Vahter K, Paavel B (2017) Mapping water quality parameters with Sentinel-3 ocean and land colour instrument imagery in the Baltic Sea. Remote Sens 9(10):1070

Usali N, Ismail MH (2010) Use of remote sensing and GIS in monitoring water quality. J Sustain Dev 3(3):228

Wardlow BD, Egbert SL, Kastens JH (2007) Analysis of time-series MODIS 250 m vegetation index data for crop classification in the US Central Great Plains. Remote Sens Environ 108(3):290–310

Influence of Poor Water Quality on Crop Productivity and Quality of Produce

Lalita Rana, Manish Kumar, Navnit Kumar, Jitendra Rajput, Sumit Sow, Shivani Ranjan, Meenu Kumari, S. N. Singh, and A. K. Singh

Abstract Quality irrigation water is crucial for sustainable crop production. The use of low-quality water for irrigation poses significant environmental constraints that can hinder crop productivity and affect broader ecosystem health. Effective management practices and strategic interventions are essential to mitigate these impacts and ensure sustainable agricultural productivity. This book chapter emphasizes on sources and types, problems associated with poor quality irrigation water, heavy metal toxicity

L. Rana (✉) · N. Kumar
Department of Agronomy, Sugarcane Research Institute, Dr. Rajendra Prasad Central Agricultural University, Pusa, Samastipur 848125, Bihar, India
e-mail: lalita@rpcau.ac.in

N. Kumar
e-mail: navnit.sri@rpcau.ac.in

M. Kumar
Metamorphic Petrology Lab, Department of Geology, Institute of Science, Banaras Hindu University, Varanasi 221005, India
e-mail: manishkumar562@gmail.com

J. Rajput
Division of Agricultural Engineering, Indian Agricultural Research Institute, Pusa, New Delhi 110012, India
e-mail: jitu.caepht@gmail.com

S. Sow · S. Ranjan
Department of Agronomy, PGCA, Dr. Rajendra Prasad Central Agricultural University, Pusa, Samastipur 848125, Bihar, India
e-mail: sumitsow19@gmail.com

S. Ranjan
e-mail: ranjanshivani54@gmail.com

M. Kumari
Department of Horticulture (Floriculture and Landscaping), Bihar Agricultural College, Sabour 813210, Bihar, India
e-mail: meenubhu08111@gmail.com

S. N. Singh
Department of Plant Pathology, Sugarcane Research Institute, Dr. Rajendra Prasad Central Agricultural University, Pusa, Samastipur 848125, Bihar, India
e-mail: snsingh@rpcau.ac.in

Dimple et al. (eds.), *Surface Water and Groundwater Quality for Sustainable Utilization*, Springer Hydrogeology, https://doi.org/10.1007/978-981-96-8189-1_9

in crops and microbial contamination, their pathogenic effect on crops and also addressed direct and indirect impacts on crop yield, quality and soil health. Additionally, implementing good agricultural practice, use of pre-treated poor-quality water for irrigation and enforcing regulatory standards for irrigation water quality can significantly reduce contamination risks. This chapter underscores the necessity of ensuring high-quality irrigation water to maintain agricultural productivity, safeguard food safety, and promote sustainable farming practices.

Keywords Agricultural productivity · Ecosystem health · Food safety · Heavy metal toxicity · Low-quality water · Microbial contamination · Sustainable farming practices

1 Introduction

Agriculture, the cornerstone of many economies worldwide, heavily relies on irrigation where rainfall is scant to sustain productive cultivation of crops. The future supply of high-quality irrigation water will decline as new water sources are not being developed quickly enough to meet the increasing demands of industries and municipalities. To feed a growing population, irrigated farms must find ways to use less water, sometimes of lower quality, while simultaneously producing more food and fibre (Suresha and Nagesh 2015). Nonpoint nutrient inputs are challenging to measure and regulate because they originate from widely dispersed land-based activities and exhibit temporal variability influenced by weather conditions (Carpenter et al. 1998; Chee-Sanford et al. 2009). Excess use of fertilizers poses a problem in river, lakes and reduces the quality of water through eutrophication (Smith and Schindler 2009). A key issue in eutrophication science is the potential for N-fixing cyanobacteria to compensate for any deficiency in biologically available N (Dodds and Smith 2016; Fergusson 1990; Garcia-Delgado et al. 2019). The findings of Malakar et al. (2019) using poor irrigation water sources, specifically those contaminated with toxic metals, pathogenic organisms (Gerba and Smith 2005; Gessel et al. 2004) and emerging pollutants, can damage soil and crop quality, endangering both agricultural and human health. Historically, issues like soil salinization have been primary concerns, but the focus has shifted to include newer contaminants that complicate water management practices and affect long-term sustainability. Over time, this can result in soil salinization, reduced microbial activity, impaired soil structure, and ultimately, a decline in agricultural productivity. Traditional waste disposal systems in modern economies have focused on soil, air and water and causing poor quality irrigation water which are contaminated with salts, heavy metals, pathogens, and

A. K. Singh
Directorate of Research, Dr. Rajendra Prasad Central Agricultural University, Pusa, Samastipur 848125, Bihar, India
e-mail: shantaanil@yahoo.com

other pollutants can lead to a range of agronomic problems, adversely affecting both the yield and quality of crops (Ayers and Westcot 1985). Agricultural run-off of fertilizers, pesticides and manure is a potential contributor of poor-quality irrigation water. When agricultural inputs applied in excess, they can leach into water bodies and contaminate irrigation sources (Tilman et al. 2002). Industrial effluents are another significant source, especially in regions with lax environmental regulations (Singh et al. 2017). Municipal wastewater, whether treated or untreated, can introduce a variety of contaminants into the irrigation systems (Toze 2006). Wastewaters are the primary sources of emerging organic contaminants (EOCs) in the environment, with surface waters typically containing the highest loads. Compared to groundwater, both wastewaters and surface waters are generally considered to harbor a greater diversity of EOCs (Ward et al. 2005). However, this perceived difference may be influenced more by the limitations of current analytical techniques and the relatively small number of groundwater studies than by actual variations in environmental occurrence (Lapworth et al. 2012). Saline groundwater or naturally occurring minerals can lead to high levels of salts in irrigation water, particularly in coastal and arid regions (Rhoades et al. 1992). Additionally, climatic factors such as reduced rainfall and increased evaporation rates can concentrate salts and other contaminants in water bodies (Conway et al. 2019). Organic contaminants (pesticides, herbicides and industrial chemicals) can disrupt plant metabolism (Jayaraj et al. 2016), and reduce crop yield and quality. These substances often persist in the environment, leading to long-term contamination issues (Pimentel 1995). Addressing the issue of poor irrigation water quality is essential not only for maintaining agricultural productivity but also for ensuring food safety. The presence of contaminants can lead to reduced crop yields, deteriorated produce quality and significant economic losses for farmers (FAO 2011). The accumulation of harmful substances in crops poses serious health risks to consumers, highlighting the critical need for stringent water quality management practices in agriculture. Furthermore, the consumption of crops irrigated with polluted water (Gorga et al. 2013), can cause a variety of health issues in humans (Jin et al. 2012), ranging from acute illnesses to chronic conditions, including neurological disorders and cancers (Guillette and Gunderson 2001; Hamelin et al. 2013), due to the accumulation of toxic substances like arsenic, lead, and cadmium. The consequences of poor water management extend beyond individual health, impacting overall food security and the resilience of agricultural systems. As the demand for food increases alongside the pressures of climate change (Zhang et al. 2016) and industrialization, understanding the full impact of irrigation water quality on crop production is essential for safeguarding agro-ecosystems through sustainable water management practices that integrate technological innovations and data-driven approaches. In this chapter, we emphasized the sources and effects of contaminants and implementing effective mitigation strategies, protection of crop yields, food quality, and public health can be ensured, thereby supporting global food security.

2 Types of Poor-Quality Irrigation Water

The quality of irrigation water act crucial role in agricultural productivity, and its deterioration is often inter-connected to the presence of various cations, anions, biological contaminants, and organic matter. Addressing the issue of poor water quality is therefore critical not only for maintaining crop yields but also for preserving the long-term health of the soil ecosystem. Saline, alkaline and wastewater are of poor-quality (Tables 1 and 2) which are used for irrigation of crops that can impose severe constraints on agricultural productivity. Poor quality water has distinct chemical compositions that negatively affect soil properties, plant health, and overall crop yield (Table 3).

2.1 *Saline Water*

Saline water contains high levels of dissolved salts, primarily sodium chloride (NaCl), but may also include other salts like calcium chloride ($CaCl_2$) and magnesium sulfate ($MgSO_4$). Groundwater in coastal and arid regions often has high concentrations of dissolved salts due to the dissolution of minerals in the soil and rock formations. Additionally, natural soil salinization can occur through the weathering of rocks and atmospheric deposition of salts (Rhoades et al. 1992). Irrigation itself can contribute to salinity. When irrigation water evaporates, it leaves behind salts, gradually increasing

Table 1 Classification of irrigation water based on salinity

Class	EC (dS/m)	Characteristics	Suitability
C_1	<1.5	Normal water	All soils
C_2	1.5–3	Low salinity water	Light and medium textured soils
C_3	3–5	Medium salinity water	Light and medium textured soils for semi-tolerant crops
C_4	5–10	Saline water	Light and medium textured soils for tolerant crops
C_5	>10	High salinity water	Not suitable

Source Zaman et al. (2018)

Table 2 Classification of irrigation water based on SAR class

Sodium hazard class	SAR value	Characteristics of irrigation water
S_1	0–10	Excellent (little or no hazard)
S_2	10–18	Good (can be used with appropriate management)
S_3	18–26	Unsatisfactory (unsatisfactory for most of the crops
S_4	>26	Unsuitable (unsatisfactory for all the crops)

Source Zaman et al. (2018)

Table 3 Impacts of poor quality irrigation water on various crops

Impacts	Cause	Specific effects on crop yield	Specific effects on crop quality	References
Heavy metal contamination	Industrial effluents, polluted water sources	Reduced plant growth, inhibited photosynthesis, lower crop yield	Heavy metals accumulation in edible parts, health hazards	Shahid et al. (2017), Nagajyoti et al. (2010)
Pathogen contamination	Use of untreated or poorly treated wastewater	Plant diseases, reduced growth, lower yield	Presence of pathogens in produce, potential health risks	Toze (2006), Hamilton et al. (2020)
Nutrient imbalance	Use of low-quality water with poor nutrient content	Nutrient deficiencies, reduced plant growth, lower yield	Poor nutritional quality of crops, reduced shelf life	Bhargava and Bhargava (2019), Zhang et al. (2020)
High pH levels	Alkaline water sources	Reduced nutrient availability, stunted growth, lower yield	Poor crop quality, susceptibility to diseases	White (2013), Flores et al. (2016)
Low pH levels	Acidic water sources	Nutrient toxicity, inhibited root growth, lower yield	Poor crop quality, increased heavy metal uptake	Sumner (1995), Bolan et al. (2021)
High levels of suspended solids	Use of turbid water	Clogging of irrigation systems, reduced water availability to crops	Poor crop quality, potential contamination with harmful particles	Verma et al. (2018), Lamm and Ayars (2009)
Chemical contaminants (e.g., pesticides, herbicides)	Agricultural runoff, contaminated water sources	Phytotoxicity, reduced growth, lower yield	Residual chemicals in produce, potential health risks	Carvalho (2017), Mostafalou and Abdollahi (2017)

soil salinity. The use of poor-quality groundwater for irrigation, which often contains high levels of dissolved salts, can exacerbate this problem (Oster 1994). Industries, especially those involved in mining, chemical manufacturing, and food processing, can discharge saline effluents into water bodies. These effluents can contaminate surface and groundwater sources used for irrigation (Rengasamy 2006). Urban run-off, containing salts from road de-icing, wastewater discharge, and other sources, can flow into rivers and lakes, increasing the salinity in irrigation water (Kaushal et al. 2005).

2.2 Alkaline Water

Alkaline water typically has a high pH (above 8.5) and elevated levels of bicarbonates (HCO_3^-) and carbonates (CO_3^{2-}). High soil pH can lead to the precipitation of essential nutrients like iron, manganese and phosphorus, making them unavailable to plants. This results dispersion of soil particles, leading to poor soil structure, reduced infiltration, aeration, nutrient deficiencies and poor plant growth. High pH levels can damage root systems, impairing their ability to absorb water and nutrients.

2.3 Industrial and Municipal Wastewater

Major source of poor-quality irrigation water is Industrial and Municipal wastewater. An effluent released from industries (Schwarzenbach et al. 2006), and households (Pettygrove and Asano 1985) goes to groundwater, rivers and ponds can contain contaminants such as heavy metals or other toxic substances like lead, cadmium, and mercury and used for irrigation purpose leading to toxic accumulation in edible parts of the crop. This not only reduces crop quality but also poses serious health risks to consumer, significant challenges and opportunities when it comes to its impact on crop development. The composition varies widely depending on the industry, containing a mix of beneficial nutrients and harmful contaminants. This dual nature necessitates careful management to harness its potential benefits while mitigating its adverse effects. Irrigation using treated municipal wastewater is regarded as an environmentally sustainable alternative to direct discharge into surface or groundwater bodies. This practice not only aids in wastewater disposal but also contributes to water conservation and nutrient recycling in agricultural systems (Mojiri 2011).

3 Problems Associated with Poor Quality Irrigation Water

3.1 Salinity Problems

Salinity in irrigation water is a critical issue affecting agricultural productivity. High levels of dissolved salts (viz., sodium chloride, calcium, and magnesium salts) can lead to significant agronomic challenges. In developing countries about one-third of the irrigated land is affected or at serious risk of salinity hazards (Stockle 2001). In India, approximately 6.73 million hectares of land have been degraded due to salinity and sodicity, primarily as a result of irrigation practices in the western regions of the country (Singh 2009). Irrigating crops with saline water significantly elevates both agricultural and environmental hazards. It leads to soil salinization, which reduces soil fertility and crop productivity, and impairs water quality by elevating the concentration of dissolved salts. This not only threatens sustainable farming practices but also affects downstream ecosystems (Minhas et al. 2020). Salts negatively affect plant growth by increasing soil osmotic pressure and disrupting plant nutrient uptake. High concentrations of salts in the soil solution reduce the plant's ability to absorb water, a phenomenon known as the osmotic or water-deficit effect of salinity. This stress leads to physiological drought, even when water is present in the soil. Crop damage occurs when salt concentrations reach levels that begin to inhibit growth and development (Machado and Serralheiro 2017). High salt concentrations in irrigation water can lead to soil salinization and affects the agricultural development which needs to be managed to keep the root zone at an optimal salt balance for economically viable agriculture. In Australia, excessive use of freshwater for intensive agriculture and other activities threatens nearly 33% of the land with salinization (Farber et al. 2005). Recent research suggests that economic losses could reach USD 1,604 per hectare as salinity rises from low to medium, and USD 2,748 per hectare if it increases from medium to high. Unfortunately, the potential profit from mitigation efforts is often too low, discouraging farmers from adopting sustainable irrigation practices (Naifer et al. 2011). Salinity results into osmotic stress, where plants struggle to absorb water which is a consequence of the decreased osmotic potential of the soil solution. This creates a physiological drought condition, even when soil moisture is adequate. Plants under osmotic stress exhibit reduced growth rates and lower biomass production (Munns 2002). Specific ions such as sodium (Na^+) and chloride (Cl^-) can accumulate in plant tissues to toxic levels, disrupting cellular functions and causing physiological damage (Fig. 1). Sodium can replace potassium in various physiological processes, leading to deficiencies and impaired metabolic functions. Chloride toxicity can cause leaf burn and defoliation (Shannon et al. 2008). Excess salts in the soil can lead to nutrient imbalances by reducing the availability and uptake of essential nutrients. High levels of sodium (Na^+) in saline soils compete with potassium (K^+) for absorption sites in plant roots, which can result in potassium deficiencies. Likewise, Calcium ($Ca^{2+)}$ and magnesium (Mg^{2+}) can be displaced from the soil exchange sites by sodium, leading to reduced availability of these nutrients for plant uptake. For example, elevated sodium levels can interfere with potassium uptake, leading to potassium deficiency, which is essential for enzyme activation and protein synthesis in plants (Grattan and Grieve 1998).

3.1.1 Effects of Salinity on Crop Yield and Quality

High salt levels create osmotic stress around the root zone forces plants to expend more energy to absorb water, leading to energy deficits resulting in poor aeration and drainage, which are critical for healthy root development, plant development, vigour, and soil structure degradation can lower the soil's capacity to hold on to water and nutrients.

High concentrations of certain ions, especially sodium (Na^+) and chloride (Cl^-) in saline water, can be harmful to plants causing leaf burn, chlorosis (yellowing of leaves), and necrosis (tissue death) and often show reduced yields and impaired the taste, texture, and marketability of produce, particularly in crops like fruits and

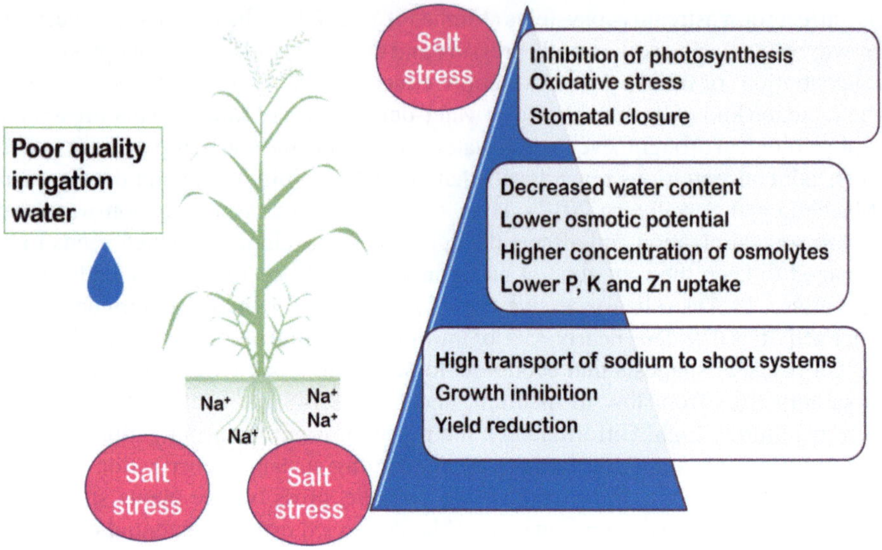

Fig. 1 Effect of salinity in crops irrigated with poor-quality water

vegetables due to poor water quality. Poor crop establishment can be caused by high saline levels, which can also hinder early seedling growth and seed germination (Table 4). According to Katerji et al. (2003), seedlings are especially vulnerable to osmotic stress and ion toxicity. High salinity levels not only reduce crop yields but can also degrade soil structure and fertility over time (Rengasamy 2006). Lee and colleagues (2008) examine the implications of salt water for irrigation on chrysanthemums. Rameshwaran et al. (2015) used a drip irrigation system to study the effects of different irrigation regimes and salinity treatments on two pepper types in a greenhouse in Antalya, Turkey. Plants under salt stress often exhibit stunted growth, reduced leaf area, and lower overall biomass production. This is due to both the osmotic and ionic effects of salinity, which limit water uptake and cause cellular damage (Flowers and Yeo 1995). Salt negatively impacts plant growth through mechanisms such as reduced osmotic potential, ion toxicity, and nutritional imbalances (Lu et al. 2023).

Salt stress triggers stomatal closure, reducing carbon dioxide availability in the leaves and consequently lowering photosynthesis rates (Kamran et al. 2020). Ahmad and Sharma (2008); Khorobrykh et al. (2020) opined that The light reactions in chloroplasts are a major source of ROS (reactive oxygen species viz. superoxide, hydrogen peroxide, hydroxyl radicals, and singlet oxygen) Research shows that abiotic stressors like salinity cause substantial yield losses in major crops, particularly during their reproductive stages (Ehtaiwesh and Rashed, 2020). The class I high-affinity potassium transporter (HKT) family plays a key role in regulating sodium ion homeostasis by preventing Na^+ accumulation in leaf blades under saline conditions. This regulation affects both grain dry matter and the K^+/Na^+ ratio, which

Table 4 Impacts of salinity on crop yield and quality

Impact	Specific effects	Mechanism	Crop examples	References
Reduced seed germination	Lower germination rates	High salt concentrations inhibit water uptake	Wheat, Rice, Barley	Panuccio et al. (2014), Munns and Tester (2008)
Stunted plant growth	Reduced plant height and biomass	Osmotic stress, ion toxicity	Tomato, Soybean, Maize	Zorb et al. (2019), Acosta-Motos et al. (2017)
Lower yield	Decreased fruit/seed production	Reduced photosynthesis, nutrient imbalance	Cotton, Potato, Sugar beet	Shrivastava and Kumar (2015), Munns (2002)
Poor fruit quality	Reduced size and weight, altered taste	Ion toxicity, osmotic stress	Citrus, Strawberry, Melon	Ashraf and Foolad (2007), Gupta and Huang (2014)
Leaf chlorosis and necrosis	Yellowing and death of leaves	Ion toxicity, nutrient deficiency	Beans, Lettuce, Spinach	Negrão et al. 2017, Gupta and Huang (2014)
Flower and fruit abortion	Premature dropping of flowers and fruits	Osmotic stress, hormonal imbalance	Tomato, Bell Pepper, Grapevine	Zorb et al. (2019), Mahajan and Tuteja (2005)
Root growth inhibition	Reduced root length and mass	Osmotic stress, ion toxicity	Carrot, Onion, Radish	Cuartero et al. (2006), Tavakkoli et al. (2011)

are closely associated with the rate and duration of grain filling under salt stress (Poustini and Siosemardeh 2004). Crops such as cereals and vegetables often experience reduced grain and fruit production in saline environments (Maas and Hoffman 1977). Sabagh et al. (2021) reported that salinity reduces grain production by 39.1, 24.3, and 13.4% at various developmental stages (anthesis, early booting, and mid-grain filling) respectively. Additionally, Grieve et al. (2012) noted that salinity can negatively affect produce quality, impacting taste, texture, and nutritional value. In high-salt environments, plants face difficulty absorbing water, despite the soil being saturated with moisture. Maggio et al. (2007) found that increasing salinity by 1 dS m^{-1} resulted in a 6% decrease in plant dry mass up to 9 dS m^{-1}. After that, the yield decline decreased to 1.4% per dS m^{-1}. Similarly, Al-Harbi et al. (2009) found that saline water irrigation reduced both total dry biomass and fresh tomato fruit output, This aligns with the findings of Jha and Dubey (2012). Elevated salt and chloride levels in the soil can also result in increased concentrations of these ions in plants, reducing flavour and marketability.

3.1.2 Mitigation Strategies

Effective management of salinity contamination in irrigation water requires a multi-faceted approach. A variety of technology developed for desalinization of water which can be used for irrigation purpose and can be helpful for mitigating the salinity problems as indicated in Fig. 2. One of the primary strategies is to improve the quality of irrigation water. This can involve blending saline water with freshwater to dilute the salt concentration (Qadir and Oster 2004). Additionally, desalination techniques, such as reverse osmosis and electrodialysis, can be used to remove salts from irrigation water, although these methods can be cost-prohibitive for large-scale agricultural use (Xu et al. 2010).

Soil management measures are critical to alleviating the negative impacts of salinity. Potassium (K) is vital for many functions in plants; however, elevated concentrations in soil solutions can still cause clay dispersion, though to a lesser extent than sodium (Na). This can result in reduced hydraulic conductivity and decreased aggregate stability (Marchuk et al. 2014). Caution is needed when using irrigation wastewater with high K concentrations, as it may negatively affect soil structural stability. The impact of K on soil structure varies depending on the soil type, particularly the mineral composition of the soil clay (Marchuk and Marchuk 2018). Leaching is a widely used method for managing soil salinity, where excess water is applied to flush salts out of the root zone. This process moves the excess salts deeper into the soil profile (Rhoades 1999). However, this requires sufficient water of good quality and appropriate drainage systems to prevent waterlogging. Another soil management strategy is the use of gypsum (calcium sulfate) which replace sodium ions on the soil exchange complex with calcium ions, thereby improving soil structure and permeability, and facilitating leaching of salts (Oster 1982).

Apart from this approach, selecting salt-tolerant crop varieties is a practical approach to managing salinity. Breeding and biotechnological advances have led

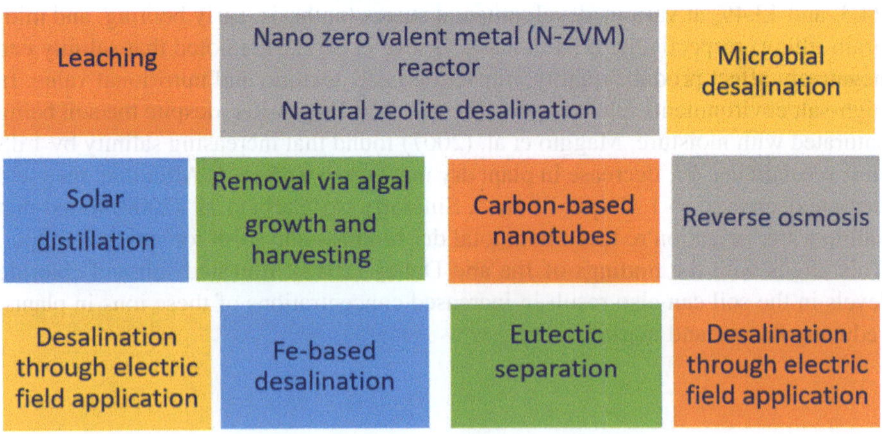

Fig. 2 Desalinization techniques of poor-quality irrigation water

to the development of crop varieties that can tolerate higher salinity levels. For example, certain varieties of barley, wheat, and rice have been developed to grow in saline conditions (Munns et al. 2006). Crop rotation and diversifying crops can also help to maintain the soil salinity. Growing salt-tolerant crops in rotation with less tolerant ones can help reduce the buildup of salts in the soil. Additionally, proper timing and application of fertilizers can minimize the risk of exacerbating salinity problems (Grattan and Grieve 1998). An integrated approach combining various management practices can be more effective. For example, employing precision irrigation techniques like drip or subsurface irrigation can reduce the amount of water applied, thereby minimizing salt accumulation in the root zone (Ravikovitch 1992). Monitoring soil and water salinity regularly using electrical conductivity meters and soil testing can help farmers make informed decisions about irrigation and soil management practices (Ayers and Westcot 1985). Implementing controlled drainage systems can also help manage the water table and prevent salinization of the root zone (Skaggs and van Schilfgaarde 1999).

3.2 Accumulation of Toxic Heavy Metals in Crops Irrigated with Poor Quality Irrigation Water

In recent years, scarcity of water forces to use of poor-quality irrigation water including waste water. Nowadays, 35 elements are designated as heavy metals, with 23 being the most dangerous metals (Hasanpour and Hatami 2020) posing severe risks to agricultural productivity and environmental health as depicted in Fig. 3. Heavy metal toxicity is determined by a variety of factors, Heavy metals, which can infiltrate water systems from industrial and consumer waste as well as acidic rain, play a significant biological role and vary in their impact throughout the life cycle of organisms. Acidic rain can break down soils, releasing these metals into streams, lakes, rivers, and groundwater (Nafchi et al. 2021). When industrial effluent containing heavy metals is used for irrigation, these metals accumulate in the soil, posing health risks to individuals who consume contaminated crops. The excessive deposition of heavy metals in agricultural soils due to such irrigation practices raises serious concerns for food safety and endangers public health, since polluted crops and vegetables constitute a serious problem for food safety and quality (Radwan and Salama 2006). Singh et al. (2010) and Nagajyoti et al. (2010) have highlighted the negative impacts of heavy metals on crop development and soil health, emphasizing the significance of frequent (Leaching 2003). Moreover, Khan et al. (2008) revealed that the consumption of crops contaminated with heavy metals poses severe health risks to humans, (kidney damage, neurological disorders, and cancer). Heavy metals may drastically influence fundamental physiological functions in plants, such as photosynthesis, respiration, and enzyme activity. Plant physiology results heavy metal uptake, with larger amounts detected in leaves and roots than in stems and fruits

(Yusuf et al. 2003; Sinha et al. 2005). For example, cadmium can replace zinc in critical biochemical pathways, interfering with plant metabolism and growth (Sanita di Toppi and Gabbrielli 1999). Heavy metal-contaminated irrigation water might pose significant health concerns to both humans and animals (Ostad-Ali-Askari et al. 2017). For example, Arefin et al. (2016) examined surface water from the Turag River in Bangladesh and discovered chromium (Cr) concentrations that above the maximum allowed limit (MAL) of 0.1 mg/L. The Food and Agriculture Organization (FAO) specifies a MAL of 5 mg/L for lead (Pb) in irrigation water (Ayers and Westcot 1985). Ngweme et al. (2020) examined irrigation water in the peri-urban districts of Kinshasa, Democratic Republic of Congo, and found lead (Pb) concentrations below the FAO's maximum acceptable limit (MAL) of 5 mg/L. Similarly, Ahmed et al. (2019) and Sagagi et al. (2022) recorded Pb levels in irrigation water at 2.04 mg/L, which is still within the MAL as set by FAO. Moreover, Giller et al. (1998) investigated that heavy metals can alter soil microbial communities and reduce soil fertility. They can bind to soil particles, making them less available for plant uptake but potentially more harmful in the long term as they accumulate in the soil. High salinity levels in industrial wastewater can lead to soil degradation, affecting its structure and ability to retain water and nutrients. Grattan and Grieve (1998) discussed how excessive salinity can cause osmotic stress in plants, reducing their growth and productivity.

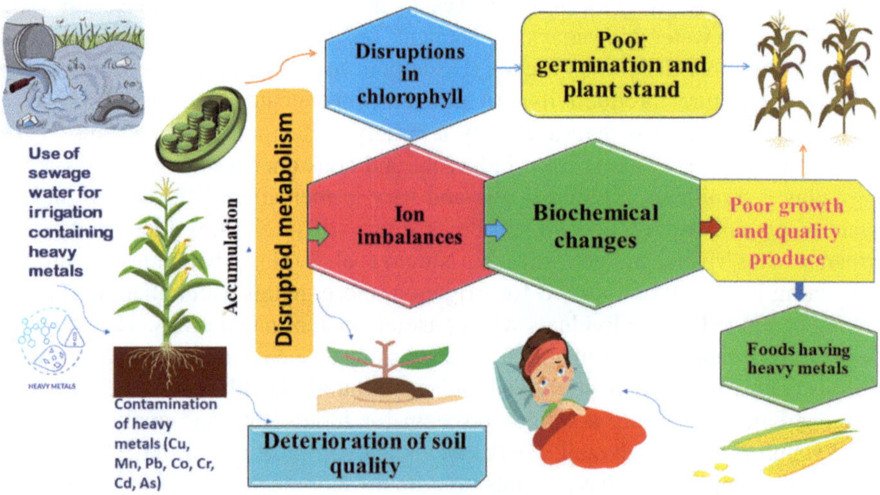

Fig. 3 Effect of heavy metals in crops irrigated with poor quality irrigation water

3.2.1 Effects of Heavy Metals Contaminated Poor Quality Irrigation Water on Crop Yield and Quality

Multiple studies in India have indicated that using wastewater for irrigating crops results in elevated heavy metal concentrations in both the soil and the agricultural products grown in that soil (Rattan et al. 2005; Arora et al. 2008) and can accumulate over time, leading to elevated levels in the soil (McLaughlin et al. 1999). As plants absorb water from this contaminated soil, they also take up these heavy metals, resulting in their presence in the crops themselves. The lasting presence of heavy metals in the soil can negatively impact soil health and its capacity to promote healthy plant growth over time, as it can inhibit seed germination and early seedling development (Fig. 3). Plants subjected to high concentrations of heavy metals frequently show signs of stunted growth, chlorosis, and necrosis (Shahid et al. 2017), which can lead to significant reductions in crop yields. For example, toxicity from cadmium and lead can hinder root development and nutrient absorption, resulting in decreased biomass and grain production (Khan et al. 2015a, b).

According to Orosun et al. (2023), the average cadmium concentration in spinach and cabbage was 5.2 and 2.4 times higher than the WHO (2011) threshold value of 0.2 mg kg^{-1} for vegetable samples. Tiwari et al. tested samples of vegetables continuously irrigated with effluents elevates heavy metal concentrations (mg kg) in soil (Cr, Fe, As, Cu, Mn, Ni, Cd, Pb and Zn (12.8, 278, 9.6, 18.4, 94, 15, 19.2, 13, 85) respectively. This is because soil that is watered with tube wells has a lower concentration of metals than land that is irrigated with wastewater. The accumulation of heavy metals in crops can affect their quality by altering taste, texture, and nutritional content. Furthermore, crops contaminated with heavy metals can pose significant health risks to consumers, raising food safety concerns (Alloway 2013). The accumulation of contaminants like heavy metals can compromise crop quality, rendering them unsafe for consumption. Muchuweti et al. (2006) investigated the heavy metal content in vegetables irrigated with mixtures of wastewater and sewage sludge in Zimbabwe, highlighting the health risks associated with consuming such produce. Additionally, Tiwari et al. found that plants watered with tube well water in the Vadodara had higher transfer factors for metals than effluent irrigated crops.

3.2.2 Remediation Strategies

Remediation of contaminated soils and water is essential for long-term management of heavy metal contamination. Sorption, precipitation, complexation, or metal balance reduction are methods for removal of heavy metals as suggested by Rana et al. (2021). Phytoremediation (Phang et al. 2024) by using hyperaccumulator plants for removal of heavy metals from contaminated soils is an effective and environmentally friendly remediation strategy. These plants can be harvested and disposed of safely, reducing heavy metal levels in the soil over time (Salt et al. 1995). Turkmen et al. (2022) revealed that commercial and polymeric adsorbents shown a good capacity for absorbing and eliminating heavy metal ions from wastewater. Different adsorbent is identified by researchers which are prepared from waste materials having high heavy metals absorbance capacity such as Orange peel (Ajmal et al. 2000), Pecan shells activated carbon (Bansode et al. 2003), Corn cope and husk (Igwe et al. 2005), Coconut shell charcoal (Babel and Kurniawan 2004), Spirogyra (green alga) (Gupta et al. 2006), Modified rice hull (Tang et al. 2003), Oedogonium species (Gupta and Rastogi 2008), Ecklonia maxima marine alga (Feng and Aldrich 2004) etc. Cryogels (Turkmen et al. 2022), modified biopolymers such as crosslinked chitosan, Alumina/chitosan composite, Crosslinked starch gel (Crini 2005) and ultrafiltration (Barakat 2011), modified natural materials (modified zeolite (Nah et al. 2006); calcined phosphate (Moufliha et al. 2005); Clay/poly(methoxyethyl) acrylamide (Şolener et al. 2008), photocatalysis are effective approach for the removal of heavy metal ions. Numerous studies have explored cryogels as innovative bioadsorbents for the removal of heavy metals from wastewater. For instance, Purohit et al. (2020) developed a low-cost, water-stable transition metal oxide-doped crosslinked chitosan cryogel created specifically for the very sensitive removal of arsenic ions from drinking water. Soil washing is also effective remedial measures which involves using of chemical solutions to extract heavy metals from contaminated soils. This method can be effective for heavily contaminated sites, although it may require subsequent soil amendment to restore soil health (Zheng et al. 2022). Mixing contaminated soil with binding agents to reduce the mobility of heavy metals can be effective approach for heavy metal removal. In, solidification contaminants convert into a stable solid form, minimizing their bioavailability and environmental impact (Conner and Hoeffner 1998).

3.3 Pathogenic Microorganisms in Irrigation Water

Pathogen contamination in irrigation water is a critical issue that poses significant risks to both agricultural productivity and public health. Pathogens (bacteria, viruses, protozoa, and helminths) can contaminate irrigation water through various sources and pathways and through untreated or inadequately treated wastewater (Fig. 4). Untreated home wastewater has significant amounts of microbes and pathogens. Freshwater and estuarine sediments are vital microbial habitats that can play a significant role in water contamination; however, they have received far less attention than warranted. Their potential impact on water quality highlights the need for more comprehensive research and monitoring (Pachepsky et al. 2011). Kay et al. (2008) found 7.6×10^{10} cfu per 100 mL of untreated wastewater, while Howard et al. (2004) found average concentrations of *Cryptosporidium* and *Salmonella* at 2.6–3.2×10^2 per 100 mL. These pathogens can infect plants, reducing their vigor and productivity, and present a direct health risk to consumers (Fig. 4), particularly when crops are consumed raw (Shannon et al. 2008). Pathogen contamination in irrigation water poses significant challenges to agricultural productivity and food safety. Pathogens can adhere to the surfaces of fruits and vegetables or infiltrate plant tissues through

Fig. 4 Pathogenic contamination of poor-quality irrigation water (waste water)

water uptake. This contamination can occur during irrigation, especially with overhead systems that directly wet the plant surfaces (Steele and Odumeru 2004) rrigating with pathogen-contaminated water can result in the buildup of pathogens in the soil. These pathogens may remain in the soil for long periods, posing continuous risks to crops cultivated in contaminated areas (Brookes et al. 2004). Consumption of raw or minimally processed contaminated produce can lead to foodborne illnesses. Pathogens (E. coli, Salmonella, and norovirus) are common culprits in foodborne outbreaks linked to contaminated irrigation water (Berger et al. 2010).

3.3.1 Effects of Pathogenic Microorganisms of Poor Quality Irrigation Water on Crop Yield and Quality

Certain pathogens can cause diseases in plants, leading to reduced growth and yield. For example, waterborne bacterial pathogens can cause wilts, blights, and rots in various crops (Timmer et al. 2000). Pathogen contamination can lead to spoilage and reduce the marketability of fresh produce. The presence of pathogens can also necessitate more rigorous washing and processing, potentially affecting the texture and flavor of the produce (Beuchat 1996). Outbreaks of foodborne illnesses linked to contaminated irrigation water can lead to significant economic losses due to product recalls, loss of consumer confidence, and legal liabilities (Painter et al. 2013).

3.3.2 Management Strategies

Treating irrigation water to remove pathogens can include filtration, chlorination, UV treatment, and other disinfection methods (WHO 2006). Implementing GAPs, such as ensuring proper manure management, preventing runoff, and maintaining clean irrigation equipment, can help minimize pathogen contamination (FDA 2015). Regular monitoring of water quality and adherence to regulatory standards for pathogen levels in irrigation water can help prevent contamination (EPA 2012).

3.4 Nutrient Toxicity/Insufficiency

Nutrients are vital for plant growth, however, their excessive presence in irrigation water can lead to serious environmental issues. Essential nutrients are beneficial in suitable quantities but can act as contaminants when their concentrations surpass the optimal levels for crop uptake. Excessive nutrients in irrigation water can affect crop production and environmental quality through various mechanisms. Elevated nutrient levels, particularly nitrogen (Vitousek et al. 1997), and phosphorus (Smith and Schindler 2009), can result in the eutrophication of water bodies. This process triggers excessive growth of algae and aquatic plants, resulting in depletion of oxygen levels in the water, and negatively impacting marine life and deteriorating water quality (Smith et al. 1999a, b). Continuous irrigation with nutrient-rich water can lead to nutrient imbalances in the soil. Excessive nitrogen, for example, can lead to soil acidification, while high phosphorus levels can cause nutrient lock-up, reducing the availability of other essential nutrients (Sharpley et al. 2003). Overabundance of certain nutrients can cause toxicity in plants leads to lush growth at the expense of fruit and flower development, making plants more susceptible to pests and diseases (Marschner 2011).

3.4.1 Effects of Nutrient Toxicity/Insufficiency in Crop Yield and Quality

Nutrient contamination can have both positive and negative effects on crop yield and quality, depending on the levels and balance of nutrients. Initially, increased nutrient levels may enhance crop growth and yields. However, sustained excessive nutrient levels can lead to nutrient imbalances, reducing overall crop productivity. For example, too much nitrogen can promote vegetative growth while inhibiting fruiting, leading to lower yields of crops that are grown for their fruits or seeds (Fageria 2009). Excessive nutrients can affect the quality of agricultural produce. High nitrogen levels can reduce the quality of fruits and vegetables by affecting their taste, texture, and nutritional content. For instance, excessive nitrogen can decrease the sugar content in fruits, making them less palatable (FAO 2013).

3.4.2 Management Strategies

Utilizing precision agriculture techniques can help optimize the application of fertilizers, ensuring that nutrients are applied by following 4 R principles to minimize runoff (Bongiovanni and Lowenberg 2004). Using SSNM techniques can improved the nutrient imbalance problems. Establishing buffer zones with vegetation around agricultural fields can help absorb excess nutrients before they reach water bodies, reducing the risk of nutrient contamination (Mankin et al. 2007). Treating irrigation water to remove excess nutrients through processes such as constructed wetlands or bioreactors can help reduce nutrient loads (Vymazal 2007).

3.5 Organic Contaminants in Irrigation Water

Organic contaminants in irrigation water, including pesticides, herbicides, pharmaceuticals, and industrial chemicals, can significantly impact crop yield and quality.

3.5.1 Effects of Organic Contaminants Present in Irrigation Water on Crop Yield and Quality

Organic contaminants, such as herbicides, pesticides, and decomposing plant and animal tissues, are commonly associated with adverse environmental impacts (McLaughlin et al. 1999). These substances can persist in ecosystems, disrupt biological processes, and pose risks to both terrestrial and aquatic life. These contaminants can disrupt plant physiological processes, inhibit growth, and result in the accumulation of harmful residues in edible plant parts. Cedergreen (2008) studied that organic contaminants can impede plant growth by interfering with essential biochemical pathways. For example, herbicides designed to kill weeds can also harm crops if present in irrigation water at high concentrations. These chemicals can inhibit photosynthesis, respiration, and nutrient uptake, leading to stunted growth, reduced leaf area, and lower biomass production. For example, atrazine blocks photosystem II, preventing plants from converting sunlight into chemical energy. Geoffroy et al. (2004) revealed that some organic contaminants affect cellular respiration by inhibiting mitochondrial enzymes. This disruption in the energy production process can lead to decreased growth rates and vitality in plants. For instance, certain fungicides and insecticides can inhibit key enzymes involved in the Krebs cycle and electron transport chain, reducing ATP production and causing energy deficits in plant cells. Pesticides and pharmaceuticals in irrigation water can alter root membrane permeability, disrupt ion channels, and interfere with nutrient transport proteins (Mackay et al. 2006). This can lead to nutrient deficiencies, poor growth, and reduced yield (Gong et al. 2009). Additionally, contaminants can disrupt root development, further impairing nutrient and water uptake and contributing to lower yields (Smith et al. 1999a, b). Kookana et al. (2014) in their study found that organic contaminants can accumulate in crops, affecting their quality and safety. Residues of pesticides and pharmaceuticals can remain in the edible parts of plants, posing health risks to consumers (Mackay et al. 2006). This contamination can lead to stringent regulations and market rejections, affecting the economic viability of affected crops Moreover, some contaminants can alter the taste, texture, and nutritional value of crops, reducing their market appeal and consumer acceptance (Kumar et al. 2005a, b).

3.5.2 Mitigation Strategies

Addressing organic contamination in irrigation water requires a comprehensive approach that includes regulatory measures, technological solutions, and sustainable agricultural practices. Governments and environmental agencies must enforce stringent regulations to control the release of organic contaminants into the environment. Policies should mandate the proper treatment of industrial effluents and urban runoff to reduce the load of organic pollutants entering water bodies. Various technologies introduced for removal of organic contaminants from irrigation water. Advanced oxidation processes (AOPs), such as ozonation and photocatalysis, can degrade organic pollutants into less harmful substances. Biological treatment methods, such as constructed wetlands and bio filtration, use microbial activity to break down organic contaminants. These methods can be effective for treating both industrial effluents and wastewater used for irrigation (Matamoros and Bayona 2006). Improving soil health can mitigate the impact of organic contaminants on crops. Practices such as organic farming, crop rotation, and the use of organic amendments can enhance soil microbial activity, promoting the degradation of organic pollutants. Biochar, a form of charcoal added to soil, can adsorb organic contaminants, reducing their bioavailability and uptake by plants (Ahmad et al. 2014). Phytoremediation is a cost-effective and environmentally friendly strategy that can be integrated into agricultural systems (Pilon-Smits 2005) and involves using plants to remove, degrade, or stabilize soil and water organic contaminants. Certain plants have the ability to take up and break down organic pollutants, reducing their concentrations in the environment. By promoting the use of natural predators, crop rotation, and resistant crop varieties, IPM can help for minimization of the application of organic pesticides, thereby reducing their presence in irrigation water (Kogan 1998). IPM reduces reliance on chemical pesticides by using a combination of biological, cultural, and mechanical control methods.

4 Conclusion

Effective management of irrigation water quality involves a multifaceted approach integrating advanced technologies, sustainable agricultural practices, and stringent regulatory measures. Addressing the issue of poor-quality irrigation water is essential not only for maintaining agricultural productivity but also for ensuring food safety and long-term environmental sustainability. Implementing effective mitigation strategies, the agricultural sector can better adapt to the challenges posed by declining water quality, thereby supporting global food security. Educating farmers, industry stakeholders, and the public about the risks of heavy metal contamination and the importance of safe practices is crucial for effective mitigation. Training programs for farmers can provide knowledge on the safe use of fertilizers, pesticides and wastewater in agriculture.

References

Acosta-Motos JR, Ortuño MF, Bernal-Vicente A, Diaz-Vivancos P, Sanchez-Blanco MJ, Hernandez JA (2017) Plant responses to salt stress: adaptive mechanisms. Agronomy 7(1):18

Ahmad M, Lee SS, Dou X, Mohan D, Sung JK, Yang JE, Ok YS (2014) Effects of pyrolysis temperature on soybean Stover and peanut shell-derived biochar properties and TCE adsorption in water. Biores Technol 118:536–544

Ahmad P, Sharma S (2008) Salt stress and phyto-biochemical responses of plants. Plant Soil Environ 54:89–99. https://doi.org/10.17221/2774-PSE

Ahmed M, Matsumoto M, Ozaki A, Thinh N, Kurosawa K (2019) Heavy metal contamination of irrigation water, soil, and vegetables and the difference between dry and wet seasons near a multi-industry zone in Bangladesh. Water 11(3):583. https://doi.org/10.3390/w11030583

Ajmal M, Rao RAK, Ahmad R, Ahmad J (2000) Adsorption studies on Citrus reticulata (fruit peel of orange): removal and recovery of Ni (II) from electroplating wastewater. J Hazard Mater 79(1–2):117–131

Al-Harbi AR, Wahb-Allah MA, Al-Omran AM (2009) Effects of salinity and irrigation management on growth and yield of tomato grown under greenhouse conditions. Acta Hort 807:201–206

Alloway BJ (2013) Heavy metals in soils: trace metals and metalloids in soils and their bioavailability. Springer Science and Business Media

Arefin MT, Rahman MM, Wahid-U-Zzaman M, Kim JE (2016) Heavy metal contamination in surface water used for irrigation: functional assessment of the Turag River in Bangladesh. Appl Biol Chem 59(1):83–90. https://doi.org/10.3839/jabc.2016.015

Arora MB, Kiran S, Rani A, Rani B, Mittal N (2008) Heavy metal accumulation in vegetables irrigated with water from different sources. Food Chem 111:811–815

Ashraf M, Foolad MR (2007) Roles of glycine betaine and proline in improving plant abiotic stress resistance. Environ Exp Bot 59(2):206–216

Ayers RS, Westcot DW (1985) Water quality for agriculture. FAO irrigation and drainage paper no. 29 Rev. 1, Rome, 1985. Food and Agriculture Organization of the United Nations, Reprinted 1989, 1994. https://www.fao.org/3/t0234e/t0234e00.htm

Babel S, Kurniawan TA (2004) Cr (VI) removal from synthetic wastewater using coconut shell charcoal and commercial activated carbon modified with oxidizing agents and/or chitosan. Chemosphere 54(7):951–967. https://doi.org/10.1016/j.chemosphere.2003.10.001

Bansode RR, Losso JN, Marshall WE, Rao RM, Portier RJ (2003) Adsorption of metal ions by pecan shell-based granular activated carbons. Biores Technol 89(2):115–119

Barakat MA (2011) New trends in removing heavy metals from industrial wastewater. Arab J Chem 4(4):361–377. https://doi.org/10.1016/j.arabjc.2010.07.019
Berger CN, Sodha SV, Shaw RK, Griffin PM, Pink D, Hand P, Frankel G (2010) Fresh fruit and vegetables as vehicles for the transmission of human pathogens. Environ Microbiol 12(9):2385–2397
Beuchat LR (1996) Pathogenic microorganisms associated with fresh produce. J Food Prot 59(2):204–216
Bhargava A, Bhargava P (2019) Organic farming: concept, applications and perspectives. In: Organic farming. Springer, Singapore, pp 23–40
Bolan NS, Park JH, Robinson B, Naidu R, Huh KY (2021) Phytotoxicity of heavy metals in soils. In: Phytoremediation. Springer, Cham, pp 117–142
Bongiovanni R, Lowenberg-Deboer J (2004) Precision agriculture and sustainability. Precision Agric 5(4):359–387
Brookes PC, McGrath SP, Klein DA (2004) Phosphatase activity in soil for monitoring biological response to heavy metal pollution. Soil Biol Biochem 17(1):41–47
Carpenter SR, Caraco NF, Correll DL, Howarth RW, Sharpley AN, Smith VH (1998) Nonpoint pollution of surface waters with phosphorus and nitrogen. Ecol Appl 8(3):559–568
Carvalho FP (2017) Pesticides, environment, and food safety. Food Energy Secur 6(2):48–60
Cedergreen N (2008) Herbicides can stimulate plant growth. Weed Res 48(5):429–438
Chee-Sanford JC, Mackie RI, Koike S, Krapac IG, Lin YF, Yannarell AC, Maxwell S, Aminov RI (2009) Fate and transport of antibiotic residues and antibiotic resistance genes following land application of manure waste. J Environ Qual 38(3):1086–1108
Conner JR, Hoeffner SL (1998) A critical review of stabilization/solidification technology. Crit Rev Environ Sci Technol 28(4):397–462. https://doi.org/10.1080/10643389891254250
Conway D, Nicholls RJ, Brown S (2019) The effects of climate change on coastal groundwater and its implications for water resources and the coastal environment. Sci Total Environ 678:547–561
Crini G (2005) Recent developments in polysaccharide-based materials used as adsorbents in wastewater treatment. Prog Polym Sci 30(1):38–70
Cuartero J, Bolarin MC, Asins MJ, Moreno V (2006) Increasing salt tolerance in the tomato. J Exp Bot 57(5):1045–1058
Dodds WK, Smith VH (2016) Nitrogen, phosphorus, and eutrophication in streams. Inland Waters 6(2):155–164
Ehtaiwesh FA, Rashed HF (2020) Growth and yield responses of libyan hard wheat (Triticum durum Desf) genotypes to salinity stress. Zawia University Bulletin 22:33–58
El Sabagh A, Islam MS, Skalicky M, Ali Raza M, Singh K, Hossain M (2021) Salinity stress in wheat (Triticum aestivum L.) in the changing climate: adaptation and management strategies. Front Agron 3. https://doi.org/10.3389/fagro.2021.661932
EPA (2012) Recreational water quality criteria. United States Environmental Protection Agency
Fageria NK (2009) The use of nutrients in crop plants. CRC Press
FAO (2011) Food and agriculture organization of the United Nations. The State of the World's Land and Water Resources for Food and Agriculture
FAO (2013) Food and agriculture organization of the United Nations. Crop yield response to water
Farber E, Vengosh A, Gavrieli I, Marie A, Bullen TD, Mayer B, Holtzman R, Segal M, Shavit U (2005) Management scenarios for the Jordan River salinity crisis. Appl Geochem 20(11):2138–2153. https://doi.org/10.1016/j.apgeochem.2005.07.007
FDA (2015) U.S. food and drug administration. Standards for the Growing, Harvesting, Packing, and Holding of Produce for Human Consumption
Feng D, Aldrich C (2004) Adsorption of heavy metals by biomaterials derived from the marine alga Ecklonia maxima. Hydrometallurgy 73(1–2):1–10
Fergusson JE (1990) The heavy elements: chemistry, environmental impact, and health effects. Pergamon Press
Flores P, Botella MA, Martinez V, Cerda A (2016) The influence of saline stress on yield and quality of some agricultural crops. In: Plant nutrition. Springer, Dordrecht, pp 77–88

Flowers TJ, Yeo AR (1995) Breeding for salinity resistance in crop plants: where next? Aust J Plant Physiol 22(6):875–884

Garcia-Delgado C, Barba-Brioso C, Delgado J, Sierra MJ (2019) Effects of two-phase olive mill waste on the microbial activity of soil and its ability to degrade phenolic compounds. Sci Total Environ 687:398–405

Geoffroy L, Frankart C, Eullaffroy P (2004) Comparison of different physiological parameter responses in Lemna minor and Scenedesmus obliquus exposed to herbicide flumioxazin. Environ Pollut 131(2):233–241

Gerba CP, Smith JE (2005) Sources of pathogenic microorganisms and their fate during land application of wastes. J Environ Qual 34(1):42–48

Gessel PD, Hansen NC, Goyal SM, Johnston LJ, Webb J (2004) Persistence of zoonotic pathogens in surface soil treated with different rates of liquid pig manure. Appl Soil Ecol 25(3):237–243

Giller KE, Witter E, McGrath SP (1998) Toxicity of heavy metals to microorganisms and microbial processes in agricultural soils: a review. Soil Biol Biochem 30(10–11):1389–1414

Gong P, Wilke BM, Alef K (2009) Soil enzymes and microorganisms in three types of contaminated soil. Bull Environ Contam Toxicol 83(6):823–828

Gorga M, Insa S, Petrovic M, Barcelo D (2013) Occurrence and spatial distribution of EDCs and related compounds in waters and sediments of Iberian rivers. Sci Total Environ 441:92–105

Grattan SR, Grieve CM (1998) Salinity-mineral nutrient relations in horticultural crops. Sci Hortic 78(1–4):127–157

Grieve CM, Poss JA, Grattan SR (2012) Combining salinity and nutrient management in forage crops: some agronomic and physiological considerations. J Plant Nutr Soil Sci 175(6):832–846

Guillette LJ, Gunderson MP (2001) Alterations in the development of the reproductive system of wildlife exposed to endocrine-disrupting contaminants. Environ Health Perspect 109(6):631–638

Gupta B, Huang B (2014) Mechanism of salinity tolerance in plants: physiological, biochemical, and molecular characterization. Int J Genomics 701596

Gupta VK, Rastogi A (2008) Biosorption of lead (II) from aqueous solutions by non-living algal biomass Oedogonium sp. and Nostoc sp. a comparative study. Colloids Surfaces B: Biointerfaces 64(2):170–178

Gupta VK, Rastogi A, Saini VK, Jain N (2006) Biosorption of copper (II) from aqueous solutions by Spirogyra species. J Colloid Interface Sci 296(1):59–63

Hamelin S, Boulos L, Theriault M, Gaboury I (2013) Impact of urbanization on the contribution of non-point sources of phosphorus to streams. J Environ Manag 128:665–673

Hamilton AJ, Stagnitti F, Premier R, Boland AM, Hale G (2020) Pathogen and chemical transport in irrigation water: a case study from a peach orchard. Environ Monit Assess 186(1):1–14

Hasanpour M, Hatami M (2020) Application of three dimensional porous aerogels as adsorbent for removal of heavy metal ions from water/wastewater: a review study. Adv Coll Interface Sci 284:102247. https://doi.org/10.1016/j.cis.2020.102247

Howard I, Espigares E, Lardelli P, Martin JL, Espigares M (2004) Evaluation of microbiological and physicochemical indicators for wastewater treatment. Environ Toxicol 19(3):241–249

Igwe JC, Ogunewe DN, Abia AA (2005) Competitive adsorption of Zn (II), Cd (II) and Pb (II) ions from aqueous and non-aqueous solution by maize cob and husk. Afr J Biotech 4(10):1113

Jayaraj R, Megha P, Sreedev P (2016) Organochlorine pesticides, their toxic effects on living organisms and their fate in the environment. Interdiscip Toxicol 9(3–4):90–100

Jha P, Dubey SK (2012) Effect of water quality on yield and water use efficiency of irrigated wheat crop. Indian Soil Conserv 40(3):236–239

Jin H, Zhou Q, Zhang L (2012) Risk assessment of pesticide residues in honey from the People's Republic of China. Food Addit Contam Part A 29(2):231–240

Kamran M, Parveen A, Ahmar S, Malik Z, Hussain S, Chattha MS (2020) An overview of hazardous impacts of soil salinity in crops, tolerance mechanisms, and amelioration through selenium supplementation. Int J Mol Sci 21(1):148. https://doi.org/10.3390/ijms21010148

Katerji N, Van Hoorn JW, Hamdy A, Mastrorilli M (2003) Salinity effect on crop development and yield, analysis of salt tolerance according to several classification methods. Agric Water Manag 62(1):37–66

Kaushal SS, Groffman PM, Likens GE, Belt KT, Stack WP, Kelly VR, Band LE, Fisher GT (2005) Increased salinization of fresh water in the northeastern United States. Proc Natl Acad Sci 102(38):13517–13520

Kay D, Crowther J, Stapleton CM, Wyer MD, Fewtrell L, Edwards A, Francis CA, McDonald AT, Watkins J, Wilkinson J (2008) Faecal indicator organism concentrations in sewage and treated effluents. Water Res 42:442–554

Khan MN, Mobin M, Abbas ZK, Alamri SA (2015a) Fertilizers and their contaminants in soils, surface and groundwater. Encycl Anthr 5:225–240

Khan S, Cao Q, Zheng YM, Huang YZ, Zhu YG (2008) Health risks of heavy metals in contaminated soils and food crops irrigated with wastewater in Beijing, China. Environ Pollut 152(3):686–692

Khan S, Hesham AEL, Qiao M, Rehman S, He Z (2015b) Effects of Cd and Pb on soil microbial community structure and activities. Environ Sci Pollut Res 22(1):122–134

Khorobrykh S, Havurinne V, Mattila H, Tyystjarvi E (2020) Oxygen and ROS in Photosynthesis. Plants 9(1):91. https://doi.org/10.3390/plants9010091

Kogan M (1998) Integrated pest management: historical perspectives and contemporary developments. Annu Rev Entomol 43(1):243–270

Kookana RS, Williams M, Boxall ABA, Larsson DGJ, Gaw S, Choi K, Yamamoto H, Thatikonda S, Zhu YG, Carriquiriborde P (2014) Potential ecological footprints of active pharmaceutical ingredients: an examination of risk factors in low-, middle-, and high-income countries. Philos Trans R Soc b: Biol Sci 369(1656):20130586

Kumar A, Sharma S, Ghosh S (2005a) Water-holding capacity and textural properties of low-fat ground pork patties with carrageenan, oat fiber, and konjac blends. J Food Sci 70(8):C473–C478

Kumar R, Sarmah AK, Padhy PK (2005b) Water quality of the sewage-fed River Kali and its relation to phytoplankton diversity. Int J Environ Sci Technol 2(4):437–444

Lamm FR, Ayars JE (2009) Micro irrigation for crop production: design, operation, and management. Elsevier

Lapworth DJ, Baran N, Stuart ME, Ward RS (2012) Emerging organic contaminants in groundwater: a review of sources, fate and occurrence. Environ Pollut 163:287–303

Leaching EA (2003) Environmental chemistry of soils. Soil Sci Soc Am J 67(6):1980–1981

Lee MK, Van Iersel MW (2008) Sodium chloride effects on growth, morphology, and physiology of chrysanthemum. HortScience 43:1888–1891

Lu K, Guo Z, Di S, Lu Y, Muhammad IAR, Rong C (2023) OsMFT1 inhibits seed germination by modulating abscisic acid signaling and gibberellin biosynthesis under salt stress in rice. Plant Cell Physiol 64(6):674–685. https://doi.org/10.1093/pcp/pcad029

Maas EV, Hoffman GJ (1977) Crop salt tolerance-current assessment. J Irrig Drain Div 103(2):115–134

Machado RMA, Serralheiro RP (2017) Soil salinity: effect on vegetable crop growth. Management practices to prevent and mitigate soil salinity. Horticulturae 3(2):30

Mackay D, Mackay D, Paterson S (2006) Evaluating the multimedia fate of organic chemicals: a level III fugacity model. Environ Sci Technol 25(3):427–436

Maggio A, Raimondi G, Martino A, Pascal SD (2007) Salt stress response in tomato beyond the salinity tolerance threshold. Environ Exp Bot 59:276–282

Mahajan S, Tuteja N (2005) Cold, salinity and drought stresses: an overview. Arch Biochem Biophys 444(2):139–158

Malakar A, Snow DD, Ray C (2019) Irrigation water quality-a contemporary perspective. Water 11(7):1482

Mankin KR, Ngandu DM, Barden CJ, Hutchinson SL, Geyer WA (2007) Grass-shrub riparian buffer removal of sediment, phosphorus, and nitrogen from simulated runoff. J Am Water Resour Assoc 43(5):1108–1116

Marchuk A, Marchuk S, Bennett J, Eyres M, Scott E (2014) An alternative index to ESP to explain dispersion occurring in Australian soils when Na content is low. In: 2014 national soils conference: proceedings of the 2014 national soils conference soil science Australia, Melbourne, Australia

Marchuk S, Marchuk A (2018) Effect of applied potassium concentration on clay dispersion, hydraulic conductivity, pore structure and mineralogy of two contrasting Australian soils. Soil Tillage Res 182:35–44

Marschner P (2011) Marschner's mineral nutrition of higher plants. Academic Press

Matamoros V, Bayona JM (2006) Elimination of pharmaceuticals and personal care products in subsurface flow constructed wetlands. Environ Sci Technol 40(18):5811–5816

McLaughlin MJ, Parker DR, Clarke JM (1999) Metals and micronutrients–food safety issues. Field Crop Res 60(1–2):143–163

Minhas PS, Ramos TB, Ben-Gal A, Pereira LS (2020) Coping with salinity in irrigated agriculture: crop evapotranspiration and water management issues. Agric Water Manag 227. Article 105832

Mojiri A (2011) Effects of municipal wastewater on physical and chemical properties of saline soil. J Biol Environ Sci 5(13):71–76

Mostafalou S, Abdollahi M (2017) Pesticides: an update of human exposure and toxicity. Arch Toxicol 91(2):549–599

Mouflih M, Aklil A, Sebti S (2005) Removal of lead from aqueous solutions by activated phosphate. J Hazard Mater 119(1–3):183–188

Muchuweti M, Birkett JW, Chinyanga E, Zvauya R, Scrimshaw MD, Lester JN (2006) Heavy metal content of vegetables irrigated with mixtures of wastewater and sewage sludge in Zimbabwe: implications for human health. Agr Ecosyst Environ 112(1):41–48

Munns R (2002) Comparative physiology of salt and water stress. Plant, Cell Environ 25(2):239–250

Munns R, Tester M (2008) Mechanisms of salinity tolerance. Annu Rev Plant Biol 59:651–681

Munns R, James RA, Läuchli A (2006) Approaches to increasing the salt tolerance of wheat and other cereals. J Exp Bot 57(5):1025–1043

Nafchi RF, Samadi-Boroujeni H, Vanani HR, Ostad-Ali-Askari K, Khastar M, Brojeni MK (2021) Laboratory investigation on erosion threshold shear stress of cohesive sediment in Karkheh Dam. Environ Earth Sci 80:681. https://doi.org/10.1007/s12665-021-09984-x

Nagajyoti PC, Lee KD, Sreekanth TVM (2010) Heavy metals, occurrence and toxicity for plants: a review. Environ Chem Lett 8(3):199–216

Nah IW, Hwang KY, Jeon C, Choi HB (2006) Removal of Pb ion from water by magnetically modified zeolite. Miner Eng 19(14):1452–1455

Naifer A, Al-Rawahy SA, Zekri S (2011) Economic impact of salinity: the case of al-batinah in Oman. Int J Agric Res 6:134–142

Negrão S, Schmockel SM, Tester M (2017) Evaluating physiological responses of plants to salinity stress. Ann Bot 119(1):1–11

Ngweme GN, Atibu EK, Alsalah D, Muanamoki PM, Kiyombo GM, Mulaji CK, Otamonga JP, Pote JW (2020) Heavy metal concentration in irrigation water, soil and dietary risk assessment of Amaranthus viridis grown in peri-urban areas in Kinshasa, Democratic Republic of the Congo. Watershed Ecol Environ 2:16–24. https://doi.org/10.1016/j.wsee.2020.07.001

Orosun MM, Nwabachili S, Alshehri RF et al (2023) Potentially toxic metals in irrigation water, soil, and vegetables and their health risks using Monte Carlo models. Sci Rep 13:21220. https://doi.org/10.1038/s41598-023-48489-4

Ostad-Ali-Askari K, Shayannejad M, Eslamian S, Zamani F, Shojaei N, Navabpour B, Majidifar Z, Sadri A, Ghasemi-Siani Z, Nourozi H, Vafaei O, Homayouni SMA (2017) Deficit irrigation: optimization models. In: Handbook of drought and water scarcity: management of drought and water scarcity. CRC Press, pp 375–391. https://doi.org/10.1201/9781315226774-18

Oster JD (1982) Gypsum usage in irrigated agriculture: a review. Fertil Res 3(1):73–89

Oster JD (1994) Irrigation with poor quality water. Agric Water Manag 25(3):271–297

Pachepsky YA, Shelton DR, McLain JE, Patel J, Mandrell RE (2011) Irrigation waters as a source of pathogenic microorganisms in produce: a review. Adv Agron 113:73–138

Painter JA, Hoekstra RM, Ayers T, Tauxe RV, Braden CR, Angulo FJ, Griffin PM (2013) Attribution of foodborne illnesses, hospitalizations, and deaths to food commodities by using outbreak data, United States, 1998–2008. Emerg Infect Dis 19(3):407–415

Panuccio MR, Jacobsen S-E, Akhtar SS, Muscolo A (2014) Effect of saline water on seed germination and early seedling growth of the halophyte quinoa. AoB Plants 6, plu047

Pettygrove GS, Asano T (1985) Irrigation with reclaimed municipal wastewater-a guidance manual. CRC Press

Phang LY, Mingyuan L, Mohammadi M et al (2024) Phytoremediation as a viable ecological and socioeconomic management strategy. Environ Sci Pollut Res 31:50126–50141. https://doi.org/10.1007/s11356-024-34585-z

Pilon-Smits E (2005) Phytoremediation. Annu Rev Plant Biol 56:15–39

Pimentel D (1995) Amounts of pesticides reaching target pests: environmental impacts and ethics. J Agric Environ Ethics 8:17–29

Poustini K, Siosemardeh A (2004) Ion distribution in wheat cultivars in response to salinity stress. Field Crop Res 85:125–133. https://doi.org/10.1016/S0378-4290(03)00157-6

Purohit S, Kumar M, Tanmoyendu C, Kanhaiya C, Yadav L, Satapathi S (2020) Rapid removal of arsenic from water using metal oxide doped recyclable cross-linked chitosan cryogel. SN Appl Sci 2:1–10. https://doi.org/10.1007/s42452-020-2525-6

Qadir M, Oster JD (2004) Crop and irrigation management strategies for saline-sodic soils and waters aimed at environmentally sustainable agriculture. Sci Total Environ 323(1–3):1–19

Radwan MA, Salama AK (2006) Market basket survey for some heavy metals in Egyptian fruits and vegetables. Food Chem Toxicol 44(8):1273–1278. https://doi.org/10.1016/j.fct.2006.02.004

Rameshwaran P, Tepe A, Yazar A, Ragab R (2015) The effect of saline irrigation water on the yield of pepper: experimental and modelling study. Irrig Drain 64:41–49

Rana L, Kumar N, Jha CK, Kumar A, Meena SK, Nandini S, Singh AK, Kumar M (2021) Bioremediation of heavy metals polluted soil. Adv Agric Technol Plant Sci 4(2):180065

Rattan RK, Datta SP, Chhonkar PK, Suribabu K, Singh AK (2005) Long term impact of irrigation with sewage effluents on heavy metal contents in soils, crops and ground water: a case study. Agr Ecosyst Environ 109:310–322

Ravikovitch S (1992) The use of saline water for irrigation. In: Szabolcs I (eds) Soils and salinization hazard. Developments in soil science, vol 22. Elsevier, Amsterdam, pp 337–362

Rengasamy P (2006) World salinization with emphasis on Australia. J Exp Bot 57(5):1017–1023

Rhoades JD (1999) Use of saline drainage water for irrigation. In: Agricultural drainage. American Society of Agronomy, Crop Science Society of America, Soil Science Society of America, pp 615–657

Rhoades JD, Kandiah A, Mashali AM (1992) The use of saline waters for crop production. FAO Irrigation and Drainage Paper 48

Sagagi BS, Bello AM, Danyaya HA (2022) Assessment of accumulation of heavy metals in soil, irrigation water, and vegetative parts of lettuce and cabbage grown along Wawan Rafi, Jigawa State, Nigeria. Environ Monit Assess 194:699. https://doi.org/10.1007/s10661-022-10360-w

Salt DE, Blaylock M, Kumar NP, Dushenkov V, Ensley BD, Chet I, Raskin I (1995) Phytoremediation: a novel strategy for the removal of toxic metals from the environment using plants. Nat Biotechnol 13(5):468–474. https://doi.org/10.1038/nbt0595-468

Sanita di Toppi L, Gabbrielli R (1999) Response to cadmium in higher plants. Environ Exp Bot 41(2):105–130

Schwarzenbach RP, Egli T, Hofstetter TB, Gunten UV, Wehrli B (2006) Global water pollution and human health. Annu Rev Environ Resour 35(1):109–136

Shahid M, Dumat C, Khalid S, Schreck E, Xiong T, Niazi NK (2017) Foliar heavy metal uptake, toxicity and detoxification in plants: a comparison of foliar and root metal uptake. J Hazard Mater 325:36–58

Shannon MC, Grieve CM, Francois LE (2008) Whole-plant response to salinity. In: Jenks MA, Hasegawa PM, Jain SM (eds) Advances in molecular breeding toward drought and salt tolerant crops. Springer, pp 199–244

Sharpley A, Foy B, Withers P (2003) Practical and innovative measures for the control of agricultural phosphorus losses to water: an overview. J Plant Nutr Soil Sci 166(4):449–456
Shrivastava P, Kumar R (2015) Soil salinity: a serious environmental issue and plant growth promoting bacteria as one of the tools for its alleviation. Saudi J Biol Sci 22(2):123–131
Singh A, Sharma RK, Agrawal M, Marshall FM (2010) Health risk assessment of heavy metals via dietary intake of foodstuffs from the wastewater irrigated site of a dry tropical area of India. Food Chem Toxicol 48(2):611–619
Singh G (2009) Salinity-related desertification and management strategies: Indian experience. Land Degrad Dev 20:367–385
Singh J, Kalamdhad AS, Singh RP (2017) Industrial effluent management in India: practices and challenges. Environ Dev 21:79–95
Sinha S, Pandey K, Gupta AK, Bhatt K (2005) Accumulation of metals in vegetables and crops grown in the area irrigated with river water. Bull Environ Contam Toxicol 74:210–218
Skaggs RW, van Schilfgaarde J (1999) Agricultural drainage. Agronomy Monograph No. 38. American Society of Agronomy, Crop Science Society of America, Soil Science Society of America, Madison, WI
Smith VH, Schindler DW (2009) Eutrophication science: where do we go from here? Trends Ecol Evol 24(4):201–207
Smith TM, Mertz KJ, Meyer CM (1999a) Water use and management: the contribution of farm ponds in New Zealand. J Environ Qual 28
Smith VH, Tilman GD, Nekola JC (1999b) Eutrophication: impacts of excess nutrient inputs on freshwater, marine, and terrestrial ecosystems. Environ Pollut 100(1–3):179–196
Şolener M, Tunali S, Ozcan AS, Ozcan A, Gedikbey T (2008) Adsorption characteristics of lead (II) ions onto the clay/poly (methoxyethyl) acrylamide (PMEA) composite from aqueous solutions. Desalination 223(1–3):308–322
Steele M, Odumeru J (2004) Irrigation water as source of foodborne pathogens on fruit and vegetables. J Food Prot 67(12):2839–2849
Stockle CO (2001) Environmental impact of irrigation: a review. In: Proceedings of the IV international congress of agricultural engineering, Chillan, Chile, 9–11 May 2001, p 15
Sumner ME (1995) Amelioration of subsoil acidity with minimum disturbance. In: Soil management: experimental basis for sustainability and environmental quality. CRC Press, pp 179–201
Suresha KR, Nagesh MA (2015) Experimental studies on effect of water and soil quality on crop yield. Aquatic Procedia 4:1235–1242
Tang PL, Lee CK, Low KS, Zainal Z (2003) Sorption of Cr (VI) and Cu (II) in aqueous solution by ethylenediamine modified rice hull. Environ Technol 24(10):1243–1251
Tavakkoli E, Rengasamy P, McDonald GK (2011) High concentrations of Na^+ and Cl^- ions in soil solution have simultaneous detrimental effects on growth of faba bean under salinity stress. J Exp Bot 61(15):4449–4459
Tilman D, Cassman KG, Matson PA, Naylor R, Polasky S (2002) Agricultural sustainability and intensive production practices. Nature 418:671–677
Timmer LW, Zitko SE, Peever TL, Solel Z (2000) Environmentally friendly approaches to citrus disease management. J Crop Prod 3:165–197
Toze S (2006) Reuse of effluent water-benefits and risks. Agric Water Manag 80(1–3):147–159
Turkmen D, Bakhshpour M, Akgonullu S, Asir S, Denizli A (2022) Heavy metal ions removal from wastewater using cryogels: a review. Front Sustain 3:765592. https://doi.org/10.3389/frsus.2022.765592
Verma P, Sharma H, Tyagi N (2018) Management of irrigation systems in agriculture. J Agric Sci 10(2):1–13
Vitousek PM, Aber JD, Howarth RW, Likens GE, Matson PA, Schindler DW, Schindler WH, Tilman DG (1997) Human alteration of the global nitrogen cycle: sources and consequences. Ecol Appl 7(3):737–750
Vymazal J (2007) Removal of nutrients in various types of constructed wetlands. Sci Total Environ 380(1–3):48–65

Ward MH, DeKok TM, Levallois P, Brender J, Gulis G, Nolan BT, VanDerslice J (2005) Workgroup report: drinking-water nitrate and health-recent findings and research needs. Environ Health Perspect 113(11):1607–1614
White PJ (2013) Ion uptake mechanisms of individual cells and roots: short-distance transport. In: Plant nutrient acquisition. Springer, Berlin, Heidelberg, pp 7–29
WHO (2006) Guidelines for the safe use of wastewater, excreta and greywater: Volume II: Wastewater use in agriculture. World Health Organization
WHO (2011) World health organization guidelines for drinking-water. Fourth edition http://whqlibdoc.who.int/publications/2011/9789241548151_eng.pdf
Xu P, Drewes JE, Heil D (2010) Beneficial use of co-produced water through membrane treatment: technical-economic assessment. Desalination 250:1002–1009
Yusuf AA, Arowolo TA, Bamgbose O (2003) Cadmium, copper and nickel levels in vegetables from industrial and residential areas of Lagos city, Nigeria. Food Chem Toxicol 41:375–378
Zaman M, Shahid SA, Heng L, Zaman M, Shahid SA, Heng L (2018) Irrigation water quality. In: Guideline for salinity assessment, mitigation and adaptation using nuclear and related techniques. Springer, Cham, pp 113–131. https://doi.org/10.1007/978-3-319-96190-3_5
Zhang X, Davidson EA, Mauzerall DL, Searchinger TD, Dumas P, Shen Y (2020) Managing nitrogen for sustainable development. Nature 528(7580):51–59
Zhang Y, Tao S, Wu W (2016) Global atmospheric emission inventory of polycyclic aromatic hydrocarbons (PAHs) for 2004. Atmos Environ 40(4):677–688
Zheng XJ, Li Q, Peng H, Zhang JX, Chen WJ, Zhou BC, Chen M (2022) Remediation of heavy metal-contaminated soils with soil washing: a review. Sustainability 14:13058
Zorb C, Geilfus CM, Dietz KJ (2019) Salinity and crop yield. Plant Biol 21(S1):31–38

Effects of Poor Irrigation Water Quality on Soil Fertility and Its Mitigation

Immanuel Chongboi Haokip, Mercy Chinneihoi Haokip, Mayanglambam Homeshwari Devi, S. Chinnunnem Haokip, B. H. Sunil, Ampee Tasung, Shannon N. Sangma, and Sanjay Srivastava

Abstract Irrigation is crucial for enhancing agricultural productivity and attaining food security. However, due to various competing uses, the availability of good quality water for agricultural use is on the decline. So, poor, marginal and unconventional irrigation water, such as saline and sodic water, treated sewage or wastewater, desalinated water, drainage water, etc. are gradually being applied to crops. This chapter briefly discussed the detrimental effect of these types of water on the soil chemical, physical and biological attributes which have negative consequences on its fertility. Excess salt and ions in irrigation water led to their accumulation in soil, higher soil pH and ionic imbalance. In addition, continuous irrigation of poor quality

I. C. Haokip (✉) · M. H. Devi · S. Srivastava
ICAR-Indian Institute of Soil Science, Bhopal, Madhya Pradesh, India
e-mail: immanuel.ssac@gmail.com

M. H. Devi
e-mail: mdevi271@gmail.com

S. Srivastava
e-mail: sanjaysrivastava238@gmail.com

M. C. Haokip
Department of Agricultural Chemistry and Soil Science, Bidhan Chandra Krishi Vishwavidyalaya, Mohanpur, West Bengal, India
e-mail: chinneimercy@gmail.com

S. Chinnunnem Haokip
ICAR-Central Institute of Agricultural Engineering, Bhopal, Madhya Pradesh, India
e-mail: haokip7697@gmail.com

B. H. Sunil
ICAR-National Bureau of Soil Survey and Land Use Planning, Nagpur, Maharashtra, India
e-mail: sunilbh62@gmail.com

A. Tasung
ICAR Research Complex for North Eastern Hill Region, Arunachal Pradesh Centre, Basar, Arunachal Pradesh, India
e-mail: ampeetasung@gmail.com

S. N. Sangma
ICAR-Indian Agricultural Research Institute, Hazaribagh, Jharkhand, India
e-mail: shannonofwinterfell@gmail.com

Dimple et al. (eds.), *Surface Water and Groundwater Quality for Sustainable Utilization*, Springer Hydrogeology, https://doi.org/10.1007/978-981-96-8189-1_10

in areas with low rainfall and high evaporative demand could lead to increased irrigation water-induced salinization and sodification. Excess salt and extreme pH cause ionic imbalance, disrupting the availability of essential plant nutrients. Although unconventional irrigation water sources such as treated wastewater and effluents might improve soil fertility, there are associated hazards of heavy metal contamination and pathogen loading. Not only this, this poor-quality water may also cause clay dispersion, structural degradation, and clogging of soil pores, which in turn reduces the soil porosity, water infiltration and conductivity. It also affects the microbial population crucial in nutrient transformation. Effective mitigation strategies such as blending or conjunctive use of good and poor irrigation water, using amendments, crop selection, appropriate irrigation and soil management techniques, etc. are also discussed.

Keywords Saline water · Sodic water · Treated wastewater · Soil fertility · Irrigation management

1 Introduction

Water is indispensable for all life forms on earth and crucial for successful crop production in agriculture. At places where rainfall is insufficient to support crops, irrigation is used to artificially apply a controlled amount of water to crops to complete their life cycles, and approximately over 30% of the global agrarian production is attributed to irrigation (Hillel 2008). Both conventional (surface water taken from lakes, reservoirs, streams, groundwater, etc.) and unconventional (processed sewage and wastewater, desalinated water, drainage water, etc.) sources can serve as irrigation water (Karimidastenaei et al. 2022). Surface water from snow-capped mountaintop streams and lakes often has relatively low levels of salts and other impurities, while groundwater quality depends on the type of soil and underlying rock formations (Winter et al. 2000). Agriculture is the world's greatest consumer of freshwater, accounting around three-fourth of freshwater and more than nine-tenths of whole consumption globally (Siebert et al. 2010; Bruinsma 2017). As conventional water resources are quickly depleting, especially in semiarid and arid regions, and the rise in the demand for freshwater for various purposes, such as home and industrial usage; the use of unconventional water sources for agriculture has been on the rise due to rising global population (Cheng et al. 2021; Mishra et al. 2023). However, using unconventional sources for irrigation is challenging due to the high amounts of impurities such as salts and ions, excess solids, contaminants and pathogens (Coppola et al. 2004). Consequently, unconventional water sources requires appropriate treatment and filtration to reduce the load of impurities to permissible levels (Shatanawi and Fayyad 1996). Globally, it is estimated that less than 7% of treated sewage is harnessed for irrigation (Ungureanu et al. 2018). However, treated wastewater may still contain contaminants, pathogens, pesticides, pharmaceutical waste, and trace elements, depending on the treatment method and sources (Malakar et al. 2019).

Additionally, crop and soil quality suffer when water quality for irrigation requirements is not met (Rengasamy 2018; Singh et al. 2018a, b). In addition, pollutants, synthetic chemicals, and nanomaterials hazardous to human lives may bioaccumulate in the edible portion of plants (Hochella Jr et al. 2019). Therefore, a critical assessment of the impact of this subpar water on soil health is required as poor irrigation water is likely to detrimentally affect soil properties and fertility (Allende and Monaghan 2015). Therefore, this chapter seeks to provide a brief assessment of the negative impact that poor-quality irrigation water has on soil fertility, as well as some mitigation techniques to reduce those negative effects.

2 Classification of Irrigation Water

The suitability of water for irrigation is based on its inherent ability to cause salinity, sodicity/alkalinity and other toxicity hazards. This is based on the pH of the water, soluble salt, the relative concentration of sodium with calcium and magnesium, and the concentration of anions such as carbonates, bicarbonates, chlorides, nitrates, sulphates, etc. (Singh et al. 2015; Minhas et al. 2019; Zaman et al. 2018). Several additional site-specific parameters, including drainage, plant characteristics, soil type, irrigation technique, etc., also affect the safe use of water in agriculture. The different classes of irrigation water depending on their safe use for irrigation are presented in Table 1.

3 Effect on Soil Chemical Properties

All irrigation water, regardless of the source, contains different kinds of impurities; the type and quantity of these impurities may vary depending on the source and their course before use. Numerous studies have indicated that irrigation with low-quality water alters soil chemical characteristics.

3.1 Soil pH and Salt Accumulation

Irrigating normal fertile soil continuously with alkali water raises soil pH (Arora et al. 2012). Chief cations and anions responsible for such rise in soil pH are sodium, bicarbonates and chlorides (Jalali and Ranjbar 2009). A rise in soil pH above 8.4 indicates soil alkalinization (Sou/Dakouré et al. 2013). The rise in solution pH coupled with a decrease in calcium ions dramatically raises the soil SAR (Sou/Dakouré et al. 2013). The unfavourable consequence of saline irrigation water is the soluble salts and/or sodium buildup in the soil which leads to the formation of saline or sodic soil with time (Wang et al. 2023). This especially holds in hot semi-arid and dry areas

Table 1 Classification of irrigation water

Parameters	Values	Class	Remarks
pH	6.5–8.4		Safe
	0–5, >9.5		Unsafe
Total soluble salt (g L^{-1})	<0.16	Low	Safe
	0.16–0.50	Medium	Moderately safe
	0.50–1.50	High	Can be used with leaching
	1.50–3.00	Very high	Unsafe
Salinity hazard or electrical conductivity, EC (dS m^{-1})	0.0–0.25	Low	Safe
	0.25–0.75	Medium	Moderately safe
	0.75–2.25	High	Safe with proper management
	2.25–5.00	Very high	Unsafe
Sodium hazard or sodium adsorption ratio (SAR)	<10	Low	Safe
	10–18	Medium	Appreciable threat
	18–26	High	Unsafe for sensitive crops
	>26	Very high	Unsafe for all crops
Bicarbonate hazard or residual sodium carbonate (RSC) (me L^{-1})	<1.25	Low	Safe
	1.25–2.50	Medium	Safe with proper management
	>2.50	High	Unsuitable
Boron (ppm)	<0.33	Very low	Safe
	0.33–0.67	Low	Safe with proper management
	0.67–1.00	Medium	Unsuitable for most boron-sensitive crops
	1.00–1.25	High	
	>1.25	Very High	
Chloride (me L^{-1})	4	Excellent	
	4–7	Moderate	
	7–12	Slightly unsuitable	
	12–20	Unsuitable	
Soluble sodium percentage (SSP)	>60	Unsuitable	
Magnesium hazard (Ca: Mg ratio)	<50	Unsuitable	
Nitrate (me L^{-1})	>5	High	Unsuitable
Lithium (ppm)	>0.05	Toxic to citrus	

Source Richards (1954), Das (2004)

that receive little rainfall having high evaporative demand (Choudhary and Kharche 2018). This is one of the primary reasons that causes healthy vibrant soil to become saline or sodic. Arora et al. (2012) reported that irrigation water containing excessive salt and sodium led to a considerable increase in pH and EC compared with tap water, which also reduces the soil salinity and alkalinity. Likewise, Ahmed et al. (2010) reported that the pH of the soil was also increased by irrigation with both concentrated and diluted alkali water. When salt accumulate in the rooting zone, it causes osmotic stress and impairs plant growth (Fagodiya et al. 2020). Prolonged irrigation with saline water dramatically raises the pH, EC, and SAR of the soil- all of which are indicators of the salinization and alkalization of the soil (Wang et al. 2023). Singh et al. (1992) also reported higher soil sodicity and salinity with increasing SAR and EC of irrigation water. This is consistent with the findings of Wang et al. (2022), who showed that soil EC is increased with a corresponding increase in EC of the irrigation water. Thus, continuous irrigation with saline and sodic water accelerates the process of salinization and alkalization (Huang et al. 2011; Shehzad et al. 2020). This phenomenon is already evident, particularly in Northwestern India where several studies using low-quality irrigation water have documented the development of saline and sodic soil in historically fertile soil with normal pH and EC (Minhas et al. 2019).

Wastewater and brackish water irrigation have contradictory results regarding soil pH and salt accumulation. While Pang et al. (2010) saw a substantial rise in salt concentration, Tahtouh et al. (2019) found no evidence of salinity or sodicity during short-term irrigation with brackish water. Similarly, Aman et al. (2018) found an inconsistent effect of raw wastewater on soil pH while irrigation with effluents dramatically raised the soil SAR and EC. In Brazilian oxisols, long-term irrigation with treated sewage effluent showed no potential salinization and sodification due to the low SAR of the treated sewage effluent (Barbosa et al. 2022). However, untreated wastewater irrigation can induce alkalinization and sodification when it contains excess sodium (Sou/Dakouré et al. 2013). The application of acidic sewage decreased the soil pH even in soils with substantial gypsum accumulation (Navas et al. 1998). Soil sodification and salinization are also aided by a variety of factors, including low precipitation, high evaporative demand, pedological characteristics, irrigation technology, unsustainable land use, inappropriate soil and cropping management (Shrivastava and Kumar 2015).

3.2 *Soil Fertility and Organic Carbon*

Higher pH and salt concentration has serious detrimental effects on plant nutrient availability, soil fertility and the soil-plant-water continuum. Excess cations and anions cause ion toxicity, disturb the nutrient behaviour, and produce extreme ratios of sodium/calcium and sodium/potassium which hinder the availability and uptake of calcium and potassium by plants (Singh et al. 2014). Wang et al. (2023) stated that continuous saline water application decreases the soil organic carbon (SOC) and is unfavourable for carbon (C) sequestration. Similar effects of high EC and pH of irrigation water on SOC also have been reported by Dong et al. (2022). This is because of the dissolution of the soil organic matter (SOM) under saline or alkaline conditions and is observed as black deposits (black alkali) at the soil surface, indicating deterioration of soil quality (Sou/Dakouré et al. 2013). As SOM is considered the storehouse of nutrients, loss in SOM will have irreparable consequences on nutrient availability (Singh et al. 2018a, b). Similarly, Singh et al. (2021) reported that saline water application reduces the availability and uptake of primary nutrients. It also reduces essential secondary and micronutrient availability with a concomitant increase in the risk of sodium hazard (Teffera et al. 2019). These micronutrient deficiencies in plants are attributed to the decrease in micronutrient solubility under alkaline condition (Zhao et al. 2008). High soil pH and excess salt also inhibit nitrification process and accelerate volatilization and denitrification thereby causing greater nitrogen loss from the soil (Zhu et al. 2020). This nutrient imbalance exacerbates the detrimental effects of poor irrigation water (Brar and Singh 2022). The evaluation of soil quality showed that the soil quality significantly decreased by 3.7, 11.1, 16.9, 21.0, and 21.7% when irrigated with water that had an EC of 3.4, 7.1, 10.6, 14.1, and 17.7 dS m^{-1}, respectively, in comparison to control treatment (Zhang et al. 2020).

Normally, wastewater and sewage are rich sources of organic matter and nutrients (Aman et al. 2018). As a result, their application can improve soil fertility and improve crop productivity (Hao et al. 2022). Ibrahimi et al. (2022) observed that treated wastewater increases available phosphorus, potassium, and SOM by 472, 43, and 90%, respectively. Barbosa et al. (2022), Coelho et al. (2020) also reported improvement in soil fertility status after irrigation with treated sewage water. Thus, co-application of treated wastewater with chemical fertilizers can minimise our reliance on chemical fertilizers (Magwaza et al. 2020; Muscarella et al. 2024). In Jordan, continuous and prolonged treated wastewater irrigation increased the SOM and major nutrients, but the effect on micronutrient availability was inconsistent (Rusan et al. 2007). It was estimated that appropriate application of wastewater could add about 108 kg of nitrogen, 15 kg of phosphorus, and 25 kg of potassium per hectare annually, thereby lowering expensive chemical fertilizers requirements (Lal et al. 2015, 2020).

3.3 Heavy Metal Contaminants

Wastewater irrigation is an important source of heavy metal pollution in the soil and also substantially enhances its concentration in the edible portion, especially when used over extended periods (Chaoua et al. 2019). Also, the absence of adequate and proper treatment infrastructure in many cities and towns has resulted in the outflow of extremely toxic untreated wastewater and effluents into rivers and irrigation canals (Bashir et al. 2021). A multi-industry zone adjacent to farmland in Bangladesh releases wastewater into surrounding irrigation canals and has resulted in significant heavy metal contamination in the adjacent cultivated soils (Ahmed et al. 2019). A comparable finding was also documented in the Marrakech region of Morocco, where heavy metal concentration exceeds the health risk index due to wastewater irrigation (Chaoua et al. 2019). Wastewater-irrigated soil in Northern India also showed higher concentrations of heavy metals than groundwater-irrigated soil (Singh et al. 2024). The accumulation of toxic metals beyond the safe limit will negatively impact the health of residents who consume produce grown on contaminated soils (Singh et al. 2024). Abbas et al. (2023) observed that though heavy metals present in wastewater are beyond those permitted by WHO, their concentration in soil is within permissible limits and hence may not cause harmful effects on crops and humans. Rusan et al. (2007) found no adverse effect of prolonged wastewater application on soil lead and cadmium content. Similar results have been reported from a long-term investigation by Coelho et al. (2020). This is explained by the reduction of heavy metal load when it undergoes different treatment stages (Barbosa et al. 2022). This showed that, with frequent monitoring and treatment, heavy metals in wastewater and effluents can be effectively reduced and safely used for irrigation.

4 Effect of Soil Physical Properties

Poor irrigation water quality can also negatively impact the soil physical properties. It can destroy the soil structure and aggregation, and modify the bulk density, porosity, erodibility, hydrophobicity, penetration resistance, and soil moisture behaviour (Geerts et al. 2008). These changes in the soil physical behaviour have direct consequences on nutrient behaviour, water movement, and aeration, and thereby affect crop growth and productivity.

4.1 Soil Structure and Aggregation

When soil is irrigated with water having high SAR and ESP, there is a predominance of sodium ion which replaces the flocculating agents like calcium and magnesium ions in soil complex (Wu et al. 2010; Wang et al. 2022). This causes the soil particles to disperse and repel each other (Awedat et al. 2021). Thus, clay colloids separate from one another and get dispersed, the soil structure also collapses dramatically and clay particles leach down the soil pores (Bresson and Moran 2004). Because of this structural disruption, surface crusting, hardsetting and sealing arose (Vanani et al. 2017). On drying, the clay particles separate from one another and become dense, cloddy, structurally unstable and form crust at the surface. This causes the soil aggregate and structure to breakdown (Awedat et al. 2021), and is more noticeable when the soil ESP is more than 15% (Levy and Shainberg 2005). The structural deterioration and poor permeability of soils are often accompanied by yield penalty (Minhas and Gupta 1992).

Soil aggregate stability greatly depends on the ESP, SAR and EC of the irrigation water. As the ESP, SAR and EC of the irrigation water increased, macroaggregates and aggregate stability index is gradually reduced (Huang et al. 2011; Wang et al. 2023). There was a dramatic decrease in water-stable aggregate by 4.42–45.3% in soil irrigated with saline water (Zhou et al. 2023). A high concentration of magnesium relative to calcium is also detrimental to soil structure, and this is particularly true when the magnesium/calcium ratio is more than unity (Qadir et al. 2018). In some cases, the high salt content of irrigation water can mitigate the negative effects of high SAR, such as swelling and clay dispersion (Guo and Liu 2014).

With respect to the effect of wastewater irrigation on soil physical properties, the extent of deterioration is a function of its organic matter and ions present in it. Similar to saline and alkali water, wastewater water having high EC and SAR values led to dispersion and surface crust formation (Vanani et al. 2017). Excess sodium and bicarbonates raise the soil pH and causes structural disintegration at the surface (Sou/Dakouré et al. 2013; Schacht and Marschner 2015). However, when more organic matter is added through treated wastewater and sewage effluent, it improves soil aggregation and stability because organic matter is the primary cementing agent of primary soil particles (Coelho et al. 2020). Irrigation with treated olive mill effluent for fifteen years showed a significant increase in SOC content and aggregate stability (Mahmoud et al. 2012; Castor et al. 2023). On the contrary, Coelho et al. (2020) observed no change in soil aggregation after four years of fertigation with treated effluent. This beneficial impact on the structural stability of the soil demonstrated that wastewater, with the appropriate treatment, might substitute freshwater irrigation (Mandal et al. 2008).

4.2 Penetration Resistance

Soil penetration resistance controls seedling germination, root elongation and water accessibility. High penetration resistance inhibits root growth, provides a restricted flow of water and nutrients, and lowers crop yield (Colombi and Keller 2019). Minhas et al. (2021) observed penetration resistance at 15–20 cm layers in soils continuously irrigated with sodic water. This is mainly due to the structural breakdown and soil particle dispersion, especially in the subsurface, creating a compact impermeable layer (Sou/Dakouré et al. 2013). Ibrahimi et al. (2022) reported lower penetration resistance in treated wastewater irrigated plots than in control plots. In contrast, Coelho et al. (2020) reported no alteration in soil penetration resistance with the application of treated sewage water for four years in an Oxisol.

4.3 Soil Porosity

Soil pores are volumes of soil that are filled either with gases or water, and affects the exchange of water, nutrients and air, root proliferation and nutrient uptake (Nimmo 2004). Poor irrigation water, through its effect on clay dispersion and soil aggregate, alters the soil porosity (Gonçalves et al. 2010). In this scenario, soil aggregates separate into individual soil particles, which clog the soil pores and obstruct the interstitial pore space. Brackish water irrigation considerably reduced soil porosity in comparison to freshwater and saltwater irrigation treatments, particularly in the top plough layer (Ding et al. 2023a, b). According to Huang et al. (2011), soil porosity decreases when irrigation water is more saline, and the EC of the saturated paste rises, particularly in surface soil. In wastewater-irrigated soils, suspended particles from wastewater also clog the soil pores, reducing soil porosity (Schacht and Marschner 2015). The pores are additionally obstructed by clay migration that occurs during wastewater irrigation (Lado and Ben-Hur 2009). According to Sahin et al. (2002), wastewater irrigation enhances the micropores and ultramicropores but reduces the macropores and mesopores. Sou/Dakouré et al. (2013) also reported the least specific volumes from soils irrigated with recycled wastewater. In contrast, Coelho et al. (2020) noted that continuous fertigation with treated wastewater does not alter soil porosity significantly.

4.4 Bulk Density

The effect of irrigation water on soil bulk density depends on the EC, SAR, ESP and organic matter present in it. Sodic water causes clay dispersion, migration, and clogged soil pores which raise the bulk density (Choudhary et al. 2011, 2019). The increase in bulk density is highly correlated with the EC of irrigation water (Wang et al. 2023). Ding et al. (2023a, b) reported a much higher bulk density in the subsoil when irrigated with brackish water. Soil irrigated with saline-sodic water showed a higher bulk density than tube well-irrigated or rainfed (Chinchmalatpure et al. 2018). As the organic matter present in wastewater cements the soil particles, it improves aggregation and porosity thereby lowering the bulk density of the soil (Nadav et al. 2013). Abd-Elwahed (2018) stated that long-term wastewater irrigation led to a small decrease in bulk density. Similarly, continuous irrigation with untreated municipal effluent for ten years significantly reduced the soil bulk density and became more pronounced over time (Mathan 1994). Lopes et al. (2024) also reported compaction and restrictions on sugarcane root propagation to deeper soil layers due to the decreased macroporosity and higher bulk density. Coppola et al. (2004) revealed, in contrast to these findings, that reduced soil porosity as a result of wastewater irrigation can lower soil bulk density.

4.5 Soil Hydraulic Properties

Modification in soil porosity, soil particles dispersion and pore-clogging due to poor irrigation water negatively impacts soil hydraulic properties (Mandal et al. 2008). Hard setting and crusting also impair water infiltration, soil permeability and hydraulic conductivity (Wang et al. 2022). Minhas et al. (1999) saw a notable drop in hydraulic conductivity, a consequence of surface soil dispersion and disaggregation, which is exacerbated by the excess sodium present in the irrigation water. Haj-Amor et al. (2018) observed that saline water irrigation impeded water movement, soil water storage and percolation into the soil profile. There was also a considerable reduction in the final infiltration rate with prolonged sodic water irrigation (Choudhary et al. 2011).

Wastewater application effects on soil hydraulic attributes are governed by the chemical, biological and physical composition of the wastewater (Coppola et al. 2004). When treated wastewater contains excess organic matter and suspended materials, it leads to exponential growth of microbes which clog the soil pores thereby reducing the hydraulic conductivity and infiltration rate (Magesan et al. 2000). Assouline and Narkis (2011) observed that treated wastewater lowers the saturated hydraulic conductivity, sorptivity, and infiltration rates at all depths. Sou/Dakouré et al. (2013) observed that the infiltration rate in plots irrigated with recycled wastewater was, on average, 2.5-fold lesser than in normal water-irrigated soils. Gharaibeh et al. (2007) observed that rainfed plots had a higher infiltration rate than wastewater-irrigated plots, however, the cracks developed on the surface of vertisols may enhance the infiltration rate. It also causes an accumulation of suspended solid particles and dispersed clay in the soil pores, thereby lowering soil hydraulic conductivity (Adhikari et al. 2012). Copolla et al. (2004) detected a disturbed surface layer with low porosity, water-holding capacity and hydraulic conductivity when irrigated with treated municipal wastewater, which deepens over time. In contrast, Coelho et al. (2020) observed no appreciable change in the infiltration rate under continued treated sewage effluent fertigation in an Oxisol.

4.6 Soil Erodibility

Soil erodibility determines the capacity of soil to resist erosion. Soil salinization and alkalization enhance vegetation loss, fertility loss, organic matter dissolution, and breakdown of binding or flocculating agents which ultimately reduce the soil stability and aggregation (Oster and Shainberg 2001). All these factors reduce the rate of infiltration, severely limit permeability and water storage, and enhance runoff loss and erosion (Zaman et al. 2018). Higher infiltration rates and lower erodibility index showed stronger resistance to erosion, and are generally found in soil containing high SOM and structural stability (Leonard et al. 2006; Musa et al. 2017). Consequently, using irrigation water containing high sodium will result in a high erodibility index and vulnerable to erosion. Soil loss is more distinct when soil is dispersed and aggregates breakdown (Mandal et al. 2008). Thus, poor water quality negatively affects aggregate stability and enhances clay dispersion leading to high soil erodibility (Adnan et al. 2021).

4.7 Soil Hydrophobicity

Wastewater containing excess particulate organic matter when applied to soil can impart hydrophobicity due to enhanced microbial growth and microbial cells that coat the clays and aggregates (Abegunrin et al. 2016). Hydrophobic behaviour decreases initial sorptivity, promotes preferential flow, reduces the rate of wetting, infiltration rate and water storage, and increases runoff across the soil surface (Hallett 2008; Schacht et al. 2014). This hydrophobicity is determined by the fraction of the particles and aggregates coated by the hydrophobic organic materials (Vogelmann et al. 2013). Abegunrin et al. (2016) observed least hydrophobicity in rainwater irrigation, while untreated cassava effluent irrigation produced the maximum hydrophobicity followed by domestic wastewater irrigation and abattoirs. Nadav et al. (2013) reported that freshwater-irrigated soils were hydrophilic, while reclaimed wastewater-irrigated soil displayed hydrophobicity. Likewise, Schacht et al. (2014) observed that treated wastewater application considerably increases the repellency index when compared to freshwater and saline irrigation. In contrast, Ibrahimi et al. (2022) did not observe hydrophobicity in both treated wastewater-irrigated and control sites. In terms of hydrophobicity, Wallis and Horne (1992) proposed that organic matter properties are more significant than its content.

5 Effect on Soil Biological Properties

Soil is a habitat of diverse microorganisms which perform diverse functions. By controlling SOM decomposition and nutrient transformation, soil microflora contributes significantly to maintaining and improving the fertility of the soil. They control ecological function and soil fertility, and poor irrigation water such as saline, sodic, raw and treated wastewater, sewage water, etc. significantly impacts their populations and their activity (Yuan et al. 2007).

5.1 Microbial Biomass

The biomass associated with soil microbes constitutes the most labile portion of the SOM and contributes up to 3% of the SOM. Excessive salt accumulation consequent to saline water irrigation causes salt toxicity, osmotic stress and dehydration of cells causing cell death (Yuan et al. 2007). Singh et al. (2018a, b) noticed that saline water reduced the microbial biomass carbon and nitrogen by 52.9 and 70%, respectively with saline water irrigation compared with normal water. It also reduces total microbial biomass, microbial biomass carbon and nitrogen, bacterial and fungal biomass and fungal to bacterial (F/B) ratios in arid soil (Min et al. 2015). Ding et al. (2023a, b) also observed similar results in brackish water irrigated soil, although the effects of freshwater and saline water irrigation were not different statistically. Excessive boron in irrigation water also negatively impacts the microbial biomass, although the effect is time-dependent as it promotes microbial growth as a micronutrient in the initial stage and becomes toxic only at the later stage (Vera et al. 2021). Soil irrigated with treated wastewater showed improvement in soil microbial biomass which could be attributed to higher SOM and nutrient availability (Muscarella et al. 2024). This positive effect of treated wastewater irrigation could improve soil fertility since microorganisms are involved in several soil processes including nutrient cycling and SOM decomposition.

5.2 *Microbial Abundance and Diversity*

Soil microbial diversity covers the genetic variability within species, and the number and relative abundance of species and functional groups in communities (Torsvik and Øvreås 2002). Numerous studies reported contrasting effects of irrigation water quality on soil microbial diversity (Hou et al. 2021; Ding et al. 2023a, b). Yu et al. (2021) reported lower bacterial diversity in saline water-irrigated soils. Ding et al. (2023a, b) observed that saline water irrigation does not significantly alter the bacterial and fungal diversity and abundance, but the sodicity of water regulates bacteria whereas fungi were regulated by salinity. Hou et al. (2021) concluded that fungal communities respond more quickly to saline water irrigation compared to bacterial communities. Continuous brackish or saline water irrigated soils showed lower soil microbial activity and functional diversity, which may be a reflection of soil degradation (Min et al. 2015). Excess dissolved particles in irrigation water decreased the fungus biomarkers but had the reverse effect on the bacteria (Ding et al. 2023a, b). Similarly, treated wastewater had a detrimental effect on fungi but stimulates bacteria, and this is visible from the low F/B ratio (Muscarella et al. 2024). While wastewater can contain pathogens and contaminants, its application increases the population of microbes involved in organic matter decomposition and nitrogen transformation (Leonel et al. 2022). Vera et al. (2023) reported that bacterial and fungal abundances were also affected differently, with toxic boron levels harming the microbiota associated with tomato plants, and shifting the F/B ratio towards fungal dominance (Vera et al. 2023). Evidence suggested that saline water irrigation suppresses the nitrifying genes (Wang et al. 2018), and enhanced the denitrifying genes (Guo et al. 2023). Saline water also favours the growth of species that produce spores, biofilm, exopolysaccharides, antioxidant metabolites and enzymes, and osmolytes which provide tolerance to salt stress (Alsharif et al. 2020).

5.3 *Soil Enzymes Activities*

Microbes secrete enzymes such as dehydrogenase, β-glucosidase, urease, catalase, phosphatase, etc. that are crucial in nutrient transformation. They are sensitive to changes in management practices and show the fertility status of the soil. Singh et al. (2018a, b) observed a significant reduction of dehydrogenase enzyme activity in saline water-irrigated soil. Zhai et al. (2018), Zhou et al. 2023) reported lower urease, β-glucosidase, α-glucosidase and alkaline phosphatase by 45.5, 50.8 and 24.9%, respectively when water compared to non-saline irrigated soils. Subsurface drip irrigation of arid soils with treated wastewater significantly reduces the esterase, phosphatase, glucosidase, urease, protease, and dehydrogenase activities while increasing the catalase activity (Díaz et al. 2021). With the increase in salt content in irrigation water, there was a corresponding decrease in soil enzyme activity (Hou et al. 2021). The reduction in enzyme activities is ascribed to the detrimental consequence of excess salt on microbial biomass and abundance (Singh et al. 2018a, b; Díaz et al. 2021; Zhou et al. 2023). By providing organic C and nutrients, wastewater irrigation promotes a variety of organisms and metabolic pathways and stimulates the hydrolytic and proteolytic enzymes in the soil (Frenk et al. 2014). However, the excessive proliferation of microbes consequent to wastewater irrigation leads to the development of biofilms that can act as a reservoir of pathogenic microbes and may even clog soil pore spaces (Kanarek et al. 2023).

5.4 *Soil Respiration*

Soil respiration is a measure of the CO_2 released from soil, and is a combined effect of soil flora, fauna, and litter decomposition. Reports have suggested that soil respiration was reduced by up to 30% when irrigated with saline water which may be associated with the cations and anions present in it (Zhou et al. 2023). Min et al. (2015) also reported irrigation with brackish water or saline water effectively retard soil basal respiration by 15–28%, mainly attributed to osmotic stress. Ibrahimi et al. (2022) observed higher basal respiration in treated wastewater-irrigated soils indicating improvement in soil microbial activity.

6 Management of Poor Irrigation Water Quality

Effective management practices are those techniques that minimize the concentration of salt around the root and reduce salt stress. The detrimental effect of saline, sodic and brackish water can be alleviated through appropriate water treatment, proper soil and water management, selection of suitable crops, etc.

6.1 Water Quality Improvement

Water quality that does not meet the standards for irrigation can be improved can be improve before field application. They can be improved by adding chemical amendments, allowing them to undergo chemical treatment or filtration before field application or conjunctive use of poor and good quality water.

6.1.1 Chemicals Amendments

The quality of sodic and saline-sodic water can be reclaimed with amendments that contain calcium such as gypsum and lime so that exchangeable sodium present in irrigation water is effectively replaced. Passing alkali water through a gypsum bed significantly reduces the RSC and improves the calcium content in the irrigation water (Singh 2013). One of the disadvantages of this method is that it requires specially designed gypsum beds to mix different types of water with varying quality. Sodic water with initial RSC of 8.0 me L^{-1} and 6.2 me L^{-1} when treated with 0.1% and 0.35% gypsum reduced the RSC up to 0.6 me L^{-1} and 1.2 me L^{-1}, respectively at the retention time of 1 h (Pradeepa et al. 2019). The gypsum required to improve the water quality is a function of the initial RSC and SAR levels, meteorological conditions, soil properties, crops grown and water requirement (Choudhary and Mavi 2023).

6.1.2 Wastewater Treatment

Raw sewage and wastewater that originated from industries and municipalities contain excess suspended solids, dissolved toxic inorganic and organic contaminants, organic matter and nutrients, pathogens, bad odour, etc. that make them unfit for direct application (Minhas et al. 2022). For safe use, it must undergo primary and secondary treatments. Primary treatment removed the gross, suspended and floating solids by passing them through screen or filter, grit chamber or sedimentation tank. This process lowers the biological oxygen demand (BOD) and suspended solids by 20–30% and 50–60%, respectively (Kesari et al. 2021). The secondary treatment removes up to 85% of the suspended solids and BOD to give a clear liquid. A tertiary treatment is also undertaken if the water still contains microbial, chemical, and metal contaminants, and this step ensures that the final product for field application is free of almost all impurities. Biswas et al. (2015) stated that a reasonably inexpensive filtration system consisting of sand, charcoal, rice husks, sawdust, and brick chips effectively eliminates pollutants from municipal wastewater for crop applications.

6.1.3 Conjunctive Use of Poor and Good Water Quality

The conjunctive use of poor and good water can be achieved by either blending or cyclic use. In the blending mode, good water is mixed with saline or sodic water in an appropriate ratio to get the desired water quality that is safe for crop as well as soil application; whereas in the cyclic mode, the crop is irrigated with waters of varying quality alternately (Singh 2014). Blending different water with high and low salinity yields lower water use efficiency and salt tolerance compared to cyclic application (Naresh et al. 1993). Cyclic application of freshwater at an early critical crop growth stage and saline water irrigation at maturity instead of reducing the number of irrigation events was observed to be an effective way of using both types of water without reducing crop productivity (Murad et al. 2018). This cyclic irrigation allows the plant to withstand the salts better and does not decrease crop yield or soil quality (Sharma et al. 1994). It also retards soil salinization due to poor irrigation management (Sang et al. 2020). In general, cyclic irrigation is advantageous as it gives better yields, especially in sensitive crops than blending; also, blending requires special facilities for mixing the two types of water (Singh et al. 2014). A key disadvantage of the concurrent use of saline and sodic water with good water is the formation of salinity zones near the roots and also, this strategy requires the availability of good-quality water (Singh et al. 2014).

6.2 *Soil and Crop Management*

6.2.1 Soil Amendment

Direct application of gypsum, calcium chloride, and H_2SO_4 which are a good source of soluble calcium or sulfur can solubilize the relatively insoluble carbonates abundant in sodic soils is also another way of reducing the detrimental effects of saline-sodic water (Jat et al. 2019; Minhas and Bajwa 2021). The acidic amendments like pyrites, H_2SO_4 and sulfur counteract the detrimental effects of sodic water, both through soil amendment and by passing through pyrite beds (Minhas and Bajwa 2021). It is easier and more economical to spread and completely mix the required amount of powdered amendment over a field that has already been levelled than to apply it with water. Applying pyrite once before wheat is sown is more effective than applying it through irrigation water (Minhas and Bajwa 2021). Both methods effectively neutralize and alleviate the residual carbonates present in irrigation water. Murtaza et al. (2006) found that combining saline-sodic water with gypsum and farmyard manure significantly improved the infiltration rate, bulk density, and crop yield while significantly reducing salt concentration. Hafez et al. (2015) reported higher salt removal was achieved when water was added at frequent intervals after gypsum treatment.

6.2.2 Leaching

Salt removal from the maximum root zone by leaching with good-quality water is an effective way of reducing salt from the root zone. This requires the availability of an ample amount of water and an excellent drainage system. The aim of any leaching process is to bring down the salt concentration to a level which allows crop growth without salt or osmotic stress, which is generally <4 dS m^{-1} for sensitive and <8 dS m^{-1} for tolerant crops (Hillel 2005). Adopting an appropriate leaching method is vital to leach salt from the root zone successfully. From a column study, Yang et al. (2023) reported that continuous ponding was more effective in light-textured soils whereas intermittent ponding and unsaturated application was more efficient in heavy-textured soils. Leaching out of salt is more efficient when the water application rate is kept below the soil infiltration rate (Chu et al. 2020). Generally, the average water requirement for leaching is greater for soil with higher clay content (Jat et al. 2019). Salt removal through leaching with a sprinkler is more effective than drip leaching (Yang et al. 2019). The depth of subsurface drainage affects the leaching efficiency. Feng et al. (2017) achieved maximum desalination when the drainage depth is 0.8 m, as opposed to no drainage or drainage depth of 1.2 m.

6.2.3 Nutrient Management

Continuous irrigation with saline and saline-sodic water consequently disturbs the nutrient availability and hence requires effective nutrient management strategies. As some irrigation water may contain high levels of nutrients, water analysis before application to determine the nutrient content should be carried out, and fertilizer doses adjusted accordingly to avoid over-fertilization (Murtaza et al. 2006). Application of organic nutrients such as farmyard manure improves nutrient availability, prevents nutrient leaching and acts as a soil conditioner (Choudhary et al. 2021). Application of farmyard manure, gypsum, and pressmud with chemical fertilizer with sodic groundwater irrigation improved the sustainability of the intensively cultivated cropping systems in India (Yaduvanshi and Swarup 2005). Choosing the right nitrogenous fertilizer source is also crucial, as applying urea and monoammonium phosphate with saline water results in high ammonia volatilization (Zhu et al. 2020). As sole gypsum application reduces the phosphorus solubility, it is recommended to use it in conjunction with organics to improve the phosphorus availability (Sundha et al. 2020). In North-Western India, combined application of farmyard manure, vermicompost, biogas slurry, and 5% recommended fertilizers, and seed treatment with *Azotobacter*, *Azospirillum*, and phosphate solubilizers, effectively improved crop yield when irrigated with saline water (Rani et al. 2019).

6.2.4 Crop Selection

High salinity levels negatively impact many crops due to osmotic effects, leaf surface expansion, primary C metabolism, and nutritional imbalance as well as oxidative stress. Since many crops are sensitive to excess salt, irrigation with saline and sodic water negatively affects crop growth, leading to loss of productivity and potential profits (Kim et al. 2016). To efficiently use low-quality water such as saline or sodic water, selecting crops and varieties that can survive soil salinity or sodicity is vital (Oster 1994). Also, crops and varieties that require large amounts of water should be avoided as the use of huge quantities of poor water may lead to salt accumulation (Dimple and Singh 2022). This allows efficient crop production in areas where good water is scarce.

6.2.5 Land Shaping

Land shaping and levelling improves water distribution, infiltration, and salt leaching and facilitates drainage (Kaledhonkar et al. 2019). This also allows rain harvesting and avoids water stagnation during the rainy season (Sharma et al. 2016). Shaping land into different shapes, ridge formation allows uniform distribution of water and nutrients and reduces salt accumulation. To accomplish uniform salt leaching, precise land levelling and efficient drainage systems are essential (Sonbol et al. 2007). For example, building a permanently raised bed and furrow provides drainage of leached water, minimizes salt concentration, and maintains suitable soil water availability for crops (Velmurugan et al. 2016). Precision land levelling has a better leaching efficacy than traditional levelling (El-Mowelhi et al. 1999). This provides better water availability, higher water and uniform nutrient availability and efficiency, less salt concentration around the roots, and increased crop production (Bai et al. 2013).

6.3 Method of Irrigation

Irrigation methods also influenced the concentration and flux of salt in soil. When saline water is applied through flooding and sprinklers, the salt is concentrated at the top layer. In contrast, when drip irrigated, salt concentration is maximum at the topsoil around the wetting soil zone, and least close to the water source and between two drippers (Zaman et al. 2018). In subsurface drip irrigation, greater salinity at the top layer is due to capillary motion from the buried lateral drip lines. While drip and sprinkler irrigation offer more control over salt and water distribution, they are more expensive to install, and the latter can cause leaf scorching especially in crops that are sensitive to salt (Choudhary et al. 2011). Jha et al. (2017) observed that tuber production was significantly lower in furrow irrigation than in drip-irrigated plots when the same saline water was used. Sonbol et al. (2007) reported that shorter furrow lengths in surface irrigation are effective in reducing irrigation-induced salinity. Furthermore, drip irrigation produces 22–23% less salt buildup than surface irrigation and requires 38–42% less water for leaching it (Fagodiya et al. 2020). Subsurface drip irrigation outperformed conventional drip irrigation in terms of yields, improved moisture distribution and mitigated osmotic stress (Oron et al. 2002). A study found that a nano irrigation system allows the safe usage of water having soluble salt up to 6 dS m^{-1} for irrigation without significant short-term salinity issues (Mouhdi et al. 2022). Saline water irrigation through sprinklers before emergence led to low soluble salt concentrations during the germination stage and better crop establishment at a later growth stage (Sharma and Minhas 2003).

7 Conclusion

The expanding global demand for food, feed, and fibre is predicted to drive up the demand for water in agriculture even further. The shortage of good water quality necessitates the utilization of poor water quality sources, especially from non-conventional sources. However, there are certain drawbacks to this, and further research into the negative impacts of various water sources on soil characteristics and agricultural output is essential. Irrigation water rich in sodium, bicarbonates, and chlorides not only increases the soil pH, and accumulation of excess salt, it limits the availability of essential plant nutrients and has caustic effect on organic matter. It also increases the risk of sodium and boron toxicity in soil. Their reckless usage destroys the soil aggregates, clogs the pores, changes the pore distribution, increasing bulk density and compaction which negatively impacts profile water redistribution. Irrigation water quality also influences soil microbial diversity and community structure. Poor irrigation water inhibits enzymes necessary for nutrient and organic matter cycling. reduces microbial biomass, which directly affects soil quality and fertility. To mitigate the negative effects on the soil and crop production, proper management and choosing appropriate soil and crop are required. Chemical treatment of poor-quality water, or its concurrent use with good-quality water, may mitigate the negative effects of prolonged use of poor-quality water. Wastewater, if properly treated, has the potential to improve soil fertility but may become a source of contaminants if applied untreated. Thus, irrigation with poor water quality necessitates extreme caution, frequent inspection, and continuous monitoring.

References

Abbas MT, Wadaan MA, Ullah H et al (2023) Bioaccumulation and mobility of heavy metals in the soil-plant system and health risk assessment of vegetables irrigated by wastewater. Sustainability 15(21):15321

Abd-Elwahed MS (2018) Influence of long-term wastewater irrigation on soil quality and its spatial distribution. Ann Agric Sci 63(2):191–199

Abegunrin TP, Awe GO, Idowu DO, Adejumobi MA (2016) Impact of wastewater irrigation on soil physico-chemical properties, growth and water use pattern of two indigenous vegetables in southwest Nigeria. CATENA 139:167–178

Adhikari P, Shukla MK, Mexal JG (2012) Spatial variability of hydraulic conductivity and sodium content of desert soils: Implications for management of irrigation using treated wastewater. Trans ASAE 55(5):1711–1721

Adnan TA, Al-Madhhachi AS, Mohammed EA (2021) Relationships of soil erodibility parameters and water quality indices along Tigris Riverbanks, Baghdad City, Iraq. Cogent Eng 8(1):1917330

Ahmed BAO, Inoue M, Moritani S (2010) Effect of saline water irrigation and manure application on the available water content, soil salinity, and growth of wheat. Agric Water Manag 97:165–170

Ahmed M, Matsumoto M, Ozaki A, Thinh NV, Kurosawa K (2019) Heavy metal contamination of irrigation water, soil, and vegetables and the difference between dry and wet seasons near a multi-industry zone in Bangladesh. Water 11(3):583

Allende A, Monaghan J (2015) Irrigation water quality for leafy crops: a perspective of risks and potential solutions. Int J Environ Res Public Health 12(7):7457–7477

Alsharif W, Saad MM, Hirt H (2020) Desert microbes for boosting sustainable agriculture in extreme environments. Front Microbiol 11:1666

Aman MS, Jafari M, Reihan MK, Motesharezadeh B, Zare S (2018) Assessing the effect of industrial wastewater on soil properties and physiological and nutritional responses of Robinia pseudoacacia, *Cercis siliquastrum* and *Caesalpinia gilliesii* seedlings. J Environ Manage 217:718–726

Arora NK, Chaudhari SK, Farooqi JA, Basa N (2012) Effect of poor quality water on the chemical properties of the salt affected soils and performance of rice. J Soil Sal Water Qual 4(2):114–121

Assouline S, Narkis K (2011) Effects of long-term irrigation with treated wastewater on the hydraulic properties of a clayey soil. Water Resour Res 47(8):1–12

Awedat AM, Zhu Y, Bennett JM, Raine SR (2021) The impact of clay dispersion and migration on soil hydraulic conductivity and pore networks. Geoderma 404:115297

Bai G, Du S, Yu J, Zhang P (2013) Laser land leveling improve distribution of soil moisture and soil salinity and enhance spring wheat yield. Trans Chin Soc Soc Agric Eng 29(8):125–134

Barbosa AM, Faria RT, Saran LM et al (2022) Impact of treated sewage effluent on soil fertility, salinization, and heavy metal content. Bragantia 81:e0222

Bashir S, Gulshan AB, Khan MJ et al (2021) The role of different organic amendments to improve maize growth in wastewater irrigated soil. J King Saud Univ Sci 33(7):101583

Biswas GC, Sarkar A, Rashid MH et al (2015) Assessment of the irrigation feasibility of low-cost filtered municipal wastewater for red amaranth (*Amaranthus tricolor* L cv. *Surma*). Int Soil Water Conserv Res 3(3):239–252

Brar HS, Singh P (2022) Pre-and post-sowing irrigation scheduling impacts on crop phenology and water productivity of cotton (*Gossypium hirsutum* L.) in sub-tropical north-western India. Agric Water Manag 274:107982

Bresson LM, Moran CJ (2004) Micromorphological study of slumping in a hardsetting seedbed under various wetting conditions. Geoderma 118(3–4):277–288

Bruinsma J (2017) (ed) World agriculture: towards 2015/2030: an FAO study. Earthscan Publications Ltd, London

Castor CP, Gastélum-Strozzi A, Sammartino S et al (2023) Long-term (>90 years) wastewater irrigation effect on the pore characteristics and stability of soil aggregates. Geoderma 434:116469

Chaoua S, Boussaa S, El Gharmali A, Boumezzough A (2019) Impact of irrigation with wastewater on accumulation of heavy metals in soil and crops in the region of Marrakech in Morocco. J King Saud Univ Sci 18(4):429–436

Cheng M, Wang H, Fan J et al (2021) Crop yield and water productivity under salty water irrigation: a global meta-analysis. Agric Water Manag 256:107105

Chinchmalatpure AR, Kumar S, Rao GG et al (2018) Impact of irrigation on soil characteristics of saline Vertisols of Bara tract under Sardar Sarovar canal command of Gujarat. J Indian Soc Soil Sci 66(4):381–385

Choudhary DR, Singh A, Panghal VP, Kumar V (2021) Effect of FYM and gypsum on growth and yield attributing traits on radish (*Raphanus sativus*) using RSC water. Curr Hort 9(2):56–59

Choudhary OP, Bhalla M, Sharma S et al (2019) Long-term impact of cyclic use of sodic and canal water for irrigation on soil quality and wheat yield in cotton-wheat cropping system. J Indian Soc Soil Sci 67(1):34–43

Choudhary OP, Ghuman BS, Thuy N, Buresh RJ (2011) Effects of long-term use of sodic water irrigation, amendments and crop residues on soil properties and crop yields in rice-wheat cropping system in a calcareous soil. Field Crops Res 121(3):363–372

Choudhary OP, Kharche VK (2018) Soil salinity and sodicity. In: Soil science: an introduction, Indian Society of Soil Science (ISSS), New Delhi

Choudhary OP, Mavi MS (2023) Advances in reclamation and management of sodic waters for irrigation: management of sodic waters for irrigation. J Soil Salin Water Qual 15(1):1–4

Chu LL, Kang YH, Wan SQ (2020) Effects of water application intensity of micro-sprinkler irrigation and soil salinity on environment of coastal saline soils. Water Sci Eng 13(2):116–123

Coelho AP, Silva MF, Faria RT et al (2020) Long-term impact of fertigation with treated sewage effluent on the physical soil quality. Environ Pollut 266:1–10

Colombi T, Keller T (2019) Developing strategies to recover crop productivity after soil compaction-A plant eco-physiological perspective. Soil Tillage Res 191:156–161

Coppola A, Santini A, Botti P et al (2004) Methodological approach for evaluating the response of soil hydrological behavior to irrigation with treated municipal wastewater. J Hydrol 292(1–4):114–134

Das DK (2004) Introductory soil science. Kalyani publishers

Díaz FJ, Sanchez-Hernandez JC, Notario JS (2021) Effects of irrigation management on arid soils enzyme activities. J Arid Environ 185:104330

Dimple MHK, Singh PK (2022) Groundwater quality parameters for irrigation utilization: a review. Indian J Agric Sci 92(7):803–810

Ding B, Bai Y, Guo S et al (2023a) Effect of irrigation water salinity on soil characteristics and microbial communities in cotton fields in Southern Xinjiang, China. Agronomy 13(7):1679

Ding B, Bai Y, Guo S, He Z, Wang B, Liu H, Zhai J, Cao H (2023b) Effect of irrigation water salinity on soil characteristics and microbial communities in cotton fields in Southern Xinjiang, China. Agronomy 13(7):1679

Dong X, Wang J, Zhang X et al (2022) Long-term saline water irrigation decreased soil organic carbon and inorganic carbon contents. Agric Water Manag 270:107760

El-Mowelhi NM, Shams El-Din HA, Abo El-Soud MA, Saied MM (1999). Water management for improving irrigation efficiencies at North Delta. In: Third conference of on farm irrigation and agroclimatology. Jan. 25–27, 1999, Dokki, Egypt, pp 37–52

Fagodiya RK, Meena BL, Kaledhonkar MJ et al (2020) Salt accumulation and leaching requirement for saline irrigation: a comparative study of surface and drip irrigation methods. J Soil Sal Water Qual 12(1):89–94

Feng G, Zhang Z, Wan C, Lu P, Bakour A (2017) Effects of saline water irrigation on soil salinity and yield of summer maize (*Zea mays* L.) in subsurface drainage system. Agric Water Manag 193:205–213

Frenk S, Hadar Y, Minz D (2014) Resilience of soil bacterial community to irrigation with water of different qualities under Mediterranean climate. Environ Microbiol 16:559–569

Geerts S, Raes D, Garcia M et al (2008) Could deficit irrigation be a sustainable practice for quinoa (*Chenopodium quinoa* Willd.) in the Southern Bolivian Altiplano? Agric Water Manag 95(8):909–917

Gharaibeh MA, Eltaif NI, Al-Abdullah B (2007) Impact of field application of treated wastewater on hydraulic properties of vertisols. Water Air Soil Pollut 184:347–353

Gonçalves RA, Gloaguen TV, Folegatti MV et al (2010) Pore size distribution in soils irrigated with sodic water and wastewater. Rev Bras Cienc Solo 34:701–707

Guo K, Liu X (2014) Dynamics of meltwater quality and quantity during saline ice melting and its effects on the infiltration and desalinization of coastal saline soils. Agric Water Manag 139:1–6

Guo X, Du S, Guo H, Min W (2023) Long-term saline water drip irrigation alters soil physicochemical properties, bacterial community structure, and nitrogen transformations in cotton. Appl Soil Ecol 182:104719

Hafez EM, Abou El Hassan WH, Gaafar IA, Seleiman MF (2015) Effect of gypsum application and irrigation intervals on clay saline-sodic soil characterization, rice water use efficiency, growth, and yield. J Agric Sci 7(12):208

Haj-Amor Z, Hashemi H, Bouri S (2018) The consequences of saline irrigation treatments on soil physicochemical characteristics. Euro-Mediterr J Environ Integr 3:1–2

Hallett PD (2008) A brief overview of the causes, impacts and amelioration of soil water repellency–a review. Soil Water Res 3(1):521–528

Hao X, Shi X, Khan A et al (2022) Industrial organic wastewater through drip irrigation to reduce chemical fertilizer input and increase use efficiency by promoting N and P absorption of cotton in arid areas. Agriculture 12(12):2007

Hillel D (2008) Soil in the environment. Academic Press, London

Hillel D (ed) (2005) Encyclopedia of soils in the environment. Elsevier, Amsterdam, The Netherlands

Hochella Jr MF, Mogk DW, Ranville J et al (2019) Natural, incidental, and engineered nanomaterials and their impacts on the Earth system. Science 363(6434):eaau8299

Hou Y, Zeng W, Hou M, Wang Z, Luo Y, Lei G, Zhou B, Huang J (2021) Responses of the soil microbial community to salinity stress in maize fields. Biology 10(11):1114

Huang CH, Xue X, Wang T et al (2011) Effects of saline water irrigation on soil properties in northwest China. Environ Earth Sci 63:701–708

Ibrahimi K, Attia KB, Amami R, Américo-Pinheiro JH, Sher F (2022) Assessment of three decades treated wastewater impact on soil quality in semi-arid agroecosystem. J Saudi Soc Agric Sci 21(8):525–535

Jalali M, Ranjbar F (2009) Effects of sodic water on soil sodicity and nutrient leaching in poultry and sheep manure amended soils. Geoderma 153(1–2):194–204

Jat HS, Singh J, Yadav RK et al (2019) Good agriculture practices (GAP) for salt affected soils. ICAR-CSSRI/Karnal/Technical Bulletin/2019/07. ICAR-Centrail Soil Salinity Research Institute, Karnal

Jha G, Choudhary OP, Sharda R (2017) Comparative effects of saline water on yield and quality of potato under drip and furrow irrigation. Cogent Food Agric 3(1):1369345

Kaledhonkar MJ, Meena BL, Sharma PC (2019) Reclamation and nutrient management for salt-affected soils. Indian J Fert 15(5):566–575

Kanarek P, Breza-Boruta B, Rolbiecki R (2023) Microbial composition and formation of biofilms in agricultural irrigation systems-a review. Ecohydrol Hydrobiol 24(3):583–590

Karimidastenaei Z, Avellán T, Sadegh M et al (2022) Unconventional water resources: global opportunities and challenges. Sci Total Environ 827:154429

Kesari KK, Soni R, Jamal QM et al (2021) Wastewater treatment and reuse: a review of its applications and health implications. Water Air Soil Pollut 232:1–28

Kim H, Jeong H, Jeon J, Bae S (2016) Effects of irrigation with saline water on crop growth and yield in greenhouse cultivation. Water 8(4):127

Lado M, Ben-Hur M (2009) Treated domestic sewage irrigation effects on soil hydraulic properties in arid and semiarid zones: a review. Soil Tillage Res 106(1):152–163

Lal K, Minhas PS, Yadav RK, Rosin KG (2020) GHG mitigation from sewage effluents: C sequestration after reuse in different urban cropping systems. Carbon Manag 11:281–290

Lal R (2015) The nexus approach to managing water, soil and waste under changing climate and growing demands on natural resources. In: Ardakanian R (ed) Kurian M. Governing the nexus, Springer, pp 39–60

Leonard J, Ancelin O, Ludwig B, Richard G (2006) Analysis of the dynamics of soil infiltrability of agricultural soils from continuous rainfall-runoff measurements on small plots. J Hydrol 326:122–134

Leonel LP, Bize A, Mariadassou M, Midoux C, Schneider J, Tonetti AL (2022) Impacts of disinfected wastewater irrigation on soil characteristics, microbial community composition, and crop yield. Blue-Green Syst 4(2):247–271

Levy GJ, Shainberg I (2005) Sodic soils. In: Hillel D (ed) Encyclopaedia of soils in the environment. Elsevier, Amsterdam, NL, pp 504–513

Lopes SOP, Santos LN, Teixeira MB et al (2024) How does irrigation with wastewater affect the physical soil properties and the root growth of sugarcane under subsurface drip? Agronomy 14(4):788

Magesan GN, Williamson JC, Yeates GW, Lloyd-Jones AR (2000) Wastewater C: N ratio effects on soil hydraulic conductivity and potential mechanisms for recovery. Bioresour Technol 71(1):21–27

Magwaza ST, Magwaza LS, Odindo AO et al (2020) Partially treated domestic wastewater as a nutrient source for tomatoes (*Lycopersicum solanum*) grown in a hydroponic system: effect on nutrient absorption and yield. Heliyon 6(12):e05745

Mahmoud M, Janssen M, Peth S et al (2012) Long-term impact of irrigation with olive mill wastewater on aggregate properties in the top soil. Soil Tillage Res 124:24–31

Malakar A, Snow DD, Ray C (2019) Irrigation water quality-A contemporary perspective. Water 11(7):1482

Mandal UK, Bhardwaj AK, Warrington DN et al (2008) Changes in soil hydraulic conductivity, runoff, and soil loss due to irrigation with different types of saline–sodic water. Geoderma 144(3–4):509–516

Mathan KK (1994) Studies on the influence of long-term municipal sewage-effluent irrigation on soil physical properties. Bioresour Technol 48:275–276

Min W, Guo H, Zhang W, Zhou G, Ma L, Ye J, Liang Y, Hou Z (2015) Response of soil microbial community and diversity to increasing water salinity and nitrogen fertilization rate in an arid soil. Acta Agric Scand B Soil Plant Sci 66(2):117–126

Minhas PS, Bajwa MS (2021) Use and management of poor quality waters for the rice-wheat based production system. In: The rice-wheat cropping system of South Asia. CRC Press, pp 273–306

Minhas PS, Bali A, Bhardwaj AK, Singh A, Yadav RK (2021) Structural stability and hydraulic characteristics of soils irrigated for two decades with waters having residual alkalinity and its neutralization with gypsum and sulfuric acid. Agric Water Manag 244:106609

Minhas PS, Gupta RK (1992) Quality of irrigation water: assessment and management. Indian Council of Agricultural Research, Publication Section, New Delhi, p 123

Minhas PS, Qadir M, Yadav RK (2019) Groundwater irrigation induced soil sodification and response options. Agric Water Manag 215:74–85

Minhas PS, Saha JK, Dotaniya ML et al (2022) Wastewater irrigation in India: current status, impacts and response options. Sci Total Environ 808:152001

Minhas PS, Singh YP, Chhabba DS, Sharma VK (1999) Changes in hydraulic conductivity of a highly calcareous and non-calcareous soil under cycles of irrigation with saline and simulated rainwater. Irrig Sci 18:199–203

Mishra S, Kumar R, Kumar M (2023) Use of treated sewage or wastewater as an irrigation water for agricultural purposes-environmental, health, and economic impacts. Total Environ Res Themes 6:100051

Mouhdi K, Attaoui A, Bouizgarne B et al (2022) Effect of saline irrigation water on growth and productivity growth of sugar beet (*Beta vulgaris* L.) under nano irrigation (case of moistube). Environ Sci Proc 16(1):72

Murad KF, Hossain A, Fakir OA et al (2018) Conjunctive use of saline and freshwater increases the productivity of maize in saline coastal region of Bangladesh. Agric Water Manag 204:262–270

Murtaza G, Ghafoor A, Qadir M (2006) Irrigation and soil management strategies for using saline-sodic water in a cotton–wheat rotation. Agric Water Manag 81(1–2):98–114

Musa JJ, Anijofor SC, Obasa P, Avwevuruvwe (2017) Effects of soil physical properties on erodibility and infiltration parameters of selected areas in Gidan kwano. Niger J Technol 12(1):10–20

Muscarella SM, Alduina R, Badalucco L, Capri FC et al (2024) Water reuse of treated domestic wastewater in agriculture: Effects on tomato plants, soil nutrient availability and microbial community structure. Sci Total Environ 928:172259

Nadav I, Tarchitzky J, Chen Y (2013) Induction of soil water repellency following irrigation with treated wastewater: effects of irrigation water quality and soil texture. Irrig Sci 31:385–394

Naresh RK, Minhas PS, Goyal AK, Chauhan CP, Gupta RK (1993) Conjunctive use of saline and non-saline waters. II. Field comparisions of cyclic uses and mixing for wheat. Agric Water Manage 23(2):139–48

Navas A, Bermudez F, Machin J (1998) Influence of sewage sludge application on physical and chemical properties of Gypsisols. Geoderma 87:123–135

Nimmo JR (2004) Porosity and pore size distribution. In: Hillel D (ed) Encyclopedia of soils in the environment. Elsevier, London, pp 295–303

Oron G, DeMalach Y, Gillerman L, David I, Lurie S (2002) SW—soil and water: effect of water salinity and irrigation technology on yield and quality of pears. Biosyst Eng 81(2):237–247

Oster JD (1994) Irrigation with poor quality water. Agric Water Manag 25(3):271–297

Oster JD, Shainberg I (2001) Soil responses to sodicity and salinity: challenges and opportunities. Soil Res 39(6):1219–1224

Pang HC, Li YY, Yang JS, Liang YS (2010) Effect of brackish water irrigation and straw mulching on soil salinity and crop yields under monsoonal climatic conditions. Agric Water Manag 97(12):1971–1977

Pradeepa K, Lavanya G, Inbarasan S et al (2019) Reclamation of sodic bore well water for irrigation through gypsum treatment at Anbil Dharmalingam Agricultural College and Research Institute, Trichirapalli District, Tamil Nadu. Agric Sci Digest 39(4):286–329

Qadir M, Schubert S, Oster JD, Sposito G, Minhas PS et al (2018) High-magnesium waters and soils: emerging environmental and food security constraints. Sci Total Environ 642:1108–1117

Rani S, Satyavan KA, Beniwal S (2019) Integrated nutrient management as a managerial tool for applying saline water in wheat crop cultivated under sub-tropic and semi-arid conditions of North-Western India. J Plant Nutr 43(4):604–620

Rengasamy P (2018) Irrigation water quality and soil structural stability: a perspective with some new insights. Agronomy 8(5):7

Richards LA (ed) (1954) Diagnosis and improvement of saline and alkali soils. USDA, US Government Printing Office, Washington, DC

Rusan MJ, Hinnawi S, Rousan L (2007) Long term effect of wastewater irrigation of forage crops on soil and plant quality parameters. Desalination 215(1–3):143–215

Sahin U, Anapali O, Ercisli S (2002) Physico-chemical and physical properties of some substrates used in horticulture. Gartenbauwissenschaft 67:55–60

Sang H, Guo W, Gao Y et al (2020) Effects of alternating fresh and saline water irrigation on soil salinity and chlorophyll fluorescence of summer maize. Water 12(11):3054

Schacht K, Chen Y, Tarchitzky J, Lichner L, Marschner B (2014) Impact of treated wastewater irrigation on water repellency of Mediterranean soils. Irrig Sci 32:369–378

Schacht K, Marschner B (2015) Treated wastewater irrigation effects on soil hydraulic conductivity and aggregate stability of loamy soils in Israel. J Hydrol Hydromech 63(1):47–54

Sharma DK, Singh A, Sharma PC (2016) Role of ICAR-CSSRI in sustainable management of salt-affected soils—achievements, current trends and future perspectives 4th International Agronomy Congress (November 22–26), pp 91–103

Sharma DP, Rao KV, Singh KN et al (1994) Conjunctive use of saline and non-saline irrigation waters in semi-arid regions. Irrig Sci 15:25–33

Sharma DR, Minhas PS (2003) Management options and policy guidelines for use of poor quality groundwater in agriculture. In: Sharma, Bharat R, Villholth Karen G, Sharma KD (eds) Groundwater research and management: Integrating science into management decisions. International

workshop on creating synergy between groundwater research and management in South and Southeast Asia, -9 February 2005, Roorkee, India
Shatanawi M, Fayyad M (1996) Effect of Khirbet As-Samra treated effluent on the quality of irrigation water in the Central Jordan Valley. Water Res 30(12):2915–2920
Shehzad I, Sarwar G, Manzoor MZ et al (2020) Effect of saline water irrigation on chemical properties and fertility status of soil. Pakistan J Agri Res 33(3):527–534
Shrivastava P, Kumar R (2015) Soil salinity: a serious environmental issue and plant growth promoting bacteria as one of the tools for its alleviation. Saudi J Biol Sci 22(2):123–131
Siebert S, Burke J, Faures JM et al (2010) Groundwater use for irrigation–a global inventory. Hydrol Earth Syst Sci 14(10):1863–1880
Singh A (2014) Conjunctive use of water resources for sustainable irrigated agriculture. J Hydrol 519:1688–1697
Singh G, Mavi MS, Choudhary OP, Gupta N, Singh Y (2021) Rice straw biochar application to soil irrigated with saline water in a cotton-wheat system improves crop performance and soil functionality in north-west India. J Environ Manag 295:113277
Singh M, Kumar J, Singh VP, Prasad SM (2014) Plant tolerance mechanism against salt stress: the nutrient management approach. Biochem Pharmacol 3(165):2167–2501
Singh P, Choudhary OP, Mavi MS (2018a) Irrigation-induced salinization effects on soil chemical and biological properties under cotton-wheat rotation on loamy sand soil in northwest India. J Indian Soc Soil Sci 66(4):386–391
Singh R, Singh PK, Madheshiya P et al (2024) Heavy metal contamination in the wastewater irrigated soil and bioaccumulation in cultivated vegetables: assessment of human health risk. J Food Compos Anal 128:106054
Singh RB, Minhas PS, Chauhan CP, Gupta RK (1992) Effect of high salinity and SAR waters on salinization, sodication and yields of pearl-millet and wheat. Agric Water Manag 21(1–2):93–105
Singh S, Ghosh NC, Gurjar S et al (2018b) Index-based assessment of suitability of water quality for irrigation purpose under Indian conditions. Environ Monit Assess 190:1–4
Singh SK, Srivastava PK, Singh D et al (2015) Modeling groundwater quality over a humid subtropical region using numerical indices, earth observation datasets, and X-ray diffraction technique: a case study of Allahabad district, India. Environ Geochem Health 37(1):157–180
Singh YP (2013) Gypsum bed for improving quality of alkali water. Indian Farm 63(1):16–17, 28
Sonbol HA, El Hadidi EM, Saied MM, El-F A (2007) Effect of surface irrigation methods and land leveling on salt distribution and moisture uptake by cotton plant in salt affected soils at North Delta. J Agric Sci Mansoura Univ 32(9):7999–8015
Sou DMY, Mermoud A, Yacouba H, Boivin P (2013) Impacts of irrigation with industrial treated wastewater on soil properties. Geoderma 200:31–39
Sundha P, Basak N, Rai AK, Yadav RK, Sharma PC, Sharma DK (2020) Can conjunctive use of gypsum, city waste composts and marginal quality water rehabilitate saline-sodic soils? Soil Tillage Res 200:104608
Tahtouh J, Mohtar R, Assi A et al (2019) Impact of brackish groundwater and treated wastewater on soil chemical and mineralogical properties. Sci Total Environ 647:99–110
Teffera ZL, Li J, Astatkie T, Gebru ZM (2019) Effects of irrigation systems on physicochemical properties of soil at different depths: a case study at a farm near Ziway Lake, Ethiopia. Irrig Drain 68(2):165–175
Torsvik V, Øvreås L (2002) Microbial diversity and function in soil: from genes to ecosystems. Curr Opin Microbiol 5(3):240–245
Ungureanu N, Vlăduţ V, Dincă M, Zăbavă BŞ (2018) Reuse of wastewater for irrigation, a sustainable practice in arid and semi-arid regions. In: Proceedings of the 7th international conference on thermal equipment, renewable energy and rural development, vol 31, Drobeta-Turnu Severin, Romania, pp 379–384
Vanani HR, Toudeshki AR, Shayannejad M et al (2017) Flow hydraulic investigation of the wastewater on the soil and magnetic field effects in this field. Int J Civ Eng 3(3):1–5

Velmurugan A, Swarnam TP, Ambast SK, Kumar N (2016) Managing waterlogging and soil salinity with a permanent raised bed and furrow system in coastal lowlands of humid tropics. Agric Water Manag 168:56–67
Vera A, Bastida F, Patiño-García M, Moreno JL (2023) The effects of boron-enriched water irrigation on soil microbial community are dependent on crop species. Appl Soil Ecol 181:104677
Vera Ayala A, Moreno-Ortego JL, Siles Martos JA, López Mondejar R et al (2021) Interactive impacts of boron and organic amendments in plant-soil microbial relationships. J Hazard Mater 408:124939
Vogelmann ES, Reichert JM, Prevedello J, Awe GO, Mataix-Solera J (2013) Can occurrence of soil hydrophobicity promote the increase of aggregates stability? CATENA 110:24–31
Wallis MG, Horne DJ (1992) Soil water repellency. Adv Soil Sci 20:91–146
Wang H, Feng D, Zhang A et al (2022) Effects of saline water mulched drip irrigation on cotton yield and soil quality in the North China Plain. Agric Water Manag 262:107405
Wang H, Gilbert JA, Zhu Y, Yang X (2018) Salinity is a key factor driving the nitrogen cycling in the mangrove sediment. Sci Total Environ 631:1342–1349
Wang H, Zheng C, Ning S et al (2023) Impacts of long-term saline water irrigation on soil properties and crop yields under maize-wheat crop rotation. Agric Water Manag 286:108383
Winter TC, Harvey JW, Franke OL, Alley WM (2000) Groundwater and surface water: a single resource. U.S. Geological Survey, Virginia, USA
Wu AM, Li CQ, Xu YZ, Liu JR, Luo GZ (2010) Key issues influencing sustainable groundwater utilization and its countermeasures in North China Plain. South-to-North Water Transf Water Sci Technol 8(6):110–113
Yaduvanshi NPS, Swarup A (2005) Effect of continuous use of sodic irrigation water with and without gypsum, farmyard manure, pressmud and fertilizer on soil properties and yields of rice and wheat in a long term experiment. Nutr Cycl Agroecosyst 73:111–118
Yang T, Cherchian S, Liu X et al (2023) Effect of water application methods on salinity leaching efficiency in different textured soils based on laboratory measurements and model simulations. Agric Water Manag 281:108250
Yang T, Šimůnek J, Mo M et al (2019) Assessing salinity leaching efficiency in three soils by the HYDRUS-1D and-2D simulations. Soil Tillage Res 194:104342
Yu X, Jin Z, Wang H (2021) Effect of saline water for drip irrigation on microbial diversity and on fertility of aeolian sandy soils. Diversity 13(8):379
Yuan BC, Li ZZ, Liu H, Gao M, Zhang YY (2007) Microbial biomass and activity in salt affected soils under arid conditions. Appl Soil Ecol 35(2):319–328
Zaman M, Shahid SA, Heng L (2018) Guideline for salinity assessment, mitigation and adaptation using nuclear and related techniques. Springer Nature Switzerland
Zhai HM, Cao CY, Liu MY (2018) Impact of irrigation with saline water on soil enzyme activity and soil enzyme kinetics. J Agric Res Arid Areas 36(1):95–101
Zhang A, Li K, Sun J et al (2020) Effects of a 10-year irrigation with saline water on soil physico-chemical properties and cotton production. J Soil Water Conserv 75(5):629–639
Zhao YK, Zhang WS, Wang YN et al (2008) Research progress in physiology and molecular biology of plant responses to high pH. Chin J Eco-Agric 3:783–787
Zhou S, Gao Y, Zhang J, Pang J, Hamani AK, Xu C, Dang H, Cao C, Wang G, Sun J (2023) Impacts of saline water irrigation on soil respiration from cotton fields in the North China plain. Agronomy 13(5):1197
Zhu H, Yang J, Yao R et al (2020) Interactive effects of soil amendments (biochar and gypsum) and salinity on ammonia volatilization in coastal saline soil. Catena 190:104527

Crop Selection Criteria for Different Quality of Irrigation Water and Mitigation Techniques

Udit Saxena, Vinod Kumar, Yogendra Kumar, Arvind Singh Tomar, and Dinesh Kumar Vishwakarma

Abstract The implementation of effective crop selection and management strategies becomes an essential undertaking when the quality of water is not ideal for irrigation. The quality of water varies by location and is independent of whether the source is surface or sub-surface water. Water quality highly impacts crop productivity, soil health, nutrient availability, and microbial contamination. The presence and concentration of salinity, sodicity, toxic elements, pH, heavy metals, and microbial communities in water determines the standard of water for irrigation use. The quality of irrigation water is mainly assessed for EC, TDS, SAR, and RSC and classified into different categories proposed by various agencies and researchers. These classifications aid in crop selection and deciding management techniques to improve the standard of irrigation water to meet regional agricultural needs. The assessment of specific toxic ions mainly includes sodium (Na), chloride (Cl), and boron (B) ions present in irrigation water. The irrigation water quality index (IWQI) reduces the complexity of the analysis of various parameters and assigns a single value to evaluate the status of water for irrigation quality. This classification is accountable for water use restrictions imposed on soil and plants having different properties and tolerances. Strategies for managing crop selection are very essential and include practices such as soil testing to determine nutrient levels, crop rotation to prevent soil depletion, and irrigation management for optimal use of water. These practices improve crop yields and soil health that were depleted over time. The process of applying leaching water to reduce salinity and reduction of sodicity using gypsum are widely employed technique to mitigate the adverse effects of poor-quality irrigation water. The utilization of geographic information system (GIS) in analyzing and interpreting irrigation water quality is also discussed in this chapter.

Keywords Microbial contamination · Salinity · Sodicity · Geographic information system · Electrical conductivity · Total dissolved solids

---

U. Saxena (✉) · V. Kumar · Y. Kumar · A. S. Tomar · D. K. Vishwakarma
Department of Irrigation and Drainage Engineering, Govind Ballabh Pant University of Agriculture and Technology, Pantnagar, Udham Singh Nagar, Uttarakhand, India
e-mail: uditsaxena14@gmail.com

Dimple et al. (eds.), *Surface Water and Groundwater Quality for Sustainable Utilization*, Springer Hydrogeology, https://doi.org/10.1007/978-981-96-8189-1_11

1 Introduction

Water is one of the most important fundamental for proper functioning of the ecosystem which does not have any substitute. Domestic activities, agricultural processes, and industrial operations are the most water-consuming sectors and never-ending processes. The water demand is not only related to human-related activities as the whole ecosystem relies on it. The realm of nature gives birth to this pivotal fluid and also uses it to nourish itself. Forests, plants, trees, animals, and marine life are major assets that depend on water for their existence. The hydrological cycle governs the circulation of water across, over, and under the surface of the earth, depending on the environmental conditions, seasonal variations, climate change effects, and various other factors.

Over the past few decades, heavy importance has been laid upon the standard of surface and sub-surface water. Researchers have pointed out the possibility of water being polluted to such an extent that it cannot reversed back to its standard circumstances. This can only be identified by carrying out proper water quality assessment techniques. The agriculture sector depends on various components such as fertile land, seeds, chemicals, and most importantly water. A decline in water quality will lead to a decline in food production. This will not only create an effect on the agricultural sector but also the global economy. The United States of America, China, India, and Russia are the top producers of agricultural goods, and the downfall in the economy of these countries will create an economic imbalance in the whole world. Therefore, the requirement of water quality assessment must not be approached softly. This assessment can create a domino effect, setting involvement of various aspects such as identification of sources of contamination, concentration of salts, public health and safety, demand and supply of water, policy making, and various religious and social issues.

The processes involved in improving the surface and sub-surface water quality should be carried out without compromising the current supply and demand of crops. Improving water quality is not a short process and at some locations, it is very difficult to restore the quality of water due to the unavailability of resources. Therefore, crop selection becomes a crucial task in sowing only those crops that can withstand the current quality of water supply so that crop production can be maintained. The result of the water quality assessment can be subsequently analyzed along with the soil fertility, climate, and additional significant factors to determine the suitability of the crop. By integrating water quality data, farmers can optimize crop selection to enhance crop productivity and resource efficiency.

Factors like salinity, sodium adsorption ratio (SAR), and the presence of toxic ions or heavy metals usually determine the quality of irrigation water. For example, cotton and barley can tolerate elevated levels of salts in irrigation water, and can be cultivated under highly saline water conditions, while, crops that are more delicate, like citrus and beans, need high-quality irrigation water. The negative effects of suboptimal water quality can be reduced by using mitigation strategies such as blending low-quality water with higher-quality sources, adding gypsum to soil to prevent sodicity, and implementing sophisticated irrigation techniques like drip irrigation. Various water quality parameters and mitigation techniques are discussed in this chapter. Water quality index method involved in simplifying the result of the analysis of various parameters is also discussed.

2 Crop Selection

Crop selection can be utilized as a strategic plan to eliminate the risk of hazards caused to the plants due to the rise in toxicity in soil water. Salinity and toxicity, both are major hindrances in crop growth development. To mitigate the detrimental effects of salts accumulating in the root zone and affecting crop yield; opting for crops that are more resistant to salinity, specific ion toxicity, heavy metal concentrations, and other pollutants should be preferred rather than those that are less tolerant to these hazardous conditions. It is crucial to remember that these are just approximations and can be altered due to local farming conditions.

Crops with high tolerance, such as sugar beet and barley, deal with salt stress by preserving high potassium levels and promoting osmotic adjustment by accumulating suitable solutes such as proline and glycine betaine. On the other hand, crops with moderate tolerance, such as soybeans and wheat, use techniques to limit the build-up of salt and chloride in their leaves while allowing them to absorb enough water. Soybeans and kinds of wheat regulate the selective uptake of ions, which includes the preference of potassium ions over sodium and chloride ions.

Furthermore, crop rotation and diversification can also play a significant role in managing soil salinity and toxicity. Rotating salt-tolerant crops with less-tolerant ones allows for the natural remediation of soil conditions, as certain crops help improve soil structure, organic matter content, and microbial activity. For example, legumes in crop rotations can enhance nitrogen fixation and improve soil fertility, reducing the overall impact of salinity on subsequent crops. Additionally, integrating cover crops between growing seasons can prevent the accumulation of salts by promoting water infiltration and reducing evaporation, which further mitigates the risk of salt build-up in the root zone. This holistic approach ensures long-term soil health and sustainable crop production in challenging environmental conditions.

3 Irrigation Water Sampling

3.1 General Sampling Procedures

Various sampling procedures have been assorted by different agencies around the world for carrying out water sample collection. However, the basic functions that are carried out while water sampling are precisely the same while collecting water samples for assessment. A simple approach has been designed to understand the general sampling procedure as described below.

i. A sampling and assessment strategy (SAT) has to be mobilized that includes the source and location of sampling points, and the amount of water to be collected for testing of various parameters. Sometimes samples are collected from random locations, in this case, the exact geographical coordinates and location of that sampling point should be recorded at the time of sampling.
ii. Before collecting samples, it is necessary to ensure that the chemicals, instruments, and various sampling and testing apparatus are available in the laboratory, which will be needed for water quality assessment of different parameters.
iii. Collect all the equipment that is to be taken on the site of the sampling location. Sampling bottles must be rinsed with distilled water carefully. All the equipment required for the analysis of water quality parameters that are to be tested on-site must be in proper working conditions and a record book must be carried to note the location and name of the sample, date and time of sample collection, and method of testing and result of analysis in case of on-field tests.
iv. Collect samples in an environment devoid of significant dust, rain, snow, or any other potential sources of contamination.
v. To ensure the removal of any pre-existing elements in the sampling bottle, it should be rinsed with the water to be collected at least three times.
vi. If the water is collected from wells, piezometers, or test holes, the containers from which the water is being extracted should not be subjected to corrosion and free from any foreign particles. In the case of handpumps, taps, or any source that consists of any faucet, it should be considered that the material should not be in a state of disrepair, or else collect water from any other location.
vii. On-field tests should be carried out very carefully, especially for surface water quality assessment. Proper distance should be maintained from the water body, or in special cases, proper safety equipment should be carried and used wisely.
viii. Carry the samples to the laboratory carefully so that the holding time requirements are met, which begins at the time of sample collection and ends at the end of analysis.
ix. Field kits, sampling bottles, and other materials should be returned to the laboratory for proper disposal or reuse.

3.2 Sources of Water Collection

Water quality assessment involves the collection of water samples from various sources and analyzing their physical, chemical, and biological characteristics. Some surface and sub-surface water sources are mentioned in Table 1. Each source is subjected to unique challenges and problems, assessing the quality of water from each source in an area aids researchers in assessing the overall water quality and making informed decisions for water resource management and protection.

Table 1 Sources of water for quality assessment

Category	Source
Surface water	Rivers and streams Lakes and reservoirs Ponds and wetlands Canals and ditches Urban stormwater runoff
Sub-surface water	Monitoring wells Water wells Piezometers Spring outflows Test holes Seepage or surface water interaction zones

4 Assessment of Quality of Irrigation Water and Suitability of Crops

Several physicochemical parameters are engaged in assessing irrigation water quality, consisting of several anions and cations, heavy metals, microbial contaminants, and organic matters, that are mostly found in surface and sub-surface water applied for irrigation use. These parameters are not just simply evaluated but then are further studied to justify whether they are within the acceptable range or in an amount that leads to a decline in grade. Although many indicators are available for irrigation water quality assessment, researchers mostly study fundamental measures that aid in identifying the quality of irrigation water.

4.1 Salinity Hazard

The occurrence of salts in water depends on the geologic conditions, as most of the salts observed in water are naturally occurring salts, governed by the rocks surrounding water. Nevertheless, chemicals used in farming, and polluted water flowing from various point and non-point sources are a few other factors that are to be considered while studying about accumulation of salts. Elevated magnitude of salts mostly leads to yield losses and to counter this, the salt levels must be controlled below a certain concentration. A very common situation arises in the irrigation field, that plants tend to wilt even in the presence of adequate water. This happens because the osmotic pressure created by the salts prevents water molecules from moving freely and forms a barrier that hinders the roots from absorbing water.

Salts soluble in irrigation water are mostly evaluated by measuring their EC (in $\mu S\ cm^{-1}$) and TDS (in mg/l), and utilized to describe the presence of salinity in irrigation water. The relative crop salinity tolerance ratings approved by Mass (1984) are presented in Table 2.

Table 2 Tolerance of agricultural crops to salinity (Mass 1984)

Salinity tolerance	Soil salinity threshold for initial yield loss (dS/m)	Crops
Sensitive	<1.30	Bean, sesame, carrot, guayule, onion, okra, parsnip
Medium sensitive	1.30–3.00	Rye, broccoli, sesbania, celery, alfalfa, siratro, cabbage, cauliflower, Brussels sprouts
Medium tolerant	3.00–6.00	Soybean, barley, triticale, cowpea, sorghum, safflower, oats, wheat, artichoke
Tolerant	6.00–10.00	Kallar grass, jojoba, alkali sacaton, date palm, asparagus, cotton, sugarbeet
Unsuitable for most crops	>10.00	

4.1.1 Electrical Conductivity

Distilled water has low electrical conductivity as compared to natural water bodies due to the deficiency of dissolved salts and minerals, which gives rise to electrical conductivity in the form of ions. Higher electrical conductivity values reveal the concentration of dissolved ions such as sodium, chloride, calcium, magnesium, and sulfate. These ions contribute to the overall electrical conductivity, providing an indication of the ionic composition and quality of water. Elevated conductivity levels can signal higher concentrations of these ions, which might affect water suitability for various uses.

4.1.2 Total Dissolved Solids

TDS is the total concentration of dissolved organic and inorganic matter in water which may or may not occur naturally. The dissolved concentration of minerals, salts, and nutrients is mostly due to the geologic conditions, land use practices, and treatment processes applied, moreover, higher TDS levels sometimes indicate the contamination occurred due to industrial discharges, return flow from agricultural fields, waterlogging and other sources consisting wastewater effluents. As a result of dissolved contaminants in the water, TDS can govern the taste, odor, and appearance of water.

The classification of salinity hazard is shown in Table 3, as suggested by Follett and Soltanpour (2002) and Ayers and Westcot (1985).

Table 3 Salinity hazard classification based on EC and TDS levels

References	EC unit	EC	TDS (ppm)	Effects on crops
Follett and Soltanpour (2022)	$\mu S\ cm^{-1}$	750	500	None
		750–1500	500–1000	Some
		1500–3000	1000–2000	Moderate
		3000–7500	2000–5000	Severe
Ayers and Westcot (1985)	$dS\ m^{-1}$	<0.7	<450	None
		0.7–3.0	450–2000	Slight to moderate
		<3.0	>2000	Severe

4.2 *Sodium Hazard*

A rise in sodium concentration in irrigation water diminishes soil permeability, disrupts infiltration rates, and accelerates the deterioration of soil structure. This phenomenon occurs when exchangeable sodium in the soil displaces calcium and magnesium ions. Research indicates that when calcium and magnesium dominate as the primary cations in the soil, it tends to be more workable, with a granular and permeable structure. However, as sodium ions gradually replace calcium and magnesium, the soil structure begins to deteriorate, becoming dense and compact, leading to decreased permeability and slower infiltration rates. Soils with fine textures, especially those with high clay content, are particularly susceptible to this degradation process.

4.2.1 Sodium Adsorption Ratio (SAR)

SAR is a calculated index employed to assess the influence of Na^+ ions on Ca^{2+} and Mg^{2+} ions within the soil matrix. It provides insight into the potential risks of sodium-induced soil degradation, particularly with respect to soil structure and permeability. Irrigation water with an SAR value exceeding 10 $(\text{mmole l}^{-1})^{0.5}$ is generally deemed unsuitable for agricultural purposes due to the risk of sodium accumulation, which can lead to soil compaction, reduced infiltration, and poor crop performance. However, certain exceptions arise, particularly when the soil contains sufficient levels of gypsum. In such scenarios, even water with an SAR value above 10 $(\text{mmole l}^{-1})^{0.5}$ may be deemed acceptable, as the presence of gypsum helps mitigate sodium's adverse effects by providing a source of calcium, which counters the destabilizing influence of sodium on soil particles.

SAR quantifies the relative concentration of sodium in comparison to divalent cations like calcium (Ca^{2+}) and magnesium (Mg^{2+}), which are crucial for maintaining soil structure. This metric can be determined using the formula proposed by Richards (1954),

Table 4 Sodium adsorption ratio limits as suggested by USSL (Richards 1954) and FAO (Ayers and Westcot 1985)

References	Class	Range (dS/m)
USSL (Richards 1954)	Low	<10
	Medium	10–18
	High	18–26
	Very high	>26
FAO (Ayers and Westcot 1985)	Low	<3
	Medium	3–9
	High	>9

$$SAR = \frac{Na^{+}}{\sqrt{\frac{Ca^{2+} + Mg^{2+}}{2}}} \quad (1)$$

The amount of Na^{+}, Ca^{2+} and Mg^{2+} in milligrams per litre (mg l^{-1}) is divided by their molecular mass to obtain the concentration in milliequivalent per litre (meq l^{-1}). The formula reflects the complex interactions between these ions, where higher sodium levels relative to calcium and magnesium increase the risk of sodium replacing calcium and magnesium on soil exchange sites, leading to soil dispersion and reduced permeability. Classification of sodium adsorption ratio as suggested by USSL and FAO is given in Table 4.

4.2.2 Calcium and Magnesium

The occurrence of these ions in water arises from the presence of dominant rocks like calcite, dolomite, marble, gypsum, limestone, and fluorite. Calcium and magnesium are the indicators of hardness in water and higher levels of hardness in water indicate the elevated level of these minerals. The proportion of calcium to magnesium in water generally lies between 4:1 and 2:1, when the total dissolved solid is beyond the permissible range of 500 mg per litre (mg l^{-1}). The ethylenediaminetetraacetic acid (EDTA) titration technique is the most prevalent procedure employed to quantify the concentration of metal ions present in aqueous samples. This method relies on the ability of EDTA, a chelating agent, to form stable complexes with multivalent metal cations. This procedure is favored due to its precision, reproducibility, and applicability in diverse water testing scenarios.

4.2.3 Residual Sodium Carbonate

Determination of RSC is a subsequent method to study the additional sodium hazard in irrigation water. This method is associated with the determination of calcium carbonate ($CaCO_3$) ions and magnesium carbonate (MgO_3) ions. RSC can be calculated by the equation given by Eaton (1950).

Table 5 Evaluation of irrigation water based on residual sodium carbonate (Eaton 1950; Wilcox et al. 1954)

RSC (milliequivalents per liter)	Water category
<1.25	Safe for irrigation
1.25–2.50	Marginal use of irrigation water
>2.50	Unsuitable for irrigation

$$RSC = \left(CO_3^{2-} + HCO_3^-\right) - \left(Ca^{2+} + Mg^{2+}\right) \quad (2)$$

The concentration of calcium, magnesium, carbonates, and bicarbonates are in milliequivalent per litre (meq l^{-1}). The classified values of residual sodium carbonate (RSC) given by Eaton (1950) are presented in Table 5.

4.3 Specific Ion Toxicity

Soil and water consist of various numbers of ions, which may be present in low, medium, or high concentrations. Toxicity arises when plants absorb and accumulate excessive amounts of these ions, resulting in harm to crops or vegetation and diminished yield. This toxicity is not limited to perennial crops subjected to high concentrations of ions, but at low concentrations, toxicity causes damage to sensitive crops. Almost all crops are inferior to toxicity and will be damaged if subjected to higher concentrations. The higher accumulations can be observed on the leaf tips and edges where water loss is maximum due to transpiration and it is evident that toxicity is firstly recognized by symptoms such as marginal leaf burn and interveinal chlorosis. When water is applied to a field, firstly it may not impose any harmful effect on the plant but after regular irrigation periods, ions accumulate to a toxic level gradually. The major ions causing toxicity are chloride, sodium, and boron. The problem of toxicity can arise when these ions are present in small amounts.

4.3.1 Chloride Toxicity

The amount of chloride concentration required by a plant for photosynthesis, osmosis, and plant metabolism is very subtle. As chloride is neither adsorbed nor restrained by the soil, it is readily taken up by plants with the soil water. Chloride deficiency and chloride toxicity are both very intense processes, but, as most of the water sources and fertilizers aid in fulfilling the requirement of chloride, it is very rare to see chloride deficiency in crops or plants, until water requirements are not met properly. After taking soil water, crops transfer chloride to the leaves through a transpiration stream. The effect of various levels of chloride concentration on crops is shown in Table 6. As reported by Ayers and Westcot (1985), the effect of excess chloride concentration

Table 6 Chloride concentration and crop selection (Bauder et al. 2011)

Chloride concentration		Effect on crops
(meq/l)	(ppm)	
<2.0	<70.0	Reliable for all crops
2.0–4.0	70.0–140.0	Sensitive plants typically exhibit minor to medium damage
4.0–10.0	141.0–350.0	Crops with moderate tolerance display minimal to significant damage
>10.0	>350.0	Can lead to significant issues

develops symptoms in crops, such as leaf scorching or desiccation of leaf tissue. These symptoms initially occur at the tip of the plants and then start moving toward the edges due to an increase in toxicity.

4.3.2 Sodium Toxicity

Sodium is required in small amounts for photosynthesis and metabolism of plants and facilitates in opening and closing of stomates to normalize internal water balance. Sodium can be partially used as a substitute for potassium. Sodium toxicity is mainly identified by leaf burns and dead tissues which mainly occur at the edges of the leaves. Sodium toxicity can be reduced by the presence of calcium in soil. Various studies have shown that most of the crops are dealing with calcium deficiency rather than sodium toxicity. Typically, foliar tissue examination is employed to identify sodium-induced phytotoxicity; however, a comprehensive assessment incorporating soil, water, and vegetative tissue provides a more precise and reliable diagnosis. By analyzing these three components in tandem, one can better ascertain the extent of sodium accumulation and its impact on plant health, leading to a more holistic understanding of the environmental factors contributing to ionic imbalances within the ecosystem. Exchangeable Sodium Percentage levels, instead of SAR, indicate the tolerance of various crops as shown in Table 7.

Table 7 Crop tolerance to exchangeable sodium percentage in irrigation water (Pearson 1960; FAO-UNESCO 1973; Abrol 1982)

Exchangeable sodium percentage	Class	Tolerance of selected crops
<15	Sensitive	Avocado, peas, tangerine, orange, grapefruit, maize, mung, mash, lentil, gram, cowpeas, groundnut, peach
15–40	Semi-sensitive	Carrot, vetch, rye, tomato, sorghum, lettuce, sugarcane, onion, radish, dallisgrass, rice, bersem, spinach, benji, bajara, wheat, oat
>40	Tolerant	Alfalfa, cotton, bermuda grass, rhodes grass, beet (sugar), barley, paragrass, karnal grass

4.3.3 Boron Toxicity

Boron is a fundamental component required for plant growth and is engaged in various operations such as maturation, division, meristematic growth cellular differentiation, and elongation. Boron toxicity is identified by marginal leaf chlorosis and necrosis, progressing to scorching, burning, and premature shedding of inner foliage. This chronic defoliation causes various stresses in plants and leads to deterioration. Due to the weathering of rocks, boron is released into the soil in the form of boric acid, and in this form, it can leach easily through the soil. The pH of soil water is capable of affecting the availability of boron. Boron toxicity is often linked to irrigation water from wells, particularly in saline soils. The subsequent boron uptake by plants is contingent upon various soil factors, including texture and its capacity to adsorb boron. The effect of various levels of boron concentration in irrigation water on the crops is presented in Table 8. Water with a boron concentration of <0.5 ppm is satisfactory for all crops. Wilcox also gave a classification for boron toxicity in plants, which consists of three classes, namely tolerant, semi-tolerant, and sensitive. Maas (1984) reviewed various techniques for the presence of boron toxicity in crops and gave a classification for the relative tolerance of crops to boron which is shown in Table 9.

4.4 United States Salinity Laboratory (USSL) Diagram

The United States Salinity Laboratory (USSL) diagram is a technique for evaluating irrigation water quality based on salinity and sodium risks. The salinity risk is quantified using electrical conductivity (EC, in μS/cm), while the sodium risk is determined through the sodium adsorption ratio (SAR). The USSL diagram is constructed by graphing SAR values against the EC values for the respective water samples. Both the salinity and sodium risks are categorized into four levels: low, moderate, high, and very high, as detailed in Table 10, which includes the corresponding EC and SAR measurements. The USSL diagram is segmented into various classifications, as depicted in Fig. 1.

Table 8 Effect on crops due to various levels of boron concentration in irrigation water (Follett and Soltanpour 2002)

Boron concentration (ppm)	Effect on crops
<0.5	Satisfactory for all crops
0.5–1.0	Satisfactory for most crops
1.0–2.0	Satisfactory for semi-tolerant crops
2.0–4.0	Satisfactory for tolerant crops only

Table 9 Boron tolerance levels in agricultural crops (Maas 1984)

Boron concentration (ppm)	Class	Crops
<0.50	Very sensitive	Blackberry, lemon
0.50–0.75	Sensitive	Orange, walnut, apricot, avocado, bean, persimmon, plum, onion, cherry, peach, cowpea, grapefruit, grape
0.75–1.00	Less sensitive	Strawberry, sweet potato, garlic, peanut, lupine, sunflower, wheat, sesame, barley, artichoke
1.00–2.00	Medium sensitive	Radish, potato, cucumber, pea, carrot, pepper (red)
2.00–4.00	Medium tolerant	Oats, celery, tobacco, artichoke, lettuce, cabbage, turnip, maize, mustard, muskmelon
4.00–6.00	Tolerant	Tomato, parsley, sugarbeet, alfalfa, sorghum
6.00–15.00	Very tolerant	Asparagus, cotton

Table 10 Different levels of salinity and sodium hazards caused due to irrigation water (USSL Staff 1954)

Hazard	Standard	Class	Limit
Salinity hazard	EC (μS/cm)	Low (C1)	100–250
		Medium (C2)	250–750
		High (C3)	750–2250
		Very high (C4)	>2250
Sodium hazard	SAR $(\text{meq l}^{-1})^{0.5}$	Low (S1)	<10
		Medium (S2)	10–18
		High (S3)	18–26
		Very high (S4)	>26

4.4.1 Salinity Hazard Classes

Salinity hazard is evaluated by measuring the EC of the water. The different classifications of salinity hazard are discussed below:

(a) Low salinity water (100–250 μS/cm)

Low-salinity water can be safely used for irrigating a wide range of crops on different soils, posing only a minimal risk of increasing soil salinity. Apart from soils with very low permeability, leaching can be adopted under normal irrigation practices.

(b) Medium salinity water (250–750 μS/cm)

Medium-salinity water is suitable for use in irrigation when moderate leaching is practiced. This type of water allows for the cultivation of plants with medium salt

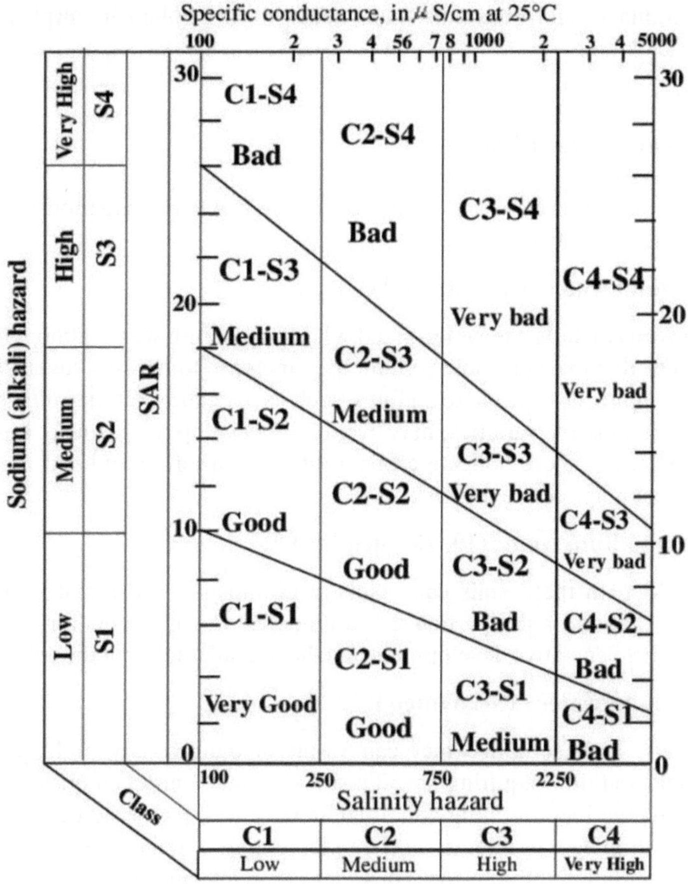

Fig. 1 USSL diagram (USSL Staff 1954)

tolerance without the need for stringent salinity control measures. Crops that thrive in medium-salinity conditions often include certain varieties of vegetables, grains, and forage crops.

(c) High salinity water (750–2250 μS/cm)

Soils with inadequate drainage cannot be treated with high-salinity water. Significant salinity control management is necessary even with sufficient drainage, and utilization of salinity-tolerant crops should be considered.

(d) Very high salinity water (>2250 μS/cm)

Although this water can be used occasionally for salt-tolerant crops it should not be practised under normal conditions. The exceptions to using this water for irrigation should meet a few requirements, mainly, soils should have good permeability and sufficient drainage, and irrigation water needs to be applied in surplus to ensure significant leaching.

4.4.2 Sodium Hazard Classes

The sodium hazard classes are illustrated by the SAR of irrigation water. These classes are described below:

(a) Low sodium water (<10 (meq l^{-1})$^{0.5}$)

Almost all types of soils can be fertilized with low-sodium water, although there are a few chances that exchangeable sodium may increase to harmful levels. Nevertheless, crops that are susceptible to sodium, such as avocados, citrus fruits, stone fruit trees, and certain berry varieties, may eventually develop dangerously high sodium concentrations. For these delicate crops, routine monitoring and soil management techniques can help reduce this risk.

(b) Medium sodium water (10–18 (meq l^{-1})$^{0.5}$)

Sodium buildup in these soils can result in diminished soil architecture, heightened disaggregation of soil particles, and hindered water percolation, all of which adversely affect vegetative development and the overall vitality of the soil ecosystem.

(c) High sodium water (18–26 (meq l^{-1})$^{0.5}$)

In most soils, high sodium water can result in dangerously high quantities of exchangeable sodium, requiring specific soil management techniques. These practices may involve incorporating soil amendments such as the addition of gypsum, overcoming drainage problems, and implementing regular leaching to prevent sodium accumulation. Opting for crops that can withstand high salinity water and monitoring the sodium content of the soil is also essential for reducing the possible negative impacts of high sodium irrigation water.

(d) Very high sodium water (>26 (meq l^{-1})$^{0.5}$)

Irrigation with very high sodium water should be avoided as it could result in elevated levels of exchangeable sodium in the soil. Soil amendments like gypsum can help mitigate sodium hazards by enhancing soil structure and mitigating the negative effects of sodium on crops. Even with gypsum treatment, the use of very high sodium water should be limited to only highly salt-tolerant crops and requires careful monitoring to prevent long-term soil salinization.

4.5 Water Quality Index

The water quality index is a crucial tool that simplifies the intricacy of datasets and returns a single unit to identify the condition of water. While carrying out water quality assessment, numerous parameters are analyzed and it is not necessary that all the parameters either lie within or above the desirable range suggested by various agencies. Therefore, it can be difficult to determine the quality of water based on numerous parameters, and to overcome this challenge, WQI is evaluated, which ascertains the quality of water by integrating the effects of each parameter and giving a single value for its classification. Researchers have proposed several water quality indices and are still being utilized in ongoing studies around the world.

4.5.1 Irrigation Water Quality Index (IWQI)

IWQI, formulated by Meireles et al. (2010), serves as an extensive framework for evaluating and categorizing the suitability of water for irrigation purposes. This index integrates five critical parameters, namely EC, SAR, Na^+, Cl^-, and HCO_3^- levels. By analyzing these variables, IWQI provides a holistic understanding of how water quality might affect soil properties, crop health, and long-term agricultural sustainability. Each parameter is assigned a water quality rating (Q_i) and a weight (W_i), which reflect its relative importance and impact on water quality. The IWQI is computed by summing the products of these ratings and weights for all five parameters. By integrating multiple indicators, the IWQI offers a more nuanced understanding of water quality, helping to identify potential issues and guide appropriate management practices for maintaining optimal irrigation conditions.

$$IWQI = \sum_{i=1}^{n} Q_i W_i \tag{3}$$

The water quality rating is determined by employing the tolerance limits of each parameter, as shown in Table 11, in the equation given below.

$$Q_i = Q_{imax} - \left[\frac{x_{ij} - x_{inf}}{x_{amp}}\right] \times Q_{iamp} \tag{4}$$

where, Q_{imax} denotes the utmost Q_i measure for the class, x_{ij} represents the perceived value of respective parameter, x_{inf} is the value equivalent to the diminished boundary of the class to which the parameter relates, x_{amp} signifies the class amplitude for the parameter, and Q_{iamp} is the amplitude of the class.

To determine x_{amp} for the concluding category of each factor, the upper threshold was set as the peak magnitude derived from the physicochemical assessment of the water specimens.

The value of the relative weight (W_i) of each parameter is given in Table 12.

Meireles et al. (2010) defined classes for IWQI based on the likelihood of salinity problems, a decrease in soil water infiltration potential, and plant toxicity. These classes for irrigation water quality index along with their characteristics are described in Table 13.

Table 11 Tolerance limits for calculating irrigation water quality rating (Meireles et al. 2010)

Q_i	EC ($dS\ m^{-1}$)	SAR ($(meq\ l^{-1})^{1/2}$)	Na^+ ($meq\ l^{-1}$)	Cl^- ($meq\ l^{-1}$)	HCO_3^- ($meq\ l^{-1}$)
85–100	$0.20 \leq EC < 0.75$	$2.0 \leq SAR < 3.0$	$2.0 \leq Na < 3.0$	$1.0 \leq Cl < 4.0$	$1.0 \leq HCO_3 < 1.50$
60–85	$0.75 \leq EC < 1.50$	$3.0 \leq SAR < 6.0$	$3.0 \leq Na < 6.0$	$1.0 \leq Cl < 4.0$	$1.50 \leq HCO_3 < 4.50$
35–60	$1.50 \leq EC < 3.00$	$6.0 \leq SAR < 12.0$	$6.0 \leq Na < 9.0$	$1.0 \leq Cl < 4.0$	$4.50 \leq HCO_3 < 8.50$
0–35	$0.20 > EC \geq 3.00$	$2.0 > SAR \geq 12.0$	$2.0 > Na \geq 9.0$	$1.0 > Cl \geq 10.0$	$1.0 > HCO_3 \geq 8.50$

Table 12 Relative weights (W_i) for determining IWQI (Meireles et al. 2010)

Parameter	EC	SAR	Na^+	Cl^-	HCO_3^-	Total
Weight (W_i)	0.211	0.189	0.204	0.194	0.202	1

Table 13 IWQI characteristics (Meireles et al. 2010)

IWQI	Restrictions	Plants recommendations
85–100	Not required	Most vegetation is unlikely to experience toxicity
70–85	Low restrictions	Avoid use with plants sensitive to salts
55–70	Moderate restrictions	Plants with intermediate salt tolerance can be cultivated effectively
40–55	High restrictions	Suitable for plants with moderate to high salt tolerance, with careful salinity management, except when the water has minimal Na^+, Cl^-, and HCO_3^- concentrations
0–40	Severe restrictions	Use exclusively for plants with high salt tolerance, except when the water has very low Na^+, Cl^-, and HCO_3^- concentrations

5 Mitigation Strategies for Managing Crop Selection

Different crops require different soil conditions to grow, and the amount of nutrients and water for all the crops also varies depending on the specific growing season and different climatic conditions. To increase crop yield and conserve water and other resources, it is essential to choose the most suitable crop under favorable conditions. Resolving issues with poor soil and water quality requires the use of efficient crop selection and management strategies. A few approaches to improve crop selection are discussed below:

5.1 Comprehensive Soil and Water Testing

As soil and water are the fundamental physical components required for crop production, the degradation of these two essential elements is an issue that must not be overlooked or taken lightly. Analysis of various parameters aids in determining the salinity, toxicity, sodicity, heavy metal concentration, and presence of other pollutants present in irrigation water. Electrical conductivity determines the salinity levels and an increase in electrical conductivity can inhibit plant growth. The assessment of specific ions causing toxicity in soils is also important, as these ions can be very hazardous if present at high levels. Sodium (Na), chloride (Cl), and boron are usual toxic ions that are mostly found at alarming levels. The assessment of these selected parameters and various other parameters gives a supporting testament to the farmers regarding crop selection, such as growing barley or sugar beet in high-salinity soils.

5.2 Crop Rotation

Crop rotation is a crucial dynamic of sustainable agriculture that promotes crop production, improves soil health, and aids in managing pests and disease cycles. Implementing crop rotation with vegetation that is tolerant to the local soil conditions, and climatic conditions and resistible to irrigation water that is subjected to high levels of sodicity, salinity, or toxicity. If the source of irrigation is groundwater, it is significant to highlight that the quality of groundwater during the pre-monsoon season is generally lower compared to the post-monsoon season. As the groundwater level during pre-monsoon season is lower, the concentration of salts is higher and crops that are tolerant to high salt levels must be preferred. In terms of maintaining soil structure, deep-rooted crops like alfalfa, carrot, oak trees, etc., should be considered in areas that are highly vulnerable to droughts throughout the season. These deep-rooted crops possess an ability to extract nutrients and water from deeper soil layers, contributing to improved soil aeration and soil health. This comprehensive crop rotation strategy supports sustainable farming practices and a stronger soil ecology.

5.3 Soil and Water Amendments

The rate of crop production is primarily determined by the rating of water and soil fertility. Therefore, to maintain or increase the performance of crop production and contribution towards sustainable agriculture, it becomes necessary to adapt soil and water amendments. These amendments comprise compounds and treatments that target diverse soil and water-related concerns that amplify the overall productivity of farms. A few amendments are discussed below.

i. Gypsum

Gypsum is the most widely utilized amendment for improving soil and water conditions. Gypsum (calcium sulfate dihydrate) is widely used in sodic soils to displace sodium ions, which improves soil structure, permeability, and root growth. Granular gypsum is utilized for enhancing sodic soils, and it is applied in the field at rates ranging from 5 to 40 tons per hectare. For extremely sodic soils, gypsum is applied at a rate of 40 tons per hectare as a single application. Water infiltration issues due to low EC or high SAR values typically affect the upper soil layer. To enhance the efficacy of gypsum, it should be either left on the surface or integrated into the soil at a minimal depth, rather than deeply embedded. This approach ensures optimal interaction with the soil, promoting better results in soil structure improvement and nutrient availability. Gypsum applied on the surface may leach into the soil rapidly, therefore, repeated application of gypsum in small quantities is more effective for infiltration problems. Whereas, for sodic soil problems, a larger application of gypsum at a single time is more effective.

Gypsum (calcium sulfate) interacts with sodium ions (Na^+) in sodic soils to produce sodium sulfate (Na_2SO_4), calcium ions (Ca^{2+}), and water. This reaction helps displace sodium ions from the soil exchange sites, improving soil structure and reducing sodium toxicity.

$$CaSO_4 \cdot 2H_2O + 2Na^+ \rightarrow Na_2SO_4 + Ca^{2+} + 2H_2O$$

Gypsum is particularly advantageous when introduced to water with minimal salinity (EC < 0.5 dS/m). Since it is more difficult to administer and dissolve enough calcium into the water to adequately counteract the sodium present in water, it is significantly less effective in higher salinity water. Applying irrigation water in which gypsum is present naturally, sodic soils can be reclaimed over several years by growing salt-tolerant crops. For rapid restoration, severely sodic soil may need up to 40 t/ha of gypsum and considerable leaching. On the other hand, soil can be restored over an extended period by depending on the amount of calcium in irrigation water (generally, Ca ≥ 2–3 meq/l), cultural methods like disking, plowing, etc., and sowing crops that can withstand high sodium levels. Deep cultivation promotes accelerated reclamation and improved infiltration.

Example 1 A water treatment facility is tasked with reducing the sodium concentration in a reservoir used for industrial cooling processes. The water in the reservoir has an electrical conductivity (EC) of 4 dS/m, and a sodium concentration of 7 meq/l. The facility needs to reduce the sodium concentration to 2 meq/l by adding gypsum to the water before it is used in the cooling systems. The total volume of water in the reservoir is 9000 cubic meters. If the gypsum available for this treatment has a purity of 80%, calculate the amount of gypsum (in metric tons) required to treat the entire reservoir.

Solution

The sodium concentration needs to be reduced from 7 to 2 meq/l.

Therefore, the difference in Na^+ concentration will be

$$= 7 - 2 = 5\,\text{meq/l}$$

Note: *For every 1 meq/l of* Na^+*, an equal amount of 1 meq/l of* Ca^{2+} *is needed, which corresponds to 0.08606 grams of gypsum per liter of solution.*

Hence, 5 meq/l of Na^+ will necessitate 5 multiples of 0.086065 g of gypsum per liter of the solution.

Now,

$$\text{Gypsum required} = 5 \times 0.08606 = 0.43035\ \text{grams/litre}$$

The total volume of water in the reservoir is 9,000 m^3, which is equivalent to 9,000,000 L.

Now calculating the total gypsum requirement,

$$\text{Total gypsum required} = 9{,}000{,}000 \times 0.43039{,}000{,}000$$
$$= 3{,}872{,}700 \text{ grams or } 3.87 \text{ metrictons}$$

As the gypsum provided for the treatment of reservoir water is 80% pure,

$$\text{Adjusted gypsum requirement} = 3.87 \div 0.80 = 4.84 \text{ metric tons}$$

The water treatment facility would require 4.84 metric tons of 80% pure gypsum to reduce the sodium concentration from 7 to 2 meq/l in the 9,000 cubic meters of reservoir water.

ii. Lime

Lime serves as an effective irrigation amendment by neutralizing acidic water, reducing element toxicity, and providing essential calcium. These improvements in irrigation water quality support healthier soil and plants, ultimately leading to more sustainable and productive agricultural practices.

Lime in the form of calcium carbonate ($CaCO_3$), is an efficient addition to raise the quality of irrigation water. It is especially useful when irrigation water is acidic or has large concentrations of dissolved components that can harm plants and soil. Acidic irrigation water can cause several problems, including the breakdown of harmful metals like manganese and aluminum, which can damage plant roots and impede the uptake of nutrients. The pH of the irrigation water is lowered to a neutral level by adding lime, which makes it safer and better suited for plant growth. This reaction increases the calcium ion concentration in the water, which is beneficial for plants and soil structure.

$$CaCO_3 + H_2O + CO_2 \rightarrow Ca^{2+} + 2HCO_3^-$$

The amount of lime required to be added to the irrigation water depends on the initial pH and composition of the water, and it can be directly added to the irrigation water source. A more accurate control over the quantity of lime added to irrigation water is possible if lime is added to the irrigation system using fertigation. This approach is essential for large-scale agricultural operations.

5.4 *Leaching*

Irrigation water applied to the crops consists of various salts present in it. With each irrigation, salts are added to the soil. The depth and volume of water provided to the crops directly affects salinity. If only that limited amount of water is provided to the crops that have been adsorbed by the plants or water lost through transpiration, then each irrigation will lead to the addition of more salts to that already present in the soil. It is necessary to remove these salts from the soil, as it will lead to a reduction in crop yield.

Leaching is a process of supplying additional water to the crops to avoid a buildup of salts in the root zone. The regions receiving adequate rainfall do not require leaching at a very necessary rate, as rainwater leaches out all the salts accumulated in the root zone. However, the semi-arid and arid regions are subjected to less rainfall, and salinity problems are purely taken into account to avoid any type of hazards caused to the crops due to the presence of excessive salts. Leaching is accomplished by irrigating the plant with enough water to allow some of the accumulated salts to seep down and beneath the entire root zone. The fraction of the applied water that passes through the entire root zone and drains below is known as the leaching fraction (LF) (Ayers and Westcot 1985) and is expressed as follows:

$$\text{Leaching Fraction (LF)} = \frac{\text{depth of water that percolates below the root zone}}{\text{depth of water delivered to the surface}} \tag{5}$$

A higher leaching fraction indicates a low accumulation of salts.

Additionally, Ayers and Westcot (1985) developed a method for estimating the salinity of drainage water that infiltrates below the root zone. This estimation is possible if the salinity of the irrigation water applied (EC_w) and the leaching fraction are known or can be calculated. The equation is provided below,

$$EC_{dw} = \frac{EC_w}{LF} \tag{6}$$

where, EC_{dw} refers to the salinity of the water that percolates below the root zone, EC_w represents the salinity of the irrigation water, and LF indicates the leaching fraction.

5.5 *Blended Water Supply*

Blended water is obtained when a high saline water is mixed with water from a better source, to provide a required quality of irrigation water for the crops. This process of diluting the high saline water with a good quality of water extends in reducing the SAR of water. However, there is no need to use blended water on farms where high-quality water is available in adequate amounts. But if the water from a better supply is not enough alone to fulfil the irrigation requirement, then it becomes necessary to mix the less desirable water with the better-quality water, which may eventually return to higher crop production. However, blending water is not commonly practiced, even when two different qualities of water are readily accessible. Poor quality of water is applied to the crops only if the irrigation demand is not fulfilled by the better quality of water. This alternating water supply is not an efficient way to deal with a rise in salinity levels within the root zone. The salinity problem can be elevated if low-saline waters are applied after the use of high-saline water, which may also lead to reduced infiltration rates. This technique helps in managing the availability of irrigation water during different seasons by mixing water from various sources such as surface water, groundwater, or treated wastewater.

$$\begin{aligned}&\text{Concentration of the blended water}\\&\quad=\left[\text{Concentration of freshwater}\times\text{percentage of fresh water used}\right]\\&\qquad+\left[\text{Concentration of brackish water}\times\text{percentage of brackish water used}\right]\end{aligned}\tag{7}$$

Example 2 A rice field was set to be irrigated by utilizing water from a nearby canal. Due to a shortage of water supply from the canal, comparatively poor-quality water was drawn from a nearby well. To ensure water quality, the water from both sources was mixed. The results of the water quality analysis for the canal and well water is provided below.

Parameter	Canal water	Well water
pH	7.0	8.5
EC (meq/l)	0.5	3.5
Ca (meq/l)	1.0	4.5
Mg (meq/l)	0.5	0.9
Na (meq/l)	0.3	12
HCO_3^- (meq/l)	0.1	1.5

i. Determine the ratio in which the water from the two supplies was mixed if the blended water has an EC and calcium concentration of 1.55 dS/m and 2.225 meq/l respectively.
ii. Calculate the SAR of the blended water.

Solution

i. Ratio of blended water

Given,

EC of blended water is 1.55 dS/m.

Ca of blended water is 2.225 meq/l.

Let x represent the proportion of canal water.

Let y represent the proportion of well water.

Now, using Eq. 7,

$$0.5x+3.5y=1.55\tag{8}$$

$$1x + 4.5y = 2.225 \tag{9}$$

Multiplying Eq. 9 by 0.5,

$$0.5x + 2.25y = 1.1125 \tag{10}$$

By solving Eqs. 8 and 10,

$$y = 0.35$$
$$x = 0.65$$

Therefore, the water was blended with 65% of canal water and 35% of well water.

ii. SAR of the blended water

The value of calcium concentration in the blended water is given to be 2.225 meq/l.

By using Eq. 8, the values of magnesium and sodium concentration of the blended water can be calculated,

$$\begin{aligned}&\text{Magnesium concentration of blended water}\\&= (0.5 \times 0.65) + (0.9 \times 0.35)\\&= 0.64\ \text{meq/l}\end{aligned} \tag{11}$$

$$\begin{aligned}&\text{Sodium concentration of blended water}\\&= (0.3 \times 0.65) + (12 \times 0.35)\\&= 4.395\ \text{meq/l}\end{aligned} \tag{12}$$

The SAR of the mixed water can be determined using the following equation,

$$SAR = \frac{Na^{+}}{\sqrt{\frac{Ca^{2+} + Mg^{2+}}{2}}} \tag{13}$$

Putting values of Na^{+}, Ca^{2+} and Mg^{2+} in the above equation,

$$SAR = \frac{4.395}{\sqrt{\frac{2.225+0.64}{2}}} = 3.67 \tag{15}$$

As a result, the SAR of the mixed water is 3.67.

6 GIS in Water Quality Assessment

GIS is an effective and pivotal mechanism in water quality assessment, as it aids in the illustration, examination, and interpretation of the data derived through water quality assessment. As earlier, we studied the importance of location considerations for sampling points, we discussed the effects of several locations and the distance between them. Using various interpolation methods such as inverse distance weighting, kriging, spline interpolation, or radial basis function, GIS predicts values at unknown locations by using data from nearby known points. This data is then utilized in constructing spatial distribution maps of different water quality parameters. These maps provide a clear delineation of sampling locations, facilitating in identification of safe or high-risk zones, epicentres of contamination, and other features.

GIS also works as a decision support system, as it helps the stakeholders in making policies related to water quality restoration, pollution control measures, and watershed management strategies. GIS-based modeling tools can be used to simulate the transport of various contamination-causing parameters in an aquatic system. It is used to simulate the movement of contaminants through various mediums like soil, surface water, or groundwater, to identify their behavior and distribution over a while.

GIS proves to be an indispensable tool in crop selection by evaluating a multitude of environmental, climatic, and pedological parameters to identify the most suitable cultivars for a given region. It facilitates the amalgamation of spatial datasets, such as soil characteristics like pH balance, salinity levels, texture, and nutrient composition, enabling a thorough assessment of the land's aptitude for various crops. This method empowers agricultural planners and cultivators to make empirically informed decisions regarding which crops are likely to thrive, thereby enhancing productivity and optimizing resource allocation. Additionally, GIS incorporates microclimatic information—such as temperature fluctuations, precipitation trends, and wind dynamics—further refining the crop selection process based on atmospheric suitability, thus reducing the likelihood of crop failure while promoting sustainable agronomic practices.

Furthermore, GIS offers the capability for prognostic modeling, simulating how alterations in environmental factors—such as climate variability or shifts in land use—might influence the future viability of particular crops. For instance, these models can predict changes in regional temperature and precipitation dynamics, helping farmers adapt their crop choices by opting for drought-tolerant species or varieties better aligned with evolving hydrometeorological conditions. GIS also assists in evaluating water availability and land utilization patterns, enabling the recommendation of crops that optimize hydrological efficiency. This is especially critical in regions where water scarcity or degraded water quality imposes limitations on crop options, as GIS can pinpoint species that are more resilient to such challenges, thereby fostering long-term agricultural sustainability.

References

Abrol IP (1982) Technology of chemical, physical and biological amelioration of deteriorated soils. In: Panel of experts meeting on amelioration and development of deteriorated soils in Egypt, pp 2–6

Ayers RS, Westcot DW (1985) Water quality for agriculture, vol 29. Food and Agriculture Organization of the United Nations, Rome, p 174

Bauder TA, Waskom RM, Davis JG, Sutherland PL (2011) Irrigation water quality criteria. Colorado State University Extension, Fort Collins, pp 10–13

Eaton FM (1950) Significance of carbonates in irrigation waters. Soil Sci 69(2):123–134

Follett RH, Soltanpour PN (2002) Irrigation water quality criteria. Colorado State University Publication No. 0.506

Kovda VA, Hagan RM, Berg C (1973) Irrigation, drainage and salinity: an international source book. FAO/UNESCO

Maas EV (1984) Salt tolerance of plants. In: Christie BR (ed) The handbook of plant science in agriculture. CRC Press, Boca Raton, Florida

Meireles ACM, Andrade EMD, Chaves LCG, Frischkorn H, Crisostomo LA (2010) A new proposal of the classification of irrigation water. Revista Ciência Agronômica 41:349–357

Pearson GA (1960) Tolerance of crops to exchangeable sodium, no. 216. Agricultural Research Service, US Department of Agriculture

Richards LA (1954) Diagnosis and improvement of saline and alkali soils, vol 78, 2nd edn. Washington, 154 pp

Staff USL (1954) Diagnosis and improvement of saline and alkali soils. Agric Handb 60:83–100

Wilcox LV, Blair GY, Bower CA (1954) Effect of bicarbonate on suitability of water for irrigation. Soil Sci 77(4):259–266

Crop Selection and Management Strategies for Poor Quality Irrigation Water

Shreeya Baghel and M. P. Tripathi

Abstract Optimal agricultural productivity is heavily reliant on the implementation of effective crop selection and management strategies, particularly in cases where poor-quality irrigation water is utilized. The initial crucial step involves selecting salt-tolerant crops like barley, sugarbeet, and specific varieties of tomato and alfalfa. These selected crops demonstrate the ability to endure elevated salinity levels while still achieving satisfactory yields. The incorporation of leguminous crop rotation can enhance soil health and mitigate negative effects linked to low-quality water usage, while the strategic application of gypsum serves to displace sodium ions and improve soil structure, thereby promoting better water infiltration and root growth. The utilization of treated water in drip irrigation systems proves advantageous as they precisely deliver water to the root zone, thereby reducing water wastage and minimizing salt accumulation on soil surface. Mulching serves as an effective method to mitigate salt buildup by enhancing soil moisture retention and reducing evaporation rates. It is imperative to conduct regular soil assessments and monitor water quality consistently to ensure well-informed decision-making in the realm of agricultural management. Furthermore, the incorporation of organic substances, such as compost, may significantly improve soil fertility and structure, consequently augmenting crop resilience to inadequate irrigation conditions. By combining these approaches, it is possible to effectively tackle the issues presented by low-quality irrigation water, ultimately promoting sustainable agricultural methods and consistent crop productivity. Globally, freshwater resources are limited for agricultural practices, necessitating heightened food grain production from limited natural resources to cater to the needs of a swiftly growing population. Thus, emphasis has been placed on the sustainable production capacity of agricultural crops by scientists and policymakers. The production of crops is further limited by the problematic fact that freshwater bodies are diminishing at an accelerated rate. In regions with limited freshwater resources, low-quality water could be a suitable substitute. Toxic contaminants should not be

S. Baghel (✉)
ABE Department, University of Florida, Gainesville, USA
e-mail: baghelshreeya1@gmail.com

M. P. Tripathi
SWCE, SVCAET and RS, IGKV, Raipur, Chhattisgarh, India
e-mail: mktripathi64@gmail.com

Dimple et al. (eds.), *Surface Water and Groundwater Quality for Sustainable Utilization*, Springer Hydrogeology, https://doi.org/10.1007/978-981-96-8189-1_12

present in excess of a few key thresholds. However, there are either none at all or insufficiently established critical thresholds for many contaminants as well as acceptable quality standards for certain wastewater kinds. Before being applied to a crop field, industrial effluent and water of marginal quality that are utilized in crop cultivation should be treated. Consequently, given the evolving circumstances, the secure reclamation of wastewater for food production is essential to satisfy the requirements of the burgeoning global population.

Keywords Crop rotation · Crop productivity · Salt build-up · Management strategies

1 Introduction

Irrigation has historically constituted a critical determinant in the evolution of agricultural practices. The global expanse of land subjected to irrigation has experienced significant growth, particularly during the latter part of the previous century. The widespread installation of irrigation systems during the period spanning the mid-1960s to the mid-1980s. Increased global food output by more than 50% (El-Ashry and Duda 1999). While merely approximately 17% of global arable land is irrigated, this segment is responsible for generating over one-third of the world's fibre and food output (Hillel 2000). The escalation of irrigated agricultural practices must persist in response to the rising global population; however, the annually renewable freshwater resources appear to be predominantly allocated for the foreseeable future. Certain regions may witness fluctuations in freshwater availability, influenced by variations in precipitation attributable to climate change; nonetheless, these alterations are anticipated to be minimal when juxtaposed with the escalating upcoming requirements for freshwater (Wallace 2000). Due to population increase, struggle for freshwater resources has become apparent among the municipal use factories, and farming sectors in many locations. This has caused the agricultural sector to get a smaller share of freshwater resources (Tilman et al. 2002). It is anticipated that this tendency will continue and increase, especially among less developed countries with dry climates that are already battling high rates of population increase and serious threats to the environment.

As the availability of high-quality irrigation water is projected to decline, it is imperative that existing water resources are utilized with greater efficiency (Wichelns 2002; Hatfield et al. 2001; Oweis et al. 1999), wherein a potential strategy involves the recycling of sodic and/or saline drainage waters resulting from farming with irrigation (Oster 2010; Shalhevet 1994; Rhoades 1999), or the utilization of inferior-quality water produced by urban municipalities (Bouwer 2002). This idea also holds true for salt-affected soils, and which account for over 20% of irrigation land (Ghassemi et al. 1995). In order to improve agricultural productivity, concentrated efforts must be made toward effective, economical, and sustainable ecological restoration and management (Qadir and Oster 2002). The use of these lower-quality water and soil, if mismanaged, may worsen salinity and the sodicity problems, which are currently harmful to many irrigation projects and eventually jeopardize agricultural production.

A significant challenge associated with irrigated farming pertains to its deleterious ecological ramifications. Over time, irrigated agriculture inevitably leads to negative off-site impacts due to the drainage water it generates (Van Schilfgaarde 1994). While producing drainage water through irrigation is crucial for maintaining soil salinity at levels favorable for crop growth by means of leaching, relying solely on crop salt tolerance and irrigation water salinity to set leaching requirements is no longer sufficient. Similarly, focusing solely on decreasing soil sodicity and salinity to limits that improve crop yields as the primary goals of soil restoration or remediation is inadequate. It's also important to recognize the effects that drainage waters from reclaimed land, rehabilitation, and normal irrigation methods have on the ecosystem. Where is the drainage water to be disposed of, and more importantly, the salts it contains? What influence may this have on the receiving water bodies' or soil strata's chemical structure? The crucial questions are: (1) What steps may be taken to reduce the amount of drainage water? (2) Is it appropriate to limit the removal of unusable drained water to the particular areas where the waters come from? Improving irrigation management is one practical way to deal with these issues (Wichelns 2002). This will make sure the extra water isn't provided in excess of what is required for evapotranspiration and leaching. Reusing drainage water to irrigate crops that are suitable for salt tolerance is another tactic (Rhoades 1999).

In the coming decades, there is great potential to increase agricultural yields while reducing negative environmental effects using sustainable systems of irrigation that make use of saline-sodic soils and waters. A holistic strategy for managing soil, water, and crops will be necessary for this topic. The areas of concentration need to include reclaiming newly discovered land, restoring salty and sodic lands that were irrigated in the past, increasing production per unit of water, and protecting ecosystem. The management of crops and water, which are main topics of this study, will play significant roles within this all-encompassing framework.

The features of a water supply that affect whether it is suited to a certain application—more especially, to what extent the quality that satisfies user needs—are collectively referred to as the quality of irrigation water. Certain chemical, biological, and physical characteristics define quality. Despite personal preferences like palatability can be used as a crude gauge of acceptance. In the event when two readily available sources of drinking water with similar quality are found to be more agreeable to people than the other, the more agreeable source is identified as the preferred one. When evaluating water used for irrigation, the physical and chemical features of the water are given a lot of significance, but other elements are rarely seen to be important.

The quality of water used for irrigation plays a crucial role in crop production, soil fertility, and ecosystem preservation. For instance, the types of exchangeable ions present in irrigation water significantly affect the soil's mechanical and physical properties, such as its porosity and structure (aggregate stability). Even when other factors and farming practices are optimal, the quality of irrigation water can still influence crop yields and soil physical characteristics. Additionally, different crops require specific types of irrigation water. The quality of irrigation water is determined by three categories of factors: physical, chemical, and biological. The chemical properties include the concentration of salts and related metrics, such as sodium adsorption ratio (SAR), electrical conductivity (EC), alkalinity, total dissolved solids (TDS), and hardness. Chemical analysis is the most efficient method of evaluating irrigation water condition. The main variables that affect whether or not water is suitable for farming purposes are pH, salinity danger, sodium hazard (SAR), extra trace elements, toxic anions, free-chlorine, and the presence of bicarbonates and carbonates in combination with calcium and magnesium content.

The types and concentrations of dissolved salts can greatly affect the quality of water used for irrigation. Irrigation water contains salts in comparatively trace yet significant amounts. These salts originate by dissolution or weathering actions that impact both soil and rocks. These processes include the slow-dissolving disintegration of minerals such as gypsum, lime, and different soil minerals. These salts are carried to their application locations by water. When it comes to irrigation specifically, these salts are added to water and remained in soil until the water, dries, evaporates or is absorbed into crops. The nature and severity of the resulting issues vary based on factors like crop types, soil characteristics, climate, and the experience of water user. Consequently, there is no definitive threshold for water quality; instead, the water's suitability depends on conditions of use, which influence the accumulation of water components and can impact agricultural productivity. The most common soil-related problems used as benchmarks for assessing water quality include infiltration rate, salinity, toxicity, and various other challenges.

i. Irrigation water salinity

The salinity of water utilized for irrigation is main issue with regard to its quality. Salinity denotes the cumulative concentration of salts that are dissolved within the water, although it does not specify the particular types of salts that are present. High salt concentrations in the irrigation water reduce the amount of water available to crops, mainly because of osmotic pressure, leading to lower yields. Beyond a specific threshold, the decline in crop yield exhibits a direct correlation with the escalation of salinity levels. Various crops exhibit differing degrees of tolerance to salinity, thus possessing distinct thresholds and rates of reduction of their yields. The most common indicators used to evaluate irrigation water quality in terms of salinity are Total Dissolved Solids (TDS) and Electrical Conductivity (EC). The parameters pertaining to the water quality as well as the threshold limit for salinity are outlined in Table 1.

In instances where salinity of irrigation water surpasses the established threshold for every specific crop, a decrease in yield is observed. Mathematical equations have been formulated to estimate the potential yield, taking into consideration the salinity levels of water used for irrigation.

$$\text{Percent Yield (of maximum)} = 100 - b(ECe - a)$$

Table 1 Irrigation water quality parameters

EC (ds/m)	TDS (mg/l)	Salinity levels
0.8	<500	Low
0.8–1.6	500–1000	Moderate
1.6–3.0	1000–2000	Moderate to high
>3	>2000	Very high

Here, *b* represents the percentage decrease in relative yield for each unit increase in salinity, *a* is the EC threshold the crop can withstand, and *ECe* is the electrical conductivity of the saturated soil paste, measured in the laboratory. Residual sodium carbonate (RSC), electrical conductivity (EC) and sodium absorption ratio (SAR) serve as critical variables for the classification of water into saline or alkaline categories.

In contemporary times, numerous areas experience the emergence of salt-affected soils as a consequence of human-induced economic mismanagement. The main factors leading to the presence of K^+, NO_3^-, Cl^-, and Na^+ in groundwater are human activities, such as improper fertilizer use, over-application of fertilizers, insufficient sewage treatment, and inadequate industrial waste management, as identified through multivariate statistical analysis (Shukla and Saxena 2021).

Sodium absorption ratio (SAR): It serves as a gauge for the concentration of Na or the concentration of Na in relation to other cations in farms.

$$SAR = \frac{\mathrm{Na}^+}{\sqrt{(\mathrm{Ca}^{2+} + \mathrm{Mg}^{2+})/2}}$$

(A) ***Exchangeable sodium percent (ESP)***: Quantity of Na in soil that can be exchanged for other cations.

(B) ESP = 100 [Exchangeable Na^+ ions/(CEC)] (All cations and CEC in *m*Mol (+)/kg of soil)

(i) Based on electrical conductivity (EC), which is divided into two categories, the total salt content is calculated.

(ii) For soil solutions with EC ranges from 0.1 to 5 dS m^{-1}

$$\text{Entire dissolved solids mg L}^{-1} = \text{EC dS m}^{-1} \times 640$$

(iii) For soil solutions with EC ranges from 3 to 30 dS m^{-1}

$$\text{Osmotic pressure (bars)} = \text{EC dS m}^{-1} \times (-0.36)$$

ii. Sodium hazard and irrigation water infiltration

The Sodium Adsorption Ratio (SAR) is used to evaluate the sodium hazard. This metric elucidates sodium concentration present in the water used for irrigation, relative to the levels of magnesium and calcium. Magnesium and calcium are known to mitigate the detrimental impacts of sodium. Elevated levels of SAR may lead to the disintegration of soil structure and complications regarding water infiltration. When desiccated, soil typically exhibits tendencies to seal and become increasingly hard and compacted. Paradoxically, heightened salinity serves to diminish the adverse effects of sodium on soil structure. When sodium concentrations in the soil are high relative to magnesium and calcium, resulting in a high SAR, leaching with high-quality irrigation water can worsen the issue.

iii. Toxicity of specific ions

The quality of agricultural water can also be assessed based on the toxicity of certain ions. Unlike salinity issues, which affect the soil, toxicity problems arise within the plant due to the accumulation of ions in the leaves. Almost prevalent ions that may give rise to a toxicity issue are chloride, sodium, and boron ions. Similar to salinity, various crops exhibit differing levels of sensitivity with these ions. Although boron is a necessary plant nutrient, it can be hazardous at very low doses. Therefore, it requires careful monitoring. Irrigation water that contains toxic amounts of even a single ion may not be suitable for use in agriculture. However, there are a number of management strategies that might help to lessen the harm. These methods include enough leaching, more frequent watering, avoiding overhead water supply, not using fertilizers containing boron or chloride, and choosing the right crop kinds, among others.

iv. Alkalinity and pH

Alkalinity is the total amount of hydroxide (OH^-), carbonate ($CO_3{}^{2-}$), and bicarbonate ($HCO_3{}^-$) ions in water. $CaCO_3$ is measured in milligrams per liter or meq/l. Alkaline water helps stabilize pH levels and prevent rapid changes. If the alkalinity is too low, the addition of acid fertilizers can quickly lower the pH. Similarly, ions released by plant roots in containers and hydroponic systems can cause rapid pH fluctuations if alkalinity is insufficient.

2 Current Freshwater Availability for Farming Operations

In the forthcoming decades, potable freshwater may emerge as a most expensive resources in a densely urbanized nation such as India. In addition to the issue of water shortage, the increasing peoples necessitates substantial quantity of food, thereby rendering the employment of treated sewage water for the agricultural purposes a viable alternative to conserve freshwater for drinking. For example, the Godavari, Tapi, Sabarmati, Ganges, Indus, Mahanadi, Bramhaputra, Brahmani-Baitarani, Narmada, Pennar, Krishna, and Mahi Rivers are the main sources of surface water in India. It has been calculated that the mean annual flow among these regions of rivers is 1869 km^3. The river Ganga is the most important of these major rivers, while the Godavari and Brahmaputra also contribute significantly to the restocking of surface water supplies. A significant amount of freshwater, representing 89% of all water on the surface and 92% of all groundwater, is devoted to India's agricultural crop producing industry. Over 110,000 km^3 of precipitation are observed annually on average in terrestrial areas. Approximately 5% of this precipitation is thought to be a result of rainfed agricultural techniques. On typical, 56% of this rainfall evapo-transpirates via plants and into the atmosphere. The remaining amount (39%, or 43,000 km^3 per year) flows via several natural water basins and helps recharge aquifers. This volume is progressively diminishing, with the desiccation of natural aquifers becoming increasingly pronounced in the current context. A fraction of the available freshwater is also employed in construction or in the extraction of additional water volumes through advanced mechanization techniques, referred to as withdrawal. These challenges are anticipated to exacerbate in the ensuing years (Faurès et al. 2002). Globally, freshwater withdrawal is allocated as follows: 69%, 12%, and 19% for agriculture, municipal, and industrial purposes, respectively (Fig. 1). This distribution can be further delineated, with the majority associated with crop production (27%), particularly involving maize, paddy, and wheat; followed by meat and their products, which account for 22%, and dairy activities, which utilize approximately 6–7%. According to projections, roughly 70% of water is expended on irrigation for farming, whereas 15–35% of water remains improperly utilised.

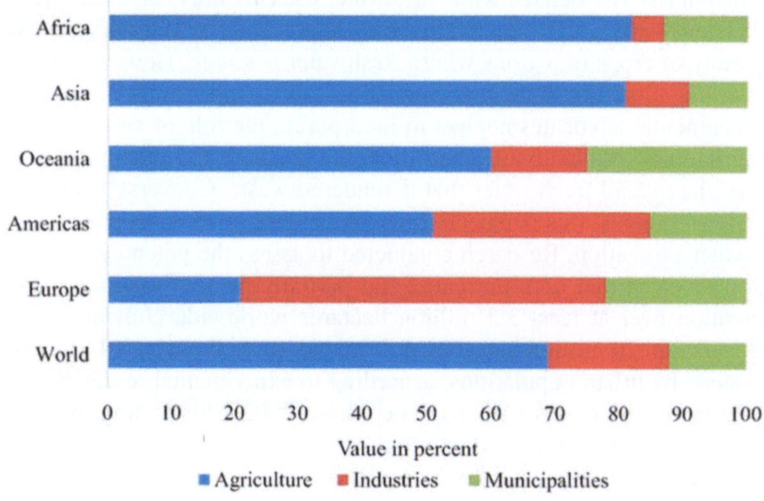

Fig. 1 Freshwater withdrawal ratios by continent. *Source* Modified from AquaStat (2015)

3 Why Is Low-Quality Water Necessary for Us to Use?

The limited availability of the potable water has compelled individuals to resort to the utilization of water of marginal quality for agricultural endeavors. A significant number of peri-urban regions are engaged in the cultivation of vegetables through the application of untreated or treated sewage. Water of inferior quality harbors a considerable concentration of the organic matter along with necessary plant nutrients. From the perspective of agricultural practitioners, the application of sewage for irrigation purposes, specifically in the range of four to five instances, effectively reduces the recommended fertilizer dosage for crops by 50% (Rajendiran et al. 2020). The practice of using sewage for irrigation promotes sewage farming in peri-urban regions. Studies have estimated that global annual wastewater production exceeds thirty million tons (Minhas and Samra 2004; Roy 2020). Specifically, urban areas generate substantial volumes of waste, consequently exacerbating the per capita daily burden on waste management systems. Merely 15–24% of fresh water is utilized efficiently, with the remainder reintegrating into the urban hydrological cycle. This indicates that approximately 75–85% of water allocated for domestic purposes ultimately manifests as wastewater (Qadir et al. 2010). Globally, more than 800 million of farmers are engaged in urban agriculture, with around 200 million of them using water of marginal quality due to the lack of better options (Qadir et al. 2010). To tackle future water scarcity effectively, it will be crucial to adopt more cost-effective and sustainable technologies and improve management practices to mitigate the health and environmental risks associated with wastewater use (Mishra et al. 2023; Qadir et al. 2007). Wastewater is widely applied in the agricultural sector; farmers use it to cultivate a variety of crops in regions where freshwater is scarce. However, there is still doubt over the reliability of the real estimations. Certain researchers, policymakers, and environmental advocates neglect to incorporate the role of sewage or wastewater in agricultural productivity. Their rationale posits that it constitutes a minimal fraction of the overall fresh water that is rendered waste. Conversely, in contemporary lifestyle contexts, this perspective stands in stark contrast to previous patterns of freshwater utilization. Research conducted to assess the potential of wastewater utilization revealed that both untreated and partially treated wastewater can facilitate irrigation over at least 3.5 million hectares worldwide (Jimenez and Asano 2004). More than 20 million hectares of farmland might be irrigated using wastewater produced by urban populations, according to experimental research conducted in 2002 under the direction of Wim van der Hoek, 2002. This finding would provide a major boost to agricultural productivity.

The utilization of wastewater that is mixed with sewage significantly enhances the availability of crucial plant nutrients, including nitrogen, phosphorus, and various micronutrients. Standard estimations indicate that sewage water provides approximately 25–50% of nitrogen and phosphorus required for optimal crop growth. When projecting crop yields, an increase of nearly 15–30% has been observed in various crops as compared to those irrigated through tube wells. A considerable number of peri-urban agriculturalists resort to the utilization of substandard water, either by necessity or by choice, to facilitate multi-crop farming. In these areas, the rate of cropping is noticeably higher (300–400%), and economic returns are said to be greater than three–four times those of groundwater-irrigated land (Minhas and Samra 2004). Beyond agricultural applications, low-quality water is also employed in the farming of aquatic organisms and for various industrial purposes in countries across Africa and Asia. In India, nearly every state engages in the utilisation of contaminated water for the farming of agricultural crops (Minhas and Samra 2004; Sia 2023). In numerous locations, substandard water is utilized for the forestry or fodder crops cultivation intended for wildlife, as well as for the restoration of degraded wetlands. A significant portion of peri-urban regions is engaged in vegetable cultivation utilizing wastewater; researchers Raschid-Sally and Jayakody (2009) have estimated that 32% of vegetables and 27% of cereal crops are grown with the poor-quality water supply.

4 Some Sources of Poor-Quality Water Used for Agriculture

a. Industrial wastewater

The tannery sector is characterized by significant consumption of freshwater resources and the subsequent discharge of chromium (Cr) contaminated effluent into aquatic ecosystems and agricultural land, which poses carcinogenic risks.

The automotive manufacturing industry releases oil and grease which causes a hydrophobic layer to grow on the water's surface, thereby inhibiting gas exchange between aquatic or terrestrial environments and the atmosphere. This form of pollution has been implicated in the mortality of aquatic organisms due to hypoxic conditions within aquatic habitats. It is imperative to employ biological processes for the biodegradation of oil constituents. Various chemicals such as oregano clay, bentonite, and alum are utilized to mitigate oil contamination and turbidity in wastewater originating from the automobile industry. Notably, organo clay has demonstrated superior efficacy compared to alum in reducing turbidity in aquatic environments (Pachepsky et al. 2011). In regions experiencing water scarcity, treated wastewater can be repurposed for agricultural production following both secondary and primary treatment protocols.

The paper and pulp sector in India is notorious for its substantial water consumption and significant pollution levels, as it releases considerable quantities of contaminants into both aquatic ecosystems and soil matrices. According to Chakrabarti (2015), industries reliant on wood as a raw material exhibit a greater demand for freshwater compared to those that utilize agro-based resources. The wastewater produced by these industries is laden with various soluble salts, including sodium carbonate, chloride, sodium sulfate, acids, charcoal dust, bleaching agents, and residual raw materials. The concentrations of these soluble salts predominantly influence both the biological and chemical oxygen demands of the wastewater. Additionally, elevated levels of soluble salts can curtail plant biomass by interfering with nutrient availability of soil. The kinetics of nutrient mineralization and amount of carbon microbial biomass in soil are closely related to dynamics of plant nutrients. The textile industry also contributes significantly to the generation of wastewater laden with pollutants, including salts, organic and inorganic dyes, paint residues, polymers, and trace metal concentrations. Such effluents are frequently applied in agricultural practices, particularly in cultivation of the vegetables within peri urban regions. Prolonged application of these wastes, especially those containing elevated concentration of soluble sodium, chloride, and sulphate, adversely impacts soil fertility and the potential biomass yield of crops. Before being dumped into agricultural fields or aquatic systems, textile waste is frequently treated using a variety of absorbents to remove soluble salts (Wu et al. 2018).

b. Sewage

Sewage, as utilized in peri-urban regions of megacities, contributes a substantial quantity of organic matter, along with essential macronutrients and micronutrients required for plant growth. A significant proportion of Indian sewage systems contain trace levels of heavy metals, accompanied by considerable organic and inorganic loads. To improve soil's fertility and encourage sustained agricultural output, the ideal interchange of crop nutrients among SOM, soil moisture, clay particles, and different soil constituents are essential.

It has been demonstrated that adding organic material to the soil enhances its general chemical properties, microbe population and variety, infiltration rates, nutrient dynamics, and soil aggregate stability. Prolonged use of sewage water has resulted in significant accumulation of heavy metals in the soil. These metals can enter the food chain and cause various physiological problems in humans and other organisms. Farmers often regard sewage water as a form of liquid fertilizer due to its elevated nutritional composition. The nutrient concentrations are estimated to be approximately 48 mg L^{-1} of nitrogen, 8 mg L^{-1} of phosphorus, 72 mg L^{-1} of potassium, and 35 mg L^{-1} of sulfur (S). In addition to these primary nutrients, sewage also contains micronutrients quantified in mg kg^{-1} (0.34 for Zn; 10.8 for Fe; 0.2 for Cu; and 0.36 for Mn) (Singh et al. 2018). If farmers utilize 5 irrigation cycles of 7.5 cm each with the use of sewage, they would contribute approximately 181 kg ha^{-1} of nitrogen, 29 kg ha^{-1} of phosphorus, 270 kg ha^{-1} of potassium, and 130 kg ha^{-1} of sulfur, that is adequate to meet the fertilizer demands of various crops (Saha et al. 2012). While this practice can augment the plant nutrients availability for agricultural production, the primary obstacle remains the concentration of heavy metals (Dotaniya et al. 2018). Employing various treatment methods—such as primary (for removing larger particles), secondary (for eliminating dissolved and smaller organic matter), and tertiary (for addressing biological loads)—can effectively lower trace metal levels in sewage water. However, to ensure proper treatment of sewage or low-quality water before its use in agriculture, ongoing monitoring and public awareness campaigns are essential. Moreover, low-quality water may be utilized for cultivating non-edible crops, fibers, and oilseed plants, such as flowers, castor, and jatropha, which can be advantageous for generating economic returns.

5 Management for Improving Irrigation Water Quality

Poor irrigation water quality adversely affects soil and crop production, making it crucial to maintain water quality through various methods, like technical, agronomic, and managerial strategies. The key features of these different approaches are:

*A. **Technical***

- Implementing land leveling techniques to facilitate the uniform application of irrigation water.
- Employing surge irrigation methods to enhance the distribution of water resources.
- Utilizing efficient sprinkler systems to achieve a more uniform distribution of water.
- Adopting low energy precision application sprinklers to minimize losses due to wind drift and evaporation.
- Implementing drip irrigation systems to diminish evaporation and various types of water loss, thereby enhancing crop productivity.

*B. **Managerial***

- Optimizing irrigation scheduling practices for improved efficiency.
- Enhancing canal operations to ensure timely water deliveries. Applying irrigation water during critical periods to maximize crop yields.
- Utilizing field preparation methodologies and water-conserving tillage.
- Ensuring better maintenance protocols for canals and irrigation instruments.
- Implementing recycling processes for tail and drainage water.

*C. **Institutional***

- Establishing water user associations to promote greater farmer engagement and facilitate fee collection.
- Reducing irrigation subsidies and/or instituting pricing mechanisms that encourage conservation.
- Creating a legal framework to support efficient and equitable water market practices.
- Promoting rural infrastructure development to enable the private sector's dissemination of advanced techniques.
- Enhancing extension and training services to improve agricultural practices.

*D. **Agronomic***

- Choosing crop types with significant yields per liter of water transpired.
- Putting intercropping techniques into practice to maximize soil moisture consumption.
- Choosing crops that are compatible with the current climate and the quality of the water resources available.
- Crop cultivation should be spaced out to optimize yield potential in situations where salt of the soil and water is a defining factor.

- Selecting varieties of crops resistant to drought in areas with irregular or scarce water supplies.
- Implementing breeding initiatives with the goal of creating crop types that are water-efficient.

E. Other management practices affecting the quality of irrigation water

(1) **Infiltration Challenges**: One important metric for evaluating irrigation water quality is the Sodium Adsorption Ratio (SAR), which makes it easier to anticipate any problems with water seeping into the soil. A variety of tactics can be used to address problems with water infiltration, such as adding organic residues, managing irrigation methods, lowering the SAR in the water supply, applying soil or water amendments, and cultivating and tillage techniques.
(2) **Irrigation Water Quality and Soil Amendments**: Soil amendments are mostly used to increase the soluble calcium content or raise the irrigation water's salinity levels in an effort to counteract the negative effects of salt.
(3) Among the many often used soil amendments are gypsum and other calcium-supplying substances. Since salt-related water penetration issues primarily affect the upper soil layers, it is better to apply gypsum in small, gradual amounts rather than in a single large dose. In cases where water salinity is very low (EC < 0.5 ds/m), gypsum can be added to irrigation water at concentrations of 1–4 meq/l of calcium.
(4) **Additional Amendments**: Some acids or amendments that produce acid can be used in situations when the soil matrix contains lime ($CaCO_3$). These additions help release calcium into the soil solution.Sulfuric acid, ferric sulfate, and elemental sulfur are a few notable instances of these modifications.
(5) **Organic Residues**: Adding organic residues improves water infiltration and soil structure by preserving soil porosity.
(6) Mixing different sources of irrigation water can enhance water infiltration by reducing the SAR. Even when calcium and magnesium levels are high, diluting the water with a source that has lower sodium content can decrease the SAR.

6 Reusing Poor Quality Water for Sustainable Crop Production

For the development of sustainable agricultural practices, irrigation water is systematically classified based on the presence and concentration of specific cations. Electrical conductivity and pH measurements gauge the activity of these ions, while sodium ion concentrations are determined using a flame photometer. This data is reported in terms of residual sodium carbonate and the sodium adsorption ratio. Generally, irrigation water is categorized into three main types: excellent (A), saline (B), and alkali (C). This classification can be further divided into three subcategories based on additional water properties (see Table 2).

Table 2 Irrigation water categorization

Group	EC (dS m^{-1})	SAR (mmol)	RSC (meq L^{-1})
A. Excellent	<2	<10	<2.5
B. Saline			
i. Moderately saline	2–4	<10	<2.5
ii. Saline	>4	<10	<2.5
iii. High saline	>4	>10	<2.5
C. Alkali water			
i. Moderately alkali	<4	<10	2.5–4.0
ii. Alkali	<4	<10	>4.0
iii. High alkali	Variable	>10	>4.0

Table 2 outlines out the fundamental concepts for assessing irrigation water quality. Most of these recommendations highlight the cumulative impact of water quality variables on crop quality and production, soil features, and general soil health in a particular agricultural environment (Singh et al. 2004). The implementation of these defined rules and guiding frameworks is generally simple and aids in the efficient classification of wastewater or irrigation during agricultural output within a particular region.

Table 2 outlines these norms, which are predicated on a few key suppositions. When using low-quality water for agricultural purposes, recommendations serve as a preventive strategy for water quality and call for a communal set of principles and actions. Poor-quality water needs to be used responsibly, and legitimate agencies must regularly assess the crop and provide techniques of application and soil adaptation.

7 Policy Guidelines for Wastewater Reuse

The guidelines set forth by international organizations, such as the US EPA, FAO, and WHO (World Health Organization), are followed by many nations. A few countries have set their own guidelines for crop reprocessing in agriculture, most notably Spain and the US. Less European nations have specific regulations and suggestions on restoration and reuse since there is less need to reuse water and enough dilution factor in the current river bodies. In many underdeveloped countries, it is forbidden from being reused for producing raw veggies. The international agencies (FAO, WHO, US EPA) that established guidelines for its agricultural reuse state that this type of limitation necessitates frequent inspection of low-quality water channels, methods of treatment, and products harvested from inferior quality water.

8 Restriction on Use

There are categories exist for the "Restriction on Use": none, moderate, and high. These categories may vary based on the kinds of crop and soil. There is no distinct border separating one division from another because the categorization is based on shared qualities and its use. The majority of the time, this kind of poor water is used for farming or forestry. Humans do not consume the edible portions of plants, nor do metals or other hazardous materials migrate from the roots to the edible portions of plants. However, the different environmental and natural resource conservation bodies indicate the metal toxicity limitations with relation to its safe use. The crop output is significantly impacted by the limit, which can be used a little bit higher or lower. It is necessary to reevaluate these guideline values considering crop, soil, and climate circumstances.

9 Crop Management Approaches for Safe Use of Poor-Quality Water

Numerous factors have influenced the strategies employed for crop management and the utilization of substandard quality water in agricultural practices, including the regional rainfall patterns, depth of the soil composition, water table, types of crops, and the associated crop management methodologies. In a detailed analysis, the various crop-based management strategies are delineated under:

a. Selection of crops

The negative effects of using poor-quality water can be reduced by implementing targeted crop management strategies and practices. Each crop possesses a distinct genetic capacity to withstand the harmful constituents present within its environment. Resilient crops exhibit a higher tolerance in comparison to crops characterized by softer stems. The lignin concentration within the wood of crop plants enhances their potential for tolerance (Meena et al. 2014). A majority of leguminous crops demonstrate susceptibility to toxic metals such as mercury (Hg), nickel (Ni), cadmium (Cd), and chromium (Cr) when juxtaposed with cereal crops. The germination of pulse crops is adversely affected by saline irrigation; specifically, pigeon pea, lentils, blackgram, and peas can endure exchangeable sodium percentages (ESP) of 10–15, cowpea can tolerate 20–25, in contrast, rice and barley show resistance to increased ESP levels.

b. Growth stages

When exposed to low-quality water—such as salt, alkaline, or heavy metal-filled water—crop development stages do not function as well. Compared to the later phases of development, the initial stages of growth are very susceptible. It has been demonstrated that chromium can prevent pigeon pea germination when added to soil at a 40 mg kg^{-1} concentration (Meena et al. 2014). Thus, it is best to avoid applying low-quality water to crops while they are germination ready.

c. Cropping sequence

This procedure is an essential management technique used to lessen the impact of water of inferior quality on agricultural output. In this strategy, pulse crops—such as combinations of pearl millet and gram or mungbean and wheat—are grown in the following growing season after cereal crops are harvested during the previous one. Pulse crop growth is adversely affected by salty irrigation, which is exacerbated by the presence of different cations and anions. This finally results in a decrease in crop production. Chloride ions exhibit a higher level of toxicity in comparison to sulfate ions; therefore, the incorporation of sulfate-rich water in irrigation practices can ameliorate the detrimental effects of chloride (Manchanda and Garg 2008; Dotaniya et al. 2019).

d. Use of horti crop sequence

Salinity and heavy metals generally have a milder impact on tree species compared to other types of plants. Integrating rows of horticultural plants within cereal crop fields achieves two main benefits: it mitigates the adverse effects of metal toxicity while concurrently providing additional income for farmers. This practice functions as a form of crop insurance in the face of unfavorable climatic conditions. Certain medicinal plants, such as isabgol, aloe, and kalmegh, thrive effectively within saline water-irrigated fields.

e. Farmyard Manure (FYM) application

The application of inferior-quality water has significantly influenced the nutrient absorption patterns of crops while concurrently diminishing the fertility levels of the soil. The introduction of FYM in irrigated soils characterized by poor quality has facilitated alterations in rhizospheric micro-climate and fostered proliferation of various organic acids, thereby augmenting the uptake of nutrients by plants and enhancing their yield (Dotaniya et al. 2013; Dotaniya et al. 2013). In reaction to these factors, plant roots release various low molecular weight organic acids (LMOAs) into soil. This release has been found to boost soil microbial biomass and accelerate nutrient transformation. Additionally, applying nitrogen fertilizers to soils with high organic matter content increases the rate of mineralization, leading to improved soil mineralization and higher crop yields (Unuofin and Siswana 2019). Moreover, using biofertilizers along with compost has been effective in reducing salinity and minimizing the toxic effects of heavy metals in ill-quality irrigated crop production systems (Sarkar and Chourasia 2017).

f. Soil amendment

The impact of using marginal quality water for agriculture on groundwater and crops depends significantly on soil texture and soil type. Soils with high permeability are more likely to contaminate groundwater, especially when using poor-quality irrigation water. Additionally, sandy or arid soils, which typically have low organic matter content, have a reduced capacity to bind metals, increasing the risk of metal leaching into deeper soil layers. These metals are subsequently extracted and utilized for crop cultivation, thereby exerting detrimental effects on both crop productivity and quality. Soils with a lighter texture exhibit superior drainage capabilities when compared to those with a heavier texture. Soils with high content of clay tend to resist the downward movement of metals due to the creation of complexes between humus and metals. The application of poor-quality water with high sodium content has been shown to adversely affect both soil fertility and productivity. After carefully examining the physicochemical characteristics of low-quality water, various counter-management strategies were developed. These include identifying high levels of heavy metals in effluents, incorporating absorbents into soil, and altering irrigation channels to minimize the presence of metals in the water used for irrigation.

A variety of absorbents are presently available in marketplace, including bentonite, bio-absorbent, nano clay, and chitosan-based bio sorbents (Wu et al. 2018). Continuous use of effluents with high sodium levels in agricultural practices results in a rise in soil alkalinity. With this scenario, applying gypsum to cultivated fields helps counteract soil alkalinity, while adding lime to fields irrigated with low-pH effluents improves crop germination rates. The activity and diversity of soil microbes boost the production of low-molecular-weight organic acids, enhancing nutrient availability during different phases of crop growth. Additionally, nanotechnology offers a cutting-edge method for reducing metal toxicity, with various nanoparticles being developed and to employ worldwide to remove trace metals from water bodies. In comparison to conventional treatment methods, magnetic microparticles offer targeted selectivity and economic efficiency (Shukla et al. 2023).

Several organic soil amendments have proven effective in mitigating the harmful effects of heavy metals in the soils (Dotaniya et al. 2018; Mwangi et al. 2019). One such example is a hydrogel synthesized from the animal waste, which has successfully reduced soil salinity and minimized water-loss. The findings indicate that the incorporation of hydrogel formulated from animal waste enhances moisture availability in the soil, with symptoms of plant wilting manifesting at a later stage. Such organic amendments have been shown to elevate the concentration of essential nutrients within the soil. A selection of both organic and inorganic amendments is employed for the immobilization of heavy metals during the cultivation of crops affected by effluents containing heavy metals or from contaminated soils.

g. Nutrients and water uptake by crops

The capacity of plants to absorb water from the soil is greatly impacted by properties of their roots and the prevailing climatic conditions in a given region. A decrease in moisture during crucial growth stages, such as flowering/pod development, can lead to a significant reduction in the crop yield. Furthermore, low moisture levels intensify soil-salinity, which negatively affects crops more severely than when moisture is maintained at field capacity. Frequent irrigation practices serve to alleviate salinity within the root zone, allowing for its redistribution to lower soil layers. An increase in the depth of the root zone correlates with a heightened concentration of salinity in well-drained soils. The use of diverse moisture measurement techniques is crucial for optimizing irrigation scheduling, which supports sustainable agricultural practices. In saline soils, a larger quantity of water is necessary to leach the salts from the root zone into deeper soil layers, preventing salt accumulation that could hinder crop yield. This practice of moving salts beyond root zone through irrigation is especially important in managing the saline sandy soils found in regions such as Haryana, Gujarat, and Rajasthan.

Plant nutrient acquisition in terrestrial environments is largely determined by soil properties, moisture levels, trace metal concentrations, biomass carbon content, and a range of climatic factors. Extended agricultural activities can lead to the buildup of significant amounts of the organic matter, vital plant nutrients, as well as harmful metalloids and metals in the soil (Dotaniya et al. 2023). The interactions among these elements have a complex impact on soil health and crop productivity, which is highly dependent on the chemical composition of the irrigation water used (Dotaniya et al. 2019).

Throughout the globe, irrigation, which refers to provision of water for the agriculture practices or cultivation of crops, is arguably the most paramount application of water, apart from potable water. Nearly 60% of the total freshwater extractions across the world are allocated for irrigation purposes. Large-scale agriculture would not be able to sustain the world's vast populations without irrigation of farmland utilizing water drawn from wells, rivers, reservoirs, and lakes. In the absence of irrigation, the cultivation of crops in the arid regions of various parts of the world would be an impossibility. This review aims to foster a comprehensive understanding of impact of the water quality on soil and on crops, as well as to aid in the selection of appropriate alternatives to address potential issues related to water quality that may hinder production under existing usage conditions. Therefore, it is crucial to implement a variety of conservation and preservation-focused management systems to alleviate pressure on the fresh natural resources. This includes maintaining available irrigation water, which in turn boosts crop production and supports the delicate balance of soil flora and fauna.

References

AquaStat FAO (2015) Aquastat

Bouwer H (2002) Integrated water management for the 21st century: problems and solutions. J Irrig Drain Eng 128(4):193–202

Chakrabarti P (2015) Purifying the river: pollution and purity of water in colonial Calcutta. Stud Hist 31(2):178–205

Dotaniya ML, Meena HM, Lata M, Kumar K (2013) Role of phytosiderophores in iron uptake by plants. Agric Sci Digest Res J 33(1):73–76

Dotaniya ML, Rajendiran S, Meena VD, Coumar MV, Saha JK, Kundu S, Patra AK (2018) Impact of long-term application of sewage on soil and crop quality in Vertisols of Central India. Bull Environ Contam Toxicol 101:779–786

Dotaniya ML, Meena VD, Meena BL (2019) Use of wastewater for sustainable agriculture. Sustain Agric Res 71–87

Dotaniya ML, Meena VD, Saha JK, Dotaniya CK, Mahmoud AED, Meena BL, Rai PK (2023) Reuse of poor-quality water for sustainable crop production in the changing scenario of climate. Environ Dev Sustain 25(8):7345–7376
El-Ashry MT, Duda AM (1999) Future perspectives on agricultural drainage. Agric Drain 38:1285–1298
Faurès JM, Hoogeveen J, Bruinsma J (2002) The FAO irrigated area forecast for 2030. FAO, Rome, pp 1–14
Ghassemi F, Jakeman AJ, Nix HA (1995) Salinisation of land and water resources: human causes, extent, management and case studies, pp xviii+–526
Hatfield JL, Sauer TJ, Prueger JH (2001) Managing soils to achieve greater water use efficiency: a review. Agron J 93(2):271–280
Hillel D (2000) Salinity management for sustainable irrigation: integrating science, environment, and economics. World Bank Publications
Jimenez B, Asano T (2004) Acknowledge all approaches: the global outlook on reuse. Water 21:32–37
Manchanda G, Garg N (2008) Salinity and its effects on the functional biology of legumes. Acta Physiol Plant 30:595–618
Meena VS, Maurya BR, Verma JP (2014) Does a rhizospheric microorganism enhance K+ availability in agricultural soils? Microbiol Res 169(5–6):337–347
Minhas PS, Samra JS (2004) Wastewater use in peri-urban agriculture: impacts and opportunities
Mishra S, Kumar R, Kumar M (2023) Use of treated sewage or wastewater as an irrigation water for agricultural purposes-environmental, health, and economic impacts. Total Environ Res Themes 6:100051
Mwangi VW, Kollongei JK, Kipkorir EC (2019) Effect of irrigation with waste water on soil characteristics and bean yield: a case study of University of Eldoret Farm
Oster JL (2010) Reconstructing precipitation variability in California using Sierra Nevada speleothems. University of California, Davis
Oweis T, Hachum A, Kijne J (1999) Water harvesting and supplemental irrigation for improved water use efficiency in dry areas, vol 7. IWMI
Pachepsky Y, Shelton DR, McLain JE, Patel J, Mandrell RE (2011) Irrigation waters as a source of pathogenic microorganisms in produce: a review. Adv Agron 113:75–141
Qadir M, Oster J (2002) Vegetative bioremediation of calcareous sodic soils: history, mechanisms, and evaluation. Irrig Sci 21:91–101
Qadir M, Oster JD, Schubert S, Noble AD, Sahrawat KL (2007) Phytoremediation of sodic and saline-sodic soils. Adv Agron 96:197–247
Qadir M, Wichelns D, Raschid-Sally L, McCornick PG, Drechsel P, Bahri A, Minhas PS (2010) The challenges of wastewater irrigation in developing countries. Agric Water Manag 97(4):561–568
Rajendiran S, Dotaniya ML, Coumar MV, Sinha NK, Singh VK, Kundu S, Patra AK (2020) Block level soil fertility status of tribal dominated Jhabua district of Madhya Pradesh, India. J Indian Soc Soil Sci 68(1):70–77
Raschid-Sally L, Jayakody P (2009) Drivers and characteristics of wastewater agriculture in developing countries: results from a global assessment, vol 127. IWMI
Rhoades JD (1999) Use of saline drainage water for irrigation. Agric Drain 38:615–657
Roy A (2020) The pandemic is a portal. Financ Times 3(4):45
Saha AK, Moses CS, Price RM, Engel V, Smith TJ, Anderson G (2012) A hydrological budget (2002–2008) for a large subtropical wetland ecosystem indicates marine groundwater discharge accompanies diminished freshwater flow. Estuar Coasts 35:459–474
Sarkar P, Chourasia R (2017) Bioconversion of organic solid wastes into biofortified compost using a microbial consortium. Int J Recycl Org Waste Agric 6:321–334
Shalhevet J (1994) Using water of marginal quality for crop production: major issues. Agric Water Manag 25(3):233–269

Shukla N, Gupta S, Rai S (2023) Potential impacts of climatic changes and human activity on water quality. In: Environmental processes and management: tools and practices for groundwater. Springer International Publishing, Cham, pp 103–111

Shukla S, Saxena A (2021) Appraisal of groundwater quality with human health risk assessment in parts of Indo-Gangetic alluvial plain, North India. Arch Environ Contam Toxicol 80(1):55–73

Sia ET (2023) A Swat-based decision support system for multipurpose reservoir operation and food-water-energy-environment trade-off analysis: case study of Selingue Reservoir. Doctoral dissertation, Université d'Ottawa/University of Ottawa

Singh KP, Malik A, Mohan D, Sinha S (2004) Multivariate statistical techniques for the evaluation of spatial and temporal variations in water quality of Gomti River (India)—a case study. Water Res 38(18):3980–3992

Singh S, Ghosh NC, Gurjar S, Krishan G, Kumar S, Berwal P (2018) Index-based assessment of suitability of water quality for irrigation purpose under Indian conditions. Environ Monit Assess 190:1–14

Tilman D, Cassman KG, Matson PA, Naylor R, Polasky S (2002) Agricultural sustainability and intensive production practices. Nature 418(6898):671–677

Unuofin FO, Siswana M (2019) Enhancing organic waste decomposition with addition of phosphorus and calcium through different sources. Int J Recycl Org Waste Agric 8:139–150

Van Der Hoek W, Hassan MU, Ensink JH, Feenstra S, Raschid-Sally L, Munir S, Matsuno Y (2002) Urban wastewater: a valuable resource for agriculture: a case study from Haroonabad, Pakistan, vol 63. IWMI

Van Schilfgaarde J (1994) Irrigation—a blessing or a curse. Agric Water Manag 25(3):203–219

Wallace JS (2000) Increasing agricultural water use efficiency to meet future food production. Agr Ecosyst Environ 82(1–3):105–119

Wichelns D (2002) An economic perspective on the potential gains from improvements in irrigation water management. Agric Water Manag 52(3):233–248

Wu Z, Wang X, Chen Y, Cai Y, Deng J (2018) Assessing river water quality using water quality index in Lake Taihu Basin, China. Sci Total Environ 612:914–922

Unseen Threats: The Impact of Emerging Contaminants on Surface Water and Groundwater Quality

Sreedeep Dey

Abstract Over the past few years, maintaining surface and groundwater quality has become major worldwide concern due to the presence of emerging contaminants (ECs), which include synthetic or naturally occurring chemicals such as pharmaceuticals and personal care products (PPCPs), pesticides, surfactants, and industrial pollutants or any microorganisms that are not commonly monitored in the environment. Commonly present in exceptionally lower concentrations, generally in parts per billion and trillion, most of these contaminants and their by-products have not yet been studied and can often evade conventional water treatment systems, leading to significant environmental and health threats to living organisms. Based on long-term monitoring data and regional comparisons, this chapter explores the origins, routes, and effects of ECs on water quality and, in relation to that, showcases the occurrence of chronic pharmaceutical pollutions in the Ganges River Basin, 16,216 ng/L of antibiotics found in the surface water of the United States, and ibuprofen analgesics in river of India reaching concentrations of 2.32 μg/L. Conversely, it reports detectable levels of triclosan and microplastics to co-exist in PPCPs almost across the world. The chapter also describes how recent technological advances, and in particular the advent of treatment technologies such as TiO_2-doped PVDF ultrafiltration membranes, eliminated upto 96% of steroidal estrogens, whereas biochar-based adsorbents obtained high adsorption capacities, ultimately eliminating ECs from water sources. In the context of other relevant literature, the findings provide important insights into the need for improved legislation, advanced detection, and remediation methods, as well as raising awareness among the public alongside influencing future studies to focus on the long-term consequences of ecosystems and well-being from EC exposure. Ultimately, adaptation of number of fast-evolving techniques that keep expanding sustainable water treatment may quantitatively diminish the level of these unseen threats in terms of attenuating their negative impact on compromising the quality of surface water and groundwater.

Keywords Emerging contaminants · Surface water · Groundwater · Water quality · Pharmaceuticals and personal care products · Water treatment

---

S. Dey (✉)
Department of Chemical Engineering, University of Calcutta, Kolkata 700009, West Bengal, India
e-mail: sreedeepdey04@gmail.com

Dimple et al. (eds.), *Surface Water and Groundwater Quality for Sustainable Utilization*, Springer Hydrogeology, https://doi.org/10.1007/978-981-96-8189-1_13

1 Introduction

ECs (also termed as "*substances of emerging concerns*") are hazardous compounds that have not yet been regulated but significantly present an overwhelming number of environmental or human health risks, most of which are far from being understood (Campo and Pico 2015; Hazra et al. 2022). Among huge varieties of ECs, pharmaceuticals, personal care products (PCPs), nanomaterials, fire retardants, pesticides, surfactants, industrial chemicals (e.g., plasticizers), disinfection by-products (DBPs), antibiotic resistant bacteria (ARB) and genes (ARG) belong to the most common categories (Lesser et al. 2018; Gutiérrez Jaber 2024). They are labeled "emerging" because they are increasingly becoming a matter of concern. Besides being absolutely indispensable for health and general life quality, as ECs are not presently being regulated by national or international law, they become a more serious threat as a consequence of almost impossible putative reduction (Mahlknecht et al. 2023). In this chapter, the most prevalent emerging pollutants detected in groundwater are marked as ibuprofen, a pain reliever and nonsteroidal anti-inflammatory drug (NSAID), and carbamazepine, a anticonvulsant medicine commonly used in epileptic treatment. Some of the ECs with the chlorinated hydrocarbons like dichlorodiphenyltrichloroethane (DDT) which are not naturally present in the groundwater may pollute the environment by bioaccumulation and biomagnification or pose a hazard to human health by propagating through the food-chain (Kibuye et al. 2019; Sharma et al. 2019; Hazra et al. 2022). Unfortunately, many ECs are not directly considered pollutants by regulatory bodies as they do not fullfill certain characteristics. ECs are present in concentrations of less than a microgram and survive regularly practiced water treatment such as sedimentation, filtration and activated sludge, designed principally for removing larger particles (Badruddoza et al. 2010; Gutiérrez Jaber 2024). The sources of ECs include agricultural fields, livestock farming, industrial activities, waste disposal bodies, and rapidly growing cities. From these sources, ECs undergo leaching to reach groundwater. Untreated effluent discharged from wastewater treatment facilities containing higher concentrations of microbial ECs facilitates unwelcoming biofilm accumulation which may cause organoleptic changes of water and harbors detrimental pathogens such as *Pseudomonas aeruginosa*, *Legionella pneumophila*, and nontuberculous *Mycobacteria* (NTM) (Alengebawy et al. 2021; Mishra et al. 2023; Osuoha et al. 2023).

In our daily practices we use large amounts of PPCPs which contaminate water systems by accumulating ECs like pharmaceutical active compounds (PhACs) and Pharmaceutical contaminants (PCs) (Badruddoza et al. 2010). These products including antibiotics, soaps and toothpaste, sunscreens, hair treatment products, and their metabolites are harder to be eradicated. Some leading researchers, have discussed that after these pollutants accumulate in water, they persist in the environment for longer periods of time due to their stable chemical structures. Some added that the problem gets worsened proportionally by growing population and uncontrolled utilization of resources, ultimately depleting the quality of groundwater and surface to subsurface waterbodies. On the otherhand, some of the popular studies, particularly those from Mora et al. (2021), highlighted the worldwide nature of PPCP contamination. Whereas a recent study has reported various sources of ECs, including stimulants like caffeine, amphetamines, antibiotics, pain relief medicinal substances (e.g., diclofenac and ibuprofen), antiepileptics, antidepressant medications, and hormones from the contraceptive pill (Corada-Fernández et al. 2017; Sharma et al. 2019; Mora et al. 2021; Kumar et al. 2023). Kibuye et al. (2019) indicated the United States reaching maximum antibiotic (Ofloxacin) concentrations of 122,700 ng/L in groundwater (Private wells and springs). The figure for countries like India's (Delhi) aquifer system for analgesic (diclofenac) compound has been found 1.3 mg/L. Triclosan, an chlorinated phenolic antibiotic found in soaps and shampoos has been found in undesirable levels in groundwaters almost all over the world (Kibuye et al. 2019; Silori et al. 2022; Gutiérrez Jaber 2024).

Over 40% of the population in Mexico consumes groundwater (Mahlknecht et al. 2023). A study in 2018 conducted at Valle del Mezquital, Hidalgo, contributes to the literature by revealing the presence of up to 65 PhACs, along with 3 human hormones in wastewater, 23 PhACs in groundwater and benzoylecgonine (a cocaine metabolite), medications like paracetamol and metformin in other water sources (Lesser et al. 2018). In 2021, a study was carried out to determine the contamination status at the Atoyac River in Puebla. It was documented that the river water contained triclosan, naproxen, diclofenac, microplastics, as well as coliform bacteria and heavy metals at intolerable levels. This study presents a compelling argument that as the irrigation rivers are getting contaminated, as an obvious consequence, it matures as a threat to the food supply chain. PPCPs in surface waters increase the incidence of cancer, lower the reproductive rates, and cause birth defects. The high level of antibiotics in untreated groundwater presents a major concern with antibiotic resistance (Mora et al. 2021). The characteristics of the ECs are important determinants for predicting their environmental fates and risk profiles which can be taken account for understanding ECs' role in affecting surface waters. Reports say, nearly 80% of the wastewater is thrown into rivers and seas without being treated, causing water pollution to the main sources of drinking water (Lin et al. 2022; Dutta et al. 2021). A study led by researchers from India found that 166 ECs are present in some of the countries' six environmentally, culturally, and geographically important locations substantially devoted to agricultural purposes and urban populations. Examples of these 36 categories ECs that have been found in surface water of these areas include medicines, caffeine and nicotine, pesticides, and polyfluoroalkyl substances (PFAS) (Philip et al. 2017). Microplastics are synthetic solid particles of less than 5 mm. These water insoluble microplastics are found in almost all aquatic habitats. Primary microplastics including microfibers, plastic glitter, and plastic pellets (also known as nurdles) are being considered a crucial threat in recent years as they are being consumed by aquatic animals, leading to trophic chain accumulation. Due to the persistent nature of ECs, even their existence in trace amounts (in parts per billion) is sufficient to pose hazards to human health and risk the environment. From that perspective, developing new processes for removal or reduction of ECs to meet safe levels in the water resources is imperatively needed. According to the researchers, scarcity of water, combined with pollution and increased population loads in addition to global warming, are catalysts aggravating the growing concern (Renau-Pruñonosa et al. 2020; Petrovic et al. 2016; Sudarsan et al. 2024).

Research has provided evidence that ECs may act as endocrine disruptors and cause perturbation (Alengebawy et al. 2021). ECs hamper hormonal production that influences reproduction, metabolism, and behavior and take care of our immune system (Arman et al. 2021; Mora et al. 2021). Some of the drugs that are intended for treating certain animals may also affect health conditions of other organisms. For instance, chemicals used in irrigation water can influence crop growth. These chemicals sometimes contain some groups of antibiotics that may cause antibiotic resistance in those plants (Alengebawy et al. 2021; Pradhan et al. 2023). Therefore, it is vital that ECs are removed from the water-bodies before being deposited on plant surfaces or absorbed through stomata. Likewise, it is not unusual for ECs to bypass most wastewater treatment plants (WWTPs) engineered to extract typical pollutants from agricultural fields. Therefore, more stringent treatment strategies should have to be followed for their complete removal. Adequate water treatment plants should be established through subsequent technological improvements ensuring safety against ECs (Diamanti-Kandarakis et al. 2009).

To preserve the water resources from ECs and ensure it stays potable for the future, many alternative approaches have been introduced till date (Corada-Fernández et al. 2017; Sharma et al. 2019; Kumar et al. 2022; Meffe and de Bustamante 2014; Sudarsan et al. 2024). However, most of these elimination practices contain considerable backlogs and demand better management approaches. Through an extensive literature review, several research gaps have been found in the recently published articles dealing with ECs and their impacts on surface water quality. Primarily, long-term effects of these ECs were not focused on in those papers. These studies mainly focused on time-based monitoring systems. These studies were conducted at the sub-route level and were limited to specific regions within each country, restricting global comparison with regional results. In spite of carrying out effect assessments regarding new contaminants, innovative approaches towards detection technologies were overlooked. Quantitative risk assessments often relied on qualitative risk measurements. Broad characterization of hazards and ecological impacts—especially those regarding non-target species and ecosystem processes—was minorly discussed. Human health impact was inadequately investigated. The linkages between ECs and targeted public health outcomes, policy/regulation analysis demonstrating currently existing scientific findings, as well as policy making were little discussed. As concerns about these chemicals and compounds affect economies across the globe, the mother target should be public awareness, education, and engagement considerations. Traditional econometric evaluations for managing pollutants from point sources were missing substantially. Therefore, there remains a scarcity of well-established reviews to date.

When compared to those prior relevant literatures, the chapter places emphasis on ECs and addresses those gaps. A Ganges River Basin study was conducted by Sharma et al. (2019) discussing persistent pharmaceutical contaminants and their environmental impacts with a hazardous risk for both the environment and humans. Bioaccumulation of these pollutants and chronic effects were also highlighted in their paper but the study was limited in its capacity to explain temporal variability and geographical coverages. In this chapter data from sites such as the Guadalete River Basin in Spain (Corada-Fernández et al. 2017) and Italy (Meffe and de Bustamante 2014) have been merged providing a global perspective of pollutants, geographical differences, as well as the influence of environmental settings along with government legislation. The development of novel and more sophisticated detection and removal technologies for ECs, such as TiO_2-doped PVDF ultrafiltration membranes that can remove steroidal estrogen (Wang et al. 2024) and biochar adsorbents, for increasing removal efficiency has also been discussed. Specific quantitative risk analyses were conducted by Kumar et al. (2022). In their study, pollutant levels as well as environmental and health impacts associated with adverse effects were predicted using probabilistic risk assessment models. This chapter not only covers the studies relating to contaminants but also reviews their impacts on freshwater ecosystems with special reference to pharmaceuticals affecting the endocrine functions of fishes (Sharma et al. 2019). ECs' effect on biodiversity and ecosystems has also been reviewed alongside ecological impact assessments. Through an exploration of epidemiological review reports, this chapter examines how pollutant exposure might influence health outcomes, like BPA-related concerns (Meffe and de Bustamante 2014). This chapter reviews the current rules and regulations widely followed by European countries (Houtman 2010). It critiques previous water treatment practices as insufficient and proposes policy changes. Initial recommended strategies to raise public awareness and involvement in contamination management include the implementation of community-based monitoring programs, public education campaigns, and involving a broad spectrum of stakeholders at various levels of stages for drafting policies, standards, or criteria. Moreover, it successfully highlights specific pathways through which communities could engage themselves to resolve major EC-related issues.

The aims of this chapter are to:

1. Categorize ECs, for example, ECs from pharmaceuticals or industrial pollutants.
2. Assess ECs' impact on the quality of surface and groundwater.
3. Evaluate existing detection and treatment technologies that deal with ECs present in water sources.
4. Compare ECs among different areas and scenarios.
5. Review risk assessment of ECs from the point of view of human health and affected ecological systems.
6. Provide practical policy and management guidance that can mitigate ECs.
7. Emphasize why public-communication campaigns are important to deal with ECs.
8. Make the readers understand the economics of managing ECs in the evaluation of costs and benefits.

2 Types of Emerging Contaminants

In recent years, several compounds have been detected as ECs in surface water and groundwater systems and the number is ever increasing. As shown in Fig. 1, they are primarily derived from sources, including medicines, cosmetics, fragrances, and chemicals that are rapidly used in industries. Other sources and pathways include agricultural runoffs, industrial discharges, urban runoffs and WWTPs. ECs can be divided into the following mentioned types. However, these classifications constantly change as new contaminants (or effects) are discovered and ECs from past years become inconsequential.

(1) Pharmaceuticals: Pharmaceuticals, such as prescribed drugs, OTC (over-the-counter) medicines, β-blockers (e.g., metoprolol), anti-lipemics, X-ray contrast media, or animal medication, are always present in the domestic sewage as a result of continual introduction. These products are generally excreted by humans or animals after being used and are disposed of inappropriately. In the Ganges River Basin, India, antibiotics and anti-inflammatory drugs were found in surface as well as subsurface waters (Hazra et al. 2022; Sharma et al. 2019). As these ECs hold the capacity to alter endocrine systems in aquatic organisms and can bypass standard wastewater treatment, they become a matter of serious concern (Mishra et al. 2023). A study by

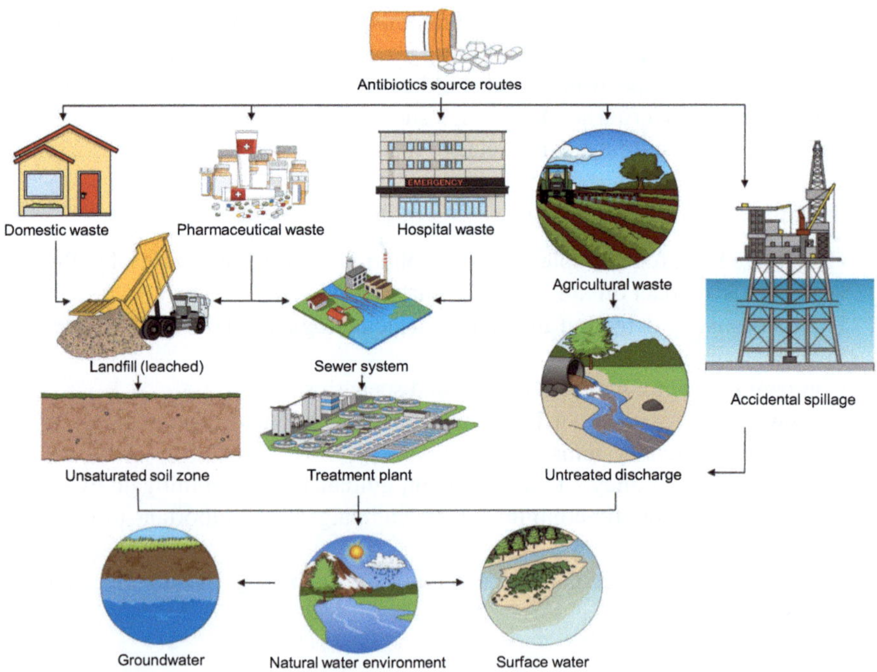

Fig. 1 Visual representation of pathways through which ECs get accumulated in groundwater and surface water

Rivera-Utrilla et al. (2013) aimed to review the literature on pharmaceutical-derived compounds in water and their removal. The primary argument presented in this study was to describe the most common pharmaceutical families found in water and analyze related water pollution issues. The study also summarized the performance of different water treatment systems in removing pharmaceuticals. These systems included conventional systems like chlorine, chlorine dioxide, WWTPs, adsorption/bioadsorption on activated carbon, advanced oxidation processes like ozonation, photooxidation, radiolysis, and electrochemical processes. Additionally, the study also discussed the effects of these treatments on pharmaceutical compounds, the advantages and disadvantages of different methodologies, and the most important parameters of these systems. Key publications on pharmaceutical removal from water were also summarized (Rivera-Utrilla et al. 2013). While significant, the study could become more valuable if it could sufficiently considered environmental and economical sustainability such as energy consumption, carbon footprint and economic feasibility. **(2) Personal Care Products**: PCPs, which include detergents, soaps, and shampoos, as well as cosmetics, UV filters (e.g., oxybenzone), plasticizers (e.g., phthalates), antimicrobials (e.g., triclosan), fragrances, and colorants are another source of EC contamination of water sources. Typically, all these products contain preservatives like parabens, which unfortunately return back to the environment. In the study of Kumar et al. (2022), PCPs have been found to be widely distributed in surface and groundwater. The presence of endosulfan is a threat to aquatic biota with acute toxicity level (LC_{50} value 96-h) found as 0.004 (0.001−0.01) mg/L for African catfish (Islam et al. 2021). PCPs also contain endocrine-disrupting chemicals (EDCs), which are mostly mixture of lipophilic exogenous chemicals and bioaccumulate in adipose tissues with longer half-lives ultimately demonstrating obesogenic and toxic effects. These characteristics hamper their release into the environment on a large scale. The study highlighted the need for new treatment technologies capable of degrading those pollutants. Osuoha et al. (2023) reviewed the current information on PPCPs in aquatic environments, assessing their risk, toxicity, and bioaccumulation. This study provided information on sources, migration, and degradation of these contaminants. According to this study, Carbamazepine is the most persistent PPCP, and a combined treatment strategy, including microbial degradation and other technologies, is recommended for improved eradication efficiency (Osuoha et al. 2023). **(3) Synthetic Chemicals and Artificial Sweeteners**: The lion's share of emerging water contaminants usually comes from industrial processes such as synthetic chemicals and artificial sweetener (ASWs) processing. However, they are also found in daily-used products. Long-lasting endocrine-disrupting chemicals like bisphenol A (BPA) and ASWs are harmful to the water environment (Wang et al. 2024). Meffe and de Bustamante (2014) summarized a case report of Italy highlighting the occurence of 298 emerging organic compounds (EOCs) in surface water and ground water. In this report it was seen that synthetic chemicals as well as their byproducts directly affect water quality. Corada-Fernández et al. (2017) detected the presence of synthetic organic substances and ASWs on the surface of groundwater. A negative association was reported where heavy rainfall influenced the EC concentrations in the Guadalete River Basin of Spain (Xu et al. 2009; Corada-Fernández et al. 2017). **(4)**

Industrial and Agricultural Pollutants: Heavy metals, pesticides, and chemical residues contaminate the water system. This can be seen as a result of industrial processes taking place within urban and agricultural lands (Alengebawy et al. 2021). In Europe, Houtman (2010) investigated the relevance of those contaminants with a special focus on drinking water quality. Pesticides & heavy metals in water bodies are serious problems. They may threaten human health, aquatic life, as well as the ecosystem. Kumar et al. (2022) have also investigated some sustainable technologies based on recent studies that can remove these contaminants. They proposed that in addition to MBRs, reverse osmosis, ultrafiltration or nanofiltrations may considered as sustainable processes for improving water treatment whilst addressing detrimental environmental impacts of ECs. Aside from these principle types, other types of EC include cyanotoxins, industrial chemicals like 1,4-dioxanes, perfluorooctane sulfonate (PFOS) and perfluorooctanoic acid (PFOA), Illicit drugs, non-controlled drugs, organic solvents MTBE and ETBE, complexing agents (e.g., EDTA) and nanomaterials (e.g., carbon-based materials, metal oxides, metals, and quantum dots).

3 Impacts of Emerging Contaminants on Surface Water Quality

ECs like PPCPs, and synthetic chemicals pose a widespread impact on surface water quality with severe consequences. These impacts, in turn, result in alterations of inherent aquatic environments, pose risks to human safety, and slow down the drinking water processing and operations (Arman et al. 2021). This section features the impacts of ECs in detail. Figure 2 portrays the maximum concentrations of some of the most concerning ECs. It is understood, the graphs have been contextualized considering both surface and ground water.

3.1 Effects on Aquatic Ecosystems

Emerging pollutants are environmentally unfriendly to aquatic ecosystems. In a meta-study conducted in 2019, it was found that the Ganges River Basin contained high levels of pharmaceuticals, such as antibiotics and anti-inflammatories. The researchers found that some of these changes had negative effects on aquatic animals by preventing or slowing the rate of their reproduction. For instance, antibiotics are known to decrease fertility and interfere with the development of fish embryos, while anti-inflammatories have been reported to cause abnormal behaviors in fish along with altered feeding responses (Sharma et al. 2019; Subaramaniyam et al. 2023). Meffe and de Bustamante (2014) reported the effects of PFASs and BPA in marine ecosystems with loss in aquatic biodiversity as one of the fatal consequences. This study evidenced almost all fauna species present in the environment to be roughly

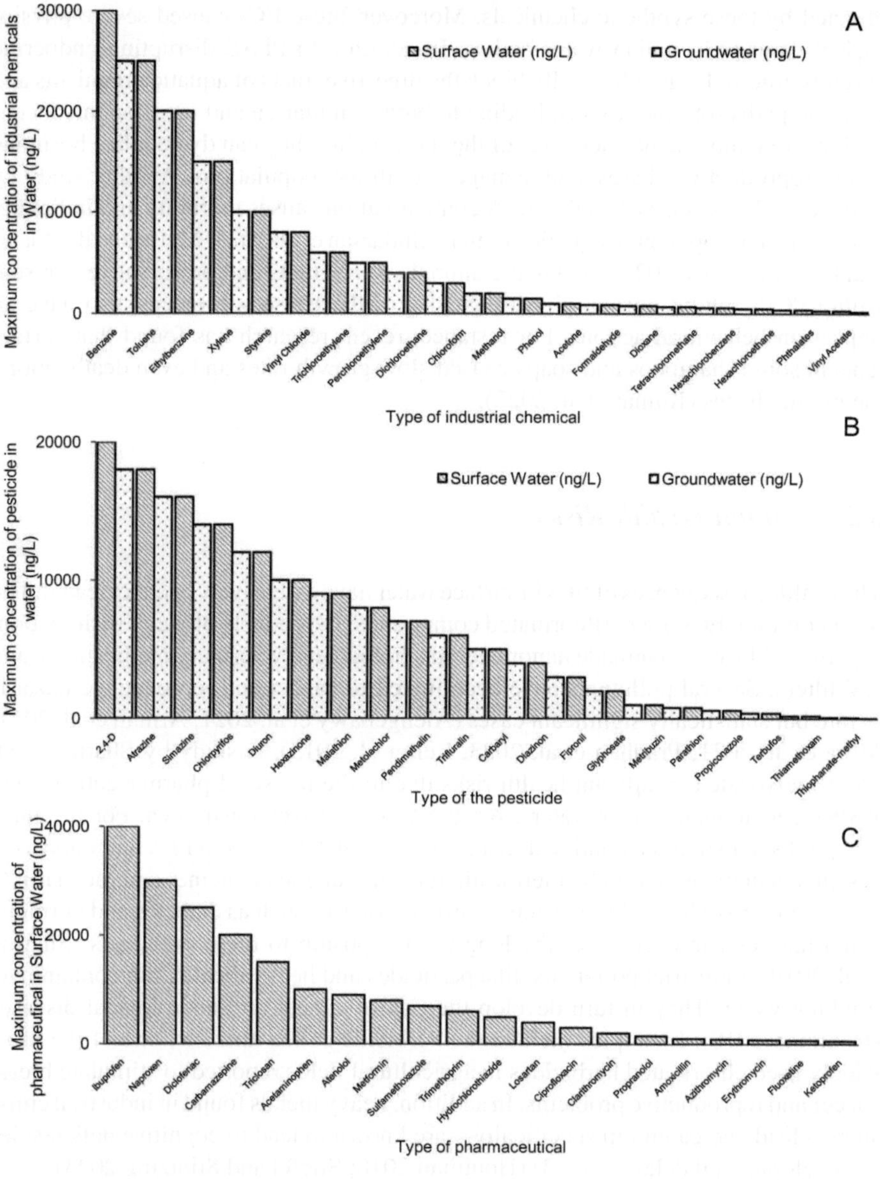

Fig. 2 Column representations displaying maximum concentrations (in ng/mL) of various ECs in surface water and groundwater

affected by these synthetic chemicals. Moreover, these ECs caused severe physiological damage in Italian water bodies. In addition to PFAS disrupting endocrine systems, microplastics physically block the digestive tracts of aquatic organisms and clear the paths for other toxins, leading to bioaccumulation and increase in concentration while moving up each level of the food chain. They can dysregulate hormone levels, reproductive phases, and damage overall fish populations. Another study by Ersoy (2022) also described that BPA contamination causes mortality and indirectly affects the development of aquatic animals (Sudarsan et al. 2024; Meffe and de Bustamante 2014; Ersoy 2022). In aquatic animals, surfactants and preservatives present within PCPs inhibit normal physiological growth. These substances also have an impact on behavioral actions. For instance, recent research has found that surfactants in some shampoos and soaps caused slow growth rates and even death among the marine fishes (Kumar et al. 2022).

3.2 Human Health Risks

The health consequences of ECs in surface water have extensively been reviewed. The most common ECs are perfluorinated compounds, byproducts of water disinfection, gasoline additives, manmade nanomaterials, human and veterinary medications, and UV filters. Several pollutants have been linked to cancer and reproductive hazards in rare but statistically significant cases (Alengebawy et al. 2021; Arman et al. 2021; Mora et al. 2021; Pradhan et al. 2023; Lei et al. 2015). A study by Sharma et al. (2019) advocated significant health risks due to the intake of pharmaceuticals and ASWs contaminated river water from the Ganges River Basin area. For instance, this published study has outlined disturbance of metabolic activities and endocrine systems in humans caused by metformin (an antidiabetic medicine) and saccharin (a renowned ASW) in drinking waters. Chronic diseases such as diabetes and hormone abnormalities can occur over the long-term exposure to these pollutants (Sharma et al. 2019). Industrial pollutants, like pesticides and heavy metals, can contaminate drinking water. They in turn develop the risk of cancer and neurological diseases (Houtman 2010). Examples include the herbicide called atrazine, one of the most widely used chlorinated herbicides in agricultural fields reported to stimulate breast cancer and reproductive problems. In addition, heavy metals found in industrial emissions—lead and cadmium specifically—are known to lead to cognitive deficiencies or developmental delays (Fig. 3) (Houtman 2010; Stiefel and Stintzing 2023).

3.3 Impact on Water Treatment Processes

The ECs directly interfere with the conventional water treatment plants and often require new directions of hybrid technologies for effective EC removal. Mathew

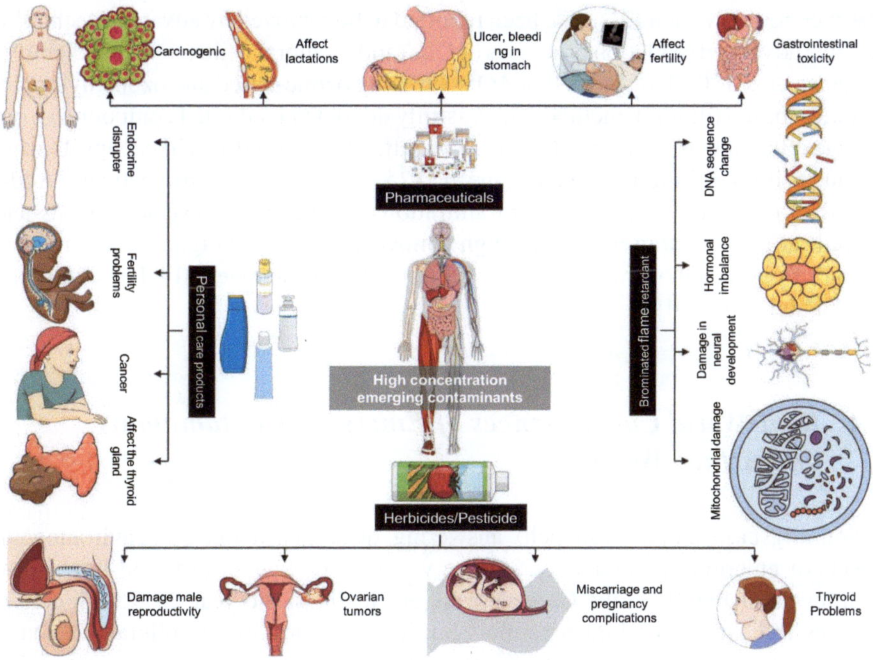

Fig. 3 Simplified diagram depicting the negative health effects of high-concentration ECs on humans

and Kanmani (2020) provided an unique perspective on the efficiency of traditional treatment methodologies including the use of activated carbon and biological treatments. One thing they found was that most of the impurities were absorbed by activated carbon. However, PPCPs were not entirely removed. Recalcitrant compounds (e.g., lignin, an aromatic heteropolymer consisting of condensed phenylpropane units), a class of synthetic pollutants can be found resistant to degradation by conventional biological treatment methods such as activated sludge or microbial treatment. These compounds can only be degraded by only a few groups of specialized fungi over long time periods (Reif et al. 2008; Patel et al. 2019; Mathew and Kanmani 2020; Zabel and Morrell 2020). ECs pose a global concern, particularly pharmaceuticals like antibiotics, which spread antimicrobial resistance. Non-antimicrobials also interact with microorganisms in bulk phase and biofilms. A review by Gomes et al. (2020) aims to answer questions about ECs exposure, tolerance to antimicrobials, biofilm function, and public health concerns. While some ECs have significant effects on microbial behavior, the conclusions of different studies remain controversial. Standardized tests are needed for accurate assessment of ECs' effects on water systems' microbiome. The main challenge is overcoming variability in experimental conditions, mixtures of ECs, and lack of information on non-pharmaceutical ECs (Gomes et al. 2020). Previously, Meffe and de Bustamante (2014) argued that new water treatment strategies are needed to solve these problems, however, recently such ECs have been reported to be removed by advanced technologies like advanced oxidation processes (AOPs) and membrane filtration. For example, Farzaneh et al. (2020) revealed that AOPs utilizing ozone-based and ozone/hydrogen peroxide-based filtration methods successfully degraded medicinal compounds from sewage effluent with 80% and 100% removal efficiencies for drugs like ibuprofen and gemfibrozil (at 1.5 mg/L ozone dosage; 1:0.25 O_3:H_2O_2 ratio), respectively. Additionally, membrane filtration and ultrafiltration in the form of reverse osmosis and nanofiltration were praised for their high removal efficiency of organic contaminants at a more significant operational cost (Meffe and de Bustamante 2014; Farzaneh et al. 2020; Patel et al. 2019).

3.4 Real-World Consequences of Emerging Contaminants on Surface Water

Examining case studies can help answering questions about the environmental effects of emerging pollutants on surface water. In particular, overland flow led to high levels of PPCPs in the Guadalete River, Spain, in the year 2017 with consequences indicating that higher rainfall can increase the loss of pollutants through runoff welcoming water pollution and ecological disruption via sources of EC contaminated surface water. Figure 4 helps understand the contamination pathway with simplified visual representation (Corada-Fernández et al. 2017). According to Ranjan et al. (2022), in the Cooun river near Chennai, 2.32 μg/L ibuprofen has been detected. Around the Ganges River Basin, PPCPs were detected by Sharma et al. (2019). Their study revealed that the pollutants, like ciprofloxacin and triclosan, are toxic to aquatic life due to their persistence in the environment and can raise concern compromising safety profiles of potable water. Feng et al. (2023) mentioned that antibiotics in surface water of the United States are widespread, with a maximum concentration of 16,216 ng/L. Furthermore, Houtman (2010) elucidated increasing pollutant loads in the rivers of Europe highlighting again that better monitoring as well as treatment protocols are needed to secure water quality and human health (Sharma et al. 2019; Houtman 2010).

4 Impacts of Emerging Contaminants on Groundwater Quality

Emerging pollutants in groundwater present serious risk due to their durability and chronic behavior. The insect repellent DEET, Pharmaceuticals such as carbamazepine and clopidol, PPCPs, and synthetic chemicals are often detected in groundwater systems which not only diminish overall water qualities but also pose serious threats

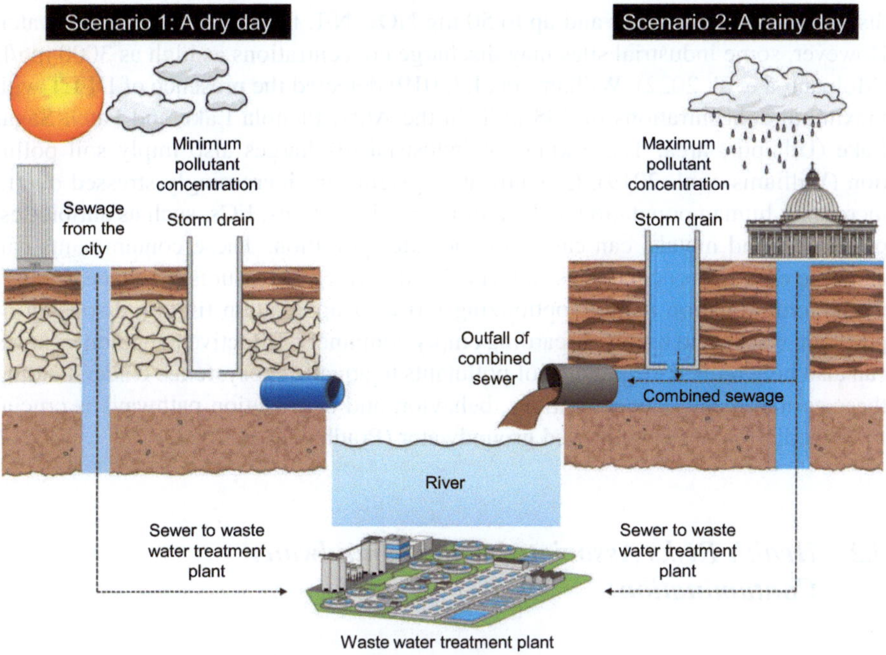

Fig. 4 Diagram depicting two different scenarios. Dry weather allows for more controllable sewage management, whereas developing pollutants cause minimal water pollution. On the other hand, severe rainfall impairs the city's sewage management system, resulting in combined sewer overflows. This ultimately pollutes the water with the highest concentrations of emerging pollutants

to public health (Swain et al. 2022; Sudarsan et al. 2024). This section reflects influences of different contaminants on the groundwater.

4.1 Groundwater Contamination Sources and Mechanisms

There are various sources from which the groundwater may get contaminated. Agrochemicals and discharged effluents from industries get mixed in fresh aquatic systems. Domestic waste leakage is another concerning problem as it leach through the soil and contaminate underground water sources (Mathew and Kanmani 2020; Kumar et al. 2022; Pradhan et al. 2023). In a review, it was stated that nitrate content in the groundwater of South-Indian semi-arid region, may rise up to 78 mg/L (post-monsoon) (Ramalingam et al. 2022). Nitrate content is an important parameter to determine environmental health. High concentrations of nitrate—above 10 mg/L—cause adverse effects in humans, including diseases like methemoglobinemia (Ward et al. 2018; Mathew and Kanmani 2020). According to Fernández-Nava et al. (2008), nitrate concentrations generally range between 10–30 mg NO_3^-N/L in industrial discharges for fresh water and up to 50 mg NO_3^-N/L for discharges into seawater. However, some industrial sites may discharge concentrations as high as 3000 mg/L (Moloantoa et al. 2022). Williams et al. (2019) detected the presence of DEET with maximum concentrations of 388 ng/L in the Ahar, Pichola Lake and Fateh Sagar Lake (Udaipur) area. These kinds of industrial discharges also imply soil pollution (Williams et al. 2019). Groundwater systems are increasingly stressed due to increasing human population and environmental concerns. ECs, such as antibiotics, pesticides, and metals, can cause groundwater pollution. These contaminants can enter the environment and cause adverse health effects. Measures to reduce groundwater contamination include optimizing fertilizer application timing, disposing of pesticide waste, and timely discarding empty containers. Effective irrigation systems can also help reduce the passage of pollutants to aquatic ecosystems. Understanding these contaminants' concentrations, behavior, and degradation pathways is crucial for remediation of contaminated groundwater (Pradhan et al. 2023).

4.2 Health Risks Associated with Groundwater Contamination

Emerging pollutants in groundwater are serious health hazards. Studies have found PPCPs, along with synthetic sweeteners, in groundwater. The concentration of acesulfame was BQL 0.73 μg/L for urban groundwater system of Mexico (Monterrey metropolitan area). These ECs cause endocrine disruption and stimulate antibiotic resistance (Petrovic et al. 2016; Ramos et al. 2022). 500–750 mg of ciprofloxacin is a frequently prescribed broad-spectrum antibiotic. This quinolone antibiotic has been found to develop the disease-causing bacteria resistant against several treatments through mutational changes affecting the target site of the drug (mutating GyrA subunit of DNA gyrase and topoisomerase IV) or through efflux pumps reducing drug accumulation within the bacteria (Sharma et al. 2019; Shariati et al. 2022). Houtman (2010) emphasized the threat of synthetic chemicals such as perfluorinated compounds (PFCs). In his study, PFCs were detected in drinking water up to 519 ng/L (Ruhr–Rhine area). PFCs pose severe health effects like liver damage and hamper maternal chromosome development. PFCs last longer in groundwater, worsening public and environmental health and are often referred to as "forever chemicals" (Houtman 2010).

4.3 Impact on Groundwater Properties and Resources

Emerging pollutants spoil the natural taste of potable water and alter its physiochemical structure, negatively affecting its usability. Meffe and de Bustamante (2014) detected industrial and pesticide EOCs in groundwater at concentrations up to 4.78 × 0^5 ng/L. Water samples were found to be contaminated with perfluoroalkyl substances. Agricultural wastes such as cattle excreta negatively affect water quality by releasing elevated levels of nitrogen, phosphorus, and pathogens, as well as the ARGs. These kinds of water cannot be directly used for drinking purposes and agricultural practices as excessive levels of nutrients may cause eutrophication (Meffe and de Bustamante 2014; Lei et al. 2015; Kumar et al. 2023; Wang et al. 2024). Corada-Fernández et al. (2017) studied the impact of intense rain on groundwater pollution. They discovered that intense rain increases the concentration of pollutants such as pesticides and other ECs in groundwater by up to 50%. This study illustrated how natural disasters hamper groundwater quality management systems (Corada-Fernández et al. 2017).

4.4 Need of Preventive Measures for Protecting Groundwater

Groundwater is the major source of drinking water for a large portion of global population. Many communities from developing countries, specially from arid and semi-arid regions particularly rely on groundwater. As groundwater is directly related with the ecosystem support (sustains rivers, lakes and wetlands; supplies water to terrestrial ecosystems), food production, and flood and drought resilience, protection of this natural reservoir is essential and thus, the groundwater pollution and depletion should be mitigated by using efficient and viable technological solutions (Arman et al. 2021; Ersoy 2022; Wang et al. 2024). Advanced filtration techniques such as activated carbon and reverse osmosis have been studied to remove ECs from groundwater. As an instance, the use of activated charcoal decreases concentrations of pharmaceutical substances from groundwater. Pollutants like triclosan (used in toothpaste as an antimicrobial agent) can be removed via reverse osmosis (Kumar et al. 2022). Mathew and Kanmani (2020) reviewed that porphyrin compounds mixed with other ECs are a concerning threat for aquatic environments. They demonstrated that through regular sampling and routine analysis, it was possible to measure the toxic substances in groundwater samples and acquire suitable solutions accordingly. Effective risk assessment and developmental planning were the major objectives of their research.

5 Emerging Contaminants Mitigation and Management Strategies

The previously mentioned issues highlight the need for mitigation strategies as well as remedial management to tackle ECs in both surface and groundwater by utilizing effective water pollutant removal techniques. However, there are still some challenges that increase the importance of developing new legislative pollution guidelines (Jurado et al. 2019). This section covers a number of remediation technologies to reduce both surface and groundwater pollution and depletion.

5.1 Sophisticated Treatment Technologies

Although various physical, chemical and biological water treatment methods have been implemented till date, even with advanced treatment technologies, such as nanostructured materials and hybrid systems, it is indeed challenging to remove ECs like PRPs (Sudarsan et al. 2024). Kumar et al. (2022) have enlisted a range of technologies in their paper, such as adsorption, advanced oxidation processes, and membrane filtration while proposing removal of acetaminophen with 99% removal efficiency using nitrification and denitrification (Kumar et al. 2022). Similarly, ozonation, photocatalysis, and AOPs have also been proven effective in some cases whereas,within almost 05 seconds residence time, modified TiO_2-doped PVDF ultrafiltration membranes along with a UV light source (365 nm), have been found suitable to remove steroidal estrogens (E2) > 96% at 100ng/L feed concentration, a flux of 60 L/m^2h, 25 mW/cm^2, and 6.5 % Ti (Liu et al. 2023). Ambaye et al. (2021) reviewed that biochar-based adsorption is effective in removing 46% toxic metals, 39% organic pollutants, 13% phosphorus and nitrogen from water.

5.2 Policy and Regulatory Measures

Policy and regulatory actions must be strict while dealing with ECs in groundwater. Sharma et al. (2019) underlined the requirement of suitable release strategies. Universal maximum contaminant levels (MCLs) have to be proposed for antibiotics such as ciprofloxacin and carbamazepine. To drive accountability for policy change, Houtman (2010) advocated the need for regular water quality monitoring programs capable of detecting even low levels of EC compounds. Furthermore, he recommended that assessing treatment removal efficiency should have to be considered as integral parts of EC monitoring programs.

5.3 Best Management Practices

Source control and pollution prevention may be regarded as the most cost-effective solutions promoting sustainability. Using integrated pest management (IPM) strategies may reduce pesticide runoff over the surfaces of the land (Corada-Fernández et al. 2017). Best management practices (BMPs) for industrial discharges include wastewater treatment and hazardous EC reduction. In addition to source reduction and prevention of ECs, BMPs effectively reduce water pollution through structural practices (erosion control, storm water management, and wastewater treatment), and managerial practices (nutrient management, irrigation management, education and awareness) (Mathew and Kanmani 2020).

5.4 Public Awareness and Education

Public awareness is a key component of the EC reduction strategy in groundwater treatment. Kumar et al. (2022) underscored the importance of public education and their potential for safe disposal mechanisms. Common people who are aware of safe disposal mechanisms know how to avoid the spreading of pharmaceutical waste. They know that flushing drugs down through their wash basins indirectly poses problems for themselves. Sharma et al. (2019) also stressed the importance of community monitoring and reporting practices.

5.5 Research and Innovation

Advanced nanomaterials and bioreactors (e.g., MBRs, AnMBRs, BFRs, MBBRs) besides enhancing water treatment efficiency, may improve effluent quality suitable for reuse or discharge, and offer compact footprint (specifically MBRs) compared to activated sludge treatments. Such nanomaterial-based approaches have already been reported to trap emerging pollutants and are favored for NPs' high surface areas, reactivities and functionalizibilities. For example, graphene-based filter materials adsorb a wide range of ECs, including pharmaceuticals and their by-products (Meffe and de Bustamante 2014; Anjum et al. 2019). Interdisciplinary research may generate novel ideas and new approaches, making the future more sustainable and greener (Corada-Fernández et al. 2017). However, potential risks of NPs and membrane fouling, scalability and economic viability issues of bioreactors have to be taken account and thus, future studies should be concerned about possible optimization.

6 Possible Implications of Emerging Contaminant Management Strategies

Some real-world scenarios that have possibilities to implement EC management strategies for successful elimination of ECs to some extends from surface and groundwater have been discussed briefly in the following sections.

6.1 Emerging Contaminants Assessment in the Ganges River Basin, India

The Ganges Basin (The largest river basin in India, covering approximately 861,404 square kilometers) receives thousands of tons of waste every year, including varieties of pharmaceuticals, PCPs and ASWs. In this area, at one or more sampling sites, Sharma et al. (2019) detected unacceptably higher levels of 54.7–826 ng/L, 743 ng/L, & 0.2–102 ng/L of $\sum PPCPs_{RW}$, caffeine & $\sum ASWs_{RW}$ in river water and 34–293 ng/L & 0.5–25 ng/L of $\sum PPCPs_{GW}$ & $\sum ASWs_{GW}$ in groundwater, respectively. Tertiary treatments such as ozonation and activated carbon adsorption, tailored microbial consortia and constructed wetlands can be recommended to treat ECs in these regions. In addition, source control measures may reduce the release of pharmaceuticals in surface waters and upgrade the design of the WWTPs (Sharma et al. 2019).

6.2 Extreme Rainfall Events in Spain

The Guadalete River Basin usually gets contaminated with surfactants, and PPCPs after an incidence of heavy rain. Corada-Fernández et al. (2017) detected the presence of OTNE, galaxolide and ibuprofen up to 6540, 2748 and 1747 ng/L in surface water (post December 2010 heavy rains) and OTNE up to 1285 ng/L in groundwater, respectively. In this situation, if TiO_2-doped PVDF ultrafiltration membranes, electrocoagulation, electrooxidation and AOPs could be combined together, it could deliver high concentrated EC and residual contaminants removal efficiencies. Over and above that, intricate focus on integrated floodwater management systems may lower combined sewer overflows (CSOs) and intercept contamination in the waterbodies of Spain (Corada-Fernández et al. 2017).

6.3 Pharmaceuticals in Italy

Meffe and de Bustamante (2014) reported EOCs, pharmaceuticals, estrogens and illicit drugs up to 15×10^6, 3.59×10^3 ng/L and <50 ng/L in surface water and josamycin >100 ng/L in groundwater, respectively in some of the cities in Italy. As recommended by Sosalla et al. (2024), secondary treatment of tunnel wash water (TWW) may found suitable to remove total EC content, total suspended solids (TSS) and PAHs from sewage water. They proposed that more sophisticated monitoring systems and stricter government supervision are needed to address those fundamental issues (Meffe and de Bustamante 2014; Sosalla et al. 2024).

6.4 Contaminants in Europe

According to a study, undetectable concentrations of pesticides (<100 ng/L) and ethynylestradiol (up to 1.4 ng/L) were found in Llobregat River and Southern Germany drinking waters. Granular activated carbon and ozonation have been used as a combined removal strategy showing remarkable removal rates for last few decades and can be utilized in these cases. The study also discussed that water quality should be maintained through continuous, routine checkups ensuring that the system provides the same efficiency after every treatment program (Houtman 2010).

6.5 Sustainable Technologies in India

In 2022, a review was published on sustainable EC removal technologies. With the aim to study microbial contaminants in India, it emphasized biochar-based and AOP approaches as well as nano-metal-oxide BC used as a biochar adsorbent. It highlighted prior studies that performed hydrogen peroxide mediated photocatalysis achieving 99–100% pesticides and certain antibiotics removal efficiency. This study described that using the mixed liquor-activated sludge techniques groundwaters contaminated with acetaminophen, caffeine and digoxigenin can be treated with 90%, 90% and 95% removal efficiencies, respectively. The researchers identified 83% hydrochlorothiazide removal and 63% diazinon removal from contaminated water using microorganisms based treatments (Kumar et al. 2022). Furthermore, the review documented that AOPs, notably UV-based systems and photocatalysis techniques, may be used as potential processes for treating over-contaminated waters. In addition, pilot plant-3-stage MBBR systems are able to remove more than 90% of propranolol, acetylsulfadiazine and ibuprofen (Kumar et al. 2022). The study suggested that additional treatment infrastructure (e.g., MBR with submerged hollow-fiber ultrafiltration membrane) may be incorporated into these technologies for improving the removal of ECs.

6.6 Limitations of Emerging Contaminants Management Strategies

Controlling emerging pollutants in surface and groundwater is always a matter of concern. The key challenges are mentioned below.

(1) Detection and Monitoring: Detection and monitoring of ECs require advanced analytical methods, especially when they are present in very low levels. These lead to the need for further developments of sensor technologies and methods (Sudarsan et al. 2024). **(2) Scalability and Cost Efficiency**: A reverse question arises that, however these treatment technologies are pretty attractive, they are not possibly scalable and cost-effective. These technologies are likely to be modified and improved in the near future (Sosalla et al. 2024). **(3) Policy and Legislation**: Government laws, rules, and regulations play important roles in mitigating environmental pollution. The partnership between scientists and regulators or industries is necessary to imply viable rules (Wang et al. 2018). **(4) Education and Public Awareness**: Transparent understanding of consequences builds desirable behavior in humans. With proper education and public awareness programs, we can reduce the EC loads from our water sources (Zulkifli et al. 2018; Wang et al. 2018).

7 Policy Frameworks for the Management of Emerging Contaminants

Emerging pollutant management involves quantitative strategies as per the country's policy framework. EC monitoring, treatment, and prevention approaches are commonly used in both the cases of surface and groundwater. This section talks about the existing policies and also where they are falling short. Last but not least, a few suggestions are given to make the regulations stronger (Sanganyado 2022).

7.1 Existing Policy Paradigms

The updated toxicity guidelines "Guidelines for Drinking-water Quality by World Health Organization (WHO)" provide a handful of parameters regarding polluted drinking water systems. Waterbody protection policies have become more imperative in many countries. The Water Framework Directive (WFD) is becoming popular within the EU. Other ECs prevention acts are also formulated to impose strict rules on water pollution monitoring (Petrovic et al. 2016). The US The Environmental Protection Agency (EPA) has established a Contaminant Candidate List (CCL) as well as an Unregulated Contaminant Monitoring Rule to monitor ECs (Korostynska et al. 2013).

7.2 Gaps in Current Policies

However, there are several gaps existing in the current global policies affecting the current regulations. More frameworks are needed to regulate emerging pollutants. The gaps in current policies are as follows. **(1) Narrow Contaminant Focus**: Most of the guidelines recommend a small number of ECs. Promising contaminants like PPCPs are overlooked in some of the policies (Sharma et al. 2019). **(2) Insufficient Detection Technologies**: Many regulations do not include the use of modern detection methods that can effectively handle newly identified contaminants. This hampers the overall oversight and does not provide cent percent assurance that existing controlling strategies are adequate (Meffe and de Bustamante 2014). **(3) Scattered Approaches**: A large number of global and national agendas do not communicate with each other. This leads to the scattering of regulatory mandates (Kumar et al. 2022).

7.3 Suggestions for Policy Improvement

Based on the preceding discussion, a series of recommendations can be proposed to improve the risk management of ECs. Some of the suggestions that may be found helpful in policy improvement are, **(1) Broaden Regulatory Scope**: Government bodies should be aware of modifying regulations that cover a wider range of ECs. The policy would be applicable to handling pharmaceutical wastes, consumer care products, and industrial chemicals. (Mathew and Kanmani 2020). **(2) Studies on Advanced Detection Technologies**: Funds should be raised to carry research on new methods and technologies for detecting ECs. Tracking and assessing of the risks should be done with better analytic approaches (Corada-Fernández et al. 2017). **(3) Strengthen International Cooperation**: Regulators should facilitate the development of internationally recognized standards and best practices for handling ECs. These regulations should follow proper jurisdictions. Global action has to be taken for managing water pollution and respecting standard harmonization (Kumar et al. 2022). **(4) Increase Public Awareness and Participation**: Studies say that the awareness regarding the issue of ECs is not prevalent within the commoner. High EC contamination levels have been recorded in marine waters where recreational activities are commonly practiced. Future plans should take actions to improve public awareness and involve community stakeholders more effectively. Proper management of wastewater and regulation is only possible with proper public education (Sharma et al. 2019; Wang et al. 2018).

7.4 Policy Implementation in Practice

The EU's WFD and the associated regulations on monitoring and managing emerging pollutants represent an example of a well-established policy implementation. The EU is preparing a large set of pollution parameters accompanied by limitation values. These parameters will help the regulators to control ECS and report the level of contamination. These regulations have brought real advancements in water quality. These adaptable regulations could easily be adapted from one site to another (Houtman 2010). However, developing countries may have a scarcity of necessary resources and infrastructure to quickly implement these regulations (Sharma et al. 2019).

8 Future Prospects and Research Needs

Further research is needed to better understand the emerging pollutants, their impacts on surface waters as well as groundwater, and develop novel methods of water quality management. Future researches should focus on the following strategies. **(1) Improved Detection Technologies**: Future research should concentrate on the generation of detection tools (e.g., Biosensors, voltammetry, potentiometry, amperometry, GC-MS and LC-MS, ELISA, colorimetry and spectrophotometry, and SERS). In particular, these tools have to be cost-effective, more sensitive, and EC-specific to overcome current restrictions. Refinements in high-resolution mass spectrometric methods and biosensors may lead to increased sensitivity of the physiochemical measurement capabilities. These high-resolution techniques can detect ECs even if they are present in trace levels and provide real-time monitoring (Meffe and de Bustamante 2014; Kadadou et al. 2022). **(2) Creative Treatment Modalities (CTM)**: CTMs are needed to effectively remove ECs. Emerging technologies like In-situ H_2O_2 production for Fenton oxidation, AOPs, membrane filtration (MF), Microalgae and nanomaterials biotechnology, and nanomaterial-based adsorbents such as CNTs are some examples of potent alternatives. Studies showed that the removal performance of TiO_2-doped membranes and biochar composites for steroidal estrogens and heavy metals from groundwater is considerably better than other conventional methods (Wang et al. 2024; Kumar et al. 2022; Cardoso et al. 2021). **(3) Strategies for Risk Assessment and Management**: Future research must focus on chemical hazard evaluation systems that could reveal the molecular connection existing between EC toxins and their breakdown products. Environmental factors should not be overlooked. Decision-support systems that incorporate these frameworks will probably facilitate better management of contamination risk and stimulate the strategic mitigation (Sharma et al. 2019; National Academies of Sciences, Engineering, and Medicine 2017). **(4) Adaptive Weather Forecast and Pre-treatment Risk Assessment**: In EC affected regions such as Spain, new pollutant management systems have to be implemented for dealing with major environmental problems like the Guadalete River basin contamination where heavy rainfall negatively impact on the natural balance of water system and dysregulate the route of pollution distribution control. Proactive management of wastewater treatment systems with adaptive management measures for weather variability and weather forecasts are required for optimizing wastewater treatment (predicting stormwater overflows, adjusting treatment processes, and managing storage capacity), reducing CSOs (integrating weather radar data with real-time monitoring, and dynamic adjustment of infrastructure) and preventing erosion and channel changes (controlling stormwater runoff) (Corada-Fernández et al. 2017). Research conducted in the Ganges River Basin of India shows how direct discharge into a river has led to increasing contamination. The focus of this study was on PPCPs, resulting in environmental health risks. It was revealed that our daily-used products, like shampoos or medicines, can have significant negative effects when released as waste into waterway systems. The results underscore concern about insufficient pre-treatment risk assessment of these chemical products. Thus pre-treatment risk assessment (e.g., source water assessment, treatment system evaluation, distribution system review, source protection, treatment optimization, system maintenance and repair, emergency response planning, and regular water quality testing) in those pinpoint areas may prevent EC contamination to some extends (Sharma et al. 2019).

9 Conclusive Remarks

Taken together, this chapter shows that due to their persistent nature and hazardous properties, ECs have become a major concern in public health and aquatic ecosystems. ECs include the wide ranges of PPCPs, synthetic chemicals, and industrial pollutants. PPCPs get into waterbodies via improper disposal routes. Antibiotics, analgesics, and hormones have been detected at high levels on the surface and groundwater. This chapter implies that artificial molecules like BPA and ASWs remain in the environment and are hard to remove using traditional water treatment methods. Heavy metals and pesticides from industrially contaminated effluents cause serious water pollution. In the chapter, TiO_2-doped PVDF ultrafiltration membranes and biochar-based adsorbents have been reviewed for detection and removal of ECs. However, lower concentrations of ECs make their elimination from groundwater difficult. The chapter pointed out some areas where further research is needed for addressing the mostly concerning challenges. Long-term outcome measures and regional and global comparisons of existing standards stimulate quantitative risk assessments. With the help of cases like the Ganges River Basin and the Guadalete River Basin, a comprehensive review is given to understand the effects of ECs from different locations. This broader approach of this chapter included an ecological perspective of ECs with human health assessments along with policy analysis and the important role of public awareness campaigns for EC management. Although this review is helpful, it would be more valuable if it provided a framework for establishing long-term monitoring schemes and cross-regional studies to identify EC impacts more clearly.

Developing advanced treatment technologies for more generalized removal of EC advocate regulatory frameworks. Economic evaluations quantify benefits as well as costs. In the context of other relevant literature, this chapter's findings suggest that further future research should be focused on real-life testing of new technologies, understanding their consequences in the long run, and connecting scientific innovators to policymakers as a collaborative effort preserving sustainable water resources.

Declarations

Conflict of Interest The author declares no competing interests.

References

Alengebawy A, Abdelkhalek ST, Qureshi SR, Wang M-Q (2021) Heavy metals and pesticides toxicity in agricultural soil and plants: ecological risks and human health implications. Toxics 9(3):42. https://doi.org/10.3390/toxics9030042

Ambaye TG, Vaccari M, van Hullebusch ED, Amrane A, Rtimi S (2021) Mechanisms and adsorption capacities of biochar for the removal of organic and inorganic pollutants from industrial wastewater. Int J Environ Sci Technol 18(10):3273–3294. https://doi.org/10.1007/s13762-020-03060-w

Anjum M, Miandad R, Waqas M, Gehany F, Barakat MA (2019) Remediation of wastewater using various nano-materials. Arab J Chem 12(8):4897–4919. https://doi.org/10.1016/j.arabjc.2016.10.004

Arman NZ, Salmiati S, Aris A, Salim MR, Nazifa TH, Muhamad MS, Marpongahtun M (2021) A review on emerging pollutants in the water environment: existences, health effects and treatment processes. Water 13(22):3258. https://doi.org/10.3390/w13223258

Badruddoza AZM, Hazel GSS, Hidajat K, Uddin MS (2010) Synthesis of carboxymethyl-β-cyclodextrin conjugated magnetic nano-adsorbent for removal of methylene blue. Colloids Surf A 367:85–95. https://doi.org/10.1016/j.colsurfa.2010.06.01889

Campo J, Picó Y (2015) Advanced mass spectrometry for food safety and quality emerging contaminants. Elsevier, pp 515–578

Cardoso IMF, Cardoso RMF, Esteves da Silva JCG (2021) Advanced oxidation processes coupled with nanomaterials for water treatment. Nanomaterials 11(8):2045. https://doi.org/10.3390/nano11082045

Corada-Fernández C, Candela L, Torres-Fuentes N, Pintado-Herrera MG, Paniw M, González-Mazo E (2017) Effects of extreme rainfall events on the distribution of selected emerging contaminants in surface and groundwater: the Guadalete river basin (SW, Spain). Sci Total Environ 605–606:770–783. https://doi.org/10.1016/j.scitotenv.2017.06.049

Diamanti-Kandarakis E, Bourguignon JP, Giudice LC, Hauser R, Prins GS, Soto AM, Zoeller RT, Gore AC (2009) Endocrine-disrupting chemicals: an Endocrine Society scientific statement. Endocr Rev 30(4):293–342. https://doi.org/10.1210/er.2009-0002

Dutta D, Arya S, Kumar S (2021) Industrial wastewater treatment: current trends, bottlenecks, and best practices. Chemosphere 285:131245. https://doi.org/10.1016/j.chemosphere.2021.131245

Ersoy N (ed) (2022) New trends in agriculture, forestry and aquaculture sciences. Duvar Publishing, Izmir

Farzaneh H, Loganathan K, Saththasivam J, McKay G (2020) Ozone and ozone/hydrogen peroxide treatment to remove gemfibrozil and ibuprofen from treated sewage effluent: factors influencing bromate formation. Emerg Contam 6:225–234. https://doi.org/10.1016/j.emcon.2020.06.002

Fernández-Nava Y, Maranon E, Soons J, Castrillón L (2008) Denitrification of wastewater containing high nitrate and calcium concentrations. Bioresour Technol 99(17):7976–7981. https://doi.org/10.1016/j.biortech.2008.03.048

Gomes IB, Maillard JY, Simões LC, Simões M (2020) Emerging contaminants affect the microbiome of water systems—strategies for their mitigation. NPJ Clean Water 3:39. https://doi.org/10.1038/s41545-020-00086-y

Gutiérrez Jaber I (2024) Emerging contaminants: a growing threat to global water quality. Tecscience. https://tecscience.tec.mx/en/health/groundwater-contamination/. Accessed 06 June 2024

Hazra A, Mondal A, Paul S, Bej S, Mondal U, Nag S, Banerjee P (2022) Chemosensing technology for rapid detection of emerging contaminants. In: Emerging contaminants in the environment. Elsevier, pp 407–464

Houtman CJ (2010) Emerging contaminants in surface waters and their relevance for the production of drinking water in Europe. J Integr Environ Sci 7(4):271–295. https://doi.org/10.1080/1943815X.2010.511648

Islam Md, Amin SN, Brown CL, Juraimi AS, Uddin MK, Arshad A (2021) Determination of median lethal concentration (LC50) for endosulfan, heptachlor and dieldrin pesticides to african catfish, Clarias gariepinus and their impact on its behavioral patterns and histopathological responses. Toxics 9(12):340. https://doi.org/10.3390/toxics9120340

Jurado A, Walther M, Díaz-Cruz MS (2019) Occurrence, fate and environmental risk assessment of the organic microcontaminants included in the Watch Lists set by EU Decisions 2015/495 and 2018/840 in the groundwater of Spain. Sci Total Environ 663:285–296. https://doi.org/10.1016/j.scitotenv.2019.01.270

Kadadou D, Tizani L, Wadi VS, Banat F, Alsafar H, Yousef AF, Barceló D, Hasan SW (2022) Recent advances in the biosensors application for the detection of bacteria and viruses in wastewater. J Environ Chem Eng 10(1):107070. https://doi.org/10.1016/j.jece.2021.107070

Kibuye F, Preisendanz H, Elkin K, Swistock B, Veith T, Watson J, Elliott H (2019) Occurrence, concentrations, and risks of pharmaceutical compounds in private wells in Central Pennsylvania. J Environ Qual 48. https://doi.org/10.2134/jeq2018.08.0301

Korostynska O, Mason A, Al-Shamma'a AI (2013) Monitoring pollutants in wastewater: traditional lab based versus modern real-time approaches. In: Mukhopadhyay S, Mason A (eds) Smart sensors for real-time water quality monitoring. Smart sensors, measurement and ınstrumentation, vol 4. Springer, Berlin, Heidelberg. https://doi.org/10.1007/978-3-642-37006-9_1

Kumar M, Sridharan S, Sawarkar AD, Shakeel A, Anerao P, Mannina G, Sharma P, Pandey A (2023) Current research trends on emerging contaminants pharmaceutical and personal care products (PPCPs): a comprehensive review. Sci Total Environ 859(1):160031. https://doi.org/10.1016/j.scitotenv.2022.160031

Kumar R, Qureshi M, Vishwakarma DK, Al-Ansari N, Kuriqi A, Elbeltagi A, Saraswat A (2022) A review on emerging water contaminants and the application of sustainable removal technologies. Case Stud Chem Environ Eng 6:100219. https://doi.org/10.1016/j.cscee.2022.100219

Lei M, Zhang L, Lei J, Zong L, Li J, Wu Z, Wang Z (2015) Overview of emerging contaminants and associated human health effects. Biomed Res Int 2015:404796. https://doi.org/10.1155/2015/404796

Lesser LE, Mora A, Moreau C, Mahlknecht J, Hernández-Antonio A, Ramírez AI, Barrios-Piña H (2018) Survey of 218 organic contaminants in groundwater derived from the world's largest untreated wastewater irrigation system: Mezquital Valley, Mexico. Chemosphere 198:510–521. https://doi.org/10.1016/j.chemosphere.2018.01.154

Lin L, Yang H, Xu X (2022) Effects of water pollution on human health and disease heterogeneity: a review. Front Environ Sci 10:880246. https://doi.org/10.3389/fenvs.2022.880246

Liu S, Véron E, Lotfi S, Fischer K, Schulze A, Schäfer AI (2023) Poly (vinylidene fluoride) membrane with immobilized TiO_2 for degradation of steroid hormone micropollutants in a photocatalytic membrane reactor. J Hazard Mater 447:130832. https://doi.org/10.1016/j.jhazmat.2023.130832

Mahlknecht J, Aguilar-Barajas I, Farias P, Knappett PSK, Torres-Martínez JA, Hoogesteger J, Lara RH, Ramírez-Mendoza RA, Mora A (2023) Hydrochemical controls on arsenic contamination and its health risks in the Comarca Lagunera region (Mexico): implications of the scientific evidence for public health policy. Sci Total Environ 857(1):159347. https://doi.org/10.1016/j.scitotenv.2022.159347

Mathew R, Kanmani S (2020) A review on emerging contaminants in Indian waters and their treatment technologies. Nat Environ Pollut Technol 19:549–562. https://doi.org/10.46488/NEPT.2020.v19i02.010

Meffe R, de Bustamante I (2014) Emerging organic contaminants in surface water and groundwater: a first overview of the situation in Italy. Sci Total Environ 481:280–295. https://doi.org/10.1016/j.scitotenv.2014.02.053

Mishra RK, Mentha SS, Misra Y, Dwivedi N (2023) Emerging pollutants of severe environmental concern in water and wastewater: a comprehensive review on current developments and future research. Water-Energy Nexus 6:74–95. https://doi.org/10.1016/j.wen.2023.08.002

Moloantoa KM, Khetsha ZP, Van Heerden E, Castillo JC, Cason ED (2022) Nitrate water contamination from industrial activities and complete denitrification as a remediation option. Water 14(5):799. https://doi.org/10.3390/w14050799

Mora A, García-Gamboa M, Sánchez-Luna MS, Gloria-García L, Cervantes-Avilés P, Mahlknecht J (2021) A review of the current environmental status and human health implications of one of the most polluted rivers of Mexico: the Atoyac River, Puebla. Sci Total Environ 782:146788. https://doi.org/10.1016/j.scitotenv.2021.146788

National Academies of Sciences, Engineering, and Medicine; Division on Earth and Life Studies; Board on Environmental Studies and Toxicology; Committee on Incorporating 21st Century Science into Risk-Based Evaluations (2017) Using 21st century science to ımprove risk-related evaluations. National Academies Press, Washington DC. https://www.ncbi.nlm.nih.gov/books/NBK424976/. Accessed 22 July 2024

Osuoha JO, Anyanwu BO, Ejileugha C (2023) Pharmaceuticals and personal care products as emerging contaminants: need for combined treatment strategy. J Hazard Mater Adv 9:100206. https://doi.org/10.1016/j.hazadv.2022.100206

Patel M, Kumar R, Kishor K, Mlsna T, Pittman CU Jr, Mohan D (2019) Pharmaceuticals of emerging concern in aquatic systems: chemistry, occurrence, effects, and removal methods. Chem Rev 119:3510–3673. https://doi.org/10.1021/acs.chemrev.8b00299

Philip JM, Aravind UK, Aravindakumar CT (2018) Emerging contaminants in Indian environmental matrices—a review. Chemosphere 190:307–326. https://doi.org/10.1016/j.chemosphere.2017.09.120

Petrovic M, Sabater S, Elosegi A, Barceló D (eds) (2016) Emerging contaminants in river ecosystems: occurrence and effects under multiple stress conditions. Springer, Cham

Pradhan B, Chand S, Chand S, Rout PR, Naik SK (2023) Emerging groundwater contaminants: a comprehensive review on their health hazards and remediation technologies. Groundw Sustain Dev 20:100868. https://doi.org/10.1016/j.gsd.2022.100868

Ramalingam S, Panneerselvam B, Kaliappan SP (2022) Effect of high nitrate contamination of groundwater on human health and water quality index in semi-arid region, South India. Arab J Geosci 15(3), 242. https://doi.org/10.1007/s12517-022-09553-x

Ramos E, Padilla-Reyes D, Mora A, Barrios-Piña H, Kant S, Mahlknecht J (2022) Assessment of artificial sweeteners as wastewater co-tracers in an Urban groundwater system of Mexico (Monterrey Metropolitan Area). Water 14(20):3210. https://doi.org/10.3390/w14203210

Ranjan N, Singh PK, Maurya NS (2022) Pharmaceuticals in water as emerging pollutants for river health: a critical review under Indian conditions. Ecotoxicol Environ Saf 247:114220. https://doi.org/10.1016/j.ecoenv.2022.114220

Reif R, Suárez S, Omil F, Lema JM (2008) Fate of pharmaceuticals and cosmetic ingredients during the operation of a MBR treating sewage. Desalination 221:511–517. https://doi.org/10.1016/j.desal.2007.01.111

Renau-Pruñonosa A, García-Menéndez O, Ibáñez M, Vázquez-Suñé E, Boix C, Ballesteros BB, Hernández García M, Morell I, Hernández F (2020) Identification of aquifer recharge sources as the origin of emerging contaminants in ıntensive agricultural areas. La Plana de Castellón, Spain. Water 12(3):731. https://doi.org/10.3390/w12030731

Rivera-Utrilla J, Sánchez-Polo M, Ferro-García MÁ, Prados-Joya G, Ocampo-Pérez R (2013) Pharmaceuticals as emerging contaminants and their removal from water: a review. Chemosphere 93(7):1268–1287. https://doi.org/10.1016/j.chemosphere.2013.07.059

Sanganyado E (2022) Policies and regulations for the emerging pollutants in freshwater ecosystems: challenges and opportunities. In: Dalu T, Tavengwa NT (eds) Emerging freshwater pollutants. Elsevier, pp 361–372. https://doi.org/10.1016/B978-0-12-822850-0.00007-7

Shariati A, Arshadi M, Khosrojerdi MA, Abedinzadeh M, Ganjalishahi M, Maleki A, Heidary M, Khoshnood S (2022) The resistance mechanisms of bacteria against ciprofloxacin and new approaches for enhancing the efficacy of this antibiotic. Front Public Health 10:1025633. https://doi.org/10.3389/fpubh.2022.1025633

Sharma BM, Bečanová J, Scheringer M, Sharma A, Bharat GK, Whitehead PG, Klánová J, Nizzetto L (2019) Health and ecological risk assessment of emerging contaminants (pharmaceuticals, personal care products, and artificial sweeteners) in surface and groundwater (drinking water) in the Ganges river basin, India. Sci Total Environ 646:1459–1467. https://doi.org/10.1016/j.scitotenv.2018.07.235

Silori R, Shrivastava V, Singh A, Sharma P, Aouad M, Mahlknecht J, Kumar M (2022) Global groundwater vulnerability for pharmaceutical and personal care products (PPCPs): the scenario of second decade of 21st century. J Environ Manag 320:115703. https://doi.org/10.1016/j.jenvman.2022.115703

Sosalla N, Uhl W, Vistnes H et al (2024) Tunnel wash water in a cold climate: characteristics, ecotoxicological risk, and effect of sedimentation. Res Sq. https://doi.org/10.21203/rs.3.rs-4681682/v1

Stiefel C, Stintzing F (2023) Endocrine-active and endocrine-disrupting compounds in food—occurrence, formation and relevance. NFS J 31:57–92. https://doi.org/10.1016/j.nfs.2023.03.004

Subaramaniyam U, Allimuthu RS, Vappu S, Ramalingam D, Balan R, Paital B, Panda N, Rath PK, Ramalingam N, Sahoo DK (2023) Effects of microplastics, pesticides and nano-materials on fish health, oxidative stress and antioxidant defense mechanism. Front Physiol 14:1217666. https://doi.org/10.3389/fphys.2023.1217666

Sudarsan JS, Dogra K, Kumar R, Raval NP, Leifels M, Mukherjee S, Trivedi MH, Jain MS, Zang J, Barceló D, Mahlknecht J, Kumar M (2024) Tricks and tracks of prevalence, occurrences, treatment technologies, and challenges of mixtures of emerging contaminants in the environment: with special emphasis on microplastic. J Contam Hydrol 265:104389. https://doi.org/10.1016/j.jconhyd.2024.104389

Swain S, Taloor AK, Dhal L et al (2022) Impact of climate change on groundwater hydrology: a comprehensive review and current status of the Indian hydrogeology. Appl Water Sci 12:120. https://doi.org/10.1007/s13201-022-01652-0

Wang F, Xiang L, Leung KS-Y, Elsner M, Zhang Y, Guo Y, Pan B, Sun H, An T, Ying G, Brooks BW, Hou D, Helbling DE, Sun J, Qiu H, Vogel TM, Zhang W, Gao Y, Simpson MJ, Luo Y, Chang SX, Su G, Wong BM, Fu T-M, Zhu D, Jobst KJ, Ge C, Coulon F, Harindintwali JD, Zeng X, Wang H, Fu Y, Wei Z, Lohmann R, Chen C, Song Y, Sanchez-Cid C, Wang Y, El-Naggar A, Yao Y, Huang Y, Law JC-F, Gu C, Shen H, Gao Y, Qin C, Li H, Zhang T, Corcoll N, Liu M, Alessi DS, Li H, Brandt KK, Pico Y, Gu C, Guo J, Su J, Corvini P, Ye M, Rocha-Santos T, He H, Yang Y, Tong M, Zhang W, Suanon F, Brahushi F, Wang Z, Hashsham SA, Virta M, Yuan Q, Jiang G, Tremblay LA, Bu Q, Wu J, Peijnenburg W, Topp E, Cao X, Jiang X, Zheng M, Zhang T, Luo Y, Zhu L, Li X, Barceló D, Chen J, Xing B, Amelung W, Cai Z, Naidu R, Shen Q, Pawliszyn J, Zhu Y, Schaeffer A, Rillig MC, Wu F, Yu G, Tiedje JM (2024) Emerging contaminants: a One Health perspective. Innovation 5(4):100612. https://doi.org/10.1016/j.xinn.2024.100612

Wang L, Zhang L, Lv J, Zhang Y, Ye B (2018) Public awareness of drinking water safety and contamination accidents: a case study in Hainan Province, China. Water 10(4):446. https://doi.org/10.3390/w10040446

Ward MH, Jones RR, Brender JD, De Kok TM, Weyer PJ, Nolan BT, Villanueva CM, Van Breda SG (2018) Drinking water nitrate and human health: an updated review. Int J Environ Res Public Health 15(7):1557. https://doi.org/10.3390/ijerph15071557

Williams M, Kookana RS, Mehta A, Yadav SK, Tailor BL, Maheshwari B (2019) Emerging contaminants in a river receiving untreated wastewater from an Indian urban centre. Sci Total Environ 647:1256–1265. https://doi.org/10.1016/j.scitotenv.2018.08.084

Xu J, Chen W, Wu L, Green R, Chang AC (2009) Leachability of some emerging contaminants in reclaimed municipal wastewater-irrigated turf grass fields. Environ Toxicol Chem 28:1842–1850. https://doi.org/10.1897/08-471.1

Yan F, An L, Xu X, Du W, Dai R (2024) A review of antibiotics in surface water and their removal by advanced electrocoagulation technologies. Sci Total Environ 906:167737. https://doi.org/10.1016/j.scitotenv.2023.167737

Zabel RA, Morrell JJ (2020) The decay setting: some structural, chemical, and moisture features of wood features of wood in relation to decay development. In: Wood microbiology. Elsevier, pp 149–183

Zulkifli SN, Rahim HA, Lau WJ (2018) Detection of contaminants in water supply: a review on state-of-the-art monitoring technologies and their applications. Sens Actuators B Chem 255:2657–2689. https://doi.org/10.1016/j.snb.2017.09.078

Unseen Threats: The Impact of Emerging Contaminants on Surface and Groundwater Quality

Lalita Rana, Manish Kumar, Navnit Kumar, Jitendra Rajput, Sumit Sow, Sarvesh Kumar, Anil Kumar, S. N. Singh, and Devendra Singh

Abstract Emerging contaminants, have become significant concerns for water quality management. The increasing detection of emerging contaminants (ECs) in surface water and groundwater raising substantial concerns about water quality, pose challenges to conventional water treatment processes and can lead to adverse ecological and health effects. These contaminants, which often escape regulation and conventional water treatment processes, persists in the ecosystems, resulting in potentially long-term impacts and their potential for bioaccumulation necessitate a comprehensive understanding of their behavior and impacts. Disruptions to aquatic ecosystems, poor quality produce due to contaminated irrigation water, proliferation of antibiotic-resistant bacteria, threat to drinking water sources and health risks are common problems poses by emerging contaminants. Addressing these challenges requires improved monitoring techniques, advanced water treatments and integrated management strategies to mitigate the impact of emerging contaminants on water resources for achieving SDGs. This chapter explains sources, behavior, fate and impacts of ECs in agro-ecosystems, public education, and stricter regulations are also crucial to protect water quality from these pervasive contaminants.

Keywords Emerging contaminants · Water quality · Sustainable Development Goals (SDGs) · Agro-ecosystems · Public health

---

L. Rana (✉) · N. Kumar
Department of Agronomy, Sugarcane Research Institute, Dr. Rajendra Prasad Central Agricultural University, Pusa, Samastipur 848 125, Bihar, India
e-mail: lalita@rpcau.ac.in

N. Kumar
e-mail: navnit.sri@rpcau.ac.in

M. Kumar
Metamorphic Petrology Lab, Department of Geology, Institute of Science, Banaras Hindu University, Varanasi 221005, India
e-mail: manish562@gmail.com

J. Rajput
Division of Agricultural Engineering, Indian Agricultural Research Institute, Pusa, New Delhi 110012, India
e-mail: jitu.caepht@gmail.com

Dimple et al. (eds.), *Surface Water and Groundwater Quality for Sustainable Utilization*, Springer Hydrogeology, https://doi.org/10.1007/978-981-96-8189-1_14

1 Introduction

Water quality is a cornerstone of environmental health and human well-being, essential for drinking, agriculture, recreation and industry. Traditionally, water quality concerns have centered around well-known pollutants such as heavy metals, organic chemicals, landfills and nutrients. However, in recent years, a new category of pollutants known as emerging contaminants (ECs) has gained attention. These contaminants encompass pharmaceuticals, personal care products, endocrine-disrupting compounds (EDCs), microplastics, and various agricultural chemicals used for killing weeds, reducing insects and disease populations (Fig. 1). Unlike traditional pollutants, ECs are not routinely monitored or regulated, yet they pose potential risks to both eco-systems and human health (Richardson and Ternes 2018). Numerous studies have detected a broad range of pharmaceuticals in surface waters and the effluents from wastewater treatment plants, underscoring the limitations of conventional treatment processes in effectively removing these compounds (Blair et al. 2013). EDCs can enter into sea, rivers, ponds, lakes, streamlines through agricultural activities, effluents discharge from industries and consumer products. Landfills also play crucial role for elevated levels of contaminates to ground and surface water. Their persistence and bioaccumulation potential make them particularly concerning (Schwarzenbach et al. 2006a, b). In other hand, micro plastics after degradation into tiny plastic particles are widespread in the environment and have been found in

S. Sow
Department of Agronomy, PGCA, Dr. Rajendra Prasad Central Agricultural University, Pusa, Samastipur 848 125, Bihar, India
e-mail: sumitsow19@gmail.com

S. Kumar
Directorate of Research, Dr. Rajendra Prasad Central Agricultural University, Pusa, Samastipur 848 125, Bihar, India
e-mail: sarvesh@rpcau.ac.in

A. Kumar
Department of Entomology, Sugarcane Research Institute, Dr. Rajendra Prasad Central Agricultural University, Pusa, Samastipur 848 125, Bihar, India
e-mail: anil.kumar@rpcau.ac.in

S. N. Singh
Department of Plant Pathology, Sugarcane Research Institute, Dr. Rajendra Prasad Central Agricultural University, Pusa, Samastipur 848 125, Bihar, India
e-mail: snsingh@rpcau.ac.in

D. Singh
Director, SRI, Dr. Rajendra Prasad Central Agricultural University, Pusa, Samastipur 848 125, Bihar, India
e-mail: dsingh@rpcau.ac.in

various water bodies even in the ground water and can be ingested by aquatic organisms, resulting in both physical and toxicological impacts (Barbosa et al. 2016). Apart from this, use of pesticides, herbicides, and fertilizers in agricultural land are major threats of ECs which can runoff into surface water and leach into groundwater. For example, nitrates from fertilizers can cause eutrophication in surface waters and pose serious health risks when present in drinking water (Lapworth et al. 2012). Other common inorganic contaminants found in groundwater include Total dissolved solids (TDS), anions and oxyanions, such as SO_4^{-2}, Cl^- and F^- and major cations, such as Ca^{2+} and Mg^{2+}. The environmental behavior of ECs is influenced by their chemical properties, including solubility, volatility, and persistence. This persistence means that even low concentrations of ECs can have significant cumulative effects over time. For instance, the presence of pharmaceuticals in surface waters has been linked to the development of antibiotic-resistant bacteria, posing a substantial risk to both environmental and public health (Luo et al. 2014). To overcome these consequences, there is a need for more comprehensive monitoring programs that include a wider range of ECs to better understand their distribution and impacts (Verlicchi et al. 2012). So, keeping in view, we explored the sources and types of emerging contaminants, their fate and environmental behavior and their impacts on surface and groundwater quality, crop growth and quality, soil health, as well as mitigation strategies. Additionally, we provided future recommendations and perspectives regarding contaminants in surface and groundwater.

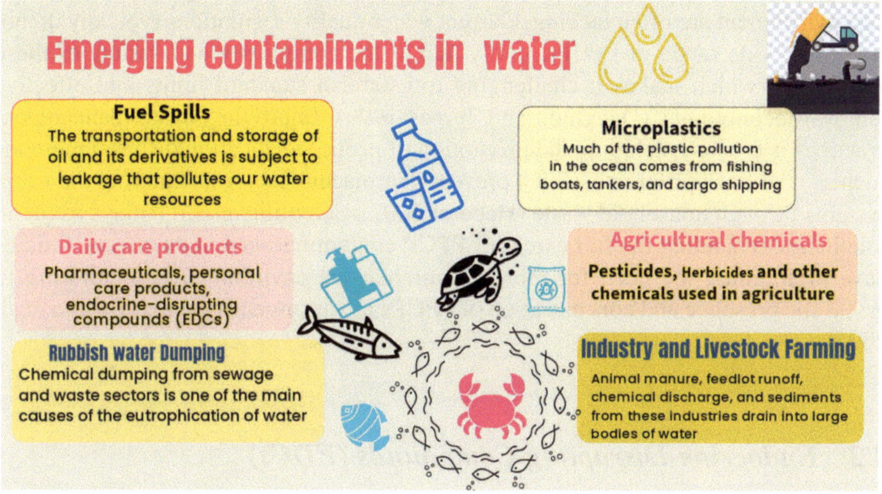

Fig. 1 Emerging contaminants (ECs) in surface and ground water

2 Sources and Types of Emerging Contaminants

2.1 *Pharmaceuticals and Personal Care Products (PPCPs)*

Pharmaceuticals and Personal Care Products (PPCPs) represent a diverse array of chemical substances utilized for health, hygiene, and cosmetic purposes. Pharmaceutical industries contribute significantly to environmental pollution. Many pharmaceutical products intended to improve human and animal health are not fully metabolized within the body. These unmetabolized residues are excreted and eventually enter aquatic ecosystems through wastewater. Wastewater treatment plants are not specifically designed to fully remove pharmaceuticals and personal care products (PPCPs), allowing these substances to persist in surface water, groundwater, and even drinking water supplies (Santos et al. 2010; Boxall et al. 2012). Marine ecosystems, in particular, are vulnerable as pharmaceuticals can disrupt the endocrine systems of aquatic organisms, affecting their reproduction and behavior. For example, exposure to certain synthetic hormones used in contraceptives has been linked to altered sex ratios in fish populations (Fent et al. 2006). Furthermore, the widespread use of antibiotics in the atmosphere can promote the emergence of antibiotic-resistant bacteria by drastically reducing antibiotic effectiveness in treating illnesses (Kümmerer 2009). Despite the growing recognition of these issues, the regulatory framework for managing PPCPs is still in its infancy in many regions. Comprehensive guidelines for monitoring and controlling PPCP residues in the environment are often lacking. Current water quality regulations typically do not cover the wide range of PPCPs, partly due to the complexity and variety of these substances, which makes it challenging to establish standard limits and effective removal technologies (Daughton and Ternes 1999). Improving waste management practices is crucial, including the promotion of proper medication disposal methods such as take-back programs, which prevent pharmaceuticals from entering the water systems through household waste (Heberer 2002). Scientific research plays a crucial role in understanding the full extent of PPCP contamination and its effects. Studies have highlighted the need for more comprehensive environmental monitoring to assess the presence and concentration of PPCPs in various ecosystems (Boreen et al. 2004).

2.2 *Endocrine-Disrupting Compounds (EDCs)*

Endocrine-Disrupting Chemicals (EDCs) are a group of substances capable of disrupting the endocrine (hormonal) systems of living organisms, potentially causing harmful developmental, reproductive, neurological, and immune effects (Fig. 2). Among the various endocrine-disrupting chemicals (EDCs) present in the environment, micropollutants like natural hormones, perfluorooctanesulfonic acids (PFOS), insecticides, and nonylphenols have been detected. These substances often originate

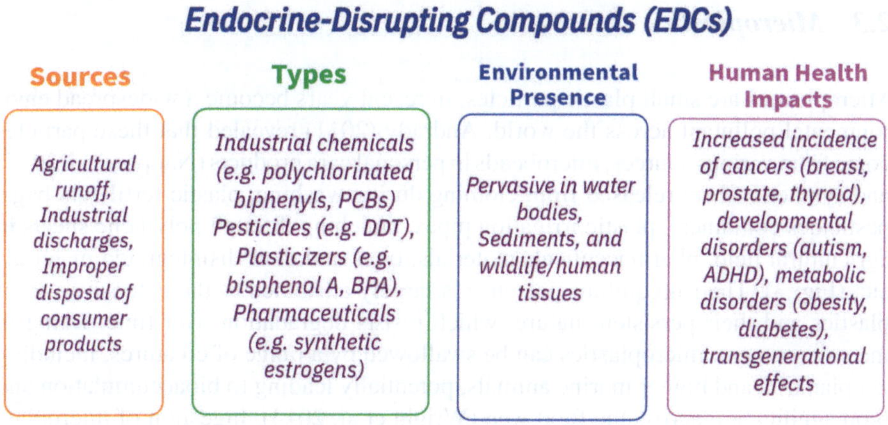

Fig. 2 Endocrine-disrupting compounds (EDCs) and their impacts

from plastic products and flame retardants. These chemicals can imitate, obstruct, or modify the normal activity of hormones, posing considerable health hazards even at minimal concentrations (Gore et al. 2015a, b). Exposure to EDCs has been linked to reproductive abnormalities in fish, amphibians, and mammals, including humans. These abnormalities can manifest as altered sex ratios, reduced fertility, and developmental malformations (Colborn et al. 1993).

In humans, endocrine-disrupting compounds (EDCs) have been linked to a higher incidence of certain cancers, including breast, prostate, and thyroid cancers, as well as developmental disorders like autism and attention deficit hyperactivity disorder (ADHD) (WHO/UNEP 2013). Additionally, EDCs are suspected of contributing to metabolic disorders like obesity and diabetes, and their effects can be transgenerational, affecting not just individuals but also their offspring (Vandenberg et al. 2012). The complexity and variety of EDCs pose challenges for their regulation and management. Traditional risk assessment methods often fall short in evaluating the low-dose effects and non-linear dose–response relationships characteristic of EDCs. Moreover, current regulatory frameworks may not adequately address the cumulative and synergistic effects of multiple EDCs present in the environment (Kortenkamp et al. 2009). Consumers can reduce their exposure by opting for products free of known EDCs, such as BPA-free plastics and organic foods. Policymakers and industry stakeholders need to prioritize research and development of safer chemicals and promote policies that limit the use of harmful substances (Zoeller et al. 2012). Their ability to interfere with hormone systems at low concentrations poses risks to wildlife and humans, contributing to a range of health problems from reproductive abnormalities to chronic diseases.

2.3 Microplastics

Microplastics are small plastic particles, in recent years become a widespread environmental pollutant across the world. Andrady (2011) revealed that these particles come from various sources, microbeads in personal care products (Napper et al. 2015) and synthetic fibers released from clothing during washing, plastic fertilizers bags, pesticides containers, plastic irrigation pipes, mulching through polythene sheets in agricultural field, pharmaceuticals materials, tyres and their disintegrated materials etc. (Fig. 3). Their ubiquitous presence is largely attributed to the extensive use of plastics and their persistent nature, which resists degradation over time. Being in the environment, microplastics can be swallowed by a range of creatures, including zooplankton and bigger marine animals, potentially leading to bioaccumulation and biomagnification across the food web (Wright et al. 2013). Ingestion of microplastics by marine creatures causes severe ecological and health hazards. For example, research has demonstrated that microplastics can physically harm marine creatures' digestive tracts, limit feeding efficiency, and affect growth and reproduction (Cole et al. 2013). Microplastics can then be conveyed to organisms through consuming food potentially resulting in toxicological consequences (Teuten et al. 2009). Furthermore, the prevalence of microplastics in seafood has generated worries about human health, as these contaminants may enter the body via contaminated marine products (Galloway and Lewis 2016). Mitigating microplastic contamination requires a multidimensional approach that includes legislation, industry innovation, and public awareness. Innovations in textile manufacturing aim to create fibers that shed fewer microplastics during washing, while improved wastewater treatment technologies can help capture microplastics before they enter water bodies. Scientific research continues to play a critical role in understanding the full extent of microplastic pollution and its impacts. Advanced analytical techniques are being developed to detect and quantify microplastics in various environmental matrices, improving our ability to monitor their distribution and abundance (Hidalgo-Ruz et al. 2012).

2.4 Industrial and Household Chemicals

Industrial and household chemicals constitute a diverse group of substances widely used in various sectors, including manufacturing, agriculture, and daily domestic activities. These chemicals range from solvents, detergents, and disinfectants to pesticides, heavy metals, and flame retardants. Their extensive use is crucial for modern living and economic activities, yet they pose significant environmental and health risks due to their persistence, bioaccumulation potential, and toxicity (UNEP 2019). Industrial chemicals, such as polychlorinated biphenyls (PCBs) and dioxins, and can enter ecosystems through industrial discharges, agricultural runoff, and improper disposal of household products have long been recognized for their harmful effects. For example, PCBs, which were widely used as dielectric and coolant fluids, have

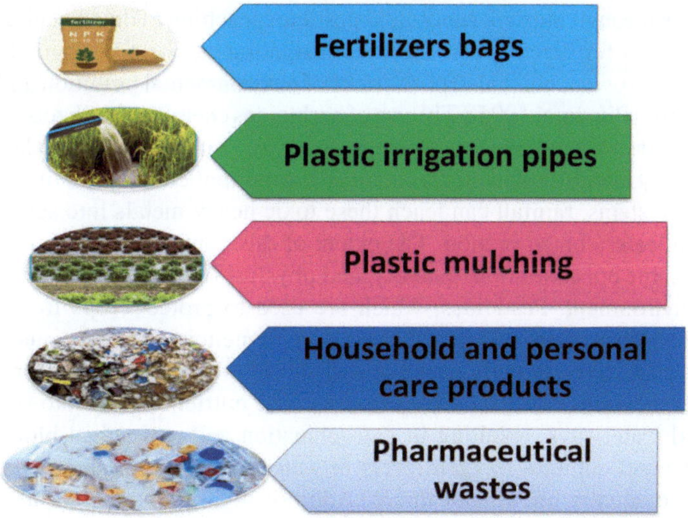

Fig. 3 Sources of microplastics in surface and ground water

been linked to cancer, immune system suppression, and endocrine disruption (Breivik et al. 2002). Similarly, household chemicals, including cleaning agents and personal care products, contribute to indoor pollution and have been associated with respiratory issues, skin irritation and endocrine disruption (Weschler 2009). The environmental pathways of these chemicals are diverse. Once released, they often persist in the environment, affecting water quality, soil health, and air purity. For instance, pesticides such as organophosphates and neonicotinoids can leach into waterways, impacting aquatic life and potentially entering the human food chain through contaminated water and crops (Aktar et al. 2009). Additionally, the improper disposal of household chemicals, can lead to the contamination of water bodies with substances that wastewater treatment plants are not designed to remove, thus posing a continuous threat to aquatic ecosystems and public health (Daughton and Ternes 1999). In recent past years, green chemistry offer promising alternatives by designing safer chemicals and processes that reduce the environmental footprint and enhance biodegradability (Anastas and Eghbali 2010).

2.5 Agricultural Chemicals

Agricultural activities encompass a wide array of practices designed to cultivate crops and raise livestock to sustain the global food supply. While agriculture is essential for human survival and economic development, use of chemicals in the field is increasing day by day and results into significant environmental challenges. Intensive farming practices often lead to soil degradation, loss of biodiversity, and pollution of water bodies through nutrient runoff and pesticide leaching (Tilman et al. 2002). The extensive use of fertilizers, pesticides, and irrigation can help mitigate the harmful effects of soil erosion, but it also poses risks of environmental pollution and potential health hazards (Pimentel 1995). The use of herbicides chemical fertilizers and organic sources of nutrients has been steadily increasing, and many of these products contain traces of heavy metals. Agricultural chemicals when applied in agricultural fields near aquatic ecosystems, rainfall can leach these toxic heavy metals into surface waters, resulting in water contamination. The extent of this contamination is influenced by the bioavailable concentration, quantity, and physical state of the chemicals released into the environment. Pesticides, which are used to protect crops from pests and diseases, can also harm non-target organisms (beneficial insects, aquatic animals and birds). Fertilizers, notably nitrogen and phosphorus, are essential for increasing crop yields; nevertheless, excessive use can cause nutrient runoff into rivers, lakes, and coastal waterways, resulting in eutrophication and toxic algal blooms (Smith et al. 1999) and depletes oxygen levels in water bodies, through creation of dead zones where aquatic life cannot thrive. Additionally, the overuse of antibiotics in animal production fosters the emergence of antibiotic-resistant bacteria, leading to significant public health concerns (Marshall and Levy 2011).

3 Environmental Behavior and Fate

It refers to the processes and pathways through which chemicals and pollutants move, transform, and persists. Understanding the environmental behaviour and fate of chemicals becomes essential to identifying their possible effects on ecosystems and human wellness. Chemicals exposed into the environment may go through several kinds of physical, chemical, and biological alterations, which affect their dispersion, duration, and toxicity. Pesticides used on farmland, for example, can evaporate into the sky, leach into groundwater, or wash off into surface waterways, where they can degrade by photolysis, hydrolysis, or microbial activity (Kookana et al. 1998). The fate of these chemicals is determined by their inherent qualities, such as solubility, volatility, and susceptibility to degradation, as well as environmental factors such as temperature, pH, and the presence of other substances. One critical aspect of environmental behavior is the partitioning of chemicals between different environmental compartments-air, water, soil, and biota. This partitioning is influenced by factors such as the chemical's octanol–water partition coefficient (Kow), which predicts its affinity for organic matter versus water, and its Henry's law constant, which indicates its potential to volatilize from water to air (Mackay et al. 2006). For instance, hydrophobic chemicals with high Kow values tend to accumulate in soils and sediments and can bioaccumulate in the fatty tissues of organisms, leading to biomagnification through the food chain (Gobas et al. 1993). Conversely, chemicals with high volatility may disperse widely through the atmosphere, potentially leading to long-range transport and deposition far from their original source. The environmental fate of pollutants is also shaped by degradation processes, which can be biotic or abiotic. Biotic degradation, primarily mediated by microorganisms, plays a significant role in breaking down organic pollutants into less harmful compounds. This process can be influenced by the availability of nutrients, oxygen, and suitable microbial communities (Alexander 1999). Abiotic degradation, on the other hand, involves chemical reactions such as hydrolysis, photolysis, and oxidation–reduction reactions. For example, sunlight can drive the photodegradation of certain pesticides in surface waters, reducing their persistence and toxicity (Zepp and Cline 1977). The fate of chemicals is further complicated by the formation of transformation products, which can sometimes be more toxic or persistent than the parent compounds. These transformation products can arise from metabolic processes in organisms or abiotic reactions in the environment (Schwarzenbach et al. 2006a, b).

4 Impacts of Contaminants

4.1 Impact on Water Quality of Surface and Ground Water

Surface water, comprising rivers, lakes, streams, and reservoirs, is a critical component of the Earth's hydrological cycle and an essential resource for ecological systems, human consumption, agriculture, and industry. The quality and quantity of surface water are influenced by a variety of natural processes and anthropogenic activities. Precipitation, runoff, and groundwater discharge naturally contribute to surface water levels, while human activities such as agriculture, urbanization, and industrial operations significantly affect both the volume and quality of these water bodies (Meybeck 2003). Current surface water pollution can serve as an early warning sign for future groundwater contamination. Contaminants from surface sources may gradually seep into aquifers, threatening water quality. Monitoring surface water quality is essential to predict and prevent long-term impacts on groundwater resources and ensure sustainable water management (Stuart et al. 2012). Pollution from agricultural runoff, which often contains fertilizers, pesticides, and sediments, is a major concern. Nutrient loading, primarily from nitrogen and phosphorus, leads to eutrophication (Fig. 4), which causes harmful algal blooms and hypoxic conditions detrimental to aquatic life (Smith and Schindler 2009). Furthermore, urban runoff adds contaminants including heavy metals, lubricants, and untreated sewage into surface water systems, deteriorating water quality and endangering human health and the environment (Paul and Meyer 2001). Industrial discharges also contribute to the contamination of surface waters with hazardous chemicals and emerging contaminants (Kolpin et al. 2002). These substances can persist in the environment, bioaccumulate in aquatic organisms, and pose long-term ecological and health risks. Surface water contamination by such pollutants is particularly concerning due to their potential to disrupt endocrine systems, cause reproductive and developmental issues in wildlife, and lead to chronic health problems in humans (Diamanti-Kandarakis et al. 2009).

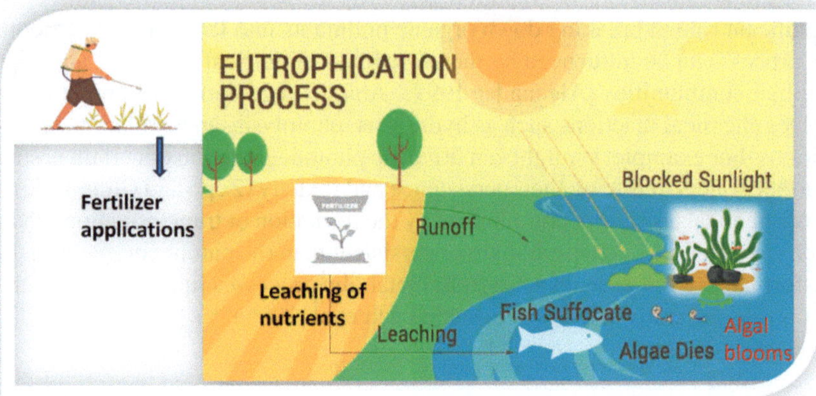

Fig. 4 Eutrophication from agricultural runoff and leaching in water

The impact of global warming exacerbates the difficulties affecting surface water resources by changing the distribution of precipitation, escalating the frequency and severity of catastrophic storms, and modifying water temperatures. These changes may lead to more frequent droughts and floods, compromising water availability and quality (IPCC 2014). Higher temperatures can exacerbate eutrophication and increase the metabolic rates of microorganisms, potentially leading to more rapid degradation of water quality (Paerl and Huisman 2008). Restoration projects aimed at improving riparian buffer zones, wetland conservation, and watershed management play vital roles in mitigating pollution and enhancing water quality (Bernhardt et al. 2005).

Groundwater is a vital resource that faces significant threats from natural processes and human activities. In other hand, groundwater, stored in the pores and fractures of underground aquifers, is a crucial resource for drinking water, irrigation, and industrial uses. It constitutes about 30% of the world's fresh water supply and is essential for sustaining ecosystems and human populations, particularly in arid and semi-arid regions (Foster and Chilton 2003). The quality and availability of groundwater are influenced by natural and anthropogenic factors. Naturally occurring processes, such as the dissolution of minerals from geological formations, can impact groundwater quality, introducing substances like arsenic, fluoride, and radon that pose health risks (Smedley and Kinniburgh 2002). Human activities, including agriculture, industrial processes, and urban development, significantly exacerbate groundwater contamination. Agricultural practices, for example, contribute to nitrate pollution through the excessive use of fertilizers, which can leach into aquifers and lead to conditions such as methemoglobinemia or "blue baby syndrome" (Spalding and Exner 1993). Similarly, pesticides used in farming can infiltrate groundwater, posing long-term health risks and challenging water treatment processes (Squillace et al. 2002). Industrial activities is the main cause of groundwater pollution having heavy metals, VOCs (volatile organic compounds), and POPs (persistent organic pollutants). For instance, the improper disposal of industrial waste or accidental spills can result in the release of hazardous substances like benzene, toluene, and chlorinated solvents into groundwater supplies, which are difficult to remediate and pose significant risks to human health (Pankow and Cherry 1996). Urbanization also contributes to groundwater pollution through the infiltration of contaminants from septic systems, landfills, and urban runoff, introducing pathogens, pharmaceuticals, and household chemicals into aquifers (Lapworth et al. 2012). The over-extraction of groundwater for agricultural, industrial, and domestic use is another critical issue, leading to the depletion of aquifers and a decline in water levels. This can cause land subsidence, reduced water availability, and the deterioration of water quality due to the concentration of contaminants (Konikow and Kendy 2005). Additionally, excessive groundwater extraction in coastal areas can induce saltwater intrusion, contaminating freshwater aquifers and rendering them unsuitable for consumption and irrigation (Barlow and Reichard 2010). Climate changes can exacerbate the stress on already overdrawn aquifers, reducing groundwater availability and affecting the sustainability of water supplies (Taylor et al. 2013). Promoting the use of efficient irrigation techniques, enhancing groundwater recharge through artificial methods, and protecting recharge zones from contamination (Gleeson et al. 2012) (Table 1).

4.2 Impacts on Human Health

Emerging contaminants (ECs) pose significant risks to human health due to their persistence in the environment, potential for bioaccumulation, and diverse sources. These contaminants include pharmaceuticals, endocrine-disrupting compounds (EDCs), microplastics, and agricultural chemicals. Their presence in water systems can lead to a variety of health issues (Fig. 5), often through complex and long-term exposure pathways. Pharmaceuticals are among the most frequently detected ECs in water bodies. These substances, including antibiotics, analgesics, and psychiatric medications, enter the environment through human excretion, improper disposal, and agricultural runoff. One of the most pressing concerns is the development of antibiotic-resistant bacteria. Antibiotics in water promote the proliferation of resistant strains, making bacterial infections increasingly difficult to treat and posing a significant public health threat (Kümmerer 2009). The presence of antibiotics in drinking water sources has been linked to the emergence of multidrug-resistant pathogens, which are responsible for various infections that do not respond to conventional treatments (Martínez 2009). Endocrine-disrupting compounds (EDCs) are another major category of ECs that significantly impact human health. These chemicals, including bisphenol A (BPA), phthalates, and certain pesticides, can mimic or interfere with the body's hormonal systems. EDCs have been associated with health issues, including reproductive disorders, developmental problems in children, and an increased risk of cancers. BPA, commonly found in plastics and food containers, has been linked to hormonal disruptions that can lead to infertility, obesity, and cardiovascular diseases (Rubin 2011). Similarly, phthalates, which are used as plasticizers

Table 1 Impact of emerging contaminants on surface and ground water with respect to climate

Emerging contaminant	Source	Impact on surface water quality	Impact on groundwater quality	References
Pharmaceuticals (e.g., antibiotics, painkillers)	Wastewater treatment plants, agricultural runoff	Development of antibiotic-resistant bacteria, toxicity to aquatic life	Contamination of drinking water sources, potential health risks to humans	aus der Beek et al. (2016), Wilkinson et al. (2022)
Personal care products (e.g., triclosan, parabens)	Household products, wastewater effluents	Disruption of aquatic ecosystems, bioaccumulation in wildlife	Persistence in groundwater, potential endocrine disruption	Sanchez et al. (2021), Luo et al. (2014)
Per- and polyfluoroalkyl substances (PFAS)	Industrial processes, firefighting foams	Long-term persistence, bioaccumulation, toxic effects on wildlife	Persistent contamination, potential carcinogenic effects	Sunderland et al. (2019)
Pesticides (e.g., atrazine, glyphosate)	Agricultural runoff, urban runoff	Toxicity to aquatic organisms, disruption of reproductive systems	Leaching into groundwater, potential health risks to humans	Kwiatkowski et al. (2020), Tang et al. (2021), Silva et al. (2019)
Microplastics	Urban runoff, wastewater effluents	Physical and chemical harm to aquatic organisms, potential for chemical adsorption	Potential for transport of contaminants, unknown health effects	Li et al. (2020), Enyoh et al. (2019)
Endocrine disrupting chemicals (EDCs) (e.g., bisphenol A)	Industrial effluents, household products	Alteration of reproductive systems in aquatic life, developmental effects	Persistence and mobility in groundwater, potential health risks to humans	Kabir et al. (2020)
Heavy metals (e.g., lead, mercury, arsenic)	Industrial discharge, mining activities	Bioaccumulation, toxic effects on aquatic life	Long-term persistence, potential neurotoxic effects on humans	Liu et al. (2021), Kumar et al. (2019)
Hormones (e.g., estrogen, testosterone)	Agricultural runoff, wastewater treatment plants	Disruption of endocrine systems in aquatic life, potential feminization of fish	Contamination of drinking water sources, potential endocrine disruption	Rehman et al. (2021), Anjum et al. (2021), Aris et al. (2014)

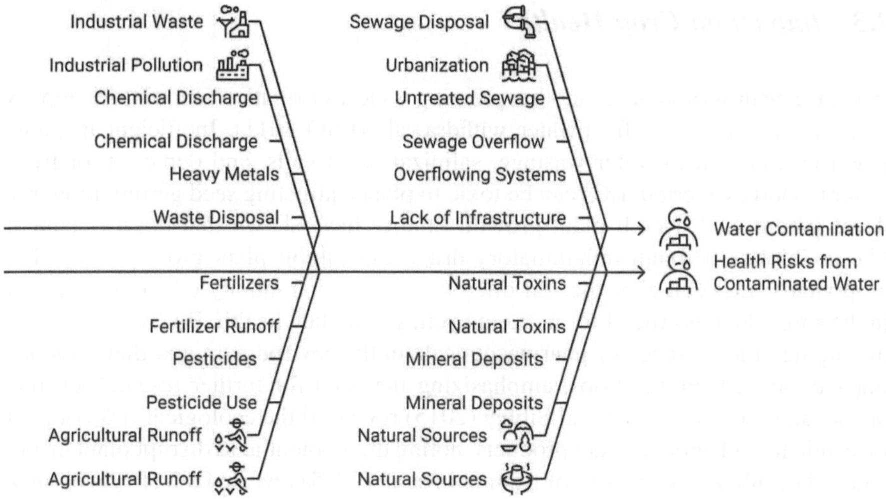

Fig. 5 Contaminated water and human health hazards

in numerous consumer products, have been shown to affect the development of the reproductive system, particularly in male fetuses and infants (Gore et al. 2015a, b). Microplastics have been detected in both surface water and groundwater. The ingestion of microplastics by aquatic organisms can lead to physical harm and toxicological effects, and these particles can enter the human food chain through seafood consumption (European Chemicals Agency 2016) and can cause physical damage to tissues, induce inflammatory responses, and serve as vectors for other harmful chemicals and pathogens (Smith et al. 2018).

Agricultural chemicals, particularly pesticides, herbicides, and fertilizers, are significant contributors to ECs in water systems. Pesticides like atrazine and glyphosate are commonly used for suppressing pests and weeds, but they may cause adverse consequences on human health. Long-term exposure to these chemicals has been linked to endocrine disruption, cancers, and developmental delays in children (Mnif et al. 2011). Atrazine, for instance, is known to disrupt hormonal activity and has been implicated in reproductive issues and cancers (Hayes et al. 2011). Glyphosate, the active ingredient in many herbicides, has been classified as a probable human carcinogen and is associated with non-Hodgkin lymphoma (IARC 2015). Nitrate contamination from agricultural runoff is another critical health concern. Nitrates in drinking water can cause methemoglobinemia, sometimes known as "blue baby syndrome," a condition in which newborns' blood's ability to carry oxygen is weakened. Other health risks linked to high nitrate levels pertain to thyroid disorders and some malignancies (Ward et al. 2005). The widespread use of nitrogen-based fertilizers has led to significant nitrate pollution in many regions, making it a persistent public health challenge. The widespread presence of emerging contaminants in water systems necessitates urgent attention due to their potential to cause significant health issues.

4.3 Impact on Crop Health

Water usage in agriculture is another pressing issue, as irrigation accounts for approximately 70% of global freshwater withdrawals (FAO 2011). Inefficient irrigation practices can lead to water wastage, salinization of soils, and depletion of freshwater resources. Certain ECs can be toxic to plants, affecting seed germination, root development, and overall plant growth. Studies have shown that pharmaceuticals like antibiotics and anti-inflammatory drugs can inhibit plant growth and reduce crop yields. ECs can alter the soil microbiome, the community of microorganisms in the soil, which is crucial for nutrient cycling and plant health. Boxall et al. (2012) highlighted the presence of pharmaceuticals in the environment and their potential impacts on agricultural crops, emphasizing the need for further research on their uptake and effects. Prosser and Sibley (2015) reviewed the ecological risks of pharmaceuticals and personal care products, noting their potential to disrupt plant growth and soil health. Investigations of Bonmatin et al. (2015) revealed the effects of neonicotinoid pesticides on non-target organisms, including plants, and found significant adverse effects on plant health and soil microorganisms. Another researcher Kumar et al. studied in their experiment found that long-term exposure of ECs could lead to reduced crop yields and altered soil properties. Rillig (2012), de Souza Machado et al. (2018) discussed the emerging issue of microplastics in the soil and their potential effects on soil structure and plant growth. For instance, Jones and de Voogt (2019) reviewed that the environmental presence of per- and polyfluoroalkyl substances (PFAS) and their potential impacts on crop health, including uptake and phytotoxicity.

5 Detection and Monitoring

Detection and monitoring of environmental pollutants are crucial components in managing and mitigating the impact of contaminants on ecosystems and human health. These processes involve a variety of methods and technologies designed to identify and quantify pollutants in different environmental media, including air, water, soil, and biota. Advanced analytical techniques, such as gas chromatography-mass spectrometry (GC–MS) and high-performance liquid chromatography (HPLC), are widely used for detecting high sensitive and specific organic and inorganic pollutants (Zuloaga et al. 2012). These methods allow for the precise quantification of trace levels of contaminants, providing critical data for environmental assessment and regulatory compliance. Additionally, field-portable instruments and sensor technologies, such as portable X-ray fluorescence (XRF) analyzers and real-time air quality monitors, have enhanced the ability to conduct in situ measurements, enabling more immediate and localized monitoring (Huang et al. 2016). Biological monitoring, or biomonitoring, is another important approach for detecting and assessing the impact of pollutants on living organisms. This method involves measuring the levels of contaminants in tissues of plants, animals, or human populations to evaluate exposure and potential health effects (CDC 2009). Biomonitoring can provide an integrated measure of bioavailable pollutants over time, reflecting actual environmental conditions more accurately than point measurements. For example, the use of bioindicator species, such as mussels or lichens, helps assess the health of aquatic and terrestrial ecosystems, respectively, by indicating the presence and effects of pollutants like heavy metals and air pollutants (Rainbow 2002). Remote sensing technologies have also revolutionized environmental monitoring, offering large-scale and long-term observation capabilities. Satellite-based sensors and aerial drones equipped with multispectral and hyperspectral imaging can detect changes in land use, vegetation health, and water quality, providing valuable data for tracking pollution sources and trends over time (Asner 2009). These technologies are particularly useful for monitoring large or inaccessible areas, such as remote forests, oceans, and urban regions, where traditional ground-based methods might be impractical or too costly. The integration of big data analytics and machine learning with environmental monitoring systems has further enhanced the ability to detect and predict pollution events. These advanced computational techniques can analyze vast amounts of environmental data from diverse sources, identifying patterns and trends that may not be immediately apparent through conventional analysis (Jordan and Mitchell 2015). Predictive modeling and real-time data analytics enable proactive management strategies, such as early warning systems for air pollution or water contamination incidents, thereby reducing the potential impact on public health and the environment (Castell et al. 2015). The use of sophisticated analytical techniques, portable instruments, biomonitoring, remote sensing, and advanced data analytics has significantly improved the ability to identify, quantify, and manage contaminants.

6 Mitigation Strategies

6.1 Advanced Water Treatment Technologies

Advanced water treatment technologies have become increasingly essential to meet the growing demand for clean and safe water amid rising pollution and resource scarcity. These technologies encompass a range of processes designed to remove contaminants from water to make it suitable for various uses, including drinking, industrial processes, and environmental protection. Membrane filtration technologies, such as reverse osmosis (RO), nanofiltration (NF), ultrafiltration (UF), and microfiltration (MF), have revolutionized water treatment by offering high efficiency in removing dissolved solids, pathogens, and organic pollutants (Shannon et al. 2008). Reverse osmosis, in particular, is widely used in desalination plants to convert seawater into potable water, playing a crucial role in addressing water scarcity in arid regions (Elimelech and Phillip 2011). Another significant advancement in water treatment is the development of advanced oxidation processes (AOPs), which involve the generation of highly reactive species like hydroxyl radicals to degrade organic contaminants, including pharmaceuticals, personal care products, and industrial chemicals, that are resistant to conventional treatment methods (Glaze et al. 1987). Common AOPs include ozone (O_3) treatment, hydrogen peroxide (H_2O_2) combined with ultraviolet (UV) light, and photocatalysis using titanium dioxide (TiO_2). These processes can effectively break down complex organic molecules into smaller, less harmful compounds, improving water quality and safety (Esplugas et al. 2007). Biological treatments, such as membrane bioreactors (MBRs), integrate biological degradation processes with membrane filtration to efficiently remove organic matter and nutrients from wastewater. MBRs combine the benefits of activated sludge treatment with the high retention capabilities of membranes, resulting in superior effluent quality and reduced footprint compared to conventional wastewater treatment plants (Judd 2008).

These systems are particularly advantageous in treating municipal and industrial wastewater, where space and stringent discharge standards are critical considerations. Additionally, adsorption technologies using advanced materials like activated carbon, biochar, and novel nanomaterials such as graphene oxide and carbon nanotubes have shown promise in removing a wide range emerging contaminants (Qu et al. 2013). These materials offer high surface area and tunable surface properties, enhancing their capacity to adsorb various contaminants effectively. Ion exchange resins and advanced coagulation-flocculation techniques also play vital roles in removing dissolved ions and particulates, further improving water quality (Wang et al. 2014). Electrochemical water treatment technologies, including electrocoagulation, electrooxidation, and capacitive deionization, leverage electric fields to remove contaminants through processes such as coagulation, oxidation, and ion removal. These methods are gaining attention for their efficiency, versatility, and potential for decentralized water treatment applications (Rajeshwar et al. 2008). Advanced water treatment technologies are critical in addressing the complex challenges of water pollution and scarcity. Adopting water-efficient technologies, such as drip irrigation, and implementing integrated water management strategies, is crucial in order to save water and maintain agricultural productivity in the midst of increased water limitation (Postel et al. 2001). Climate change further exacerbates the challenges faced by agriculture, as it influences temperature, precipitation patterns, and the frequency of extreme weather events resulting into disruption of crop yields, altered growing seasons, and more prevalence of pests and diseases (Lobell et al. 2008).

6.2 Source Control and Regulation

Source control and regulation are pivotal strategies in managing environmental pollution and safeguarding public health and ecosystems. These approaches focus on preventing pollutants from entering the environment at their origin rather than relying solely on remediation after contamination has occurred. Source control involves implementing best practices and technologies to minimize the release of pollutants from industrial, agricultural, and urban sources. For example, in industrial settings, this can include optimizing processes to reduce waste generation, employing cleaner production techniques, and using less harmful materials (Kibert 2016). Precision farming, integrated insect management, and the use of environmentally friendly chemicals such as pesticides in agriculture may all contribute to avoiding perilous chemical pollution into water bodies (Tilman et al. 2002). Regulatory measures include the establishment of standards and guidelines for air and water quality, emission limits for industries, and requirements for waste management and treatment practices. Agencies such as the U.S. Environmental Protection Agency (EPA) and the European Environment Agency (EEA) play crucial roles in developing and enforcing these regulations (EEA 2019). For instance, the Clean Air Act and the Clean Water Act in the United States set stringent standards for air and water pollutants, respectively, and require industries to obtain permits and regularly monitor their emissions and discharges (EPA 2020a, b). Economic instruments, such as taxes, subsidies, and tradable permits, are also effective regulatory tools that incentivize pollution reduction. Carbon pricing schemes, such as carbon taxation and caps and trade systems, are intended to restrict the emissions of greenhouse gases by fixing a cost to carbon emissions (Stavins 2003). Subsidies for renewable energy and energy efficiency improvements further promote the transition to sustainable practices (Jaffe et al. 2005). Participatory approach can lead to more effective implementation and compliance with environmental regulations (Dietz and Stern 2008). Additionally, transparency and access to environmental information empower citizens to hold polluters accountable and advocate for stronger environmental protection measures (Fung et al. 2007). Continuous monitoring and enforcement are critical to the success of source control and regulatory frameworks. Advanced monitoring technologies, such as remote sensing, real-time sensors, and data analytics, enhance the ability to track pollutant levels and identify non-compliance (Zhao et al. 2018). Regulatory agencies must have the authority and resources to conduct inspections, enforce penalties, and ensure that polluters comply with environmental laws and standards (Bennear and Olmstead 2008).

7 Conclusion

Emerging pollutants are rapidly being noticed in surface and ground water globally, raising worries about their presence in treated wastewater used for irrigation. Despite rising data, there is still a lack of understanding about how these toxins behave different ecosystems even in surface and ground water, particularly in terms of crop uptake, bioaccumulation, and entry into the food chain, which could pose health hazards to individuals and ecosystems. This book chapter highlights the urgent need for regulations and strategies to regulate emerging contamination levels in water, however still there is need for improvement and development of new technologies for removal of contaminates in surface and ground water. By addressing pollution at its source, establishing robust regulatory frameworks, leveraging economic incentives, engaging the public and ensuring rigorous monitoring and enforcement, these strategies help protect environmental and human health. Continued innovation and commitment to these approaches are necessary. To tackle the ever-changing ecological concerns of a swiftly changing world.

References

Aktar MW, Sengupta D, Chowdhury A (2009) Impact of pesticides use in agriculture: their benefits and hazards. Interdiscip Toxicol 2(1):1–12

Alexander M (1999) Biodegradation and bioremediation. Academic Press

Anastas PT, Eghbali N (2010) Green chemistry: principles and practice. Chem Soc Rev 39(1):301–312

Andrady AL (2011) Microplastics in the marine environment. Mar Pollut Bull 62(8):1596–1605

Anjum S, Hashmi MZ, Malik RN, Zakria M (2021) Review of endocrine-disrupting chemicals in the environment and their impact on human health. Environ Sci Pollut Res 28(4):3907–3923

Aris AZ, Shamsuddin AS, Praveena SM (2014) Occurrence of 17α-ethynylestradiol (EE2) in the environment and effect on exposed biota: an update review. Environ Int 69:104–119

Asner GP (2009) Tropical forest carbon assessment: integrating satellite and airborne mapping approaches. Environ Res Lett 4(3):034009

aus der Beek T, Weber FA, Bergmann A, Hickmann S, Ebert I, Hein A, Küster A (2016) Pharmaceuticals in the environment—global occurrences and perspectives. Environ Toxicol Chem 35(4):823–835

Barbosa MO, Moreira NF, Ribeiro AR, Pereira MF, Silva AM (2016) Occurrence and removal of organic micropollutants: an overview of the watch list of EU Decision 2015/495. Water Res 94:257–279

Barlow PM, Reichard EG (2010) Saltwater intrusion in coastal regions of North America. Hydrogeol J 18(1):247–260

Bennear LS, Olmstead SM (2008) The impacts of the "right to know": information disclosure and the regulation of drinking water quality. J Environ Econ Manag 56(2):117–130

Bernhardt ES, Palmer MA, Allan JD, Alexander G, Barnas K, Brooks S, Sudduth E (2005) Synthesizing US river restoration efforts. Science 308(5722):636–637

Blair BD, Crago JP, Hedman CJ, Klaper RD (2013) Pharmaceuticals and personal care products found in the Great Lakes above concentrations of environmental concern. Chemosphere 93(9):2116–2123

Bonmatin JM, Giorio C, Girolami VD, Goulson DP, Kreutzweiser C, Krupke M, Liess E, Long M, Marzaro EAD, Mitchell DA, Noome N, Simon D, Tapparo A (2015) Environmental fate and exposure; neonicotinoids and fipronil. Environ Sci Pollut Res 22:35–67. https://doi.org/10.1007/s11356-014-3332-7

Boreen AL, Arnold WA, McNeill K (2004) Photodegradation of pharmaceuticals in the aquatic environment: a review. Aquat Sci 66(3):320–341

Boxall ABA, Rudd MA, Brooks BW, Caldwell DJ, Choi K, Hickmann S, Ankley GT (2012) Pharmaceuticals and personal care products in the environment: what are the big questions? Environ Health Perspect 120(9):1221–1229

Breivik K, Sweetman A, Pacyna JM, Jones KC (2002) Towards a global historical emission inventory for selected PCB congeners—a mass balance approach: 1. Global production and consumption. Sci Total Environ 290(1–3):181–198

Castell N, Dauge FR, Schneider P, Vogt M, Lerner U, Fishbain B, Bartonova A (2015) Can commercial low-cost sensor platforms contribute to air quality monitoring and exposure estimates? Environ Int 85:162–170

CDC (2009) Fourth national report on human exposure to environmental chemicals. Centers for Disease Control and Prevention

Colborn TV, Saal FS, Soto AM (1993) Developmental effects of endocrine-disrupting chemicals in wildlife and humans. Environ Health Perspect 101(5):378–384

Cole M, Lindeque P, Halsband C, Galloway TS (2013) Microplastics as contaminants in the marine environment: a review. Mar Pollut Bull 62(12):2588–2597

Daughton CG, Ternes TA (1999) Pharmaceuticals and personal care products in the environment: agents of subtle change? Environ Health Perspect 107(Suppl 6):907–938

de Souza Machado AA, Lau CW, Till J, Kloas W, Lehmann A, Becker R, Rillig MC (2018) Impacts of Microplastics on the Soil Biophysical Environment. Environ Sci Technol 452(17):9656–9665. https://doi.org/10.1021/acs.est.8b02212

Diamanti-Kandarakis E, Bourguignon JP, Giudice LC, Hauser R, Prins GS, Soto AM, Gore AC (2009) Endocrine-disrupting chemicals: an Endocrine Society scientific statement. Endocr Rev 30(4):293–342

Dietz T, Stern PC (2008) Public participation in environmental assessment and decision making. National Academies Press

EEA (2019) European Environment Agency. https://www.eea.europa.eu/

Elimelech M, Phillip WA (2011) The future of seawater desalination: energy, technology, and the environment. Science 333(6043):712–717

Enyoh CE, Verla AW, Verla EN (2019) Uptake of microplastics by plant: a reason to worry or to be happy? World Sci News 131:256–267

EPA (2020a) Laws and Regulations. United States Environmental Protection Agency. https://www.epa.gov/laws-regulations

EPA (2020b) Summary of the Clean Water Act. United States Environmental Protection Agency. https://www.epa.gov/laws-regulations/summary-clean-water-act

Esplugas S, Bila DM, Krause LGT, Dezotti M (2007) Ozonation and advanced oxidation technologies to remove endocrine disruptors from wastewater. Environ Sci Pollut Res 14(1):31–35

European Chemicals Agency (2016) Understanding REACH. https://echa.europa.eu/regulations/reach/understanding-reach

FAO (2011) The state of the world's land and water resources for food and agriculture (SOLAW)—managing systems at risk. Food and Agriculture Organization of the United Nations, Rome and Earthscan, London

Fent K, Weston AA, Caminada D (2006) Ecotoxicology of human pharmaceuticals. Aquat Toxicol 76(2):122–159

Foster S, Chilton J (2003) Groundwater: the processes and global significance of aquifer degradation. Philos Trans R Soc B: Biol Sci 358(1440):1957–1972

Fung A, Graham M, Weil D (2007) Full disclosure: the perils and promise of transparency. Cambridge University Press

Galloway TS, Lewis CN (2016) Marine microplastics spell big problems for future generations. Proc Natl Acad Sci 113(9):2331–2333

Glaze WH, Kang JW, Chapin DH (1987) The chemistry of water treatment processes involving ozone, hydrogen peroxide and ultraviolet radiation. Ozone: Sci Eng 9(4):335–352

Gleeson T, Wada Y, Bierkens MF, van Beek LP (2012) Water balance of global aquifers revealed by groundwater footprint. Nature 488(7410):197–200

Gobas FAPC, Clark KE, Shiu WY, Mackay D (1993) Bioconcentration of polybrominated dibenzo-p-dioxins and dibenzofurans in guppies: validation of a predictive model. Environ Sci Technol 27(4):748–754

Gore AC, Chappell VA, Fenton SE, Flaws JA, Nadal A, Prins GS, Zoeller RT (2015a) Executive summary to EDC-2: the Endocrine Society's second scientific statement on endocrine-disrupting chemicals. Endocr Rev 36(6):593–602

Gore AC, Chappell VA, Fenton SE, Flaws JA, Nadal A, Prins GS, Toppari J, Zoeller RT (2015b) EDC-2: the Endocrine Society's second scientific statement on endocrine-disrupting chemicals. Endocr Rev 36(6):E1–E150

Hayes TB, Case P, Chui S, Chung D, Haeffele C, Haston K (2011) Pesticide mixtures, endocrine disruption, and amphibian declines: are we underestimating the impact? Environ Health Perspect 119(4):547–552
Heberer T (2002) Occurrence, fate, and removal of pharmaceutical residues in the aquatic environment: a review of recent research data. Toxicol Lett 131(1–2):5–17
Hidalgo-Ruz V, Gutow L, Thompson RC, Thiel M (2012) Microplastics in the marine environment: a review of the methods used for identification and quantification. Environ Sci Technol 46(6):3060–3075
Huang Y, Arain A, Miller M (2016) Use of portable XRF for environmental screening of heavy metal levels in urban soils of Los Angeles, California, USA. Environ Earth Sci 75(12):1–8
International Agency for Research on Cancer (IARC) (2015) IARC monographs volume 112: evaluation of five organophosphate insecticides and herbicides. IARC monographs on the evaluation of carcinogenic risks to humans, p 112
IPCC (2014) Climate change 2014: impacts, adaptation, and vulnerability. Contribution of working group II to the fifth assessment report of the intergovernmental panel on climate change. Cambridge University Press
Jaffe AB, Newell RG, Stavins RN (2005) A tale of two market failures: technology and environmental policy. Ecol Econ 54(2–3):164–174
Jones KC, de Voogt P (2019) Persistent organic pollutants (POPs): state of the science. Environ Pollut 100(1–3):209–221. https://doi.org/10.1016/S0269-7491(99)00098-6
Jordan MI, Mitchell TM (2015) Machine learning: trends, perspectives, and prospects. Science 349(6245):255–260
Judd S (2008) The status of membrane bioreactor technology. Trends Biotechnol 26(2):109–116
Kabir ER, Rahman MS, Rahman I (2020) A review on endocrine disruptors and their possible impacts on human health. Environ Toxicol Pharmacol 40(1):241–258
Kibert CJ (2016) Sustainable construction: green building design and delivery. Wiley
Kolpin DW, Furlong ET, Meyer MT, Thurman EM, Zaugg SD, Barber LB, Buxton HT (2002) Pharmaceuticals, hormones, and other organic wastewater contaminants in US streams, 1999–2000: a national reconnaissance. Environ Sci Technol 36(6):1202–1211
Konikow LF, Kendy E (2005) Groundwater depletion: a global problem. Hydrogeol J 13(1):317–320
Kookana RS, Baskaran S, Naidu R (1998) Pesticide fate and behavior in Australian soils in relation to contamination and management of soil and water: a review. Aust J Soil Res 36(5):715–764
Kortenkamp A, Martin O, Evans R, McKinlay R, Orton F, Rosivatz E (2009) State of the art assessment of endocrine disrupters. Final report. European Commission, DG Environment
Kumar V, Parihar RD, Sharma A, Bakshi P, Sidhu GPS, Bali AS, Thukral AK (2019) Global evaluation of heavy metal content in surface water bodies: a meta-analysis using heavy metal pollution indices and multivariate statistical analyses. Chemosphere 236:124364
Kümmerer K (2009) Antibiotics in the aquatic environment—a review—Part I. Chemosphere 75(4):417–434
Kwiatkowski CF, Andrews DQ, Birnbaum LS, Bruton TA, DeWitt JC, Knappe DRU, Sunderland EM (2020) Scientific basis for managing PFAS as a chemical class. Environ Sci Technol Lett 7(8):532–543
Lapworth DJ, Baran N, Stuart ME, Ward RS (2012) Emerging organic contaminants in groundwater: a review of sources, fate and occurrence. Environ Pollut 163:287–303
Li J, Liu HP, Chen J (2020) Microplastics in freshwater systems: a review on occurrence, environmental effects, and methods for microplastics detection. Water Res 137:362–374
Liu Y, Zhou C, Zhang Z, Zhang H, Zhang S, Zhang L (2021) Occurrence and risks of synthetic musks in the aquatic environment: a review. Sci Total Environ 787:147520
Lobell DB, Burke MB, Tebaldi C, Mastrandrea MD, Falcon WP, Naylor RL (2008) Prioritizing climate change adaptation needs for food security in 2030. Science 319(5863):607–610
Luo Y, Guo W, Ngo HH, Nghiem LD, Hai FI, Zhang J, Liang S, Wang XC (2014) A review on the occurrence of micropollutants in the aquatic environment and their fate and removal during wastewater treatment. Sci Total Environ 473–474:619–641

Mackay D, Shiu WY, Ma KC, Lee SC (2006) Handbook of physical-chemical properties and environmental fate for organic chemicals. CRC Press

Marshall BM, Levy SB (2011) Food animals and antimicrobials: impacts on human health. Clin Microbiol Rev 24(4):718–733

Martínez JL (2009) Environmental pollution by antibiotics and by antibiotic resistance determinants. Environ Pollut 157(11):2893–2902

Meybeck M (2003) Global analysis of river systems: from earth system controls to anthropocene syndromes. Philos Trans R Soc B: Biol Sci 358(1440):1935–1955

Mnif W, Hassine AI, Bouaziz A, Bartegi A, Thomas O, Roig B (2011) Effect of endocrine disruptor pesticides: a review. Int J Environ Res Public Health 8(6):2265–2303

Napper IE, Bakir A, Rowland SJ, Thompson RC (2015) Characterisation, quantity and sorptive properties of microplastics extracted from cosmetics. Mar Pollut Bull 99(1–2):178–185

Paerl HW, Huisman J (2008) Blooms like it hot. Science 320(5872):57–58

Pankow JF, Cherry JA (1996) Dense chlorinated solvents and other DNAPLs in groundwater: history, behavior, and remediation. Waterloo Press

Paul MJ, Meyer JL (2001) Streams in the urban landscape. Annu Rev Ecol Syst 32(1):333–365

Pimentel D, Harvey C, Resosudarmo P, Sinclair K, Kurz D, McNair M, Blair R (1995) Environmental and economic costs of soil erosion and conservation benefits. Science 267(5201):1117–1123

Postel SL, Polak P, Gonzales F, Keller J (2001) Drip irrigation for small farmers: a new initiative to alleviate hunger and poverty. Water Int 26(1):3–13

Prosser RS, Sibley PK (2015) Human health risk assessment of pharmaceuticals and personal care products in plant tissue due to biosolids and manure amendments, and wastewater irrigation. Environ Int 75:223–233. https://doi.org/10.1016/j.envint.2014.11.020

Qu X, Alvarez PJJ, Li Q (2013) Applications of nanotechnology in water and wastewater treatment. Water Res 47(12):3931–3946

Rainbow PS (2002) Trace metal concentrations in aquatic invertebrates: why and so what? Environ Pollut 120(3):497–507

Rajeshwar K, Ibanez JG, Swain GM (2008) Electrochemistry and the environment. J Appl Electrochem 28(9):955–969

Rehman K, Fatima F, Waheed I, Akash MSH (2021) Prevalence of exposure of heavy metals and their impact on health consequences. J Cell Biochem 119(1):157–184

Richardson SD, Ternes TA (2018) Water analysis: emerging contaminants and current issues. Anal Chem 90(1):398–428

Rillig MC (2012) Microplastic in terrestrial ecosystems and the soil? Environ Sci Technol 46(12):6453–6454. https://doi.org/1021/es302011r

Rubin BS (2011) Bisphenol A: an endocrine disruptor with widespread exposure and multiple effects. J Steroid Biochem Mol Biol 127(1–2):27–34

Sanchez W, Bender C, Porcher JM (2021) A review of the use of biomarkers in environmental monitoring: application to aquatic vertebrates. Ecotoxicology 20(1):1–17

Santos LHMLM, Araújo AN, Fachini A, Pena A, Delerue-Matos C, Montenegro MCBSM (2010) Ecotoxicological aspects related to the presence of pharmaceuticals in the aquatic environment. J Hazard Mater 175(1–3):45–95

Schwarzenbach RP, Gschwend PM, Imboden DM (2006a) Environmental organic chemistry. Wiley

Schwarzenbach RP, Escher BI, Fenner K, Hofstetter TB, von Johnson CA, Gunten U, Wehrli B (2006b) The challenge of micropollutants in aquatic systems. Science 313(5790):1072–1077

Shannon MA, Bohn PW, Elimelech M, Georgiadis JG, Marinas BJ, Mayes AM (2008) Science and technology for water purification in the coming decades. Nature 452(7185):301–310

Silva V, Mol HGJ, Zomer P, Tienstra M, Ritsema CJ, Geissen V (2019) Pesticide residues in European agricultural soils—a hidden reality unfolded. Sci Total Environ 653:1532–1545

Smedley PL, Kinniburgh DG (2002) A review of the source, behaviour and distribution of arsenic in natural waters. Appl Geochem 17(5):517–568

Smith M, Love DC, Rochman CM, Neff RA (2018) Microplastics in seafood and the implications for human health. Curr Environ Health Rep 5(3):375–386

Smith VH, Schindler DW (2009) Eutrophication science: where do we go from here? Trends Ecol Evol 24(4):201–207

Smith VH, Tilman GD, Nekola JC (1999) Eutrophication: impacts of excess nutrient inputs on freshwater, marine, and terrestrial ecosystems. Environ Pollut 100(1–3):179–196

Spalding RF, Exner ME (1993) Occurrence of nitrate in groundwater—a review. J Environ Qual 22(3):392–402

Squillace PJ, Scott JC, Moran MJ, Nolan BT, Kolpin DW (2002) VOCs, pesticides, nitrate, and their mixtures in groundwater used for drinking water in the United States. Environ Sci Technol 36(9):1923–1930

Stavins RN (2003) Experience with market-based environmental policy instruments. Handb Environ Econ 1:355–435

Stuart M, Lapworth D, Crane E, Hart A (2012) Review of risk from potential emerging contaminants in UK groundwater. Sci Total Environ 416:1–21

Sunderland EM, Hu XC, Dassuncao C, Tokranov AK, Wagner CC, Allen JG (2019) A review of the pathways of human exposure to poly- and perfluoroalkyl substances (PFASs) and present understanding of health effects. J Eposure Sci Environ Epidemiol 29(2):131–147

Tang FHM, Lenzen M, McBratney A, Maggi F (2021) Risk of pesticide pollution at the global scale. Nat Geosci 14(4):206–210

Taylor RG, Scanlon B, Döll P, van Rodell M, Beek R, Wada Y, Treidel H (2013) Ground water and climate change. Nat Clim Chang 3(4):322–329

Teuten EL, Saquing JM, Knappe DR, Barlaz MA, Jonsson S, Björn A, Thompson RC (2009) Transport and release of chemicals from plastics to the environment and to wildlife. Philos Trans R Soc B: Biol Sci 364(1526):2027–2045

Tilman D, Cassman KG, Matson PA, Naylor R, Polasky S (2002) Agricultural sustainability and intensive production practices. Nature 418(6898):671–677

UNEP (2019) Global chemicals outlook II: from legacies to innovative solutions. United Nations Environment Programme

Vandenberg LN, Colborn T, Hayes TB, Heindel JJ, Jacobs DR Jr, Lee DH, Myers JP (2012) Hormones and endocrine-disrupting chemicals: low-dose effects and nonmonotonic dose responses. Endocr Rev 33(3):378–455

Verlicchi PA, Aukidy M, Zambello E (2012) Occurrence of pharmaceutical compounds in urban wastewater: removal, mass load and environmental risk after a secondary treatment—a review. Sci Total Environ 429:123–155

Wang LK, Hung YT, Shammas NK (eds) (2014) Advanced physicochemical treatment processes. Springer

Ward MH, deKok TM, Levallois P, Brender J, Gulis G, Nolan BT, VanDerslice J (2005) Workgroup report: drinking-water nitrate and health—recent findings and research needs. Environ Health Perspect 113(11):1607–1614

Weschler CJ (2009) Changes in indoor pollutants since the 1950s. Atmos Environ 43(1):153–169

WHO, UNEP (2013) State of the science of endocrine disrupting chemicals 2012. World Health Organization, Geneva

Wilkinson JL, Boxall ABA, Kolpin DW, Leung KMY, Lai RWS, Wong D, Ankley GT (2022) Pharmaceutical pollution of the world's rivers. Proc Natl Acad Sci 119(8):e2113947119

Wright SL, Thompson RC, Galloway TS (2013) The physical impacts of microplastics on marine organisms: a review. Environ Pollut 178:483–492

Zepp RG, Cline DM (1977) Rates of direct photolysis in aquatic environment. Environ Sci Technol 11(4):359–366

Zhao J, Zhao T, Li C, Ni J (2018) Remote sensing technology for lake water quality monitoring: a review. J Environ Sci 64:416–431

Zoeller RT, Brown TR, Doan LL, Gore AC, Skakkebaek NE, Soto AM, Woodruff TJ, Von Saal FS (2012) Endocrine-disrupting chemicals and public health protection: a statement of principles from The Endocrine Society. Endocrinology 153(9):4097–4110. https://doi.org/10.1210/en.2012-1422

Zuloaga O, Navarro P, Bizkarguenaga E, Iparraguirre A, Vallejo A, Olivares M, Prieto A (2012) Overview of extraction, clean-up and detection techniques for the determination of organic pollutants in biota samples. Anal Methods 4(4):859–881

Advances in Surface Water and Groundwater Quality Monitoring and Analysis Techniques

Shreeya Baghel and M. P. Tripathi

Abstract Recent developments in technology for monitoring and analyzing surface water and groundwater quality have notably bolstered our capacity to safeguard and oversee water resources. Innovations like remote sensing and satellite imagery facilitate thorough monitoring of vast water expanses, pinpointing sources of contamination and variations in water quality over time. Furthermore, autonomous underwater vehicles (AUVs) and intelligent sensors provide intricate insights into groundwater conditions. In addition, the emergence of real-time sensor networks and Internet of Things (IoT) devices allows for continuous monitoring of water attributes such as pH, temperature, turbidity, and levels of pollutants. Advanced data analytics, incorporating machine learning algorithms, are utilized to forecast trends in water quality and pinpoint potential hazards. Additionally, portable and more precise analytical tools have enhanced on-site testing, delivering prompt and accurate results. The integration of Geographic Information Systems (GIS) with water quality data permits spatial analysis and more informed decision-making. These technological strides jointly contribute to more efficient management of water quality, ensuring the provision of safe drinking water and the preservation of healthier aquatic ecosystems.

Keywords Remote sensing · Machine learning · IoT · Geographic information systems · Sensors

S. Baghel (✉)
ABE Department, University of Florida, Gainesville, USA
e-mail: heyritu96@gmail.com; baghelshreeya1@gmail.com

M. P. Tripathi
SWCE, SVCAET and RS, IGKV, Raipur, Chhattisgarh, India
e-mail: mktripathi64@gmail.com

Dimple et al. (eds.), *Surface Water and Groundwater Quality for Sustainable Utilization*, Springer Hydrogeology, https://doi.org/10.1007/978-981-96-8189-1_15

1 Introduction

1.1 *Importance of Water Quality Monitoring*

Water is a vital resource that sustains life, ecosystems, and various human activities. Maintaining the quality of water is crucial for safeguarding public health, preserving the environment, and promoting economic stability. Over time, the monitoring of water quality has undergone substantial changes, propelled by technological advancements and the increasing recognition of water's importance in sustaining ecological equilibrium. Rivers, lakes, and reservoirs, as well as aquifers that store groundwater, are the main sources of freshwater for most of the world's population (Perveen 2023). Keeping a close eye on the quality of these water sources is vital to prevent pollution, guarantee safe drinking water, and safeguard aquatic ecosystems.

1.2 *Challenges in Surface Water and Groundwater Monitoring*

Despite its significance, monitoring water quality poses several challenges. Surface water bodies are constantly changing, with water quality varying depending on factors like the time of year, human actions, and natural occurrences like floods and droughts (Kumar et al. 2023). In contrast, groundwater is typically concealed from sight, posing challenges in evaluating and tracking its levels. Groundwater contamination can persist for extended periods, posing difficulties and increasing costs for remediation efforts.

Traditional methods of monitoring water quality involve collecting samples manually and sending them to a laboratory for analysis, which is a lengthy, expensive process and only provides a single point in time assessment of water quality. The ever-changing nature of water bodies requires ongoing monitoring to accurately assess variations in water quality. In recent years, technological advancements have tackled some of these challenges, resulting in more precise, effective, and all-encompassing monitoring methods (Adjovo et al. 2023).

1.3 *Overview of Traditional and Modern Techniques*

The traditional approach to evaluating water quality entails the collection of water samples from various locations, followed by comprehensive laboratory examinations. A multitude of indicators, such as pH levels, concentrations of dissolved oxygen, turbidity measurements, and the detection of pollutants including heavy metals or pathogens, are analyzed to ascertain the quality of water. Despite their efficacy, these methodologies are constrained by their reliance on manual sampling processes and the duration required to obtain analytical results (Kumar et al. 2022).

Conversely, modern methodologies leverage technological innovations to facilitate real-time, continuous assessment of water quality. The integration of remote sensing technologies, in-situ sensors, along with the application of artificial intelligence and machine learning, has revolutionized the monitoring and evaluation of water quality. These contemporary approaches offer numerous advantages, including enhanced accuracy in image capture, expedited response times, and the ability to monitor extensive areas or difficult-to-access environments.

2 Technological Advances in Surface Water Quality Monitoring

2.1 Remote Sensing Techniques

Remote sensing has emerged as a formidable technique for the assessment of surface water quality, offering extensive spatial coverage alongside the capacity to evaluate extensive and often inaccessible regions. Technological advancements in satellite systems, Unmanned Aerial Vehicles (UAVs), and sophisticated data processing methodologies have markedly improved the ability to analyze water quality parameters from remote locations (Razman et al. 2022).

2.2 Satellite-Based Monitoring

The advent of satellite remote sensing has revolutionized the monitoring of water quality by facilitating the acquisition of data across expansive geographical areas. Satellites outfitted with sensors such as the Moderate Resolution Imaging Spectroradiometer (MODIS), the Landsat series, and the Sentinel series have been extensively utilized to monitor a variety of water quality indicators, including chlorophyll concentration, turbidity, and surface temperature (Murray et al. 2022).

2.3 Chlorophyll Monitoring

Chlorophyll concentration serves as a critical metric for detecting algal blooms, which can pose significant ecological and public health risks. Satellites such as MODIS and Sentinel-2 are equipped with sensors capable of detecting specific wavelengths of light reflected from the water's surface, thereby enabling the estimation of chlorophyll concentrations. These estimations are vital for pinpointing regions susceptible to eutrophication and the proliferation of harmful algal blooms (Murray et al. 2022).

2.4 Turbidity and Suspended Sediment Concentration

Turbidity, which serves as an indicator of the optical clarity of water, is significantly affected by the presence of suspended particulate matter within the aquatic medium. Elevated turbidity levels can impede the penetration of light, thereby exerting adverse effects on aquatic ecosystems. Remote sensing satellites utilize reflectance data to derive estimations of turbidity and suspended sediment concentration. The Landsat satellite series, characterized by its high spatial resolution, has proven particularly effective for the surveillance of turbidity in fluvial systems, lacustrine environments, and coastal regions (Perveen 2023).

2.5 Temperature Monitoring

Surface water temperature constitutes a critical parameter that modulates numerous biological and chemical processes within aquatic ecosystems. Satellite-based sensors, exemplified by those deployed on the MODIS platform, possess the capability to accurately measure surface temperature with heightened temporal resolution, thereby furnishing essential data for the examination of thermal pollution and the ramifications of climate change on aquatic systems.

2.6 UAV and Drone-Based Monitoring

Unmanned Aerial Vehicles (UAVs), colloquially referred to as drones, have witnessed a surge in popularity for the assessment of water quality owing to their versatility, high-resolution imaging capabilities, and accessibility to remote or hazardous locations. Drones outfitted with multispectral or hyperspectral imaging systems are capable of capturing intricate images of water bodies, thereby facilitating the analysis of diverse water quality metrics (Sibanda et al. 2021).

2.7 Hyperspectral Imaging

Hyperspectral imaging systems mounted on drones are capable of capturing a comprehensive spectrum of wavelengths, thus enabling the identification of specific pollutants, such as petroleum spills or chemical contaminants, through the analysis of their distinctive spectral signatures (Manjakkal et al. 2021). This innovative technology is particularly advantageous for the monitoring of pollution incidents and for evaluating their spatial distribution.

2.8 *High-Resolution Mapping*

Unmanned aerial vehicles (UAVs) possess the capability to operate at reduced altitudes, thereby acquiring high-resolution imagery that facilitates the generation of intricate cartographies delineating water quality metrics, including turbidity, chlorophyll concentration, and surface temperature. Such cartographic representations can be employed to elucidate sources of pollution, monitor the dispersion of contaminants, and evaluate the efficacy of remediation initiatives.

2.9 *Real-Time Monitoring*

The capacity to process data in situ or relay it to terrestrial stations instantaneously allows for prompt intervention concerning water quality challenges. For instance, UAVs can be rapidly mobilized to evaluate the repercussions of a pollution incident, such as a chemical spill, thereby supplying essential information for emergency response personnel (Sibanda et al. 2021).

2.10 *In-Situ Sensors and IoT Integration*

In-situ sensors are sophisticated instruments that are directly immersed in aquatic environments to perpetually assess an array of water quality metrics. The amalgamation of these sensors with Internet of Things (IoT) technology has fundamentally transformed water quality surveillance by facilitating real-time data gathering, transmission, and analytical evaluation (Jan et al. 2021).

2.10.1 Deployment of In-Situ Sensors

In-situ sensors are capable of quantifying an extensive spectrum of water quality parameters, encompassing pH, dissolved oxygen, conductivity, temperature, and concentrations of specific chemicals or nutrients. These instruments are customarily positioned in strategically significant locations within a water body, such as at water intake points for potable supplies, adjacent to wastewater discharge facilities, or in regions susceptible to algal proliferation.

- Multiparameter Probes: Contemporary in-situ sensors frequently manifest as multiparameter probes that have the capacity to concurrently gauge multiple water quality indicators. This functionality permits a thorough evaluation of water quality at a singular site, thereby diminishing the necessity for an array of sampling apparatus.
- Long-Term Deployment: Innovations in sensor technology have significantly prolonged the operational lifespan of in-situ sensors, thereby enabling sustained deployment with minimal requirements for routine maintenance or recalibration. This characteristic is particularly vital for the observation of remote or challenging-to-access locales.

2.10.2 IoT Integration for Real-Time Monitoring

The amalgamation of in-situ sensors with IoT frameworks facilitates the real-time surveillance of water quality. Sensors convey data wirelessly to cloud-based repositories, from where it can be retrieved and scrutinized by water management professionals, researchers, and policy formulators (Glasgow et al. 2004).

- Real-Time Data Transmission: IoT-enabled sensors have the capability to relay data in real-time to centralized repositories or monitoring facilities. This feature empowers water quality managers to swiftly react to any identified irregularities, such as abrupt fluctuations in pH levels or the emergence of pollutants.
- Data Analytics and Predictive Modeling: The incessant flow of data accrued by in-situ sensors can be subjected to examination through advanced data analytical methodologies. Machine learning algorithms may be utilized to discern patterns within the data, forecast prospective trends in water quality, and identify preliminary indicators of pollution episodes.
- Scalability and Network Integration: IoT infrastructures can be expanded to encompass thousands of sensors throughout a given region, thereby ensuring comprehensive surveillance of aquatic systems. These networks can be assimilated with ancillary monitoring frameworks, such as meteorological stations or hydrological models, to deliver an integrative perspective on the dynamics of water quality.

2.11 Advanced Spectroscopic Techniques

Spectroscopy encompasses the examination of the interactions that occur between electromagnetic radiation and matter. In the realm of water quality assessment, spectroscopic methodologies are employed to scrutinize the chemical composition of water by evaluating the absorption, reflection, or emission of light across various wavelengths (Guo et al. 2020).

2.11.1 UV–Vis Spectroscopy

Ultraviolet–visible (UV–Vis) spectroscopy is a prevalent analytical technique in the field of water quality assessment, utilized for the detection and quantification of both organic and inorganic constituents within aqueous environments. This methodology is particularly adept at analyzing dissolved organic carbon (DOC), nitrates, and a range of environmental pollutants.

- Organic Pollutants: UV–Vis spectroscopy facilitates the identification of organic contaminants by quantifying the absorption of ultraviolet light. Substances such as phenols, pesticides, and hydrocarbons exhibit distinctive absorption maxima in the ultraviolet spectrum, thereby enabling their identification and quantification.
- Nutrient Monitoring: The concentration of nitrates present in water can be accurately ascertained through UV–Vis spectroscopy by measuring light absorption at predetermined wavelengths. Elevated levels of nitrates are frequently correlated with agricultural runoff and pose a risk of eutrophication, rendering this parameter critical for ongoing monitoring.

2.11.2 Fluorescence Spectroscopy

Fluorescence spectroscopy represents an additional robust instrument for the assessment of water quality. This technique entails the measurement of fluorescence emitted by substances present in water upon excitation by light. It is particularly advantageous for detecting trace concentrations of pollutants and evaluating the quality of dissolved organic matter.

- Dissolved Organic Matter (DOM): Fluorescence spectroscopy can be employed to investigate the composition of dissolved organic matter in aquatic systems. Various types of organic materials emit fluorescence at specific wavelengths, thereby providing insights into the sources and potential ramifications of DOM on water quality.
- Pathogen Detection: Certain pathogens and microbial metabolites can be identified through the application of fluorescence spectroscopy. For instance, *E. coli* and other bacterial species generate fluorescent compounds that can be quantified to evaluate levels of water contamination.

2.11.3 Raman Spectroscopy

Raman spectroscopy is a non-destructive technique that yields molecular signatures of substances by analyzing the scattering of light. Its application in water quality monitoring has gained traction for the detection and characterization of pollutants, particularly those which are challenging to quantify using conventional methodologies (Li et al. 2014).

- Chemical Pollutants: Raman spectroscopy has the capability to identify and quantify chemical contaminants such as heavy metals, pesticides, and hydrocarbons. The technique's heightened sensitivity to molecular vibrational modes enables the detection of even minimal concentrations of these substances.
- Water Source Fingerprinting: Raman spectroscopy can serve as a means to fingerprint various water sources through the analysis of the unique spectral attributes inherent to different aquatic bodies. This application is instrumental for tracing the origins of water in mixed-source systems or for evaluating the effects of industrial effluents.

2.12 Artificial Intelligence and Machine Learning Applications

The utilization of artificial intelligence (AI) and machine learning (ML) in the domain of water quality monitoring has facilitated novel methodologies for data analysis, predictive modeling, and strategic decision-making. These advanced technologies possess the capability to process extensive datasets, discern underlying patterns, and generate predictions that are often unattainable through conventional analytical approaches (Ahmed et al. 2019).

a. **Predictive Modeling**: Machine learning algorithms may be trained on historical datasets pertaining to water quality to construct predictive models that anticipate future water quality conditions. Such models are capable of incorporating a multitude of variables, including meteorological patterns, alterations in land utilization, and sources of pollution, thereby yielding precise forecasts.

   - **Algal Bloom Prediction**: The prediction of harmful algal blooms (HABs) constitutes a vital application of AI in the monitoring of water quality. Machine learning models can scrutinize historical datasets encompassing water temperature, nutrient concentrations, and chlorophyll levels to ascertain the probability of HAB occurrences, thus providing timely alerts to water management personnel.
   - **Pollution Source Identification**: AI algorithms can be employed to discern the origins of pollution within a water body by examining spatial and temporal trends in water quality data. This functionality proves particularly beneficial for the management of diffuse pollution emanating from agricultural runoff or urban stormwater.

b. **Automated Data Analysis**: AI-driven systems are capable of automating the analytical processes associated with water quality data, thereby diminishing the necessity for manual interpretation and reducing the likelihood of human error. These systems can assimilate data from diverse sources, including in-situ sensors, satellite imagery, and laboratory analyses, to render a holistic evaluation of water quality.

- **Anomaly Detection**: Machine learning algorithms possess the capacity to identify anomalies within water quality data, such as abrupt alterations in pH levels or the emergence of unanticipated contaminants. These anomalies can be flagged for subsequent investigation, enabling water managers to promptly address potential concerns.
- **Image Analysis**: AI can also be utilized for the analysis of satellite and drone imagery, thereby automating the detection of water quality challenges such as algal blooms, sediment plumes, or oil spills. Image recognition algorithms can categorize various features present in thc imagery, offering invaluable insights into the spatial distribution of water quality indicators.
- **Decision Support Systems**: The integration of AI and machine learning into decision support systems (DSS) can assist water managers in making evidence-based decisions regarding water quality management. These systems are equipped to analyze real-time data, predict forthcoming conditions, and propose actions based on the analytical outcomes.
- **Adaptive Management**: AI-powered DSS can facilitate adaptive management strategies by persistently analyzing water quality data and modifying management practices in response to evolving conditions. For instance, a DSS may advocate for adjustments in reservoir operations to avert downstream algal blooms predicated on real-time monitoring data and predictive modeling.
- **Risk Assessment**: AI-based risk assessment instruments are capable of evaluating the prospective effects of various pollution scenarios on water quality and public health. Such tools can aid policymakers in prioritizing interventions and optimizing the allocation of resources.

3 Technological Advances in Groundwater Quality Monitoring

3.1 Groundwater Monitoring Networks

The assessment of groundwater quality poses distinct challenges attributable to the subterranean characteristics of aquifers and the sluggish movement of groundwater. Groundwater monitoring networks, which comprise strategically positioned wells, are indispensable for evaluating the quality of groundwater across extensive regions and prolonged durations.

3.1.1 Design and Implementation of Monitoring Networks

The design of groundwater monitoring networks is intended to yield representative data regarding the quality of groundwater across various aquifers. This design process considers multiple factors, including the geological composition of the region, the spatial distribution of groundwater wells, and potential sources of contamination.

- **Well Placement**: The judicious placement of monitoring wells is paramount for obtaining precise data concerning groundwater quality. Typically, wells are situated in proximity to potential sources of contamination, such as landfills, industrial facilities, or agricultural zones, in addition to areas where groundwater is extracted for potable use or irrigation purposes.
- **Sampling Frequency**: The interval at which groundwater sampling occurs is contingent upon the specific objectives of the monitoring initiative. In regions with established contamination concerns, more frequent sampling may be necessary to monitor fluctuations in groundwater quality. Conversely, baseline monitoring in relatively unblemished areas may necessitate less frequent sampling intervals.
- **Data Integration**: Data collected from groundwater monitoring networks are frequently amalgamated with other data types, including land use, hydrological metrics, and climatic information, to furnish a holistic understanding of groundwater quality dynamics. Such integration is facilitated by Geographic Information Systems (GIS) and various data management methodologies (Machiwal et al. 2018).

3.1.2 Continuous Monitoring Technologies

Recent advancements in sensor technology have facilitated the continuous monitoring of groundwater quality, thereby providing real-time data on critical parameters. These technologies are particularly adept at identifying short-term fluctuations in groundwater quality that may elude conventional sampling techniques.

- **In-Situ Sensors**: In-situ sensors can be installed within groundwater wells to perpetually monitor parameters including pH, temperature, conductivity, and concentrations of specific contaminants. These sensors yield real-time data that can be employed to identify pollution incidents or to monitor the efficacy of remediation initiatives.
- **Telemetry Systems**: Telemetry systems facilitate the remote transmission of data from groundwater monitoring locations to centralized databases. This capability permits the ongoing monitoring of groundwater quality across expansive areas and diminishes the necessity for field visits to manually collect data.

3.2 Borehole Logging and Monitoring

Borehole logging represents a systematic methodology employed to acquire comprehensive data regarding the geological and hydrological characteristics of subsurface formations. This technique is extensively utilized in the evaluation of groundwater quality to investigate the spatial distribution of contaminants within aquifers and to pinpoint potential conduits for contamination dissemination.

3.2.1 Geophysical Logging

Geophysical logging encompasses the application of an array of instruments designed to assess the physical attributes of subsurface formations. This methodology yields critical insights into the architectural configuration of aquifers, the spatial distribution of pollutants, and the dynamics of groundwater movement.

- **Resistivity Logging**: Resistivity logging quantifies the electrical resistance exhibited by subsurface formations, which can facilitate the identification of contaminants such as saline water or hydrocarbons. This methodology is particularly advantageous for detecting the incursion of seawater into coastal aquifers or the proliferation of pollutants emanating from industrial facilities.
- **Gamma Ray Logging**: Gamma ray logging quantifies the inherent radioactivity present in subsurface formations, thereby providing insights into lithological characteristics and the distribution of specific minerals. This technique is employed to delineate layers of clay or other materials that may function as impediments to groundwater flow.

3.2.2 Chemical and Isotopic Logging

Chemical and isotopic logging entails the scrutiny of water samples procured from varying depths within a borehole. This technique furnishes intricate information regarding the chemical makeup of groundwater and the origins of contamination.

- **Isotope Analysis**: Isotopic analysis can be employed to trace the origins of groundwater pollution by examining the ratios of stable or radioactive isotopes present in the water. For instance, the detection of particular isotopes of nitrogen or oxygen can signify the provenance of nitrate contamination within groundwater.
- **Tracer Tests**: Tracer tests consist of the introduction of a non-toxic tracer substance into the groundwater and the subsequent monitoring of its trajectory over time. This methodology provides significant insights into the flow pathways and velocities of groundwater, which can be instrumental in evaluating the potential dissemination of contaminants (Lazarova et al. 2011).

3.3 Advanced Chemical Analysis Techniques

Innovations in chemical analysis methodologies have substantially enhanced the capacity to detect and quantify pollutants in groundwater systems. These methodologies provide superior sensitivity, precision, and the capability to analyze intricate mixtures of contaminants.

3.3.1 Mass Spectrometry

Mass spectrometry constitutes a robust analytical methodology employed to identify and quantify a diverse array of organic and inorganic substances present in groundwater. This technique is particularly advantageous for the detection of trace concentrations of contaminants that may elude other analytical approaches.

- **Gas Chromatography-Mass Spectrometry (GC–MS)**: GC–MS amalgamates the separation capabilities inherent in gas chromatography with the detection capabilities afforded by mass spectrometry. This methodology is extensively utilized for the analysis of volatile organic compounds (VOCs) in groundwater, including substances such as benzene, toluene, and xylene.
- **Liquid Chromatography-Mass Spectrometry (LC–MS)**: LC–MS is employed to analyze non-volatile and polar compounds within groundwater, including pesticides, pharmaceuticals, and endocrine-disrupting chemicals. The heightened sensitivity and selectivity associated with LC–MS render it an indispensable instrument for the monitoring of emerging contaminants.

3.3.2 Isotope Analysis

Isotope analysis offers critical insights regarding the origins and pathways of groundwater contamination. Recent advancements in isotope ratio mass spectrometry (IRMS) have enhanced the accuracy and precision of isotope measurements, facilitating more comprehensive investigations of groundwater systems.

- **Stable Isotope Analysis**: Stable isotopes, such as oxygen-18 and deuterium, are frequently utilized to investigate the provenance and movement patterns of groundwater. Through the analysis of the isotopic composition of groundwater, researchers can delineate recharge sources, identify mixing events between disparate water bodies, and evaluate the impacts of anthropogenic activities on groundwater integrity.
- **Radioisotope Analysis**: Radioisotopes, including tritium and carbon-14, are employed to date groundwater and examine the kinetics of groundwater movement. This data is vital for comprehending the long-term dynamics of contaminants within aquifers and for devising effective remediation strategies.

3.4 Integration of Geophysical Methods

Geophysical methodologies, such as electrical resistivity tomography (ERT) and ground-penetrating radar (GPR), have emerged as pivotal instruments in the monitoring of groundwater quality. These methodologies afford non-invasive techniques for imaging subsurface formations and detecting contaminant presence.

3.4.1 Electrical Resistivity Tomography (ERT)

ERT represents a geophysical technique that quantifies the electrical resistivity of subsurface materials. It is utilized to generate detailed visualizations of the subsurface, which can elucidate the presence of contaminants, the structural characteristics of aquifers, and the dynamics of groundwater movement (Mehmood et al. 2020).

- **Contaminant Plumes**: ERT can be employed to delineate the distribution of contaminant plumes in groundwater by identifying zones of low resistivity, which are frequently indicative of the presence of dissolved contaminants. This intelligence is essential for evaluating the extent of contamination and strategizing remediation initiatives.
- **Aquifer Characterization**: ERT furnishes comprehensive data regarding the structural attributes of aquifers, including the localization of impermeable strata, faults, and fractures. This information is crucial for elucidating groundwater flow pathways and the potential for contaminant migration.

3.4.2 Ground-Penetrating Radar (GPR)

Ground-Penetrating Radar (GPR) constitutes a non-invasive methodology that employs high-frequency electromagnetic waves to visualize subsurface formations. This technique is extensively utilized in the monitoring of groundwater quality, facilitating the detection of soil moisture variations, the identification of subterranean objects, and the mapping of contaminant distribution (Hanoon et al. 2021).

- **Soil Moisture Monitoring**: GPR possesses the capability to identify fluctuations in soil moisture content, thereby yielding insights into groundwater movement and the infiltration dynamics of contaminants. This methodology is particularly advantageous for assessing the repercussions of surface spills on groundwater integrity.
- **Contaminant Detection**: The application of GPR extends to the identification of specific contaminants within the subsurface, accomplished through the recognition of alterations in the dielectric properties of the soil matrix. A salient example includes the detection of hydrocarbons, which characteristically diminish the dielectric constant of soil.

4 Data Management and Analysis

4.1 *Big Data and Cloud Computing in Water Quality Analysis*

The emergence of big data and cloud computing technologies has fundamentally transformed the domain of water quality monitoring by facilitating the storage, processing, and analytical examination of extensive datasets derived from a multitude of sources. These innovations have enhanced the capacity to discern trends, recognize anomalies, and formulate evidence-based strategies for the governance of water quality.

4.1.1 Big Data Analytics

Big data analytics denotes the application of sophisticated computational methodologies to scrutinize extensive and intricate datasets. Within the realm of water quality monitoring, big data analytics can be employed to process information from in-situ sensors, satellite imagery, laboratory assessments, and various other data sources.

- **Data Integration**: Big data platforms are adept at amalgamating data from diverse origins, thereby enabling a holistic evaluation of water quality. For instance, data acquired from in-situ sensors may be synthesized with satellite imagery and meteorological information to evaluate the repercussions of climatic events on water quality.
- **Pattern Recognition**: The deployment of machine learning algorithms on big data sets can facilitate the identification of patterns within water quality data, encompassing seasonal fluctuations, the ramifications of land use transformations, or the impacts of industrial operations. Such patterns may yield critical insights into the determinants affecting water quality (Haggerty et al. 2023).
- **Predictive Analytics**: The application of advanced big data analytics can be employed to develop predictive models that forecast imminent water quality parameters by analyzing historical data and current trends. Such models are instrumental in equipping water management experts to identify prospective challenges and execute proactive measures to preserve water quality.

4.1.2 Cloud Computing

A scalable and adaptable platform for processing and storing massive amounts of data on water quality is offered by cloud computing. Organizations can obtain sophisticated computational tools and lower the cost and complexity of data management by utilizing cloud-based services (Jha et al. 2020).

Data Storage and Access: Water quality data can be securely and scalable stored using cloud computing platforms. Because data is accessible from anywhere, academics and water managers may work together and exchange information instantly.

Computational Power: Large datasets can be processed and analyzed with the help of the robust computational resources provided by cloud computing systems. Predictive models, for instance, can be created and implemented using cloud-based machine learning services without the requirement for specific hardware.

- **Data Sharing and Collaboration**: The creation of integrated water quality management plans is made possible by cloud computing, which makes data exchange and collaboration between many companies easier. For instance, combining data from several monitoring networks can offer a thorough picture of the water quality in an area.

5 Regulatory Requirements for Monitoring

Legislative frameworks that necessitate the implementation of sophisticated monitoring methodologies can catalyze their integration by instituting a statutory obligation for the ongoing or instantaneous assessment of water quality.

Continuous Monitoring Mandates: Regulatory bodies may impose requirements on water utilities or industrial entities to establish continuous monitoring systems for designated contaminants, particularly in regions characterized by elevated pollution vulnerabilities. For instance, regulations may stipulate the deployment of in-situ sensors for the perpetual monitoring of nitrate concentrations in agricultural locales susceptible to runoff.

Real-Time Reporting Requirements: Legislative mandates may also necessitate the immediate reporting of water quality metrics to both regulatory authorities and the general populace. This stipulation promotes the utilization of Internet of Things (IoT) enabled sensors alongside cloud-based data platforms capable of real-time data transmission, thereby augmenting transparency and accountability.

5.1 Standards and Protocols for Data Management

The formulation of standards and protocols pertaining to data management is imperative for guaranteeing the integrity, comparability, and interoperability of water quality data amassed through advanced monitoring methodologies.

Data Quality Standards: Regulatory entities can institute data quality standards that delineate the accuracy, precision, and reliability criteria requisite for water quality data. These standards ascertain that data acquired through advanced monitoring methodologies are fit for their intended purposes and can be utilized in decision-making frameworks.

Interoperability Protocols: Interoperability protocols facilitate the collaboration of diverse monitoring systems and data platforms, thereby promoting data sharing and integration. Regulatory agencies may advocate for the adoption of these protocols by embedding them within regulatory frameworks or by providing technical advisement (Bailey et al. 2021).

5.2 Capacity Building and Training

Capacity building and training initiatives are crucial for ensuring that professionals in water quality possess the requisite skills and knowledge to proficiently utilize advanced monitoring techniques.

Training Programs: Governments and academic institutions can devise training curricula aimed at water quality professionals, encompassing subjects such as the employment of in-situ sensors, remote sensing methodologies, data analysis, and the implementation of artificial intelligence and machine learning in water quality surveillance (Ferrero et al. 2019).

Workshops and Seminars: Workshops and seminars can present venues for water quality professionals to acquire insights regarding the latest innovations in monitoring technologies and to exchange best practices. Such events may be organized by governmental bodies, industry associations, or scholarly institutions.

Certification Programs: Certification initiatives can ascertain that water quality professionals fulfill specific competency standards in the application of advanced monitoring techniques. Certification may be conferred by professional organizations, regulatory authorities, or academic institutions.

Successful Implementation of Advanced Monitoring Techniques

Case Study: The Utilization of Satellite Remote Sensing in the Mekong River Basin

The Mekong River Basin, encompassing multiple Southeast Asian nations, encounters considerable challenges regarding water quality attributed to rapid industrial growth, agricultural proliferation, and climate change. The application of satellite remote sensing has been effectively executed within the region to oversee water quality, particularly in extensive and isolated areas where conventional monitoring methods prove to be arduous.

Chlorophyll Monitoring for Algal Blooms: Satellite remote sensing technology has been employed to assess chlorophyll levels in the Mekong River and its tributaries, thereby providing timely alerts concerning algal blooms. These blooms are frequently associated with nutrient pollution originating from agricultural practices and can exert detrimental effects on fisheries and potable water resources.

Sediment Transport and Erosion: Data derived from remote sensing have been utilized to evaluate sediment transport and erosion within the Mekong River, which represent critical concerns stemming from deforestation and alterations in land use. By observing sediment plumes, regulatory authorities can pinpoint regions susceptible to erosion and enact strategies to mitigate sediment influx.

Cross-Border Collaboration: The application of satellite remote sensing in the Mekong River Basin has fostered collaborative efforts across national borders among the constituent countries. The collective access to remote sensing data has enhanced the coordination of initiatives aimed at managing water quality and fortified regional partnerships.

Case Study: Real-Time Groundwater Monitoring in California

California has encountered acute groundwater depletion and quality challenges due to extended periods of drought and intensive agricultural practices. In response to these adversities, the state has instituted real-time groundwater monitoring systems to effectively manage its groundwater resources.

Deployment of In-Situ Sensors: In-situ sensors have been strategically installed in groundwater wells throughout California to facilitate continuous monitoring of water levels, salinity, and nitrate concentrations. These sensors yield real-time data that is **relayed to a** centralized database, thereby enabling water resource managers to monitor groundwater conditions and address emerging issues with alacrity.

Integration with Water Management Plans: The real-time monitoring data have been assimilated into California's Sustainable Groundwater Management Plans (SGMPs), aimed at ensuring the long-term viability of groundwater resources. The acquired data inform critical decisions regarding groundwater extraction, recharge initiatives, and measures for pollution mitigation.

Public Access to Data: California has rendered groundwater monitoring data accessible to the public via online platforms, thereby enhancing transparency and empowering stakeholders to engage in groundwater management. Agriculturalists, environmental organizations, and the general populace can access real-time data and partake in discussions pertaining to water management policies.

6 Future Directions and Emerging Trends: Innovative Technologies in Water Quality Surveillance

As the complexities surrounding water quality challenges progress, the technologies employed for the monitoring and administration of water resources similarly adapt. A multitude of emerging technologies presents significant potential for the enhancement of water quality surveillance in forthcoming years.

6.1 Next-Generation Sensors

The advancement of next-generation sensors is anticipated to improve the precision, sensitivity, and scope of water quality surveillance. These sensors will possess the capability to identify a more extensive range of contaminants, including novel pollutants such as microplastics, pharmaceuticals, and personal care products.

Nanotechnology-Based Sensors: The application of nanotechnology is being pursued to engineer sensors that exhibit superior sensitivity and specificity. Such sensors are capable of identifying trace levels of contaminants at the molecular scale, thereby facilitating the monitoring of pollutants that conventional methodologies struggle to detect.

Biosensors: Biosensors, which leverage biological molecules for the detection of specific substances, are being innovated for the real-time surveillance of pathogens, toxins, and various contaminants in water. These sensors possess the potential to yield rapid and precise results, rendering them particularly suitable for deployment in remote or resource-constrained environments (Bailey et al. 2021).

6.2 Remote Sensing Advancements

Progressions in remote sensing technologies, encompassing the employment of hyperspectral imaging and LiDAR, are projected to enhance the capability to monitor water quality from extra-terrestrial and aerial platforms.

Hyperspectral Imaging: Hyperspectral imaging encompasses the acquisition of a broad spectrum of wavelengths, thereby yielding comprehensive information regarding the composition of aquatic systems. This technology can be utilized to assess a range of water quality indicators, including the identification of specific pollutants, algal blooms, and alterations in water coloration resulting from sedimentation or organic matter. LiDAR (Light Detection and Ranging): The LiDAR technology, which employs laser pulses for distance measurement, is being investigated for its potential to delineate water quality indicators such as turbidity, sedimentation, and underwater flora. LiDAR can furnish high-resolution, three-dimensional visualizations of aquatic bodies, establishing it as a pivotal instrument for monitoring transformations within aquatic ecosystems (Adjovu et al. 2023).

6.3 Artificial Intelligence and Machine Learning

Artificial Intelligence (AI) and machine learning are foreseen to assume an increasingly pivotal role in water quality surveillance, especially in the domains of data analysis, predictive modeling, and decision support.

Advanced Predictive Models: Machine learning algorithms have the capacity to formulate sophisticated predictive models that anticipate variations in water quality based on an extensive array of variables, encompassing climatic conditions, land use alterations, and anthropogenic activities. These models can assist water managers in preemptively identifying and addressing potential challenges before they escalate into critical situations.

Automated Data Analysis: AI-enhanced systems can facilitate the automation of the analysis of extensive and intricate water quality datasets, thereby diminishing the necessity for manual interpretation and augmenting the velocity and precision of decision-making. These systems are capable of recognizing trends, identifying anomalies, and generating actionable insights in real-time (Hanoon et al. 2021).

Adaptive Management Systems: AI-informed decision support frameworks can empower the adaptive management of water resources by persistently analyzing monitoring data and modifying management strategies in response to evolving conditions. For instance, an AI system may propose adjustments to water releases from a reservoir to avert downstream pollution predicated on real-time monitoring data.

7 Sustainability and Cost

Effectiveness of Advanced Techniques: As the technologies employed for water quality monitoring evolve, there is an increasing focus on ensuring their sustainability and economic viability. Achieving a balance between technological advancement and affordability, as well as accessibility, is paramount for facilitating widespread implementation, especially in contexts characterized by limited resources.

7.1 Cost-Reduction Strategies

A variety of strategies can be implemented to mitigate the expenses associated with sophisticated water quality monitoring methodologies, thereby rendering them more accessible to a broader demographic of users.

Open-Source Software and Platforms: The implementation of open-source software and platforms has the potential to substantially lower the costs associated with data analysis and management. These resources are frequently available at no cost and can be tailored to address specific requirements, rendering them particularly suitable for small-scale monitoring initiatives or nations in the developing world.

Modular and Scalable Systems: The conception of modular and scalable monitoring frameworks permits users to initiate their efforts with a fundamental configuration and progressively enhance their capabilities as resources permit. For instance, a water utility may commence operations by deploying a limited number of in-situ sensors, subsequently augmenting their quantity or upgrading to more sophisticated equipment as financial resources allow.

Public–Private Partnerships: The establishment of public–private partnerships (PPPs) can facilitate cost reduction through the collaborative sharing of financial responsibilities associated with the development and deployment of advanced monitoring technologies. Such partnerships also serve to provide access to the expertise and resources of the private sector, thereby improving the efficiency and efficacy of monitoring initiatives.

7.2 Sustainability Considerations

Guaranteeing the sustainability of water quality monitoring initiatives necessitates a comprehensive evaluation of the long-term viability of the employed technologies and methodologies, alongside their environmental, social, and economic ramifications.

Energy-Efficient Technologies: The innovation of energy-efficient monitoring technologies, such as low-power sensors and solar-powered data transmission systems, can mitigate the environmental footprint of monitoring endeavors and ensure their sustainability in remote or off-grid settings.

Durability and Maintenance: Monitoring apparatus should be engineered for durability and simplicity of maintenance, especially in adverse environments. Minimizing the necessity for frequent repairs or replacements can significantly diminish the long-term financial implications of monitoring and reduce interruptions in data collection.

Community Involvement and Ownership: The active involvement of local communities in the governance and operation of monitoring systems can bolster their sustainability by cultivating a sense of ownership and accountability. When communities are invested in the success of monitoring initiatives, they are more inclined to endorse and sustain the requisite infrastructure.

7.3 Cost–Benefit Analysis

The execution of cost–benefit analyses is instrumental in enabling decision-makers to evaluate the economic feasibility of investing in sophisticated monitoring technologies. Such analyses take into account both the direct expenditures associated with the implementation and operation of monitoring systems, as well as the prospective advantages, which may include enhanced water quality, diminished health risks, and improved ecosystem services (Ongley 2000).

Long-Term Savings: Despite the potential for advanced monitoring technologies to incur elevated initial costs, they possess the capacity to generate considerable long-term savings through the prevention of expensive water quality issues, the minimization of remediation efforts, and the avoidance of regulatory fines or penalties. Economic Valuation of Ecosystem Services: The integration of the economic valuation of ecosystem services into cost–benefit analyses can yield a more holistic evaluation of the advantages associated with water quality monitoring. For instance, safeguarding the quality of a river that underpins fisheries, tourism, and agriculture can confer substantial economic value, thereby substantiating the investment in monitoring technologies.

8 Climate Change Impact on Water Quality Monitoring

The phenomenon of climate change presents considerable obstacles to water quality monitoring, as it has the potential to modify the physical, chemical, and biological properties of aquatic environments. Comprehending and mitigating these impacts is imperative for ensuring the sustained efficacy of monitoring initiatives.

8.1 Changes in Water Temperature and Chemistry

Climate change can induce variations in water temperature, which subsequently influence the solubility of gases, the kinetics of chemical reactions, and the spatial distribution of aquatic species.

Increased Temperature and Algal Blooms: Elevated water temperatures can facilitate the proliferation of harmful algal blooms (HABs), which generate toxins that present significant threats to both human health and aquatic ecosystems. Consequently, monitoring systems must be reconfigured to identify and anticipate the emergence of HABs within the context of a changing climate (Tung et al. 2012).

Altered Dissolved Oxygen Levels: Variations in water temperature and stratification can impact dissolved oxygen concentrations, resulting in hypoxic conditions that can adversely affect fish and other aquatic organisms. Continuous monitoring of dissolved oxygen levels is crucial for the timely detection and management of these alterations.

8.2 Impact on Hydrological Cycles and Water Availability

Climate change possesses the capacity to modify precipitation patterns, thereby influencing river flow dynamics, groundwater recharge processes, and the overall availability of water resources.

Increased Frequency of Extreme Events: The heightened occurrence of extreme meteorological phenomena, including floods and droughts, may result in abrupt alterations in water quality, characterized by elevated sediment loads, nutrient runoff, and contamination stemming from sewage system overflows. It is imperative that monitoring systems are meticulously designed to capture these swift changes and furnish timely data essential for emergency response initiatives.

Groundwater Quality and Recharge: Variations in precipitation and evaporation rates can significantly impact groundwater recharge, culminating in fluctuations in groundwater levels and quality. Consequently, monitoring networks must be reconfigured to accommodate these alterations and to guarantee the sustainable stewardship of groundwater resources.

Adaptation and Resilience Strategies: Developing adaptation and resilience strategies is paramount for ensuring the continuous efficacy of water quality monitoring systems amidst the challenges posed by climate change.

Flexible and Adaptive Monitoring Networks: Monitoring networks ought to be constructed with flexibility and adaptability in mind, thereby facilitating the reconfiguration of monitoring sites, the incorporation of novel sensors, and the modulation of monitoring frequency in response to evolving conditions.

Climate-Resilient Infrastructure: The infrastructure designated for monitoring purposes should be engineered to endure the ramifications of climate change, encompassing increased flooding, storm surges, and extreme temperature fluctuations. This may necessitate the utilization of more durable materials, elevated sensor platforms, or auxiliary power supplies.

Integration with Climate Models: The amalgamation of water quality monitoring data with climate models can yield critical insights into the prospective impacts of climate change on water quality. Such information can be instrumental in formulating proactive management strategies aimed at mitigating the risks associated with climate change.

The future trajectory of water quality monitoring is imbued with significant potential, as emerging technologies and innovative methodologies are poised to confront the escalating challenges inherent in water management. Nevertheless, the actualization of this potential will necessitate sustained investment in research and development, the adoption of conducive policies and regulations, and the active engagement of all stakeholders, encompassing governments, industry representatives, academia, and the general public. As we progress, it is crucial to emphasize the sustainability, accessibility, and efficacy of water quality monitoring endeavors. By prioritizing these aspects, we can ensure that our water resources remain secure, pristine, and plentiful for generations to come.

References

Adjovu GE, Stephen H, James D, Ahmad S (2023) Overview of the application of remote sensing in effective monitoring of water quality parameters. Remote Sens 15(7):1938

Ahmed AN, Othman FB, Afan HA, Ibrahim RK, Fai CM, Hossain MS, Elshafie A (2019) Machine learning methods for better water quality prediction. J Hydrol 578:124084

Bailey ES, Beetsch N, Wait DA, Oza HH, Ronnie N, Sobsey MD (2021) Methods, protocols, guidance and standards for performance evaluation for point-of-use water treatment technologies: history, current status, future needs and directions. Water 13(8):1094

Ferrero G, Setty K, Rickert B, George S, Rinehold A, DeFrance J, Bartram J (2019) Capacity building and training approaches for water safety plans: a comprehensive literature review. Int J Hyg Environ Health 222(4):615–627

Glasgow HB, Burkholder JM, Reed RE, Lewitus AJ, Kleinman JE (2004) Real-time remote monitoring of water quality: a review of current applications, and advancements in sensor, telemetry, and computing technologies. J Exp Mar Biol Ecol 300(1–2):409–448

Guo Y, Liu C, Ye R, Duan Q (2020) Advances on water quality detection by UV-Vis spectroscopy. Appl Sci 10(19):6874

Haggerty R, Sun J, Yu H, Li Y (2023) Application of machine learning in groundwater quality modelling—a comprehensive review. Water Res 233:119745

Hanoon MS, Ahmed AN, Fai CM, Birima AH, Razzaq A, Sherif M, El-Shafie A (2021) Application of artificial intelligence models for modeling water quality in groundwater: comprehensive review, evaluation and future trends. Water Air Soil Pollut 232:1–41

Jan F, Min-Allah N, Düştegör D (2021) Iot based smart water quality monitoring: recent techniques, trends and challenges for domestic applications. Water 13(13):1729

Jha BK, Sivasankari GG, Venugopal KR (2020) Cloud-based smart water quality monitoring system using IoT sensors and machine learning. Int J Adv Trends Comput Sci Eng 9(3)

Kumar A, Palmate SS, Shukla R (2022) Water quality modelling, monitoring, and mitigation. Appl Sci 12(22):11403

Kumar L, Kumari R, Kumar A, Tunio IA, Sassanelli C (2023) Water quality assessment and monitoring in Pakistan: a comprehensive review. Sustainability 15(7):6246

Lazarova V, Emsellem Y, Paille J, Glucina K, Gislette P (2011) Water quality management of aquifer recharge using advanced tools. Water Sci Technol 64(5):1161–1168

Li Z, Deen MJ, Kumar S, Selvaganapathy PR (2014) Raman spectroscopy for in-line water quality monitoring—instrumentation and potential. Sensors 14(9):17275–17303

Machiwal D, Cloutier V, Güler C, Kazakis N (2018) A review of GIS-integrated statistical techniques for groundwater quality evaluation and protection. Environ Earth Sci 77(19):681

Manjakkal L, Mitra S, Petillot YR, Shutler J, Scott EM, Willander M, Dahiya R (2021) Connected sensors, innovative sensor deployment, and intelligent data analysis for online water quality monitoring. IEEE Internet Things J 8(18):13805–13824

Mehmood Q, Mahmood W, Awais M, Rashid H, Rizwan M, Anjum L, Hamid S (2020) Optimizing groundwater quality exploration for irrigation water wells using geophysical technique in semi-arid irrigated area of Pakistan. Groundw Sustain Dev 11:100397

Murray C, Larson A, Goodwill J, Wang Y, Cardace D, Akanda AS (2022) Water quality observations from space: a review of critical issues and challenges. Environments 9(10):125

Ongley ED (2000) Water quality management: design, financing and sustainability considerations-II. In: Invited presentation at the World Bank's water week conference: towards a strategy for managing water quality management, pp 1–16

Perveen S (2023) Drinking water quality monitoring, assessment and management in Pakistan: a review. Heliyon 9(3)

Razman NA, Ismail WZW, Kamil NAIM, Zainurin SN, Ismail I, Jamaludin J, Balakrishnan SR (2022) A review on water quality monitoring methods based on electronics and optical sensing. J Adv Res Appl Sci Eng Technol 26(2):1–7

Sibanda M, Mutanga O, Chimonyo VG, Clulow AD, Shoko C, Mazvimavi D, Mabhaudhi T (2021) Application of drone technologies in surface water resources monitoring and assessment: a systematic review of progress, challenges, and opportunities in the global south. Drones 5(3):84

Tung CP, Ting-Chuan L, Wei-Ting L, Yun-Ju C (2012) Climate change impact assessment for sustainable water quality management. TAO: Terr Atmos Ocean Sci 23(5):565

Climate Change's Impacts on Surface Water and Groundwater Quality

Gauri U. Bhagole, S. Pooja Sakthi Rama, Barnali Saha, Pavan Panzade, Athira Saji, and Pallavi Rathod

Abstract The changing climate impacts the hydrologic cycle (evaporation, groundwater recharge, temperature, runoff, and rainfall) with higher air and water temperatures. Although the impacts of climatic variability on water availability are well understood, rising air temperatures have an impact on water quality by raising water temperature, lowering ice cover, stabilising vertical stratification, and decreasing mixing in deep lakes. These changes can alter water discharge patterns, leading to higher sediment loads and increased nutrient runoff from soil erosion. Additionally, higher temperatures can decrease oxygen content and change chemical reactions, deteriorating water quality and impacting freshwater ecosystems. Climate change-induced hydrological changes have an indirect impact on the physiochemical and biological quality of water. Climate change affects the behaviour of toxic pollutants, potentially leading to more frequent episodes of droughts and floods. Poor water quality yields significant economic, social, and environmental risks, particularly in developing countries like India. India's economy, which depends heavily on agriculture, is under pressure from rising energy, freshwater, and food prices as well as population growth. To protect water quality, strategic planning must address climate change impacts and vulnerabilities. Adaptation measures include managing the level of water in wetland regions, rivers, and lakes, as well as creating plant corridors and buffer zones. Finding the best ways to give high-quality water to a future world concerned about water scarcity and recognising the elements affecting water quality due to climate change will surely be difficult.

Keywords Hydrologic cycle · Climate change · Water quality

G. U. Bhagole (✉) · S. Pooja Sakthi Rama · A. Saji
Division of Agricultural Engineering, ICAR-IARI, New Delhi, India
e-mail: bhagolegauri@gmail.com

B. Saha
ICAR-Central Institute of Agricultural Engineering, Bhopal, India

P. Panzade
Water Technology Centre, ICAR-IARI, New Delhi, India

P. Rathod
Division of Environmental Science, ICAR-IARI, New Delhi, India

Dimple et al. (eds.), *Surface Water and Groundwater Quality for Sustainable Utilization*, Springer Hydrogeology, https://doi.org/10.1007/978-981-96-8189-1_16

1 Introduction

The relationship of groundwater and surface water is an important aspect of the hydrological cycle. Hydraulic linkages and water transfer between surface water and subsurface aquifers offer a range of ecosystem services that benefit both human and environmental health (Griebler and Avramov 2015; Goldscheider 2019; Griebler et al. 2014). Surface water is defined as freshwater found in streams, lakes, and wetlands. It is conspicuous and easily accessible. Surface water can seep down and become groundwater. Groundwater is a valuable resource for humans as well as ecosystems, which accounts for more than one-third of global usage of water (Gorelick and Zheng 2015). Groundwater is crucial for maintaining the global environment in mid-latitude, arid, as well as semiarid regions where supply of surface water through rivers and reservoirs is insufficient (Gorelick and Zheng 2015), as well as meeting societal demands such as water for drinking and agricultural output (Gleeson et al. 2020).

The hydrological cycle, often known as the water cycle, is an important part of the earth's system that supports people, animals, and plants (Gupta 2010). Earth's climate and water systems are intricately linked. Water both influences and is affected by the climate. Every climatic change affects the water system, and vice versa. Climate change affects the natural hydrologic process by boosting the temperature of the air and water (Whitehead et al. 2009). Rising air temperatures have an influence on boosting water temperature, reducing ice cover, stabilising vertical stratification, and limiting mixing in deep lakes. which disrupt the water cycle by reducing groundwater and surface water supplies, increasing water demand as temperatures rise, and degrading water quality due to increased runoff and erosion. These phenomena may lead to more frequent droughts, Floods have stressed the water infrastructure. Climate change has mostly affected water availability through floods and droughts (Delpla et al. 2009; Trenberth 2005) and they can alter the behaviour of harmful pollutants, directly influencing water quality via dilution or concentration of dissolved chemicals. Temperature rises have the greatest effect on water quality at low river flow rates (Delpla et al. 2009).

Climate change-induced hydrological changes have an indirect influence on physiochemical and biological water quality (Delpla et al. 2009; Shukla et al. 2023). The physicochemical parameters fluctuate with temperature and sun radiation, usually leading to enhanced endothermic processes. As a result, greater water temperatures can improve a variety of activities, including dissolution, solubilisation, complexation, degradation, and evaporation. This often increases the concentration of chemicals that dissolve in the water while the concentrations of dissolved gases decrease, restricting flows and velocity in lakes and rivers, increasing the danger of harmful algal blooms, and decreasing dissolved oxygen levels. Furthermore, higher temperatures can reduce oxygen levels by approximately 10% for every 3 °C increase in temperature (from 10 mg/L at 15 °C) and disturb chemical processes, affecting water quality and freshwater ecosystems (Whitehead et al. 2009). Climate change, in addition to the aforementioned factors, raises serious concerns regarding the long-term availability of freshwater for living creatures. These changes can alter water flow patterns, resulting in increased sediment loads and nutrient discharge from soil erosion (Whitehead et al. 2009; Lane et al. 2007; Leemans and Kleidon 2002).

The relation between surface and groundwater resources is critical for making decisions regarding sustaining and repairing our water bodies, particularly in light of changing climatic conditions. The global water and environmental crises are inextricably connected and exacerbated by unprecedented global pressures caused by overconsumption, growing populations, trade and economy globalization, reduced development assistance, and failure to implement suitable policy, legal, and institutional changes (Duda and El-Ashry 2000). Water quality deterioration has emerged as a global crisis, as evidenced by the widespread presence of persistent chemical pollutants, escalating nutrient enrichment leading to harmful algal proliferation, and the alarming prevalence of faecal contamination linked to microbial pathogens and antibiotic resistance development. According to current projections, rising temperatures would disproportionately affect agriculture in the globe's less temperate regions (Reddy 2015), where the majority of the poor live. To counter this, agriculture will need to better manage natural resources including land, water, and genetic resources in order to become more resilient. Poor water quality poses significant economic, social, and environmental risks, especially in growing or developing countries like India (Kumar et al. 2005; Asthana and Shukla 2014). Rising energy, freshwater, and food prices, as well as demographic expansion, are putting a strain on India's agriculture-based economy. Pollution prevention and water quality management are critical for the well-being of humanity because of the increasing vulnerability of the landless and destitute inhabitants. Climate has been highlighted as one of the key causes of uncertainty in the future management of water supply and quality.

2 Climatic Impact on Disturbance in Hydrologic Cycle

The hydrological cycle, sometimes known as the word "water cycle", is a critical component of the planet's ecological system that sustains animal, plant, as well as human life. It has an impact on the way all living things work. The continuing hydrological cycle, the largest movement of matter in the system of the Earth, is a critical component of the planet's supply of water. As water is essential for life, it's vital to understand how climate change affects water resources and availability. Evidence clearly indicates that global warming, primarily driven by rising greenhouse gas levels, is impacting not just temperatures but the entire water cycle. Global water supplies are under threat from a number of factors in along with climate change, including rising numbers of people, growing urbanisation, deforestation, and pollution. Climate changes is also a major factor along with the aforementioned factor that raise the significant concern about long term freshwater supply to the living beings. Climate change disrupts the water cycle by reducing groundwater and surface water supplies, increasing water demand due to higher temperatures, and degrading water quality through more runoff and erosion. It also leads to more frequent and intense flooding, straining water infrastructure. As precipitation patterns shift and extreme weather events become more common, the hydrological cycle's stability is increasingly jeopardized. One possible reason for this extreme event is deforestation and increased fossil fuel combustion and both of which are directly connected with human activities.

2.1 Climate Change and Cause of Climate Change

The average weather conditions of a specific region over a defined time interval is called "climate". It mostly describes the weather conditions of particular region. Climate change, therefore, represents significant deviations in these conditions. The Intergovernmental Panel on Climate Change (IPCC) defines climate change as statistically discernible alterations in the mean and variability of climatic variables that last for decades or longer. Traditionally, climate is characterized by different factors namely, wind patterns, precipitation and temperature, and it generally presents a stable view of the Earth's system.

Nonetheless, throughout its geological history, Earth's climate has undergone various changes, such as the glaciation phases during the Pleistocene epoch (Huybers 2006), which ended in a more stable climate during the Holocene. These historical changes were gradual and occurred over thousands of years. In contrast, the current climate variability is distinct in both origin and speed. Unlike past natural changes, today's climate shifts are largely driven by human activities and are occurring at a much faster rate. Disruptions in the global carbon cycle are intimately associated with the fast changes that have been observed over the past century (Jepma and Munasinghe 1998; Kondrat'ev et al. 2003). The atmospheric concentration of aerosols and greenhouse gases (GHGs) has considerably grown. The possible reason behind significant increment in the concentration is land-use changes, higher fossil fuel combustion, and industrial processes that are considered as human activities. This has upset the natural carbon balance, leading to a rise in global temperatures. The primary greenhouse gases (GHGs) contribute to global warming are CO_2, N_2O, O_3, and CH_4. Additionally, deforestation and land conversion have reduced the Earth's capacity to sequester carbon, further exacerbating the problem. Increased solar radiation due to ozone layer depletion has also contributed to the current warming trend. Table 1 displays the estimated changes in the GHG concentrations in the atmosphere. Based on data made available by the US Department of Energy's Carbon Dioxide Information Analysis Centre (CDIAC), the Table 1 presents information.

Table 1 The properties of greenhouse gases in the atmosphere (Sohoulande Djebou and Singh 2016)

Green house gases	Main sources	Atmospheric concentration		Lifetime (yrs)
		(2010)	Before 1750	
Ozone (O_3)	Industry, fossil energy	337 ppb	237 ppb	hr/days
Methane (CH_4)	Agriculture, fossil energy, domestic wastes	1893 ppb	722 ppm	12
Hydrochlorofluorocarbons (HCFCs)	Foams, solvents, refrigerants	231ppt	0	11.9
Nitrous oxide N_2O	Industry, fertilizers	326 ppb	270 ppb	121
Carbon tetrachloride (CCl_4)	Refrigerants, fire extinguisher, Solvents, cleaning agents	85 ppt	0	26
Chlorofluorocarbons (CFCs)	Refrigerants, solvents, foams	236–527 ppt	0	45–100
Hydrobromofluorocarbons (HBFCs)	Refrigerants, extinguisher, cleaning agents, fire, solvents	24 ppt	0	9.2–17.2
Carbon dioxide (CO_2)	Industry, fossil energy, wildfire	395.4 ppm	280 ppm	100–300
Hydrofluorocarbons (HFCs)	Industrial processes	75 ppt	0	13.4
Bromochlorofluoromethane	Fire protection	3.3–4.1 ppt	0	16–65
Sulfur hexafluoride (SF_6)	Electricity, chemistry	7.39 ppt	0	3200

Note ppb = parts per billion, ppt = parts per trillion, and ppm = parts per million

2.2 *Climate Changes Impact on the Water Cycle*

Climate change has a major impact on all components of the hydrological cycle, although the extent of its effects varies throughout area and time, depending mostly on the hydrologic element. Significant climate change causes visible disturbances in numerous areas of the water cycle, include rainfall, infiltration, runoff, and evapotranspiration. These disturbances vary in intensity depending on the location and the specific component affected. For example, while some regions face declining precipitation, others experience extreme peaks, and some maintain stable patterns. However, the overall trend is an increase in abnormal precipitation events worldwide. This disruption to the water budget leads to a persistent freshwater crisis in many areas. Significant changes in the climate system are shown in Table 2, which was created based on analysis published in a number of sources, including the IPCC reports from 2013 and 2022.

2.2.1 Changes in Precipitation Patterns

The changing of rainfall patterns is an important part of immediate effects for the variation of climate in hydrological cycle. As the increase in global average temperature, the atmosphere's capacity to hold moisture also rises, leading to more intense and unpredictable precipitation events. While some locations see extended droughts as a result of decreased precipitation, others see considerable increases in rainfall that results to severe flooding and frequent drought. These changes in precipitation patterns are also closely associated with a rise in the number of severe weather occurrences, such as cyclones, hurricanes and tornado rain-storms.

2.2.2 Impact on Evapotranspiration

Elevated temperatures resulting from global warming quicken the rates of transpiration and evaporation, two essential water cycle processes. This rise in water vapour entering the atmosphere has the potential to worsen precipitation unpredictability and cause more extreme weather patterns. Higher evaporation rates have the potential to negatively influence agricultural and native plants in areas already under water stress by lowering the availability of surface water and soil moisture.

2.2.3 Effects on Runoff and Stream Flow

Runoff and stream flow, which are essential for restocking rivers, lakes, and reservoirs, are also impacted by variations in precipitation and increasing evaporation. In certain regions, higher rainfall amounts result in more runoff, which can overwhelm water management infrastructure and lead to sedimentation and erosion. On the other hand, in areas with less precipitation, less runoff results in less stream flow, which lowers the amount of water available for ecosystems and human usage.

2.2.4 Impact on Groundwater Recharge

Precipitation seeps into the ground, replenishing the freshwater supply that is vital to many populations. By changing precipitation patterns and increasing surface runoff, climate change can impede this process by reducing the amount of water that can seep into aquifers. Reduced groundwater recharge can result in falling water tables and increasing competition for this essential resource in areas where precipitation is decreasing or becoming more erratic.

Table 2 Significant alterations to the climate system seen in recent and previous millennia (Sohoulande Djebou and Singh 2016)

Climate factor 1: precipitation				
Features		Latitude box: 60–30°S	Latitude box: 30°S–30°N	Latitude box: 30–60°N
Changes	Period 1	1901–2008: the trend was 2.40 mm/year per decade	1901–2008: the trend was—0.48 mm/year per decade	1901–2008: the trend was 3.14 mm/year per decade
	Period 2	1951–2008: the trend was—0.51 mm/year per decade	1951–2008: the trend was—4.16 mm/year per decade	1951–2008: the trend was 1.50 mm/year per decade
Additional fact		Estimates using Global Precipitation Climatology Centre GPCC data		
Climate factor 2: temperature				
Features		Land surface air temperature	Average earth temperature	Trend of global earth temperature
Changes	Period 1	1901–1950: increased by 0.097 °C per decade	1880–1974: average earth global temperature climbed by 0.3 °C	1906–2005: warming trend is 0.06 °C/decade
	Period 2	1951–2012: increased by 0.175 °C per decade	1975–2010: average earth global temperature increased by 0.5 °C	1956–2005: warming trend is 0.13 °C/decade
Additional fact		1979–2012: increased by 0.254 °C per decade	–	–
Climate factor 3: oceans				
Features		Sea surface temperature	Global sea level	
Changes	Period 1	1901–1950: increased by 0.066 °C per decade	1961–2003: global sea level raised by 1.8 mm/year 1901–2008: the trend was 3.14 mm/year per decade	
	Period 2	1951–2012: increased by 0.071 °C per decade	1993–2003: global sea level raised by 3.1 mm/year	
Additional fact		1979–2012: increased by 0.073 °C per decade	Correlation between sea level rise and the global temperature	
Climate factor 4: cryosphere				
Features		Aretic ice packs	Perennial ice covers	Glaciers (outside Greenland and Antarctica)
Changes	Period 1	1980: annual mean thickness was 3.6 m	1980: the extent was 7.9 × 10^6 km^2	1993–2009: ice mass loss was 0.76 mm/year
	Period 2	2000: annual mean thickness declined to 2.4 m	2012: the extent was 3.5 × 10^6 km^2	2005–2009: ice mass loss was 0.83 mm/year
Additional fact		Since 1978, Arctic sea ice extent decreased by 2.7% per decade	–	–

2.2.5 Melting Glaciers and Snowpack

As natural reservoirs, glaciers and snowpack hold water in a frozen state until it gradually melts throughout the warmer months. Climate change is hastening glacier melting and reducing snowpack in many parts of the world, causing stream flow patterns to shift. Short-term availability of water may rise as a result of this, but long-term water supplies during dry seasons may decrease due to the loss of these frozen stores, especially in areas where drinking and agricultural water are dependent on glacial melt-water.

3 Impacts of Climate Change on Water Availability

Climate change is having a significant impact on available water resources. Water contamination has frequently been directly linked to human activity on farms, in industries, and in urban areas. Degradation of surface water can also be indirectly caused by climate change. It is commonly known that changes in temperature and sun radiation cause physio-chemical characteristics to change, which frequently results in an increase in endothermic reactions. Higher water temperatures can therefore encourage a number of different processes, including evaporation, complexation, dissolution, solubilisation, and degradation. As a result, the concentration of dissolved materials in water often rises while that of dissolved gases typically falls. This loss is most noticeable for dissolved oxygen, which drops from 10 mg/L at 15 °C to roughly 10% at a temperature increase of 3 °C.

In addition, when sea levels rise, there may be a larger likelihood of seasonal or permanent salt intrusion into waterways and groundwater, affecting water quality and appropriateness for use in homes, agriculture, and industry. Variations in air temperature and precipitation may alter riverine currents, affecting the movement as well as dispersion of pollutants. When paired with deteriorating water quality, rising water temperatures will have an effect on the rate of chemical reactions and thus freshwater ecosystems. Elevated river flows have the ability to alter stream power as well as sediment loads, hence changing freshwater ecosystems in lakes and stream systems. They may also alter river morphology and sediment transport to lakes. Climate change models predict more frequent summertime dry spells and flash floods, resulting in uncontrolled overflow from towns and cities into waterways and estuaries. Reduced flows as well as velocities in lakes and rivers can cause extended water residence times, increasing the risk of toxic algal blooms and reducing dissolved oxygen concentrations. In general, climate change is predicted to cause more extreme hydrologic events, such as severe flooding and droughts. Increased fertiliser use may be the result of longer growing seasons and altered land use, which increases the danger of eutrophication and biodiversity loss by increasing nitrogen runoff into aquatic bodies.

The vulnerability of the underprivileged and landless communities will be increased by all of these changes. According to recent estimates, the world's poor live mostly in lower latitudes of the planet, where agriculture will be disproportionately affected by the predicted rise in temperatures. In order to boost agriculture's resilience, better management of natural resources like water, soil, and genetic resources will be required.

3.1 Climate Change's Effects on Agriculture

Agriculture is especially sensitive to the impacts of climate change because it is so dependent on meteorological factors like temperature and precipitation. The shifting climate has a significant impact on agricultural water management. It varies quantity, quality, and demand for water, affecting agricultural yields, land suitability, and livestock output. Climate change's negative impacts on agriculture are expected to jeopardise global food security, according to recent research from international agencies including the International Food Policy Research Institute (IFPRI) as well as the Food and Agriculture Organisation (FAO).

3.2 Water Factors and Climate Change

3.2.1 Water Requirements for Agriculture

For rainfed crops, crop evapotranspiration (ET) can be met entirely or in part by precipitation and for irrigated crops, ET can be met by both irrigation and precipitation. Climate projections have an impact on water deficits in rainfed crops, which are measured as the difference between ET and effective precipitation. Rising temperatures may extend the growing season, allowing for earlier planting and later harvesting, which could increase crop water needs. For instance, in Rajasthan, an increase in temperature is projected to raise total ET demand by 14.8%, significantly affecting the already fragile water resources in this arid region.

3.2.2 Water Availability

The impact of changing climate on the resources of surface water, such as water bodies, reservoirs, and wetlands, is mostly controlled by variations in precipitation in terms of time, place, amount, and type (rain or snow). Many huge-scale irrigation networks depend on surface water, along with climate change might make it more challenging to divide water between upstream and downstream consumers within river basins. Given the abundance of snow in some areas, glacier retreat caused by global warming is expected to significantly alter the hydrological cycle. Another threat to groundwater sustainability is widespread groundwater pumping for agriculture, which has resulted in groundwater depletion. The effects of climate change hold the potential to aggravate the problem. Climate change has two main effects on water availability, such as drought and floods.

3.2.3 Water Quality

Climate change has a direct impact on water temperatures, which is likely to harm water quality. It has the potential to have a detrimental influence on the quality of water indirectly by influencing human conduct related to agriculture and other industries that contribute to wastewater discharge and water contamination. Temperature and precipitation variations are anticipated to have an impact on agricultural land usage and management techniques, such as soil cultivation and pesticide and fertiliser use, hence affecting water quality.

3.3 Climate Change Impact on Agricultural Productivity Through Water Factors

3.3.1 Impacts on Crop Yield

Crop production oscillations induced by shifting climates are influenced by a range of reasons, including variation in atmospheric carbon dioxide levels. Elevated temperatures, altered precipitation and transpiration patterns, rising in the severity along with frequency of adverse weather conditions, and an increase in weed and pest populations. Water is a crucial aspect of this dynamic; depending on the crop, the region, and the water management techniques used, its interplay with other factors can have a beneficial or detrimental impact on yields. High temperatures, in particular, are critical for crop yields in rainfed environments, where yield reductions are likely to occur with rising temperatures across most of the globe.

3.3.2 Impacts on Agricultural Land Availability

A vital component of food production and security is the availability of agricultural land. Changes in water use for agriculture management, such as transitioning from rainfed to irrigated regions or other ways around, will have an influence on land usage. Changing climate affects the availability of land for agriculture because certain climate factors, like as precipitation and temperature, determine whether an area of land is appropriate for farming. Climate change affects both air humidity and soil temperature, which are two important factors in determining land suitability. While overly high temperatures in tropical places can be harmful, increased soil temperatures in temperate zones typically help local agriculture. Climates with higher humidity levels may, in part, improve land arability.

3.3.3 Impacts on Livestock Production

Livestock production supports 1.3 billion people livelihoods globally and accounts for 40% of the world's agricultural GDP. Demand for cattle products is expected to rise as the global population grows and diets change. Because of rising temperatures and altered precipitation patterns, climate change will have significant consequences on production systems of animals, much like it does on agricultural systems. There will be changes in the incidence and severity of severe weather conditions, which will specifically affect the productivity of livestock.

3.4 Climate Change and Its Impact on Drinking Water

Disinfection by-products (DBPs) are largely influenced by a number of variables, including dissolved organic carbon (DOC), temperature, pH, concentrations of bromide, and operational parameters like dosages and contact duration of chlorine. Rainstorms have the potential to increase organic matter levels and turbidity in river waters, which impairs the effectiveness of drinking water purification procedures. Furthermore, variations in aqueous composition, pH, and temperature brought on by changing climate may have an influence on the sorption of pollutants onto mineral phases. In general, the quality of drinking water is deteriorating through source to tap water because the result of changes in the climate, causing health concerns, especially during extreme weather.

4 Climate Change's Influence on Water Quality Degradation in Both Groundwater and Surface Water Resources

A comprehensive understanding of water quality across vast regions, including the interplay between groundwater and surface water resources, is very important for making decisions about safeguarding and restoring our water bodies, especially in the face of changing land use patterns and climate circumstances. The degradation of water quality has emerged as a global crisis, evidenced by the widespread presence of persistent chemical pollutants, escalating nutrient enrichment leading to harmful algal proliferation, and the alarming prevalence of faecal contamination linked to microbial pathogens and the development of antibiotic resistance. The increasing quality of water crisis has been exacerbated by the increasing impacts of the changing climate, such as more frequent and severe storms that disturb the hydrological cycle, increase pollution loads, and create conditions conducive to the spread of toxic substances within water bodies.

4.1 Impacts of Increasing Temperature on Groundwater Quality

While the devastating effects of global warming in our ecosystems are increasingly evident, our comprehension of the specific repercussions on subterranean environment, particularly in terms of groundwater quality, remains distressingly limited. The effects of warming on groundwater resources exhibit Significant differences between various places and depths, affected by a complex interaction of geological, hydrological, as well as climatic factors, making it exceedingly challenging to accurately evaluate and forecast these changes on a broad scale. Typically, groundwater temperatures exhibit a strong correlation with prevailing land surface temperatures (Menberg et al. 2014). One of the most critical repercussions of rising groundwater temperatures in the majority of aquifer systems is the accelerated metabolic activity of microorganisms (Brielmann et al. 2011).

Aquifers with high levels of nutrient and organic matter, sometimes known as eutrophic aquifers, are subject to the accelerated effects or impacts of climate change. Rising temperatures in these systems enhance microbial activity, which facilitates water quality degradation because of its rapid uptake of nutrients as well as organic matter (Griebler et al. 2016). Intensified microbial activity can deplete oxygen, leading to anoxic conditions. This shift impacts groundwater ecosystems by altering redox potential, microbial communities, and metabolic processes, often resulting in nitrate reduction (Borch et al. 2010). Elevated temperatures also stimulate microbial breakdown of organic matter, releasing CO_2 which subsequently lowers pH. This acidic environment dissolves calcite, releasing calcium ions into the groundwater (McDonough et al. 2020). Also, this leads to the release of ions like k^+, Mn^{2+}, and F^-.

4.2 Impact of Sea Level Change on Groundwater Quality

Global warming has caused considerable increases in atmospheric temperatures during the previous two decades. As a result of this temperature rise, many of our planet's glaciers as well as polar ice have melted, causing the sea level in many parts of the country to rise year after year. This increased sea level enhances the sea water intrusion in many aquifers and affecting the ground water quality significantly. The higher pH and electrical conductivity of the water makes it not fit for drinking and in many coastal regions not even for irrigation (Sreekesh et al. 2018).

4.3 Climate Change's Impact on Surface Water

In addition to the quantitative effects on water availability, the changing climate has a significant impact on the quality of surface water. Altered rainfall patterns, change in atmospheric temperatures, and more repeated extreme weather incidents disrupt delicate aquatic ecosystems. These changes can lead to higher water temperatures, reduced dissolved oxygen levels, increased nutrient loading, and alterations in pH levels. Moreover, climate change can facilitate the proliferation of harmful algal blooms, introduce new contaminants, and exacerbate existing pollution problems. These combined factors pose significant challenges for maintaining healthy and sustainable surface water bodies. The higher temperature of the surface water reduces the pH of the water and solubility of oxygen in the water.

Droughts concentrate organic matter, increasing winter soil carbon levels. Rising temperatures accelerate nitrogen transformations, boosting nitrate production in warmer months and ammonium in winter. Warmer conditions also stimulate enzymatic activity, enhancing nutrient cycling. Notably, temperature positively correlates with nitrification, a key process in nitrogen conversion. This creates a higher nutrient load in surface water.

More intense precipitation events, often resulted by the climate change, can dramatically raise the volume and velocity of runoff, leading to heightened soil erosion. This erosion process dislodges and transports pollutants, such as sediments, nutrients, pesticides, and pathogens, from land to water bodies. The impact is particularly pronounced following drought periods when soils are compacted and devoid of protective vegetation, making them more susceptible to erosion. Consequently, intense rainfall events can rapidly mobilize accumulated pollutants, causing significant water quality degradation in rivers, lakes, and coastal areas.

With high rainfall events associate with climate change will change the fate of herbicide and pesticide in the agricultural field. The high runoff rate takes all these chemicals from the field and mixes with surface water. This results in serious degradation of quality of the water.

Fishes, diatoms and planktons are some of the major indicators of fresh water quality. Due to global warming many of the narrow temperature tolerant species are prone to extinction or moving to more favourable locations. This can lead to imbalance in the aquatic ecosystem and reducing the productivity of the system. And these species are more replaces by many wide temperature tolerant species now (Delpla et al. 2009).

5 Challenges for Management in Water Quality with Regard to Climate Change

For human society to thrive, pollution prevention and water quality management are essential. Maintaining an adequate quantity of water to fulfil the increasing needs of the twenty-first century's manufacturing industries, humans, along with ecosystems, in addition to food and power generation, is one of the most significant environmental concerns (Hajkowicz and Collins 2007). One of the primary sources of uncertainty for managing water quantity and quality in the future has been identified as climate change (Ragab and Prudhomme 2002).

5.1 Challenges in Managing Water Quality

Utilising water sources in agriculture to adapt to climate change involves both technological innovations and non-technological problems. Among the latter, the following could be considered. Three key areas need to be considered in order to overcome the uncertainties that are changing about the climate and its effects:

(1) More communication between scientific disciplines in order to develop suitable adaptation measures;
(2) Better connections between research, policy, and decision-makingboth nationally and regionally; and
(3) Addressing the social and economic problems, when addressing climate change's impacts on agriculture. Other hurdles to water management as well as control of pollution in the basin would include the increases in sediment, water, and nutrient export caused by climate change (Yang et al. 2019).

5.2 Challenges of Integrated Quality of Water Modelling

Water quality models are frequently classified depending on their level of complexity, the predicted factors affecting water quality (e.g., nutrients, sediments, dissolving oxygen), and the application domain. Models of water quality attempt to capture the temporal and spatial dynamics of the main components. Over the past seventy years, a variety of components or condition factors have been gradually added to models in accordance with escalating issues with the overall condition of the water (Borah et al. 2006). Determining some of the next directions for advancement and exploring the core of the unresolved deficiencies are essential components of integrated water quality modelling. A significant challenge in watershed-scale water quality modelling is accounting for the significant variation in spatial distribution that often characterized the catchment. The difficulty with catchment-scale quality of water modelling is the fact that biogeochemistry is nonlinear and can add a lot of error when integrated with any transport model since the scale at which reaction terms are calculated is quite different from the size at which the model is applied. Rode et al. (2010). The following are challenges in modelling water quality (Rode et al. 2010).

1. Simple linear models have to be used at larger sizes if the application size inhibits a complete non-linear biogeochemical models' evaluation.
2. We suggest a unique approach for combining field and process-based research modelling, even if scaling up process knowledge in catchment-scale modelling is still a barrier. It is anticipated that scale-dependent process studies, encompassing the entire spectrum of river sizes and watershed features, will enhance the predictive power of process-based models at larger scales.
3. When determining the temporal and spatial accuracy and model complexity, considerations should include the study's goal, the scope of applications, the understanding of the important processes, and the accessibility of the data.

5.3 Challenges in Adapting Water Resources and Infrastructure

Additionally, the issues related with projected population growth and changes in land use, water managers face a significant conceptual challenge from climate change. Future climatic changes are predicted to cause the water cycle at global level to generally intensify, with a subsequent elevation in flooding risk (Huntington 2006; Milly et al. 2002; Kundzewicz et al. 2007). Future water management concerns arising from climate change include changes to existing water both supply and demand networks, changes to water source quality, and impacts on water. Numerous results have an impact, both directly and indirectly, on the quality, quantity, and timeliness of various water sources as well as the current infrastructure for water supply. For instance, an increase in the yearly average temperature of 0.4–2 °C in 2030 and 0.6–2 °C in 2070 will result in increased evaporation, which will raise human water demand while simultaneously decreasing freshwater supplies and placing more strain on environmental water flows.

A 3–17 cm sea level rise by 2030 and a 7–52 cm sea level rise by 2070 is projected (Amitrano et al. 2007). It elevates the possibility of saltwater incursion into estuaries, rivers, bays, aquifersalong the shore, and low-lying areas. It also raises the risk of flooding for infrastructure related to wastewater treatment and reticulation, as well as flooding during intense storms (Midgley et al. 2005; Sjerp and Charteris 2009) additionally resulting in a reduction in stormwater outfall capacity (Sjerp and Charteris 2009). Increasing coastal erosion can also lead to damage to infrastructure, such as septic tanks, pump stations, rainwater drainage systems, sewage mains, and water mains (Sjerp and Charteris 2009). The International Panel on Climate Change's present-day and projected impact modelling scenarios list just a handful of the expected effects of climate change on water supplies and infrastructure. Of course, there are additional effects of climate change on current water resources that need to be continuously considered in addition to its effects. Most notably, as the population rises, there will be an increased need for water for related to agriculture, human consumption, and other industrial operations. The future land use changes is another issue that will significantly affect water availability and demand (Neal and McMahon 2007).

6 Strategies and Policies for Management of Water Quality

The management of water quality is critical for maintaining the ideal quality of water as well as preventing future degradation due to pollution. The strategies adopted for water quality management can be broadly categorized into three categories: regulation, remediation and watershed management strategies.

6.1 Regulation Strategies

These are aimed at regulating wastewater discharge, discharge from industrial plants, sewage water discharge, and discharge from nonpoint sources into the water bodies. The standards must be set for every type of discharge in the water body and should be implemented stringently. It is essential to create and enforce water quality standards and regulations at both local and regional levels. Policy interventions in the form of incentives and penalties can also aid in regulations and maintaining the water quality.

6.2 Treatment Strategies

These techniques are used to eliminate contaminating agents from the water, which might have a negative impact on its quality. There are various remediation methods such as biological, physical, and chemical mediation. These principles are used in wastewater treatment which is considered as one of the remediation strategies.

6.2.1 Wastewater Treatment

Wastewater treatment is one of the most direct and effective strategies for improving water quality. This process involves the removal of contaminants from industrial, agricultural, and domestic wastewater before it is released back into the environment. Advanced treatment methods such as membrane filtration, activated sludge processes, and the use of flocculants are employed to remove both organic and inorganic pollutants. Additionally, decentralized wastewater treatment systems can be implemented in rural or remote areas to ensure that even the smallest communities have access to pure, safe water.

6.3 Watershed Management Strategies

Some of the watershed management strategies are explained in subsequent sections.

6.3.1 Management of Land Use and Land Cover

The kind of pollutants and the overall load of pollutants in a particular region are significantly influenced by land use and land cover. Residential as well as industrial land use negatively impacts water quality because they are more likely to produce wastewater and other contaminants. Thus, controlling land cover and usage is essential to controlling the quality of the water.

6.3.2 Soil Erosion/Runoff Management

Soil erosion is one of the primary contributors of water quality decline, which transports contaminants. Its management through various techniques will check the water quality deterioration. The runoff carries with it the excess fertilizers and pesticides applied in the agricultural field. In order to improve the water quality, the judicial usage of such input will reduce their intake into the water bodies. The soil and water conservation techniques such as contouring, trenching, grading, terracing will ensure less runoff and soil erosion and thereby reducing the pollutant load on the water body.

6.3.3 Public Awareness and Policy Implementation

Educating the public on the importance of water quality and how individual actions contribute to pollution is essential. This can be supported by stringent policies that regulate pollution levels and enforce penalties for non-compliance. Furthermore, implementing water quality standards and regular monitoring ensures that water bodies meet safety criteria, reducing the risk of waterborne diseases.

7 Conclusion

Climate is therefore one of the key factors contributing to future unpredictability about the management of water availability and quality. The world's water supplies are also threatened by pollution, deforestation, urbanisation, and population growth in addition to climate change. Temperatures and the entire water cycle are being impacted by global warming, which is primarily brought about by an increase in greenhouse gas emissions. Moreover, rising sea levels may make it more likely for rivers and groundwater to become permanently contaminated by saltwater intrusion, which would lower the water's suitability for domestic, agricultural, and industrial uses. Finding the most effective means of providing safe drinking water to a future human population that is worried about both water shortages and the factors impacting water quality as a result of climate change will undoubtedly be challenging. Maintaining water quality and preventing pollution are crucial for the well-being of humans. Climate change models predict more frequent droughts during the summer and flash floods, which would cause uncontrolled overflows from towns and cities into streams and estuaries. Increased nutrient runoff into water bodies, the potential for eutrophication, and increased hydrological extremes, including stronger floods and prolonged dry spells, are all predicted effects of climate change. To solve this, better natural resource management, includes land, water, as well as genetic resources, will be required to increase agricultural resilience. One of the most important concerns of the twenty-first century is to maintain a sustainable water supply to meet the growing demands of the manufacturing sector, the human population, the environment, as well as food and energy production. More study is needed on adaption measures like as managing water levels in wetlands, lakes, as well as rivers, along with building plant corridors as well as buffer zones. For optimal water quality, strategic planning must consider climate change impacts and vulnerabilities.

References

Amitrano L, Hargreaves R, Page I, Hennessy K, Lee T, Snow M, Kjellstrom T (2007) An assessment of the need to adapt buildings for the unavoidable consequences of climate change. Report for Victorian Department of Sustainability and the Environment

Asthana V, Shukla AC (2014) Water security in India: hope, despair, and the challenges of human development. Bloomsbury Publishing USA

Borah DK, Yagow G, Saleh A, Barnes PL, Rosenthal W, Krug EC, Hauck LM (2006) Sediment and nutrient modeling for TMDL development and implementation. Trans ASABE 49(4):967–986

Borch T, Kretzschmar R, Kappler A, Cappellen PV, Ginder-Vogel M, Voegelin A, Campbell K (2010) Biogeochemical redox processes and their impact on contaminant dynamics. Environ Sci Technol 44(1):15–23

Brielmann H, Lueders T, Schreglmann K, Ferraro F, Avramov M, Hammerl V, Griebler C (2011) Shallow geothermal energy usage and its potential impacts on groundwater ecosystems. Grundwasser 16:77–91

Delpla I, Jung AV, Baures E, Clement M, Thomas O (2009) Impacts of climate change on surface water quality in relation to drinking water production. Environ Int 35(8):1225–1233

Duda AM, El-Ashry MT (2000) Addressing the global water and environment crises through integrated approaches to the management of land, water and ecological resources. Water Int 25(1):115–126

Gleeson T, Cuthbert M, Ferguson G, Perrone D (2020) Global groundwater sustainability, resources, and systems in the Anthropocene. Annu Rev Earth Planet Sci 48(1):431–463

Goldscheider N (2019) A holistic approach to groundwater protection and ecosystem services in karst terrains. Carbonates Evaporites 34(4):1241–1249
Gorelick SM, Zheng C (2015) Global change and the groundwater management challenge. Water Resour Res 51(5):3031–3051
Griebler C, Avramov M (2015) Groundwater ecosystem services: a review. Freshw Sci 34(1):355–367
Griebler C, Malard F, Lefébure T (2014) Current developments in groundwater ecology—from biodiversity to ecosystem function and services. Curr Opin Biotechnol 27:159–167
Griebler C, Brielmann H, Haberer CM, Kaschuba S, Kellermann C, Stumpp C, Lueders T (2016) Potential impacts of geothermal energy use and storage of heat on groundwater quality, biodiversity, and ecosystem processes. Environ Earth Sci 75:1–18
Gupta SK (2010) Modern hydrology and sustainable water development. Wiley
Hajkowicz S, Collins K (2007) A review of multiple criteria analysis for water resource planning and management. Water Resour Manag 21:1553–1566
Huntington TG (2006) Evidence for intensification of the global water cycle: review and synthesis. J Hydrol 319(1–4):83–95
Huybers P (2006) Early Pleistocene glacial cycles and the integrated summer insolation forcing. Science 313(5786):508–511
Jepma CJ, Munasinghe M (1998) Climate change policy: facts, issues and analyses. Cambridge University Press
Kondrat'ev KI, Krapivin VF, Varotsos C (2003) Global carbon cycle and climate change. Springer Science & Business Media
Kumar R, Singh RD, Sharma KD (2005) Water resources of India. Curr Sci 794–811
Kundzewicz ZW, Mata LJ, Arnell NW, Doll P, Kabat P, Jimenez B, Shiklomanov I (2007) Freshwater resources and their management
Lane SN, Tayefi V, Reid SC, Yu D, Hardy RJ (2007) Interactions between sediment delivery, channel change, climate change and flood risk in a temperate upland environment. Earth Surf Process Landf J Br Geomorphol Res Group 32(3):429–446
Leemans R, Kleidon A (2002) Regional and global assessment of the dimensions of desertification. In: Global desertification: do humans cause deserts?. Dahlem University Press, pp 215–231
McDonough LK, Santos IR, Andersen MS, O'Carroll DM, Rutlidge H, Meredith K, Baker A (2020) Changes in global groundwater organic carbon driven by climate change and urbanization. Nat Commun 11(1):1279
Menberg K, Blum P, Kurylyk BL, Bayer P (2014) Observed groundwater temperature response to recent climate change. Hydrol Earth Syst Sci 18(11):4453–4466
Midgley GF, Chapman RA, Hewitson B, Johnston P, De Wit M, Ziervogel G, Forsyth GG (2005) A status quo, vulnerability and adaptation assessment of the physical and socio-economic effects of climate change in the Western Cape
Milly PCD, Wetherald RT, Dunne KA, Delworth TL (2002) Increasing risk of great floods in a changing climate. Nature 415(6871):514–517
Neal BP, McMahon TA (2007) Planning non-metropolitan water supplies: can we do better? Water J Aust Water Assoc 34(7):62–65
Ragab R, Prudhomme C (2002) SW—soil and water: climate change and water resources management in arid and semi-arid regions: prospective and challenges for the 21st century. Biosys Eng 81(1):3–34
Reddy PP, Reddy PP (2015) Impacts of climate change on agriculture. In: Climate resilient agriculture for ensuring food security, pp 43–90
Rode M, Arhonditsis G, Balin D, Kebede T, Krysanova V, Van Griensven A, Van der Zee SE (2010) New challenges in integrated water quality modelling. Hydrol Process 24(24):3447–3461
Shukla N, Gupta S, Rai S (2023) Potential impacts of climatic changes and human activity on water quality. Environmental processes and management: tools and practices for groundwater. Springer International Publishing, Cham, pp 103–111

Sjerp E, Charteris A (2009) Climate change, sea level rise and coastal subsidence along the Gippsland coast

Sohoulande Djebou DC, Singh VP (2016) Impact of climate change on the hydrologic cycle and implications for society. Environ Soc Psychol 1(1):36–49

Sreekesh S, Sreerama Naik SR, Rani S (2018) Effect of sea level changes on the groundwater quality along the coast of Ernakulam District, Kerala. J Clim Change 4(2):51–65

Trenberth KE (2005) The impact of climate change and variability on heavy precipitation, floods, and droughts. Encycl Hydrol Sci 17:1–11

Whitehead PG, Wilby RL, Battarbee RW, Kernan M, Wade AJ (2009) A review of the potential impacts of climate change on surface water quality. Hydrol Sci J 54(1):101–123

Yang Q, Zhang X, Almendinger JE, Huang M, Chen X, Leng G, Li X (2019) Climate change will pose challenges to water quality management in the St. Croix River Basin. Environ Pollut 251:302–311

Surface Water and Groundwater Quality Development Policy Under Climate Changes

Shyam Datta Waghmare

Abstract This chapter presents a comprehensive bibliometric analysis to inform the development of surface water and groundwater quality policies in the context of climate change. As climate variability continues to impact water availability and quality globally, it is essential to devise evidence-based strategies that ensure the sustainability of both surface and groundwater resources. Through a review of global research trends, institutional collaborations, and influential researchers, this analysis identifies key areas for policy intervention, such as enhanced monitoring systems, integrated water resource management (IWRM), and nature-based solutions. The analysis draws from a network of authors and institutions, with significant contributions from researchers like M.M., A.S., Zhang Y., and Li H., whose work on water resource management provides crucial insights for policy formulation. Key findings emphasize the importance of strengthening institutional collaboration, promoting climate-resilient water management practices, and fostering public participation in water conservation efforts. Additionally, the analysis underscores the role of integrated approaches in balancing water use demands with the need for maintaining water quality under changing climate conditions. By leveraging bibliometric insights, this chapter proposes actionable policy recommendations to improve water governance, reduce pollution, and adapt to climate-induced water challenges. The proposed policies focus on collaboration, innovation, and sustainable management to safeguard water resources for future generations.

Keywords Water quality policy · Climate change adaptation · Integrated Water Resource Management (IWRM) · Sustainable water governance · Bibliometrics

---

S. D. Waghmare (✉)
ICAR-Central Institute of Fisheries Education, Mumbai, India
e-mail: shyamdattaw@gmail.com

Dimple et al. (eds.), *Surface Water and Groundwater Quality for Sustainable Utilization*, Springer Hydrogeology, https://doi.org/10.1007/978-981-96-8189-1_17

1 Introduction

Access to clean and safe water remains a critical issue worldwide, with more than 4 billion people facing inadequate water accessibility for at least one month each year. As climate change exacerbates water scarcity, projections indicate that by 2050, 57% of the global population will live in regions plagued by severe water shortages (Boretti and Rosa 2019). Since groundwater accounts for approximately 20% of global water use and stores 70 times more freshwater than surface water (Earman and Dettinger 2011), its quantity and quality depletion seriously threaten sustainable water resource management. In arid and semi-arid regions, where surface water is limited and precipitation scarce, groundwater is the primary water source for domestic, agricultural, and industrial purposes. Over-exploitation of these resources often leads to groundwater depletion, hindering economic development and creating challenges in health, education, and agriculture sectors. As climate change continues to affect the hydrological cycle, groundwater protection and quality management become increasingly urgent.

Climate change, primarily driven by greenhouse gas emissions, particularly carbon dioxide (CO_2), alters global temperature and precipitation patterns. Since the Industrial Revolution, atmospheric CO_2 concentrations have surged, leading to significant warming trends. Between 2010 and 2018, global temperatures rose rapidly, with the 10 warmest years on record occurring during this period. These changes are not evenly distributed, with higher latitudes experiencing more pronounced warming. Such climatic shifts influence water availability by increasing atmospheric moisture content, leading to more precipitation in wet regions and greater evaporation and aridification in dry areas (Earman and Dettinger 2011).

Groundwater is critical in the natural water cycle, acting as a vast underground reservoir replenished by surface runoffs, rivers, and lakes. Groundwater recharge and discharge, along with evapotranspiration and seepage, control water resources' quantity and quality. Climate-induced temperature changes, however, can disrupt these processes, affecting biological and geochemical reactions that alter groundwater quality (Riedel 2019). In addition to these natural processes, human activities aimed at mitigating climate change, such as land use changes and carbon sequestration, also impact groundwater systems, often leading to contamination or depletion. While research on the relationship between climate change and surface water systems is extensive, groundwater systems have received comparatively less attention, especially concerning quality rather than quantity. Previous studies have focused on physical changes, such as groundwater recharge and storage capacity (Earman and Dettinger 2011; Smerdon 2017). Research on groundwater quality has often been limited to specific regions or contaminants (Bloomfield et al. 2006; Stuart et al. 2011, Bondu et al. 2017; Lipczynska-Kochany 2018; Mao et al. 2023). Given the complex interplay between climate change, hydrological systems, and human activities, understanding the impacts on groundwater quality is essential for shaping effective water resource policies.

To gain a comprehensive understanding of how climate change affects surface water and groundwater quality, this chapter utilizes a bibliometric analysis conducted using R Studio. By analyzing trends in scientific research and policy development, this chapter provides a broad overview of the challenges and opportunities in developing effective policies for managing water quality under changing climatic conditions. This study is based upon a four-stage process proposed by Fosso Wamba and Mishra, which involves specifying the keywords and selecting the databases, conducting a preliminary data analysis, analysing bibliometric networks, and performing thematic and structure analysis.

2 Data Source

This study is based on research articles extracted from Dimensions.ai, which was selected because of the world's largest database of research insights. It represents the most comprehensive collection of linked data related to the global research and innovation ecosystem available in a single platform. Dimensions are updated daily and constantly growing. It contains over 112m linked research publications, 1.3bn + citations, 5.6m + grants worth $1.7trillion + in funding, 41m + patents, 600k + clinical trials, 100k + organizations, 65m + disambiguated researchers, and more. The data is normalized, linked, and ready for analysis.

3 Keyword Selection and Refinement

The PRISMA approach was used to identify the studies that comprehensively understand how climate change affects surface water and groundwater quality. Further exploration of studies in this aspect resulted in the selection of phrases such as 'surface water quality', 'groundwater quality', 'water pollution', 'water resource management', 'sustainable development', 'climate change', 'water scarcity', 'water treatment', 'water conservation', and 'water quality monitoring' as key terms for identifying relevant studies. Therefore, based on these cues, a search query containing keywords related to the topic was formulated. For this study, we included research works published from 2000 to 2023. The results were also restricted to only research articles published in research journals, and thereafter a language filter was applied to limit this study to articles published only in English. The search query used was—**(surface water quality OR groundwater quality) AND climate change AND (water pollution OR water scarcity) AND sustainable development AND water resource management AND (water treatment OR water conservation) AND water quality monitoring AND groundwater quality AND water pollution AND water scarcity AND water treatment AND water conservation AND water quality monitoring**.

Therefore, the thematic search on the dimension.ai database resulted in 3,555 research papers based on the theme of this chapter.

4 Data Analysis Tools

Tranfield et al. (2003) emphasized the importance of systematic reviews and meta-analyses for summarizing research findings within a specific field. Building on this, bibliometric analysis has become vital for understanding and examining large volumes of literature across interdisciplinary domains (Zupic and Čater 2015; Donthu et al. 2021). Bibliometric techniques and text-mining approaches objectively analyze research landscapes, minimizing the risk of selection bias arising from subjective reviews (van Eck and Waltman 2014). This study employed bibliometric analysis to gain critical insights into the structure and research trends related to surface water and groundwater quality under climate change. This analysis utilized various tools to visualize and quantify the relationships within the literature, such as co-authorship networks, co-occurrences of keywords, citation patterns, bibliographic coupling, and co-citation analysis. To ensure a comprehensive understanding of the research landscape, this bibliometric analysis was supported by using text-mining tools. These tools provided further depth to the analysis by identifying the thematic focus of the literature. Specifically, text mining techniques such as word cloud generation and topic modeling were applied using the *Bibliometrix* package in R Studio. Word clouds visually highlighted the most frequently occurring terms in the literature, while topic modeling helped to categorize the studies into key themes or topics. Graphs, tabulation, and frequency analysis were used to summarize the findings, providing insights into the number of publications, top-performing journals, influential publishers, and leading countries in this research domain. These approaches allowed for a systematic examination of research trends and offered a quantitative foundation for identifying gaps in the literature and future directions for policy development in surface water and groundwater quality under climate change.

5 Studies Published from 2000 to 2023

The year-wise distribution of articles shows a significant upward trend in research publications on surface water and groundwater quality development policy under climate change from 2000 to 2023, as shown in Fig. 1. From 2000 to 2004, the number of articles remained relatively low, with a sharp increase starting in 2005, which recorded 432 publications. This marked a pivotal shift, with a consistent publication rise over the following years, particularly from 2015 onwards, and average citations per year are given in Fig. 2. The number of articles nearly doubled between 2015 (109) and 2023 (618), indicating growing academic interest and focus on this topic. Notably, from 2020 to 2023, the surge is most pronounced, reflecting a heightened emphasis on this area in recent years. This upward trend suggests expanding knowledge and increasing recognition of the topic's relevance in the academic community.

The bibliometric analysis for the period 2000–2023 covers a substantial body of research on surface water and groundwater quality under climate change, comprising 3,555 documents from 980 sources. The field has experienced rapid growth, with an annual increase of 20.8%. On average, each document is 7.07 years old, receiving 36.66 citations per document, indicating the sustained relevance of the literature. The analysis reveals significant research collaboration, with 14,281 contributing authors and an average of 4.93 co-authors per document. Notably, 22.67% of publications involved international collaborations, underscoring the global nature of the research. Despite the prevalence of multi-authored studies, 423 single-authored papers were identified, reflecting both collaborative and individual contributions to the field. The data also highlights that both *Author's Keywords* and *Keywords Plus* appeared 544

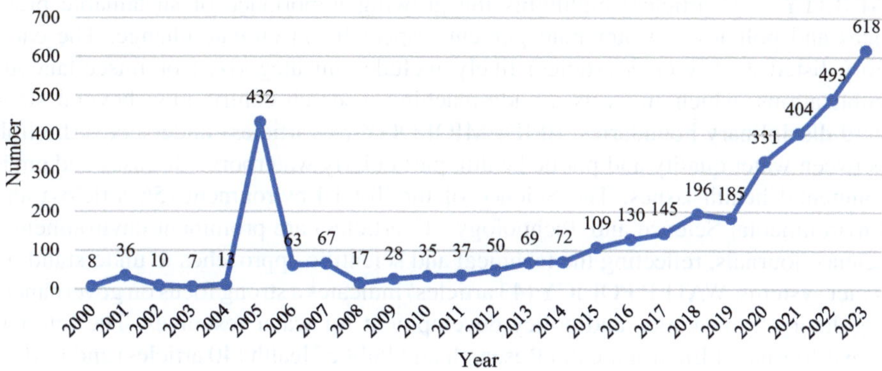

Fig. 1 Number of studies published from 2000 to 2023

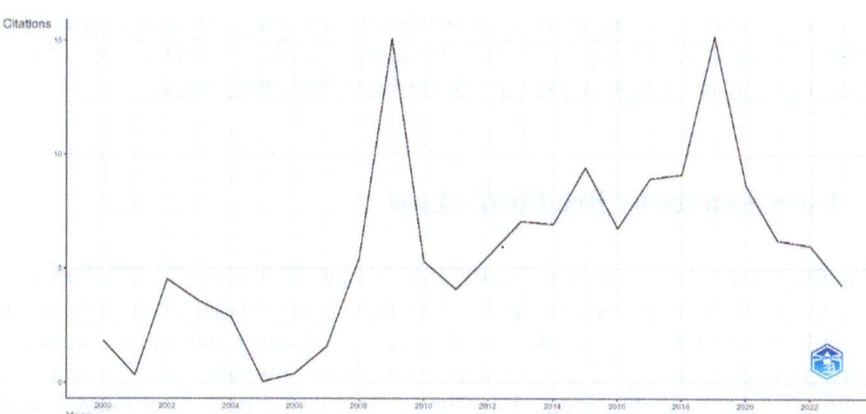

Fig. 2 Average citations per year

times, further suggesting key research trends. The data demonstrates the interdisciplinary and collaborative scope of research in this domain, with a strong emphasis on international partnerships and steady academic interest.

6 Most Relevant Sources

The analysis of the top 10 sources contributing to research on surface water and groundwater quality under climate change reveals a diverse array of journals, reflecting the interdisciplinary nature of the topic. HORTSCIENCE leads with 422 articles, indicating a strong focus on the intersection between water quality, agriculture, and horticultural practices, as shown in Fig. 3. With 285 articles, WATER emphasizes its centrality in hydrology and water resource management. SUSTAINABILITY (202 articles) highlights the growing importance of sustainable practices and policies in water management, especially in climate change. The category listed as NA (123 articles) likely includes uncategorized or miscellaneous publications, which suggests a wide-reaching research contribution beyond standard disciplinary boundaries. EPIDEMIOLOGY (99 articles) underscores the link between water quality and public health, particularly waterborne diseases and environmental health issues. The Science of the Total Environment (56 articles) and Environmental Science and Technology (43 articles) are prominent environmental science journals, reflecting the technical and scientific approaches to understanding water systems. WATER POLICY (43 articles) indicates a strong focus on governance, regulatory frameworks, and policy development for water resources. The International Journal of Environmental Research and Public Health (40 articles) and LAND (38 articles) further emphasize the connections between environmental health, land use, and water quality, demonstrating the comprehensive nature of the research in this domain. Overall, the top sources showcase the wide-ranging interdisciplinary approaches, from agricultural sciences to public health and policy, that contribute to understanding water quality issues in a changing climate. A Sankey plot in Fig. 4 summarizes relationships among top keywords, authors, and journals.

7 Core Sources by Bradford's Law

The analysis of core sources using Bradford's Law highlights the distribution of research publications on surface water and groundwater quality under climate change. According to Bradford's Law, the literature is divided into zones based on productivity showed in the Fig. 5. Zone 1 contains the core journals, which produce the majority of the research output, while Zone 2 includes the next set of journals with fewer but still significant contributions. In Zone 1, HORTSCIENCE is the most

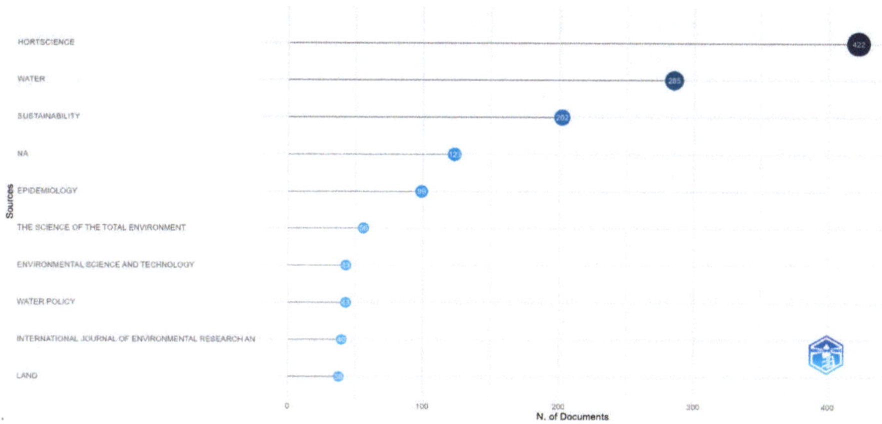

Fig. 3 Most relevant sources

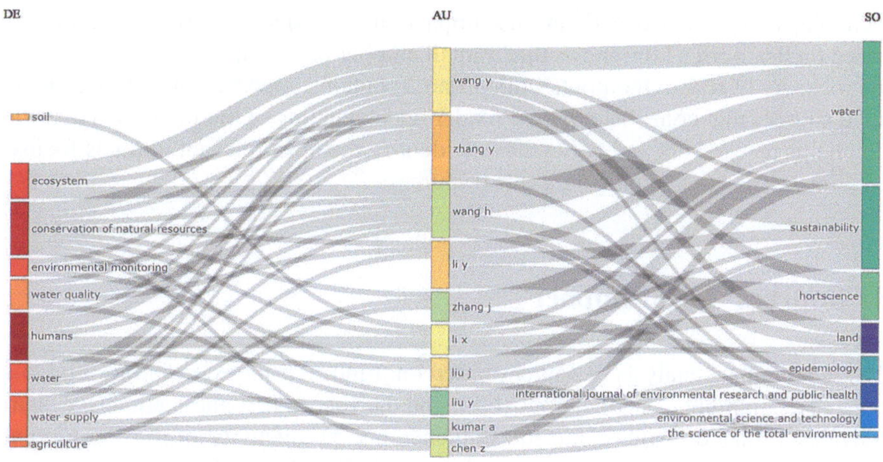

Fig. 4 Relationship among top keywords, top authors, and top journals summarized by a Sankey plot

prolific source, with 422 articles, followed by WATER (285 articles), SUSTAINABILITY (202 articles), and EPIDEMIOLOGY (99 articles). These sources represent the core journals in the field, accounting for most of the research (1,187 cumulative articles). These publications are central to the discourse on water quality, particularly in the context of climate change, with a focus on horticulture, water resource management, sustainability, and public health. Zone 2 comprises journals with slightly lower frequency but still important contributions, such as Environmental Science and Technology and Water Policy (43 articles each), and International Journal of Environmental Research and Public Health (40 articles). These journals and LAND (38 articles) extend the core topics into more specialized areas such as environmental

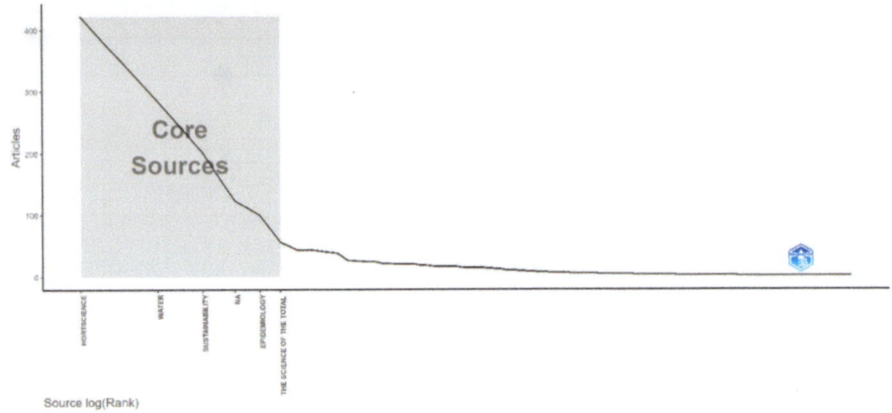

Fig. 5 Core sources by Bradford's law

technology, governance, and land-use impact on water quality. Overall, the analysis based on Bradford's Law shows that a small number of journals (Zone 1) account for the majority of research output, while Zone 2 contains important but less frequently cited sources that contribute to the broader understanding of the field. This distribution helps researchers and policymakers identify the most relevant journals focused on climate change's impact on water resources.

8 Sources' Local Impact

The local impact analysis of sources based on h-index, g-index, m-index, total citations (TC), number of publications (NP), and the year of first publication (PY_start) provides valuable insights into the influence and productivity of the top journals in the field of water and environmental research. WATER ranks highest with an h-index of 39, indicating that it has produced 39 articles cited at least 39 times. Its g-index of 63 shows that its top articles are highly influential. It has become a key water resource management research journal with over 6,115 total citations from 285 publications since 2012. Each detail of the sources' local impact by H index is given in Table 1, and the sources' production over time is shown in Fig. 6. SUSTAINABILITY follows closely with an h-index of 34 and a g-index of 60, reflecting its substantial contribution to sustainable practices in water management. Since its first publication in 2014, it has accumulated 4,614 citations from 202 articles, making it one of the most influential journals in sustainability science. The Science of the Total Environment also shows a strong impact, with an h-index of 33 and a g-index of 56. Its 56 publications since 2012 have been cited 4,001 times, highlighting its role in interdisciplinary environmental research.

Table 1 Sources' local impact by H index

Source	h_index	g_index	m_index	TC	NP	PY_start
Water	39	63	3	6115	285	2012
Sustainability	34	60	3.091	4614	202	2014
The Science of the Total Environment	33	56	2.538	4001	56	2012
International Journal of Environmental Research and Public Health	18	34	1.8	1196	40	2015
Journal of Cleaner Production	17	24	1.133	1823	24	2010
Journal of Environmental Management	16	21	1.143	1186	21	2011
Water Resources Research	16	20	0.762	2296	20	2004
Ecological Indicators	15	26	1.154	1496	26	2012
Ecology and Society	15	18	0.75	5276	18	2005
Environmental Research Letters	15	20	1.364	1889	20	2014

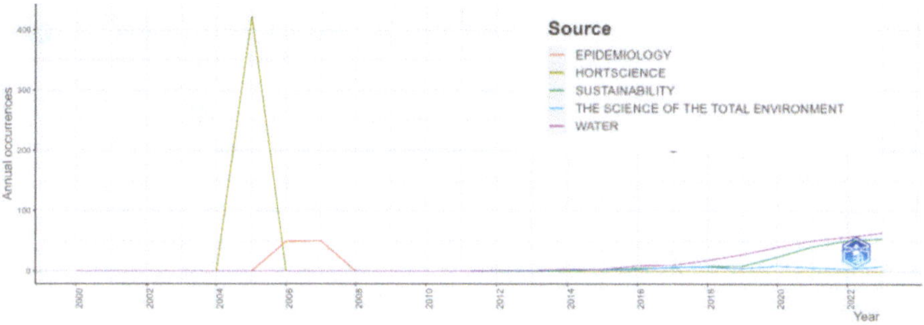

Fig. 6 Sources' production over time

Other journals, such as the International Journal of Environmental Research and Public Health (h-index 18, 1,196 citations) and the Journal of Cleaner Production (h-index 17, 1,823 citations), contribute to public health and cleaner production topics, respectively, with strong citation numbers relative to their smaller output. Despite having fewer publications (20 and 18, respectively), Water Resources Research and Ecology and Society show high impact through their total citations, particularly Ecology and Society, with 5,276 citations, indicating the deep influence of its top articles. In summary, these journals provide a significant number of publications and demonstrate high citation rates and influence within the field, particularly in water resources, sustainability, and environmental management research.

9 Most Relevant Authors

The top 10 most relevant authors in this field are led by ZHANG Y, who has published 25 articles with a fractionalized score of 4.08, indicating a significant contribution to the literature. LI Y and LIU J follow, each contributing 20 articles with a fractionalized score of 3.20. WANG Y, despite publishing 19 articles, has the highest fractionalized score (4.26), suggesting greater individual contribution or collaboration intensity. The details of most relevant authors and authors' production over time are given in Figs. 7 and 8. LI X, CHEN Z, and WANG H also feature prominently, with fractionalized scores between 2.47 and 2.59, reflecting moderate contributions relative to their total publications. ZHANG J and KUMAR A have published 15 and 14 articles, respectively, with KUMAR A standing out for a higher fractionalized score of 2.99 compared to ZHANG J's 2.15, suggesting deeper involvement in collaborative work. LIU Y rounds out the list with 13 articles and a fractionalized score of 2.43. This analysis highlights the substantial impact of these authors on the body of research in this field.

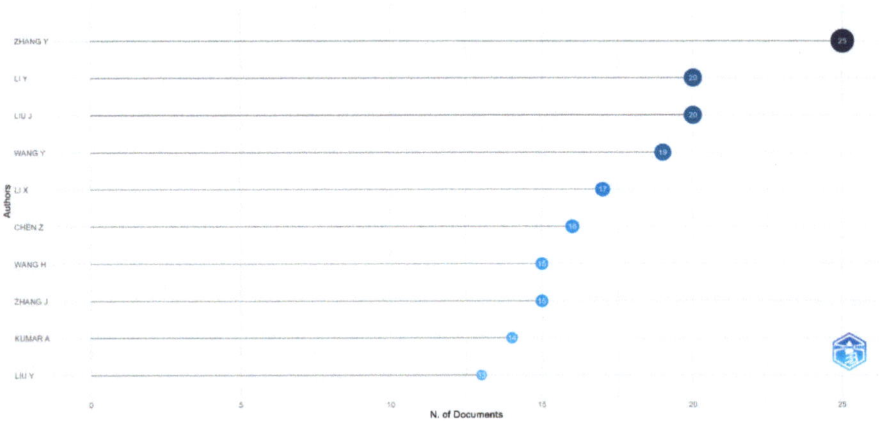

Fig. 7 Most relevant authors

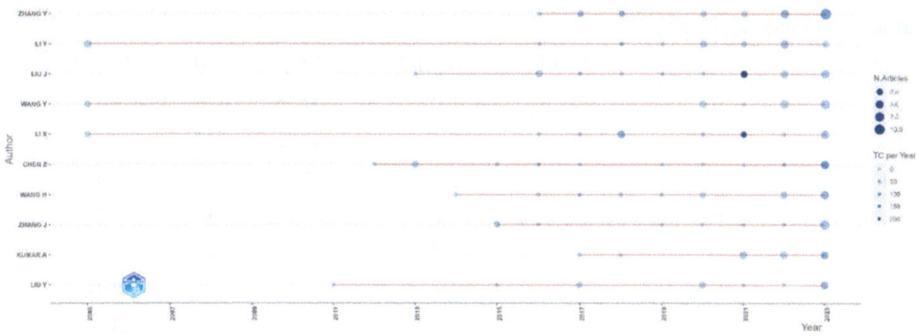

Fig. 8 Authors' production over time

10 Author Productivity Analysis Based on Lotka's Law

The data on author productivity follows Lotka's Law, which shows that most authors contribute only a single document, while a progressively smaller proportion contributes multiple documents. Specifically, 86.8% of authors (12,400) have written just one document, and only 9.1% have authored two documents. The proportion of authors contributing three or more documents declines sharply, with only 2.2% having written three documents and less than 1% contributing four or more, as shown in Fig. 9. This pattern indicates that a small number of highly productive authors are responsible for a significant portion of the literature, while the vast majority of authors have contributed just a single publication, reflecting the skewed nature of research productivity common in many scientific fields.

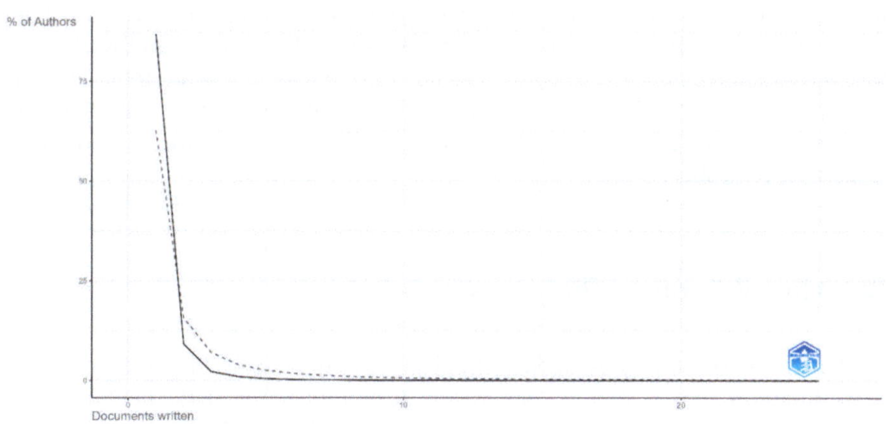

Fig. 9 Author productivity through Lotka's law

Table 2 Authors' local impact

Author	h_index	g_index	m_index	TC	NP	PY_start
Liu J	13	20	1.083	1555	20	2013
Li X	12	17	0.6	1476	17	2005
Wang H	12	15	1.091	513	15	2014
Chen Z	11	16	0.846	995	16	2012
Kumar A	11	14	1.375	328	14	2017
Zhang Y	11	25	1.222	1156	25	2016
Kumar R	10	12	1.25	308	12	2017
Li Y	10	20	0.5	621	20	2005
Angelakis AN	9	11	0.692	567	11	2012
Kumar P	9	12	1.5	858	12	2019

11 Authors' Local Impact

The local impact of the top 10 authors is measured through various indices such as the h-index, g-index, m-index, total citations (TC), number of publications (NP), and the year they started publishing (PY_start), details is given in the Table 2. LIU J leads with an h-index of 13, a g-index of 20, and total citations of 1,555 across 20 publications since 2013, reflecting a strong and consistent influence in the field. LI X, with an h-index of 12 and 1,476 citations since 2005, shows a sustained impact over a longer period. WANG H, with a high m-index of 1.091, indicates a rapid rise in influence relative to their career length. Similarly, KUMAR A and ZHANG Y have high m-indices of 1.375 and 1.222, respectively, highlighting their accelerated research contributions since they started publishing in recent years (2017 and 2016). Despite fewer publications, authors like ANGELAKIS AN and KUMAR P have made notable contributions, with h-indices of 9 and significant total citations, indicating the impactful nature of their work. Overall, this analysis reveals that while some authors have a long-term, steady impact, others have shown a rapid rise in influence over a shorter span.

12 Most Relevant Affiliations

The University of Chinese Academy of Sciences, Beijing, leads the list of most relevant affiliations with 49 articles, showcasing its prominent role in the field, as shown in Fig. 10. Several other institutions have made notable contributions, each with 32 articles, including the Department of Microbiology at The University of Hong Kong, Mahidol University, and Shanghai Municipal Center of Disease Control and Prevention. Research institutions like the Wuhan Academy of Environmental Science and the Wuhan Centres for Disease Prevention and Control are also key contributors,

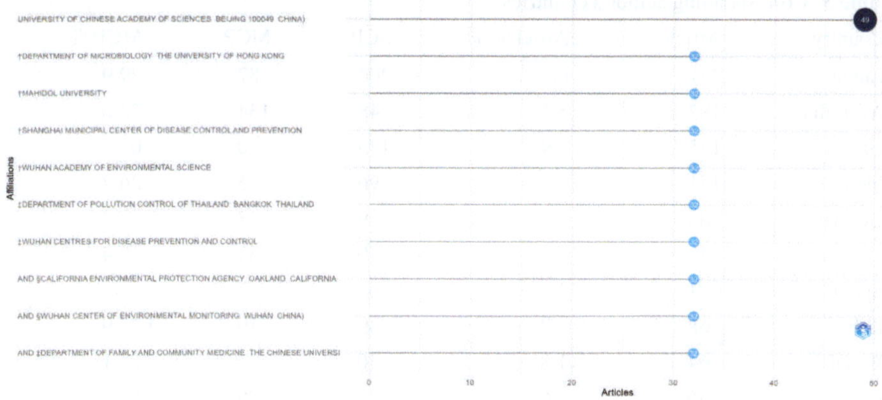

Fig. 10 Most relevant affiliations

reflecting Wuhan's strong presence in environmental and disease-related research. The California Environmental Protection Agency in Oakland, along with institutions in Thailand and China, such as the Department of Pollution Control of Thailand and the Wuhan Center of Environmental Monitoring, also appear frequently with 32 articles each. This distribution indicates a strong collaborative network of universities and environmental agencies, emphasizing their collective impact on advancing research in the field.

13 Corresponding Author's Countries

China leads as the top country with 291 articles (8.2%), with 204 single-country publications (SCP) and 87 multi-country publications (MCP), contributing 29.9% to collaborative efforts; details are given in Table 3. Australia follows with 182 articles (5.1%), strongly emphasizing international collaboration, as 73.6% of its publications are multi-country (MCP). Despite having 175 articles (4.9%), the USA has no MCPs, indicating its focus on domestic research. India, with 131 articles (3.7%), balances between single-country publications (96 SCP) and moderate international collaborations (26.7% MCP). Germany, Canada, and Brazil demonstrate substantial collaboration, with MCP percentages of 52.4%, 49.4%, and 53.1%, respectively. Spain, Italy, and Egypt contribute fewer multi-country publications, with MCP percentages below 30%. This analysis reveals that while countries like Australia and Brazil prioritize international research partnerships, others, such as the USA, focus more on independent research contributions.

Table 3 Corresponding author's countries

Country	Articles	Articles %	SCP	MCP	MCP %
China	291	8.2	204	87	29.9
Australia	182	5.1	48	134	73.6
USA	175	4.9	175	0	0
India	131	3.7	96	35	26.7
Germany	82	2.3	39	43	52.4
Canada	79	2.2	40	39	49.4
Spain	71	2	64	7	9.9
Italy	68	1.9	49	19	27.9
Brazil	64	1.8	30	34	53.1
Egypt	56	1.6	27	29	51.8

14 Countries' Scientific Production

The article's publication of the countries shows the increasing trend in recent years, as shown in Fig. 11. The USA leads in scientific production with 1,387 articles, closely followed by China with 1,323, highlighting their dominance in the research landscape, as shown in Fig. 12. India ranks third with 633 publications, showcasing its growing contribution to global research. Germany (567 articles) and Australia (459 articles) also contribute significantly, reinforcing their strong presence in the field. Spain (443 articles), Italy (364 articles), and the Netherlands (363 articles) are important European contributors. Brazil (242 articles) and Canada (232) ranked in the top 10, reflecting their active involvement in scientific production. This distribution underscores the global nature of research, with developed and emerging economies playing key roles in advancing knowledge in this domain.

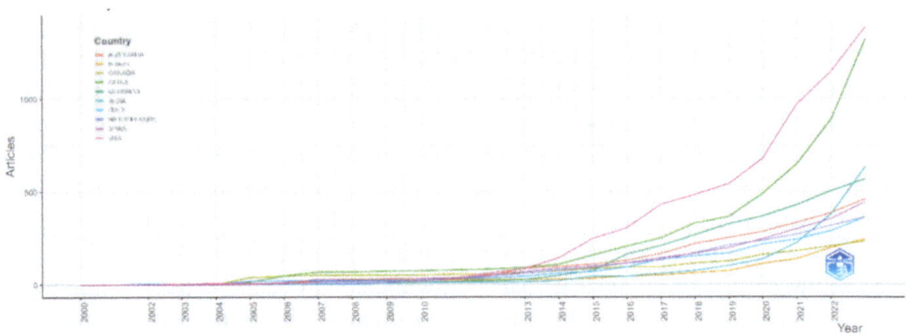

Fig. 11 Countries' production over time

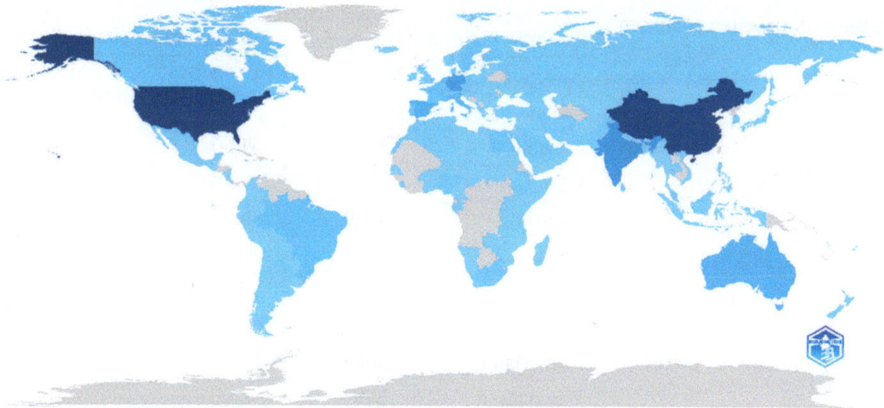

Fig. 12 Publication volume of countries

15 Most Cited Countries

Australia is the most cited country, with 26,857 citations and an impressive average of 147.60 citations per article, indicating its research's strong influence and high impact, details is given in the Table 4. China follows with 10,312 total citations and a lower average of 35.40 citations per article, suggesting that while it produces a large volume of research, the individual impact per article is comparatively modest. Canada ranks third with 5,334 citations and an average of 67.50, reflecting a solid contribution to impactful research. The USA, despite being a leading producer of articles, has 5,087 citations and a lower average of 29.10, indicating a need for more highly cited works. Austria and India demonstrate significant averages of 86.50 and 30.90 citations per article, respectively, while Germany (3,852 total citations, 47.00 average) and Belgium (3,040 total citations, 67.60 average) also maintain a strong citation presence. The Netherlands (2,664 total citations, 59.20 average) and Bangladesh (2,493 total citations, 48.90 average) round out the list, highlighting a diverse range of countries contributing to impactful scientific literature. This analysis reveals the total citation counts and the varying average citations, emphasizing the differing levels of research impact across countries.

16 Most Global Cited Documents

The most globally cited documents reflect significant contributions to their respective fields. Details are given in Table 5. Willett W's 2019 paper in The Lancet led the list with 6,462 total citations, averaging an impressive 1,077 citations per year and a normalized citation rate 71.03. Rockström J's 2009 publication in Ecology and Society follows with 4,452 total citations, averaging 278.25 citations per year

Table 4 Most cited countries

Country	TC	Average article citations
Australia	26,857	147.60
China	10,312	35.40
Canada	5334	67.50
USA	5087	29.10
Austria	4239	86.50
India	4050	30.90
Germany	3852	47.00
Belgium	3040	67.60
Netherlands	2664	59.20
Bangladesh	2493	48.90

and a normalized TC of 18.45, indicating its sustained relevance in environmental discussions. Reid AJ's 2018 paper in Biological Reviews has garnered 2,048 citations, demonstrating a strong average of 292.57 citations per year and a normalized TC of 32.07. Other notable papers include Whitmee S's 2015 article in The Lancet with 1,902 citations and a normalized TC of 20.20, and Boretti A's 2019 work in NPJ Clean Water, which received 1,536 citations, averaging 256 citations per year. Watts N's 2015 paper also holds relevance with 1,480 citations and a normalized TC of 15.72. The list continues with Thornton PK's 2010 paper (1,461 citations), Keesstra SD's 2016 article (1,173 citations), Carvalho FP's 2017 publication (1,168 citations), and Werner AD's 2013 paper (1,133 citations). This analysis highlights these documents' impact and ongoing significance within the academic community, illustrating their contributions to advancing knowledge and understanding across various disciplines.

17 Most Frequent Words Used in Abstract

The analysis of word frequency in abstracts reveals that "water" is the most commonly occurring word, with 12,661 mentions, highlighting its central role in the research context. "Management" follows with 2,861 occurrences, reflecting the frequent focus on strategies and practices in resource handling. "Quality" and "study" appear 1,952 and 1,944 times, respectively, indicating that the assessment and investigation of various aspects are core themes. Words like "environmental" (1,885 occurrences) and "resources" (1,785 occurrences) point to a strong emphasis on environmental conservation and resource management in the body of research. The term "development" appears 1,655 times, often in the context of sustainable practices or advancements. "Climate" (1,542 occurrences) and "sustainable" (1,471 occurrences) further highlight the critical importance of climate-related issues and sustainability in research.

Table 5 Most globally cited documents

Paper	DOI	Total citations	TC per year	Normalized TC
Willett W, 2019, The Lancet	https://doi.org/10.1016/S0140-6736(18)31788-4	6462	1,077.00	71.03
Rockström J, 2009, Ecology and Society	https://doi.org/10.5751/ES-03180-140232	4452	278.25	18.45
Reid AJ, 2018, Biological Reviews	https://doi.org/10.1111/BRV.12480	2048	292.57	32.07
Whitmee S, 2015, The Lancet	https://doi.org/10.1016/S0140-6736(15)60901-1	1902	190.20	20.20
Boretti A, 2019, NPJ Clean Water	https://doi.org/10.1038/S41545-019-0039-9	1536	256.00	16.88
Watts N, 2015, The Lancet	https://doi.org/10.1016/S0140-6736(15)60854-6	1480	148.00	15.72
Thornton PK, 2010, Philosophical Transactions of the Royal Society B Biological Sciences	https://doi.org/10.1098/RSTB.2010.0134	1461	97.40	18.38
Keesstra SD, 2016, The Soil	https://doi.org/10.5194/SOIL-2-111-2016	1173	130.33	19.28
Carvalho FP, 2017, Food and Energy Security	https://doi.org/10.1002/FES3.108	1168	146.00	16.35
Werner AD, 2013, Advances in Water Resources	https://doi.org/10.1016/J.ADVWATRES.2012.03.004	1133	94.42	13.35

Finally, "research" itself is mentioned 1,456 times, emphasizing the focus on scientific inquiry and exploration. This analysis demonstrates the dominance of water-related issues, resource management, and sustainability in the field, reflecting the current global concerns regarding environmental protection and climate change.

18 Word Cloud-Based Analysis

A comprehensive bibliometric analysis was conducted to assess the current state of surface water and groundwater quality development studies. As part of this, four distinct word clouds were created to visualize key terms used in different aspects of the research, as shown in Fig. 13. These visualizations provide insights into the thematic focus of the literature. The most frequent term in Keyword Plus is "humans" with 124 occurrences, followed by "conservation of natural resources" (89) and "ecosystem" (87), underscoring the focus on human–environment interactions and resource conservation, which is shown in Fig. 13a. "Environmental monitoring" (66) and "water" (65) highlight ongoing concerns with environmental oversight and water issues. "Water supply" (64), "agriculture" (56), and "water quality" (53) emphasize the role of water in agriculture and resource management. "Climate change" and "soil" (both 43) reflect the impact of climate on ecosystems and land use. In Author Keywords, "humans" again appears most frequently (124), reinforcing its importance in research. Other terms like "conservation of natural resources" (89), "ecosystem" (87), and "environmental monitoring" (66) show alignment with Keyword Plus terms, indicating shared research interests, as shown in the Fig. 13b. "Water" (65), "water supply" (64), and "agriculture" (56) continue to be crucial topics, while "water quality" (53), "climate change" (43), and "soil" (43) also remain prominent. In Title Terms, "water" is the most frequent word (1,419 occurrences), clearly establishing it as a dominant theme, as shown in Fig. 13c. "Management" (449), "review" (358), and "sustainable" (290) suggest a strong emphasis on managing and reviewing sustainability-related topics. "Quality" (256), "assessment" (237), and "climate" (230) indicate that environmental and resource evaluations are central. "Development" (211), "change" (207), and "urban" (206) reflect concerns about growth, change, and urbanization. In Abstracts, "water" leads by a large margin (12,661 occurrences), highlighting its overwhelming importance. "Management" (2,861), "quality" (1,952), and "study" (1,944) demonstrate a focus on water management and quality assessments, as shown in Fig. 13d. "Environmental" (1,885) and "resources" (1,785) show significant interest in environmental and resource issues. "Development" (1,655), "climate" (1,542), and "sustainable" (1,471) are common themes, reflecting concerns with climate change and sustainable development. "Research" (1,456) suggests the consistent emphasis on scientific inquiry. This comprehensive analysis across different categories illustrates the centrality of water, resource management, sustainability, and climate change across various research topics, with a significant focus on human impact, environmental monitoring, and resource conservation.

19 Co-occurrences Based Analysis

The co-occurrence network analysis provides valuable insights into the relationships between key topics in the research domain. The analysis was conducted on multiple clusters using "Keyword Plus," "Author Keywords," "Title," and "Abstract" to identify the most important nodes and their connections, given in Fig. 14. Below is an interpretation of the results for each set of terms.

The most prominent terms in the "Keyword Plus" category include "conservation of natural resources" and "ecosystem", both of which have the highest betweenness (85.653 and 77.363, respectively) and closeness (0.018), indicating that they are

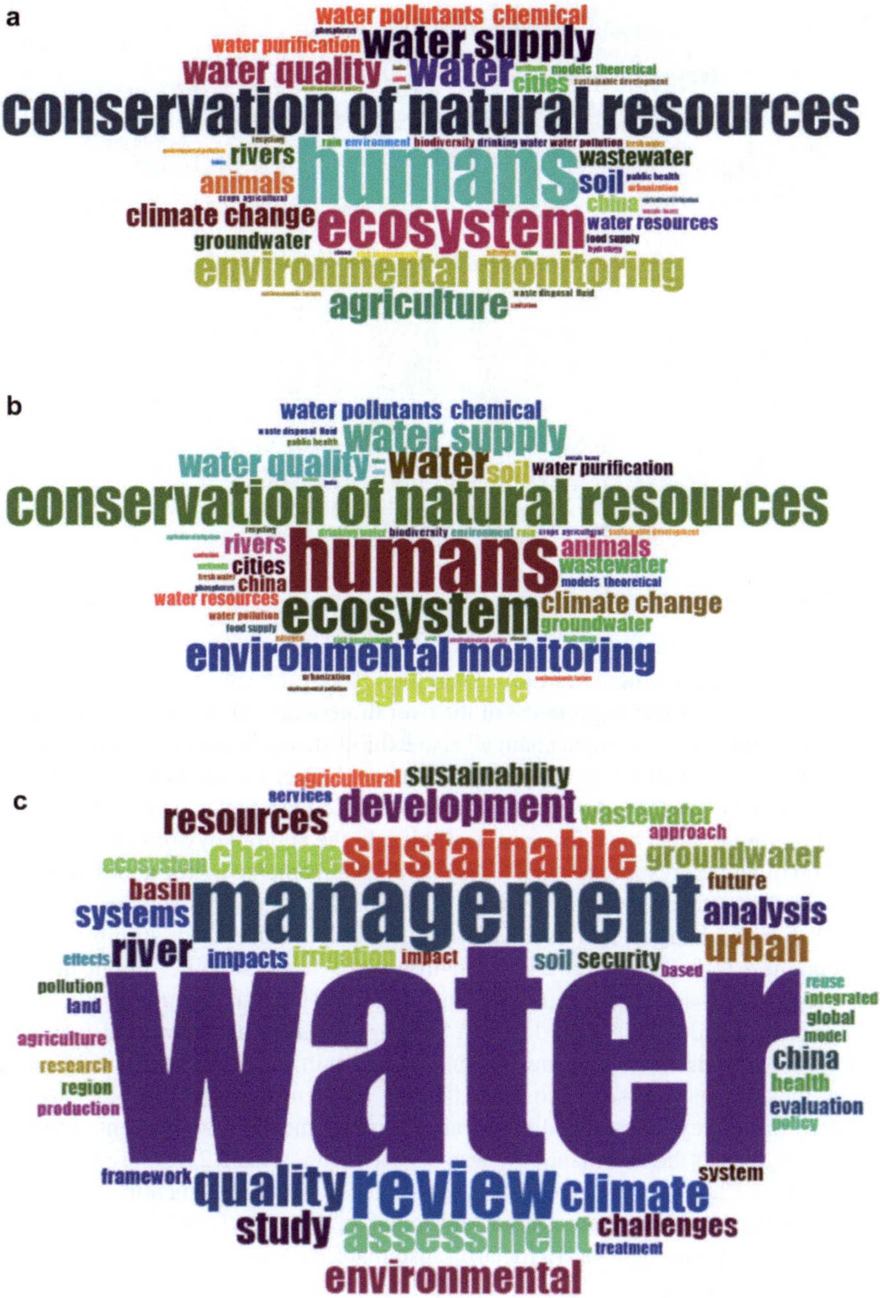

Fig. 13 Word cloud analysis

Fig. 13 (continued)

central in connecting other keywords. Their high PageRank values (0.062 and 0.059) further underscore their importance in the overall network. Other notable keywords like "agriculture" and "climate change" also exhibit strong betweenness, suggesting their relevance in interdisciplinary research areas. Lower-ranked keywords such as "soil," "animals," and "biodiversity" reflect specialized topics that contribute to environmental research but have fewer central roles. In the author keyword cluster, "conservation of natural resources" and "ecosystem" again emerge as key nodes, aligning with their importance in the "Keyword Plus" results. Keywords such as "agriculture," "climate change," and "soil" also rank highly, indicating a strong research focus on these topics. New terms like "environment," "food supply," "public health," and "fresh water" are introduced, reflecting an interdisciplinary connection between environmental and health-related issues. "Nitrogen" and "phosphorus" also appear, highlighting the significance of nutrient management in environmental sustainability discussions. Terms like "socioeconomic factors" and "environmental policy" suggest an interest in policy and social dimensions of environmental management. The title analysis highlights the dominance of "water" (betweenness: 122.218, closeness: 0.02, PageRank: 0.122) and "management" (betweenness: 13.291), indicating that water management is a critical research area. "Sustainable," "quality," and "assessment" are also important, pointing toward a focus on sustainability assessments in water-related issues. Terms like "urban," "resources," and "groundwater" reflect concerns about water security in urban and rural contexts. Keywords such as "policy," "services," and "land" emphasize the role of ecosystem services and policy frameworks in addressing climate change and environmental degradation. In the abstract section, "water" and "quality" once again dominate the network, indicating a heavy focus on water quality

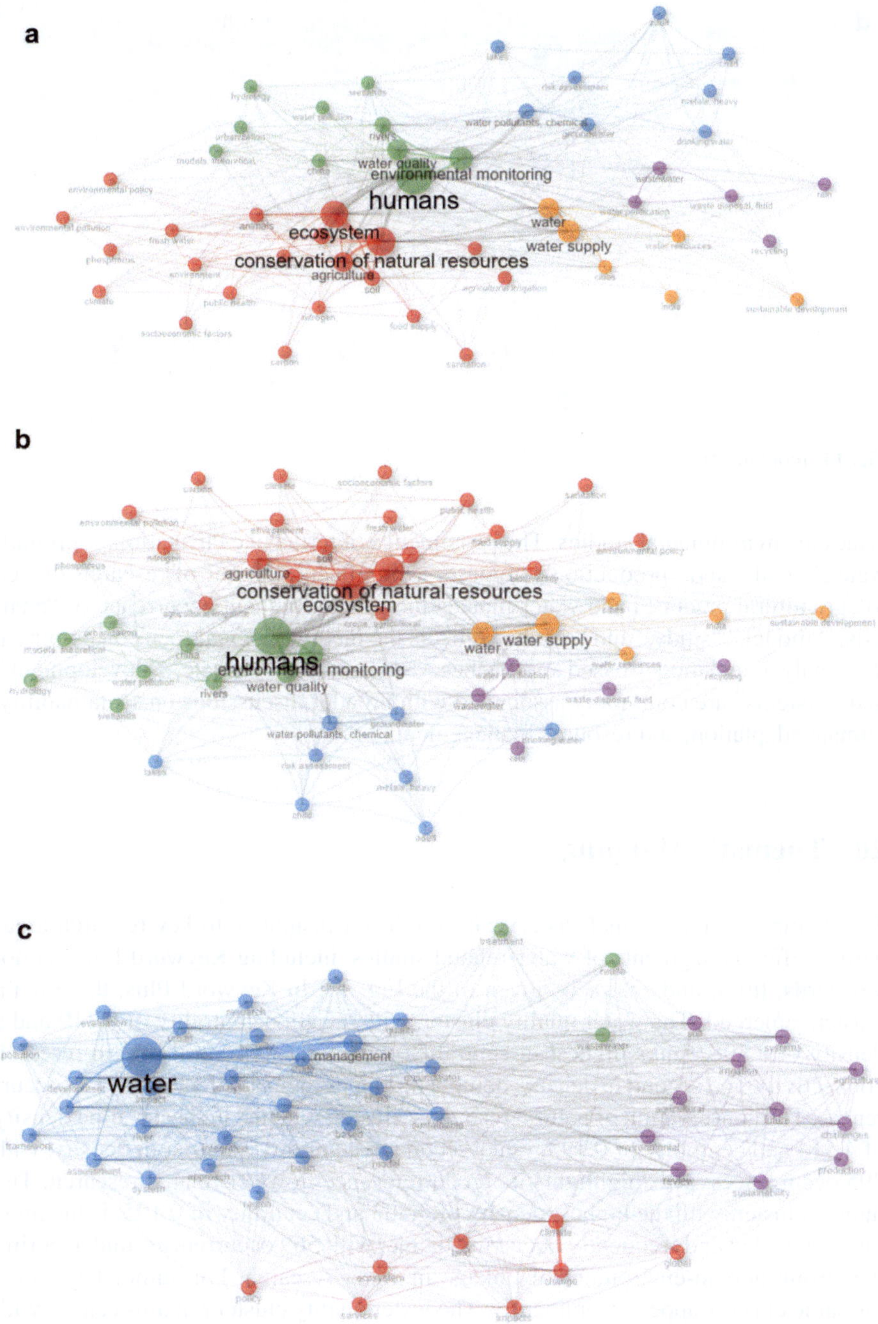

Fig. 14 Co-occurrence analysis

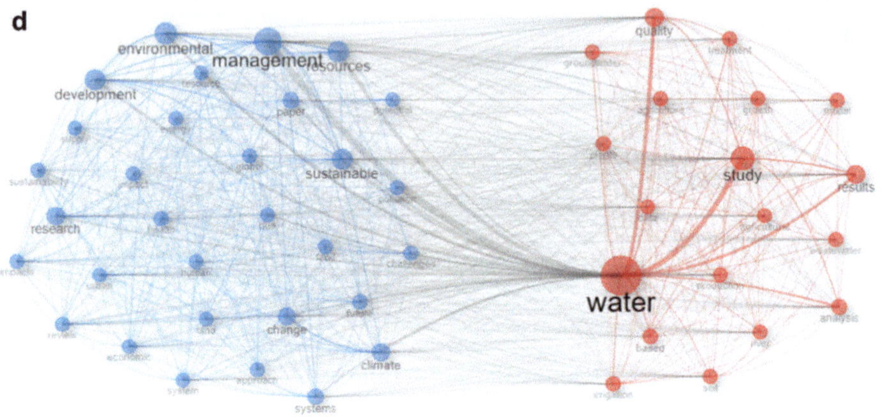

Fig. 14 (continued)

issues in environmental studies. The prominence of terms like “irrigation,” “groundwater,” “soil,” and “production” suggests a significant amount of research related to agricultural practices and water management. The frequent occurrence of “analysis,” “model,” “study,” and “results” reflects the methodological focus on empirical data analysis and model-based approaches. “Sustainable,” “climate,” “development,” and “systems” are commonly associated with broader discussions on sustainability, climate adaptation, and resource management.

20 Thematic Mapping

The thematic mapping analysis reveals significant insights into key research areas across different segments of water-related studies, including Keyword Plus, author keywords, titles, and abstracts, given in the Fig. 15. In Keyword Plus, three main clusters emerged. The water quality cluster, with a Callon centrality of 0.119 and a density of 4.598, ranks highest in centrality, indicating its central role in research connectivity, and second in density, demonstrating its thematic depth with 138 occurrences. The conservation of natural resources cluster, with the highest Callon density of 4.514 and centrality of 0.193, ranks second in centrality and first in density, with 395 occurrences. This highlights its broad relevance in water-related research. The human cluster, with the highest density of 5.503 and centrality of 0.195, is the most interconnected and frequently occurring theme, with 569 occurrences, underscoring the strong human-environmental linkage in water research. For author keywords, the same clusters appear prominently. The water quality cluster remains central with a Callon centrality of 0.119 and density of 4.598, while conservation of natural resources and humans continue to show high centrality and density, reflecting the thematic overlap with the Keyword Plus results. The human cluster maintains the highest density of 5.503, showing consistent importance across both segments. In titles, four primary clusters emerged, with the water cluster showing a notably high Callon centrality of 1.918 and density of 5.051, ranking fourth in both centrality and density. Its frequency of 9,028 occurrences indicates the foundational importance of water as a research topic. The irrigation cluster, central to agricultural water management, shows a centrality of 0.589 and a density of 4.143, ranking second in centrality and first in density with 1,319 occurrences. The quality cluster, with a Callon centrality of 1.713 and density of 4.415, ranks third and appears frequently (5,083 occurrences), highlighting its relevance. The abstract cluster, while lower in centrality (0.002), demonstrates significant density (4.545), though it is less frequently occurring (22 instances). In abstracts, four key clusters were identified. The water cluster dominates, with the highest Callon centrality of 3.077 and a density of 2.825, appearing 49,688 times, indicating its central role in research. The increased cluster follows with a centrality of 0.75 and density of 1.294, appearing 7,798 times. The development cluster, with a Callon centrality of 2.163 and density of 2.094, is significant, appearing 28,795 times. Lastly, the treatments cluster, with the lowest centrality (0.052) and density (1.07), appears 503 times but plays an important role in water treatment studies. This thematic mapping shows that core topics such as water quality, conservation, human interaction with water resources, and water management dominate the research landscape, highlighting the depth and breadth of studies in these areas.

21 Thematic Evolution

The thematic evolution analysis tracks the shifts and continuities in research topics related to water, humans, and natural resource conservation from 2000 to 2023, reflecting changes in focus and the emergence of new themes, as given in Fig. 16. Between 2000–2010 and 2011–2021, the focus on cities evolved into broader themes surrounding humans, indicating a shift from urban infrastructure to human–environment interactions. This is evidenced by a relatively low increase index of 0.5 and stability of 0.043, suggesting moderate thematic evolution with limited continuity. Similarly, the transition from cities to water supply highlights the growing concern for urban water management, particularly in relation to water purification, with an increase index of 0.5 and a slightly higher stability of 0.125. Research on the conservation of natural resources expanded to encompass themes of animals, agriculture, and ecosystems, reflecting a more holistic approach to natural resource management. This theme showed moderate continuity with a high occurrence of 8 and stability of 0.071. Some topics related to conservation persisted but were reframed within a human-centered context, indicated by an increase index of 0.431 and a low stability of 0.035. In the transition from 2011–2021 to 2022–2023, significant thematic shifts emerged. The research focus on humans transitioned toward the urgent theme of climate change, highlighting its growing relevance; this shift showed a high increase

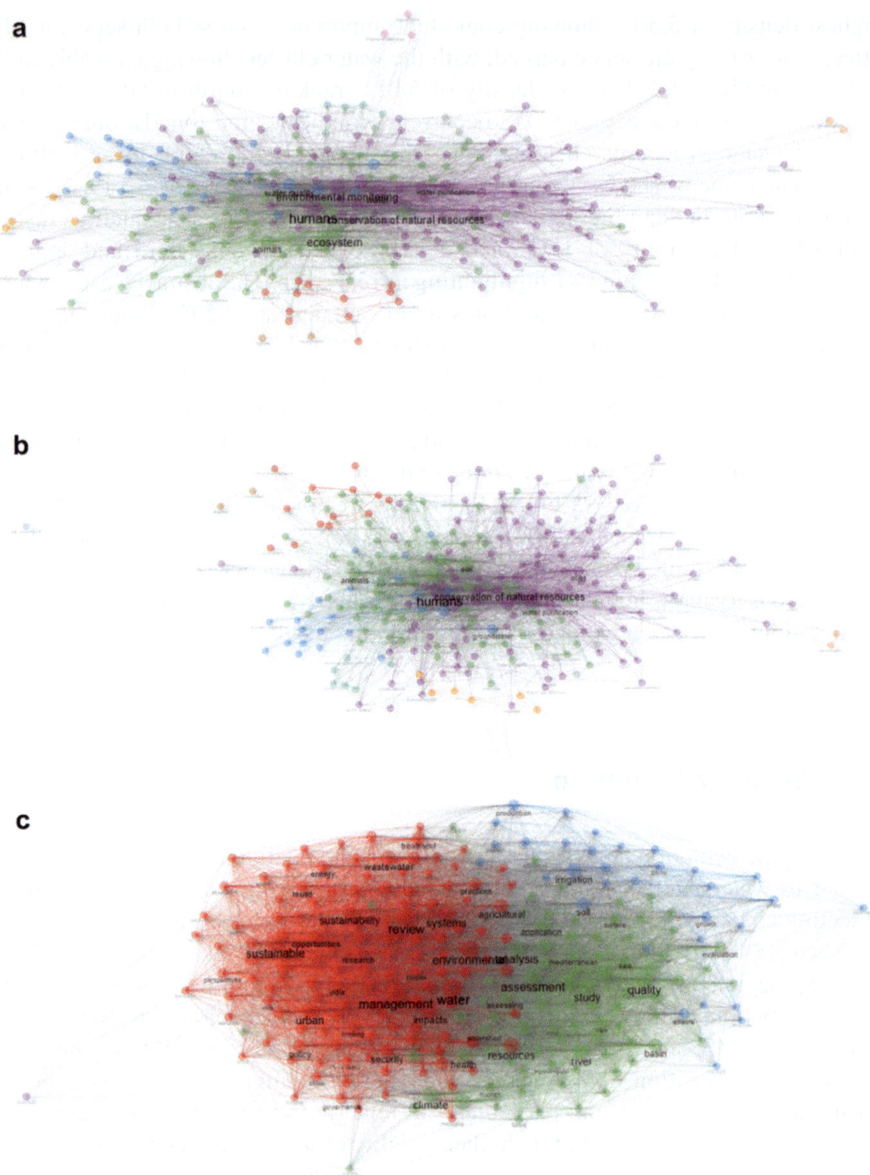

Fig. 15 Thematic mapping

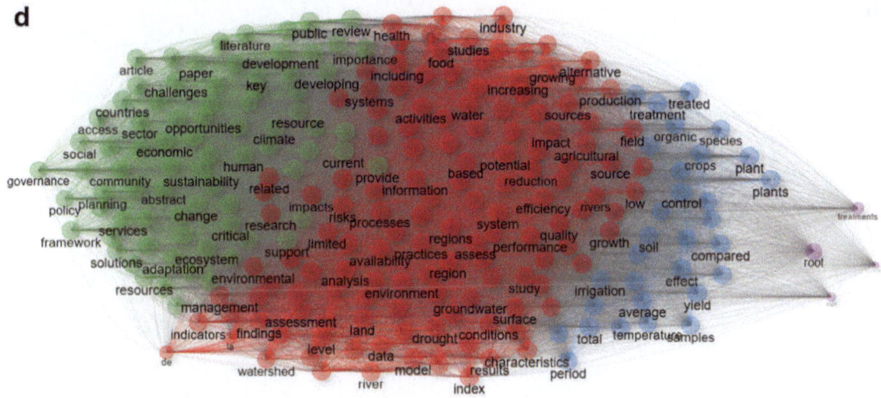

Fig. 15 (continued)

index and stability of 1.0 and 0.045, respectively. Additionally, the focus on ecosystems shifted toward soil health, suggesting a more specific concern for soil's critical role in agriculture and environmental sustainability, reflected by an increase index of 1.0 and moderate stability of 0.125. Furthermore, the transition from water supply to wastewater management underscores an increasing emphasis on issues like water purification, with a high increase index of 0.727 and stability of 0.111. Lastly, the thematic links between humans and water, including aspects of conservation of natural resources, cities, and water resources, show a growing convergence, with a high occurrence of 67 and an increase index of 0.753. In summary, the thematic evolution demonstrates a dynamic landscape of research, where early themes of urbanization and natural resource conservation have evolved into more specific focuses on climate change, wastewater treatment, and soil health, illustrating the increasing complexity and interdisciplinary nature of environmental studies.

The cluster analysis across three slices reveals significant trends in thematic focus regarding environmental studies over time. In Slice 1, the analysis identifies four clusters, with conservation of natural resources demonstrating the highest Callon Centrality (1.684), indicating its prominent role in research. It also maintains a high Callon Density of 39.54, suggesting robust interconnections with other themes. The humans cluster follows, showing a centrality of 0.518 and a density of 39.44, indicating its relevance in discussions about human interactions with the environment. The cities cluster, although less central with a Callon Centrality of 0.111, shows a noteworthy density of 44.44, indicating that while it is less prominent, it still plays a significant role in urban studies. The industry cluster, however, reflects minimal engagement, as indicated by a Callon Centrality of 0 and a frequency of only 3, suggesting limited scholarly focus. In Slice 2, the focus shifts towards the human's cluster, which gains prominence with a Callon Centrality of 0.832, indicating a strong presence in contemporary research, particularly in conjunction with themes of ecosystems (0.458) and water supply (0.654). The increase in cluster frequency for

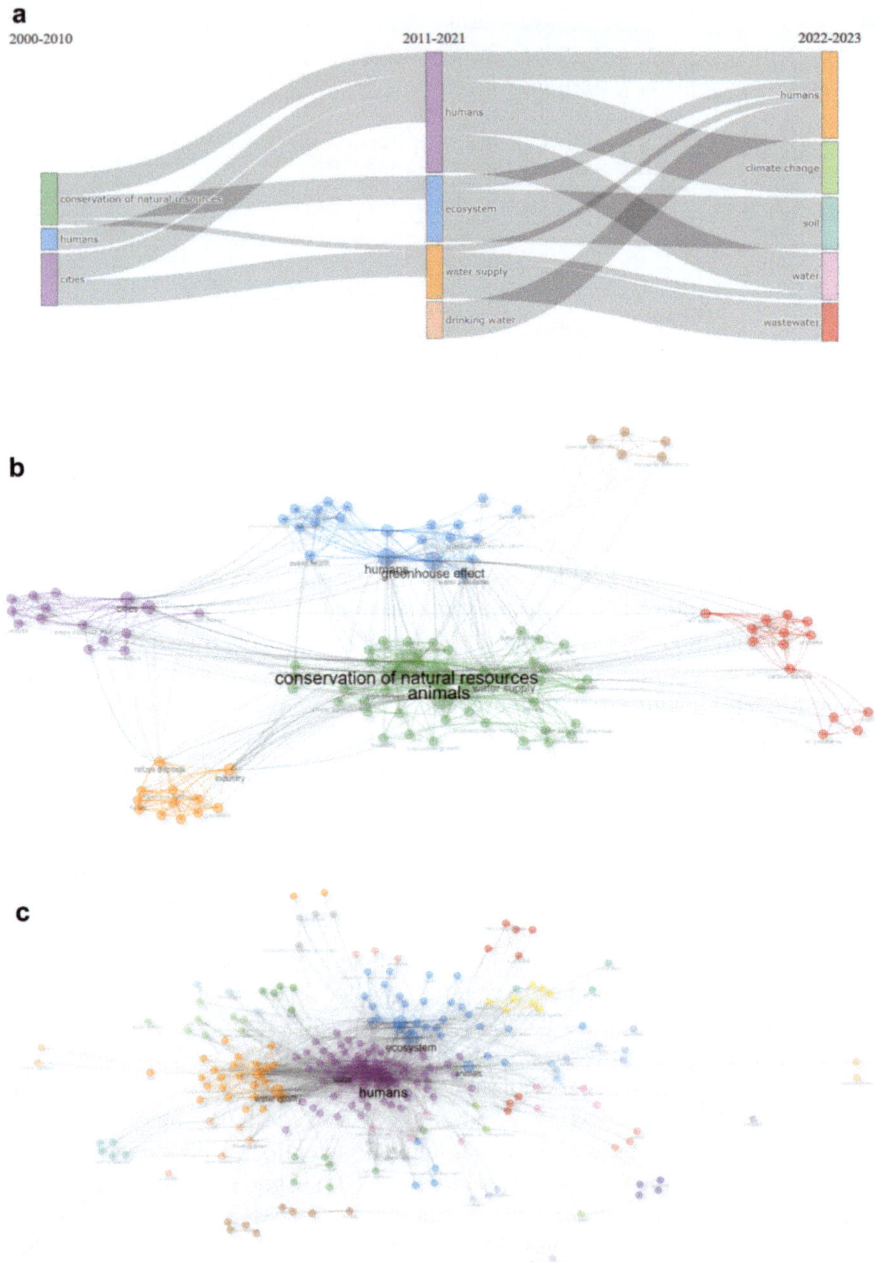

Fig. 16 **a** Thematic evolution analysis. **b** Thematic evolution analysis: slice-I. **c** Thematic evolution analysis: slice-II. **d** Thematic evolution analysis: slice-III

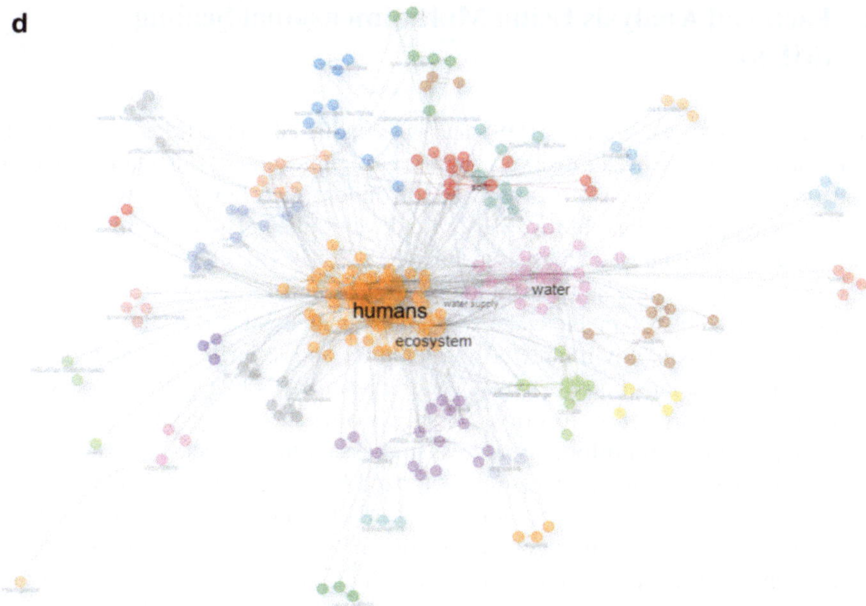

Fig. 16 (continued)

humans to 501 reflects the growing importance of human factors in environmental discussions. Conversely, the drinking water cluster remains underrepresented, with a Callon Centrality of 0.038 and a frequency of 10, indicating it is an emerging but less extensively explored theme. In Slice 3, the analysis shows a notable focus on humans, which retains a high Callon Centrality of 0.636 and a significant frequency of 239, underscoring the continued relevance of human–environment interactions. The emergence of the wastewater cluster (0.070) indicates growing concerns regarding water quality management, although its centrality remains low. The soil and water clusters also contribute to the discourse, with centralities of 0.127 and 0.502, respectively, highlighting their importance in environmental sustainability. Meanwhile, despite its lower Callon Centrality (0.064), the climate change cluster emphasizes the ongoing relevance of climate issues in environmental research. Overall, the analysis reveals a dynamic evolution in research themes, with a clear emphasis on human factors and conservation efforts, alongside growing concerns for water quality management and the impacts of climate change, reflecting the complex interplay between environmental science and societal needs.

22 Factorial Analysis Using Multidimensional Scaling (MDS)

The Multidimensional Scaling (MDS) analysis conducted on the titles and abstracts of the research corpus reveals distinct clusters, each representing a thematic grouping based on the proximity of terms across two dimensions. For the Title Analysis, the first cluster predominantly revolves around water management, quality, sustainability, and climate change themes, as given in Fig. 17a, b. This cluster includes words like water, management, quality, sustainable, climate, and development, emphasizing the critical focus on sustainable water resource management and climate resilience. Key terms like urban, environmental, agriculture, and security also highlight the interdisciplinary connections between water resources, urban planning, environmental protection, and food security. Notably, words such as systems, challenges, and evaluation highlight efforts to address and assess the impacts of these themes on global resources. Cluster two is characterized by more specific themes such as river, resources, and groundwater, indicating a shift towards localized water systems and resource management research. Terms like river, resources, and basin suggest focusing on specific water bodies and geographical regions, while evaluation and based imply analytical or data-driven research approaches. The third cluster emerges around wastewater treatment and reuse, represented by words like wastewater, treatment, and reuse, indicating a dedicated interest in water treatment technologies and sustainable reuse practices. A fourth cluster reflects broader ecosystem services, with terms like ecosystem and services suggesting the importance of natural systems in providing resources and maintaining environmental balance.

For the Abstract Analysis, the first cluster centres around water-related issues such as water quality, groundwater, soil, and irrigation, focusing on environmental challenges related to water resources, as given in Fig. 18a, b. Words like water, groundwater, soil, and irrigation reflect this focus, indicating significant research on water's interaction with natural resources and agricultural systems. Cluster two, on the other hand, focuses on management, systems, and analysis, with words like management, study, system, results, and analysis indicating a methodological focus on analyzing and managing environmental systems, particularly through studies of water resources and their sustainability. Cluster three highlights research on climate change, sustainability, and human impacts, as seen in terms like climate, sustainable, change, human, and global. This cluster captures the growing research interest in how climate change affects water resources, sustainability practices, and broader environmental impacts. It also includes terms related to economic implications (economic, challenges, and impacts), indicating interdisciplinary concerns about the socioeconomic impacts of environmental changes. The fourth cluster contains themes around agriculture, food, and health, as indicated by terms like production, agricultural, food, and health, highlighting the interconnectedness between food systems, agriculture, and human health in the context of environmental sustainability. In summary, the MDS analysis strongly emphasizes water resource management, climate change, and sustainability across both titles and abstracts, with secondary clusters highlighting specific concerns

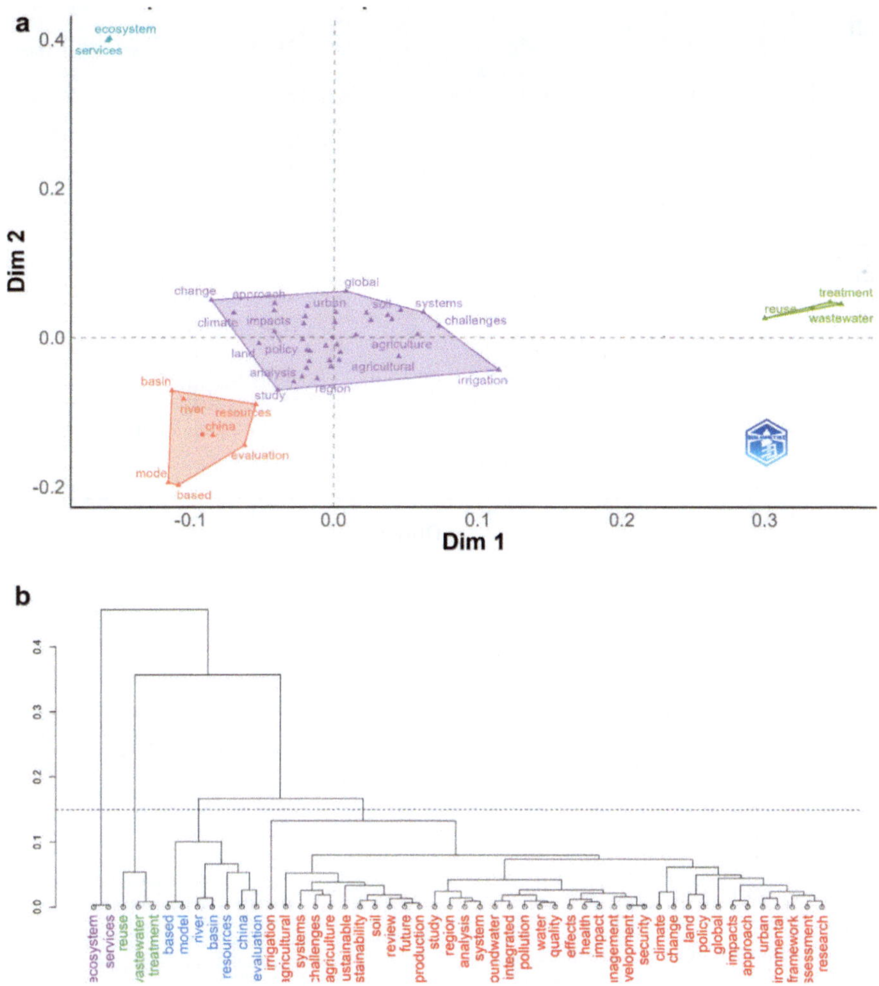

Fig. 17 **a** Conceptual structure map for the title—method MDS. **b** Hierarchical structure map for the title–method MDS

related to ecosystems, agriculture, and water treatment technologies. The clustering also reflects an interdisciplinary approach, addressing environmental, economic, and societal impacts in response to the growing challenges of sustainable development.

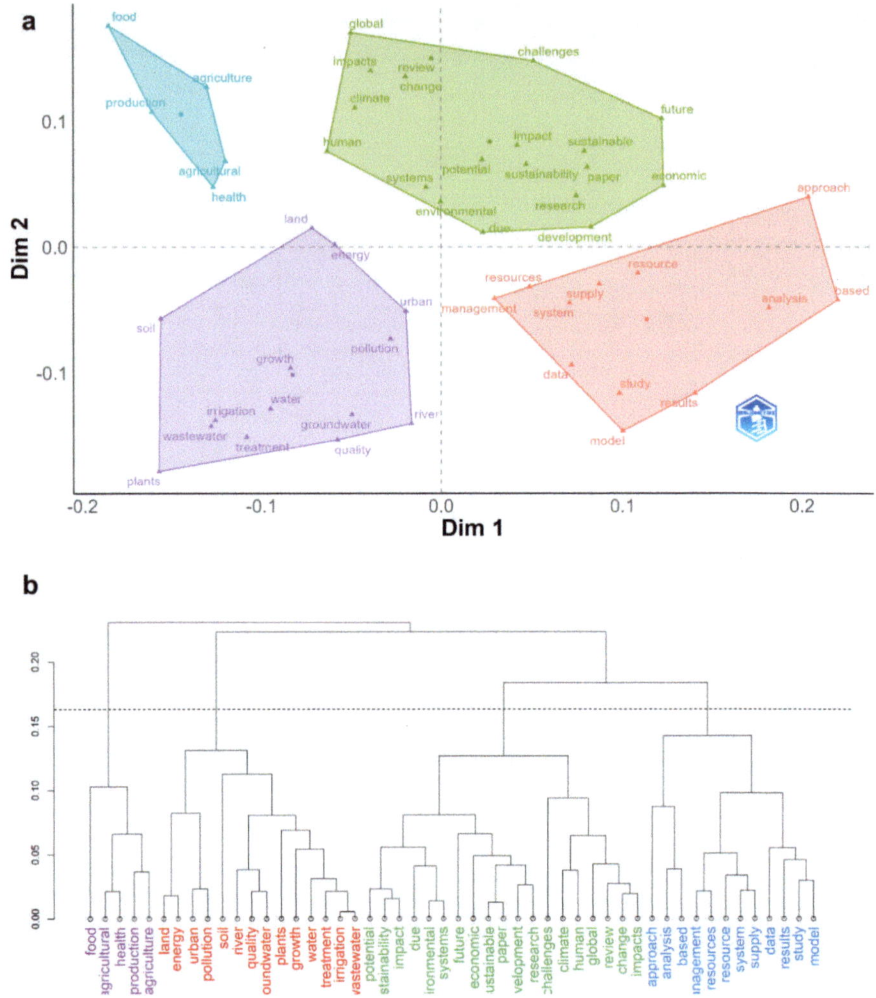

Fig. 18 **a** Conceptual structure map for the abstract—method MDS. **b** Hierarchical structure map for the abstract–method MDS

23 Author Collaboration Network

The bibliometric analysis's collaboration network output provides insight into author and institutional collaborations within the research domain; details are given in Fig. 19. For the author collaboration network, key metrics such as Betweenness, Closeness, and PageRank highlight the influence and centrality of individual authors. In Cluster 1, Kumar R stands out with a Betweenness value of 10.833, suggesting he plays a critical role in connecting different authors. In contrast, Singh S and Kumar M have lower centrality values, indicating limited inter-author connectivity. In Cluster

2, Kumar A holds significant influence with a Betweenness of 5.417, suggesting a moderate role in connecting researchers, while others in the cluster, like Kumar P and Kumar V, show lower connectivity. Cluster 3 displays a highly interconnected group, with Li H demonstrating the highest Betweenness (40.657), making him a pivotal figure in this network. Other authors such as Wang H, Wang X, and Wang J also exhibit strong centrality, showing their importance in maintaining collaboration within this cluster. In Cluster 4, Zhang Y emerges as the most influential author with a Betweenness of 54.838, closely followed by Wang Y and Chen Z, showing a dense collaboration structure. Authors in smaller clusters, such as Angelakis AN in Cluster 5, show lower connectivity but still maintain collaborative links with select researchers.

In the institutional collaboration network, institutions are divided into multiple clusters, with institutions such as M.M. and A.S. from Cluster 1 playing dominant roles, reflected in their high Betweenness values (77.015 and 80.221, respectively). These institutions seem to act as major hubs, facilitating collaborations between others. Institutions like E.S. and G.S. also exhibit significant centrality, indicating their active participation in the research network. Clusters 2 and 3 include institutions like A.L., J.B., and various organizations from Hong Kong and China, which are well connected but display lower Betweenness and Closeness values than the dominant Cluster 1 institutions. Institutions such as the University of Chinese Academy of Sciences and the State Key Laboratory of Urban and Regional Ecology are well-positioned within their respective clusters, indicating their critical role in collaborative research, especially in China-based research networks. The analysis reflects a hierarchical network with a few key authors and institutions acting as bridges between different clusters, fostering collaboration across research domains. The dense clusters, particularly Cluster 3 and 4, demonstrate an interconnected network of researchers and institutions focused on specific thematic areas.

24 Policy Recommendations for Surface Water and Groundwater Quality Development Under Climate Change

Climate change is exerting significant pressure on water resources worldwide, affecting surface and groundwater systems. The changing patterns in precipitation, increased frequency of extreme weather events, and rising temperatures contribute to water scarcity and quality deterioration. In the context of safeguarding water resources, there is a need for comprehensive policies that address surface water and groundwater quality, ensuring sustainable water management practices under climate change conditions. Bibliometric analysis of collaboration networks in water resource research provides valuable insights into global knowledge exchange, helping to guide policy development by identifying influential researchers and institutions contributing to this field. The following results from this bibliometric analysis are

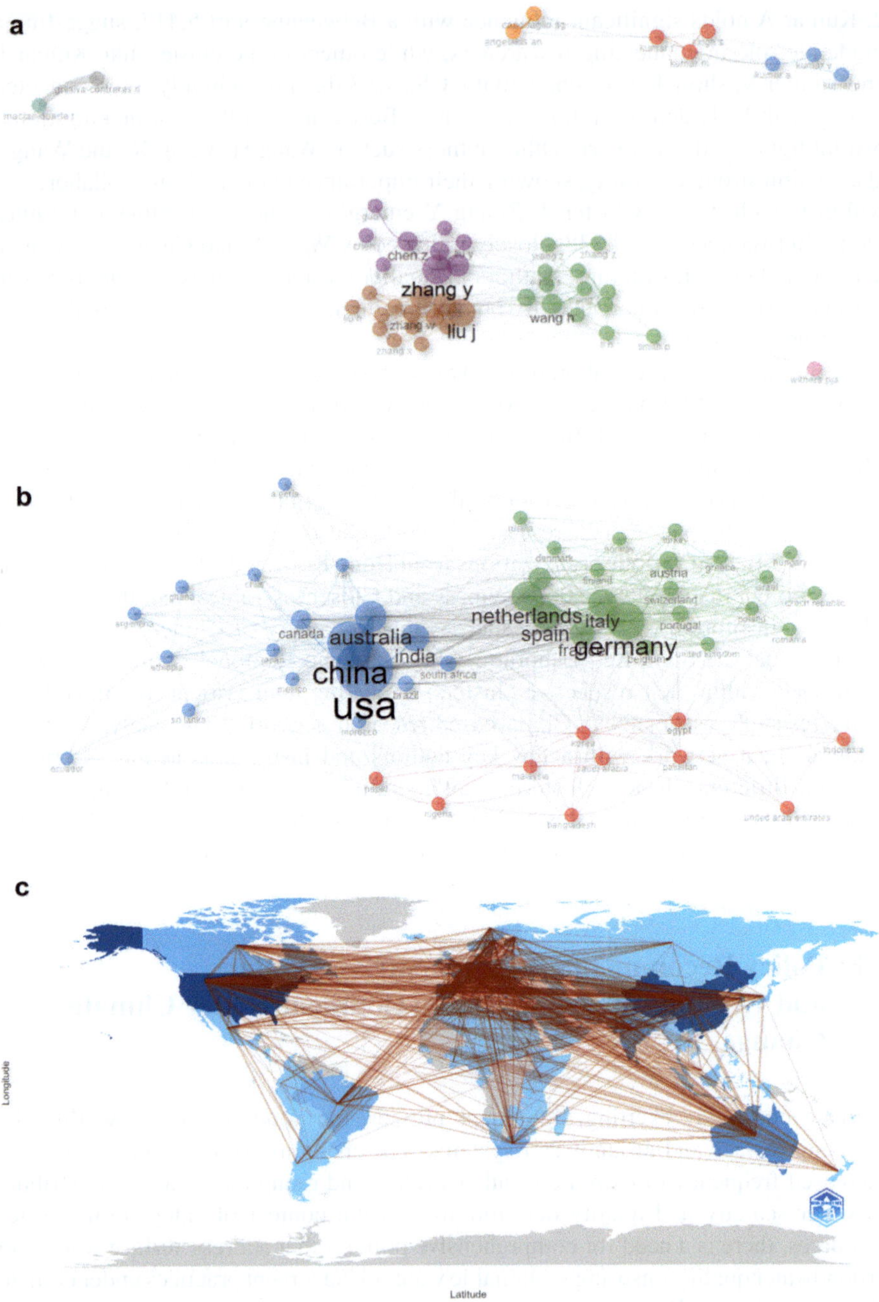

Fig. 19 Collaboration network

the policy recommendations for surface water and groundwater quality development under climate change.

1. Strengthening Institutional and Scientific Collaboration

According to the bibliometric analysis, institutions like M.M. and A.S. (with Betweenness values of 77.015 and 80.221, respectively) play pivotal roles in global water research networks, acting as major hubs for collaboration. Collaboration between scientific institutions, government agencies, and local communities must be enhanced to address water quality issues in a changing climate. This includes sharing knowledge, technological innovations, and best practices across borders to address region-specific water challenges. Strengthening the ties between institutions that exhibit high collaboration capacity, such as those in China and Hong Kong (University of Chinese Academy of Sciences and State Key Laboratory of Urban and Regional Ecology), is crucial for developing comprehensive water management strategies that are adaptable to climate variability.

2. Promoting Research in Key Areas

The bibliometric analysis highlights key research clusters contributing to understanding water resource management under climate change. For example, Cluster 3 in the author network, led by researchers like Li H and Wang H, shows a strong focus on water-related research with high Betweenness values (40.657 and 23.476, respectively). To develop effective policies, it is essential to support research in areas such as water contamination, resource efficiency, and pollution control. Government policies should fund research in these areas, particularly in understanding how climate change exacerbates issues like groundwater contamination and eutrophication in surface water bodies.

3. Enhancing Groundwater Monitoring and Regulation

Climate change increases the likelihood of groundwater depletion and contamination due to reduced recharge rates and saltwater intrusion in coastal areas. Institutions such as Shanghai Environmental Monitoring Center and Wuhan Center of Environmental Monitoring play crucial roles in water monitoring and have been identified in the institutional network analysis as key contributors to research on water quality. The policy should establish robust groundwater monitoring networks that regularly assess water quality parameters like salinity, nutrient loads, and chemical contamination. Enhanced regulation of groundwater extraction, especially in agriculture-intensive areas, must also be prioritized to prevent overexploitation.

4. Adapting Surface Water Quality Management to Climate Variability

With rising temperatures and changes in precipitation patterns, surface water bodies are vulnerable to increased nutrient loads and pollutants. Research clusters in the bibliometric analysis (Cluster 4 and Cluster 6), which include prominent authors like Zhang Y and Liu J, have contributed to the understanding of surface water pollution and its impacts on ecosystems. Policymakers must use this evidence to guide the development of adaptive management frameworks for surface water quality. This includes the introduction of buffer zones along water bodies, improving sewage treatment facilities, and encouraging the use of sustainable agricultural practices that reduce runoff. Implementing nature-based solutions, such as restoring wetlands and riparian zones, will also enhance the resilience of surface water ecosystems to climate-induced stresses.

5. Building Resilience Through Integrated Water Resources Management (IWRM)

Integrated Water Resources Management is a key approach in ensuring the sustainable use of water resources while maintaining water quality. Researchers in Cluster 2, such as Kumar A and P. Kumar, have highlighted the importance of integrated approaches in resource management. The policy should advocate for adopting IWRM frameworks at local, national, and regional levels to coordinate the management of surface water, groundwater, and associated ecosystems. IWRM should include strategies for water conservation, pollution control, and efficient water use in sectors like agriculture and industry. Climate-resilient infrastructure investments, such as constructing rainwater harvesting systems and enhancing water storage capacity, will also be critical in managing water variability.

6. Fostering Public Awareness and Community Participation

Active engagement with local communities is essential for the success of water management policies. Institutions like Mahidol University and Shanghai Municipal Center for Disease Control and Prevention, identified in Cluster 3, have been involved in community-based water management research. This policy must emphasize the role of public education in promoting water conservation, preventing pollution, and adapting to climate impacts. Public participation in water monitoring and conservation activities should be encouraged, empowering local stakeholders to take ownership of water resources. Special attention should be paid to rural and coastal areas where communities depend highly on groundwater for agriculture and drinking water.

25 Conclusion

The policy recommendations for surface water and groundwater quality management under climate change are based on an extensive review of bibliometric analyses demonstrating the importance of collaboration, research, and institutional capacity building. By promoting scientific research in key areas, enhancing collaboration among leading institutions, and implementing adaptive management strategies, the policy provides a roadmap for addressing the pressing challenges of water quality in a changing climate. Key authors and institutions identified in the bibliometric analysis, such as Zhang Y, Li H, M.M., and A.S., serve as critical knowledge hubs in the global water research community. Their contributions offer valuable insights into effective strategies for improving water quality under climate variability. Furthermore, the interconnectedness of research clusters shows the importance of multidisciplinary approaches in understanding and addressing the complex interactions between climate change and water quality.

References

Bloomfield JP, Williams RJ, Gooddy DC, Cape JN, Guha PM (2006) Impacts of climate change on the fate and behaviour of pesticides in surface and groundwater—a UK perspective. Sci Total Environ 369(1–3), 163–177

Bondu R, Cloutier V, Rosa E, Benzaazoua M (2017) Mobility and speciation of geogenic arsenic in bedrock groundwater from the Canadian Shield in western Quebec, Canada. Sci Total Environ 574:509–519

Boretti A, Rosa L (2019) Reassessing the projections of the world water development report. NPJ Clean Water 2(1):15

Donthu N, Kumar S, Mukherjee D, Pandey N, Lim WM (2021) How to conduct a bibliometric analysis: an overview and guidelines. J Bus Res 133:285–296

Earman S, Dettinger M (2011) Potential impacts of climate change on groundwater resources–a global review. J Water Clim Change 2(4):213–229

Lipczynska-Kochany E (2018) Effect of climate change on humic substances and associated impacts on the quality of surface water and groundwater: a review. Sci Total Environ 640:1548–1565

Mao H, Wang C, Qu S, Liao F, Wang G, Shi Z (2023) Source and evolution of sulfate in the multi-layer groundwater system in an abandoned mine—Insight from stable isotopes and Bayesian isotope mixing model. Sci Total Environ 859:160368

Riedel T (2019) Temperature-associated changes in groundwater quality. J Hydrol 572:206–212

Smerdon BD, Drewes JE (2017) Groundwater recharge: the intersection between humanity and hydrogeology. J Hydrol 555:909–911

Stuart ME, Manamsa K, Talbot JC, Crane EJ (2011) Emerging contaminants in groundwater

Tranfield D, Denyer D, Smart P (2003) Towards a methodology for developing evidence-informed management knowledge by means of systematic review. Br. J. Manag. 14(3):207–222

Van Eck NJ, Waltman L (2014) Visualizing bibliometric networks. In: Measuring scholarly impact: methods and practice. Cham, Springer International Publishing, pp 285–320

Zupic I, Čater T (2015) Bibliometric methods in management and organization. Organ Res Methods 18(3):429–472

Predictive Modeling of Surface and Groundwater Quality with Machine Learning Techniques

Snehil Dubey, Upasana Jha, Ashwini Suryawanshi, Radha Srivalli, and Jatoth Veeranna

Abstract Surface and ground water are primary resources used for drinking, agricultural, and commercial production purposes. However, these water sources are growingly at risk by human-made and ecological contaminants, necessitating accurately assessing their contamination intensity for robust water quality maintenance. Due to their time-consuming, labor-intensive and costly nature, traditional water quality (WQ) evaluation methods are increasingly being replaced by machine learning (ML) practices for more efficient, skillful and accurate assessments and forecasts. The review assesses ML models such as linear regression, RF, decision tree, KNN, naive Bayes, SVM and ANN for predicting the WQI based on factors like pH, turbidity, DO, TDS, and BOD. The ML model's use has surged with neural network techniques, but recent trends favor advanced techniques like deep learning and unsupervised algorithms for better accuracy. As groundwater supplies nearly 50% of worldwide drinking water, understanding water quality factors is essential for management strategies. This review evaluates surface- and groundwater water quality prediction models using unsupervised, semi-supervised, supervised, and ensemble machine learning techniques, noting the shift to data-driven approaches suited to modern computational architectures. By exploring various contaminants and a broad spectrum of algorithms, this work provides and highlights an extensive overview of ML model potential in tackling water quality forecasting, emphasizing these techniques in addressing the rising challenges in surface and groundwater resource management.

Keywords Water quality · Machine learning · SVM · Decision tree

S. Dubey (✉)
Institute of Agricultural Sciences, Banaras Hindu University, Varanasi, India
e-mail: snehildubey004@gmail.com

U. Jha
Indian Institute of Remote Sensing (IIRS), Dehradun, India
e-mail: upasana_12936@iirsddn.ac.in

A. Suryawanshi
ICAR-Research Complex for NEH Region, Basar, Arunachal Pradesh, India

R. Srivalli · J. Veeranna
College of Agricultural Engineering, Kandi, Sangareddy, Hyderabad, Telangana, India

Dimple et al. (eds.), *Surface Water and Groundwater Quality for Sustainable Utilization*, Springer Hydrogeology, https://doi.org/10.1007/978-981-96-8189-1_18

1 Introduction

Rivers are a vital source of freshwater. Current global evaluations show 8442 km^3 of water is for households, commercial production and agriculture annually (Wu et al. 2022; Singh et al. 2024a, b). 68.7% of all water is in glaciers, 1.2% is on the surface, and 30.1% is groundwater. About 1% of the world's freshwater resources can be utilized for drinking water from rivers, water reservoirs, ponds, streams and lakes (Ramaraj and Sivakumar 2023). Poorly managed Sewage Treatment Plants (STPs) release untreated sewage, along with naturally occurring minerals and chemicals such as uranium, arsenic, and radon. Additionally, local land use practices, including the application of chemicals like pesticides and fertilizers pollute water sources (Ramaraj and Sivakumar 2023). Groundwater serves as a primary drinking water source for millions, particularly in rural areas with restricted availability to surface water. Moreover, groundwater is essential for industrial and commercial activities such as manufacturing, mining, and power generation. According to the Central Ground Water Board, groundwater constitutes about 62% of irrigation, 50% of urban water supply and 85% of rural water supply. Groundwater may contain varying levels of dissolved solids. The kind and amount of these solids in water depend on things like where the water originates, the land and soil it travels through, how fast it moves, how long it stays in one area, how easily minerals can dissolve, and how much carbon dioxide is in the water. Besides human activities like sewage disposal, agricultural practices, and industrial pollution, groundwater quality is also significantly influenced. India's socio-economic development is significantly impacted by the accessibility and sustainability of its groundwater resources. For agricultural activities to be successful, maintaining good water quality is crucial. It helps decide what crops to grow, what kind of irrigation to use, and how to set up water purification for industrial use. When polluted water is used for irrigation, it can damage crops and lower their yield. It can also cause harmful substances to end up in our food, which isn't good for our health. The evaluation of water quality on surfaces and grounds is possible through generating indices based on diverse chemical and physical properties. Water Quality Index (WQI) is a widely used subjective method, which combines various water quality measurements into a single number to give an overall picture of water quality. When assessing water quality with the subjective weighting approach, we look at how important different factors are and often use the average method to weigh them. Numerous studies have been conducted following this approach. The water quality in the Santuri block's groundwater was evaluated for drinking, irrigation, and health purposes by Chakraborty et al. The assessments, using the Heavy Metal Pollution Index (HPI) and Groundwater Quality Index (GWQI), revealed a significant decline in water quality during the prior to monsoon as opposed to post-monsoon, mainly due to increased ionic activity in the summer. Advanced analytical techniques are essential for enhancing land and water management by providing accurate and efficient predictions of water quality, crucial for sustainable resource management (Suryawanshi et al. 2021, 2023; Singh et al. 2024a, b). Monitoring a large area involves substantial costs and time due to intensive fieldwork, water sampling, and laboratory analysis (Sahour et al. 2023). A lot of researchers have used ML methods like Artificial Neural Networks, XGBoost, and Random Forest in different water-related studies (Sahour et al. 2023; Elbeltagi et al. 2022; Singh et al. 2024a, b). Making predictions for both the short and long term is key to managing groundwater and figuring out what it means for the future. Machine learning models, a kind of artificial intelligence, look at data to spot patterns and learn from them, without needing a set formula. Singha et al. assessed groundwater quality in the Arang region of Chhattisgarh, India, by measuring diverse physicochemical characteristics and computing an entropy weight-based groundwater quality index (EWQI). A deep learning model was proposed and compared with other ML models: Artificial Neural Network (ANN), Random Forest (RF) also eXtreme Gradient Boosting (XGBoost). The deep learning model surpassed other models with the highest accuracy ($R^2 = 0.996$). The models ranked in performance as follows: DL > XGBoost > ANN > RF. Sahour et al. (2023) applied six different machine learning methods—eXtreme Gradient Boosting (XGB), k-Nearest Neighbors (KNN), Support Vector Machine (SVM), Random Forest (RF), Artificial Neural Networks (ANN), and Gaussian Classifier Model (GCM) to study how different factors affect the Groundwater Quality Index (GWQI). RF, SVM, and XGB performed the best, with Random Forest being the best performer. Elbeltagi et al. (2022) estimated groundwater quality in the Akot basin, Akola district, Maharashtra, India, four methods were used: Random Subspace (RSS), Support Vector Machine (SVM), Additive Regression (AR) and M5P Tree Model (M5P). The AR method turned out to be the best, with strong results and fewer errors, thanks to its use of key parameters like EC, TDS, pH, Cl, Mg, and Ca. This makes AR a reliable and precise choice for predicting groundwater quality in that area. Singh et al. (2024a, b) tested five different methods—Decision Tree, Gradient Boosted Trees, Random Forest, Support Vector Machine, and Artificial Neural Networks—to classify water quality categories for Indian rivers. They used a dataset with 2,990 records and six water quality parameters: pH, Dissolved Oxygen, Electrical Conductivity, Nitrate-Nitrogen, Biological Oxygen Demand, and Total Coliform. The Gradient Boosted Trees model performed the best, showing the highest accuracy, recall, precision, and F-score. Shah et al. examined water quality in the upper Indus River, researchers used Linear Regression Models (LRM), Artificial Neural Networks (ANN), Gene Expression Programming (GEP) and to project monthly Total Dissolved Solids (TDS) and Specific Conductivity (EC) at two points. Both GEP and ANN were effective, but GEP was the most accurate and had unique equations that made it stand out from the others. GEP's equations are straightforward for predicting TDS and EC levels each month. While ANN also performed well, its accuracy dropped with new data, probably due to its complex setup. Bedi et al. compared ANN, SVM, and XGB models for predicting nitrate and pesticide contamination in groundwater, XGB came out on top with the most accurate predictions. The author also suggests that using spatial data and mixing geostatistical methods with machine learning could make predictions even better and offer more insights into contamination patterns. This book chapter explores the potential of Machine Learning algorithms for surface and groundwater quality evaluation and forecasting.

2 Overview of Water Quality Indicators

Water quality is a key factor in maintaining environmental balance, and understanding the measures that define the quality of water is crucial for actual monitoring and control. The information offers an in-depth analysis of major water quality measures for both surface and groundwater. The importance of these indicators, the range of values they can exhibit, its effect on water quality, and the role of machine learning models in estimating these values are discussed.

Water quality is assessed by looking at several key factors, each of which can make the water unsafe for specific uses if it goes beyond certain limits. Patnaik outlines the properties of various hazardous substances that can affect water quality. To get a better handle on what matters most, the NSF surveyed 142 experts—scientists and managers working in water quality. Various factors include dissolved oxygen, fecal coliform levels, pH balance, five-day biochemical oxygen demand, temperature variation over a mile, total phosphates, nitrates, water clarity (turbidity), and total suspended solids (Tables 1 and 2).

Table 1 Parameters of water quality (*Source* Tchobanoglous et al. 1985)

No.	Water quality indicators		
	Physical indicators	Chemical indicators	Biological indicators
1	Turbidity	Acidity, alkalinity and pH	Bacteria
2	Temperature	Chloride, chlorine residual	Algae
3	Color	Sulfate, nitrogen, fluoride	Viruses
4	Taste and odor	Iron, manganese, copper and zinc, hardness and radioactive substances	Protozoa
5	Solids	Dissolved oxygen, biochemical oxygen demand (BOD), and chemical oxygen demand (COD)	–
6	Electrical conductivity (EC)	Toxic organic and inorganic substances	–

Table 2 Categorization of water based on hardness

Water category	Total hardness level in mg/L as $CaCO_3$
Softened water	<50 mg/L (expressed as $CaCO_3$)
Medium hardness	50–150 mg/L (expressed as $CaCO_3$)
High-hardness water	150–300 mg/L (expressed as $CaCO_3$)
Extremely hard	>300 mg/L (expressed as $CaCO_3$)

2.1 Surface Water Quality Indicators

Surface water quality indicators provide insight into the health of lakes, rivers, and oceans. They help in measuring the influence of natural and anthropogenic actions on aquatic ecosystems.

2.1.1 pH

A pH quantifies the acidity or basicity of water. It influences biological processes, chemical reactions, and the solubility of nutrients. 0 (highly acidic) to 14 (highly alkaline), with a neutral point at 7 (Spellman 2017). The low pH (acidic) leads to metal toxicity, harmful to aquatic life whereas high pH (alkaline) causes ammonia toxicity, reducing oxygen solubility. Checking the pH of water is really important because it affects both its chemical makeup and how living things interact with it. pH levels can vary depending on where the water comes from, like runoff from land, groundwater, or even streams coming out of forests, where organic acids and materials can cause the pH to change. Machine learning models can predict pH based on other water quality parameters using regression procedures like Random Forest Regression or Support Vector Regression (SVR).

2.1.2 Turbidity

Turbidity in drinking water occurs when tiny particles make the water look cloudy. This can happen if the water isn't filtered well, if sediment gets stirred up in the pipes, or if there are particles in the groundwater. It can also be caused by biofilm breaking loose inside the water system. Turbidity measures water clarity and is an indicator of suspended particles such as soil, algae, and microorganisms. The Turbidity measured in Nephelometric Turbidity Units (NTU) or TU. One TU equals 1 mg/L of suspended silica (APHA 2005), ranges from 0 (clear) to several hundred NTU (murky). The low turbidity indicates clear water, good for aquatic life. In other hands high turbidity can block sunlight, reducing photosynthesis, and may indicate pollution. Turbidity higher than 5 NTU may be seen in the average individual whereas turbidity in mud water, over 100 NTU (APHA 2005). Turbidity can be estimated using image processing techniques and machine learning models like Convolutional Neural Networks (CNNs).

2.1.3 Dissolved Oxygen (DO)

The DO is very important for the survival of aquatic organisms and reflects the water's ability to support life. It's an important test for water pollution (APHA 2005). The higher the concentration of DO the better the quality of water. It is measured in milligrams per liter (mg/L); typical values range from 5 to 14 mg/L. The DO may lead to hypoxia, stressing or killing aquatic life and high DO can cause gas bubble disease in fish.

The dissolved oxygen measurement methods include the colorimetric methods,

- Winkler titration technique (Quick and economical),
- Electrometric technique, and
- Traditional technique

Predictive models like Artificial Neural Networks (ANNs) can be applied to assess DO levels based on environmental data. The level of dissolved oxygen in water can vary depending on things like pressure, temperature, and how salty the water is.

2.1.4 Total Dissolved Solids (TDS)

The TDS indicates the concentration of dispersed constituents in water, with salts, minerals, also organic substances. TDS is expressed in parts per million (ppm); for drinking water it ranges from 0 to 500 ppm. Total Dissolved Solids (TDS) are the bits of substances like sodium, calcium, and chloride that stay behind when water evaporates. The low TDS is generally safe but may lack essential minerals and high TDS leads to water hardness, affecting taste and corroding pipes. Regression models can predict TDS based on other water quality parameters.

$$\text{Total solid (TS)} = \text{Total dissolved solids (TDS)} + \text{Total suspended solid (TSS)}.$$

Water is mainly grouped by the quantity of TDS/liter and is given in the following way:

For freshwater, it is below 1500 mg per liter TDS for brackish water it is 1500–5000 mg per liter TDS and for saline water it is beyond 5000 mg per liter TDS.

2.1.5 Biochemical Oxygen Demand (BOD)

The BOD is a portion of how much oxygen microbes need to break down organic stuff in the water. If the water doesn't get enough oxygen, either from natural sources or added artificially, the oxygen levels will drop as the microbe's work. This need for oxygen is known as BOD. The BOD is measured in mg/L; higher BOD indicates more organic pollution and strong sewage. The low BOD indicates clean water with low levels of organic matter. The higher BOD suggests high levels of pollution, which can lead to oxygen depletion. BOD is a time function, when $T = 0$, oxygen consumption will be zero and the BOD will be zero. As day passes by, microbes utilize oxygen and the BOD increases. Time-series analysis and regression models can also be employed to estimate BOD levels from historical data (Fig. 1).

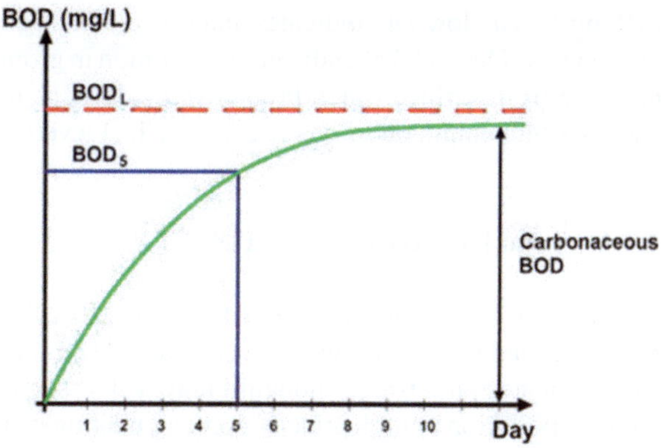

Fig. 1 BOD curve *Source* Omer (2019)

2.2 *Groundwater Quality Indicators*

Groundwater quality indicators are vital for assessing the safety also usability of underground water sources, which are critical for drinking water supply and irrigation.

2.2.1 pH

Similar to surface water, pH in groundwater affects corrosion, nutrient availability, and microbial activity. Similar to surface water (0–14), with groundwater typically slightly more alkaline. The low pH can cause leaching of toxic metals and high pH lead to scale formation in pipes and poor water taste. Groundwater pH can be predicted using multivariate analysis and regression models.

2.2.2 Turbidity

Turbidity in groundwater indicates the presence of suspended particles, often from contamination. Groundwater typically has low turbidity (<1 NTU), but higher levels may indicate contamination. Low turbidity shows clean groundwater, suitable for consumption, whereas high turbidity may indicate contamination from surface water infiltration. Predictive models can be developed using historical data to forecast turbidity levels.

2.2.3 Dissolved Oxygen (DO)

The DO in groundwater is measure of water's capacity to support aerobic microbial activity. Typically, lower in groundwater than surface water, ranging from 0 to 2 mg/L. The low DO indicates anaerobic conditions, possibly leading to harmful byproducts. The high DO indicates uncommon in groundwater but can indicate recent recharge from surface water. Predictive models can use groundwater characteristics and environmental conditions to estimate DO levels.

2.2.4 Total Dissolved Solids (TDS)

The TDS in groundwater reflects the geochemical processes and the presence of pollutants and it can vary widely; acceptable levels are generally <500 ppm. The low TDS indicates pure water, though it may lack essential minerals. The high TDS can indicate pollution or high mineral content, making water unsuitable for drinking. ML models can investigate geological and hydrological data to predict TDS levels.

2.2.5 Biochemical Oxygen Demand (BOD)

The BOD in groundwater is generally low, indicating minimal organic pollution, and its ranges typically less than 1 mg/L. The Low BOD indicates clean groundwater and high BOD uncommon but may suggest contamination from surface water or waste sources. Predictive models can estimate BOD based on surrounding land use and groundwater characteristics.

3 Machine Learning Process and Techniques

Machine learning is an advanced analytical method to handle data analysis automatically. It's part of artificial intelligence and lets systems learn from data, build models, and make decisions on their own, without needing much human help. In this case, it's used to predict and classify water quality. Machine learning is incredibly useful across different scientific fields. Researchers everywhere use methods like Random Forest (RF) and Artificial Neural Networks (ANN) to study water quality, among other things. There are also many other models out there, each suited to different tasks and needs. They give quicker and more reliable results than traditional surface and groundwater quality monitoring techniques. The inputs required for processing any machine learning model to assess water quality is shown below (Fig. 2).

3.1 Linear and Multilinear Regression Models

Linear Regression (LR) is a core quantitative method used to understand how different water quality indicators relate to the overall WQI. It's particularly useful in environmental studies where relationships are often linear or nearly linear. LR seeks

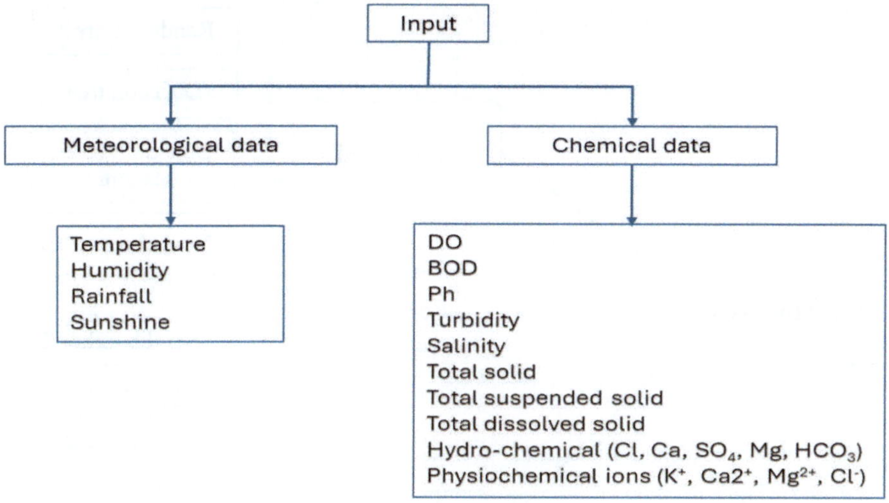

Fig. 2 Inputs categorization for machine learning model processing for surface and groundwater monitoring

to discover the best-fit line or hyperplane which shows how input variables impact an output variable. For water quality monitoring, LR can be applied in two ways: Simple Linear Regression (SLR) and Multiple Linear Regression (MLR). SLR inspects the link between one independent variable, like a contaminant level, and the dependent variable, such as WQI. MLR, on the other hand, considers multiple independent variables like pH, turbidity, and dissolved oxygen. To use LR for monitoring water quality, you start by identifying relevant variables, including dependent variables like WQI, and independent variables such as chemical and physical indicators. After collecting and cleaning the data, you build the LR model, either with SLR for single indicators or MLR for multiple indicators (Irwan et al. 2023). The model is validated by splitting the data into training and test portions, estimating coefficients, and evaluating performance analyzed through metrics like R-squared, Mean Squared Error (MSE), and Root Mean Squared Error (RMSE). Positive coefficients suggest better water quality, while negative ones indicate poorer quality. The validated model helps predict water quality trends and supports informed water management decisions.

However, LR has limitations. It assumes linear relationships, which may oversimplify complex, nonlinear interactions in environmental systems. LR models are also very sensitive to multicollinearity and outliers, that can skew results. Additionally, they assume constant variance in errors and don't naturally account for interactions between variables. Despite these challenges, LR is a useful tool for water quality monitoring, and combining it with other advanced techniques could further improve its accuracy (Fig. 3).

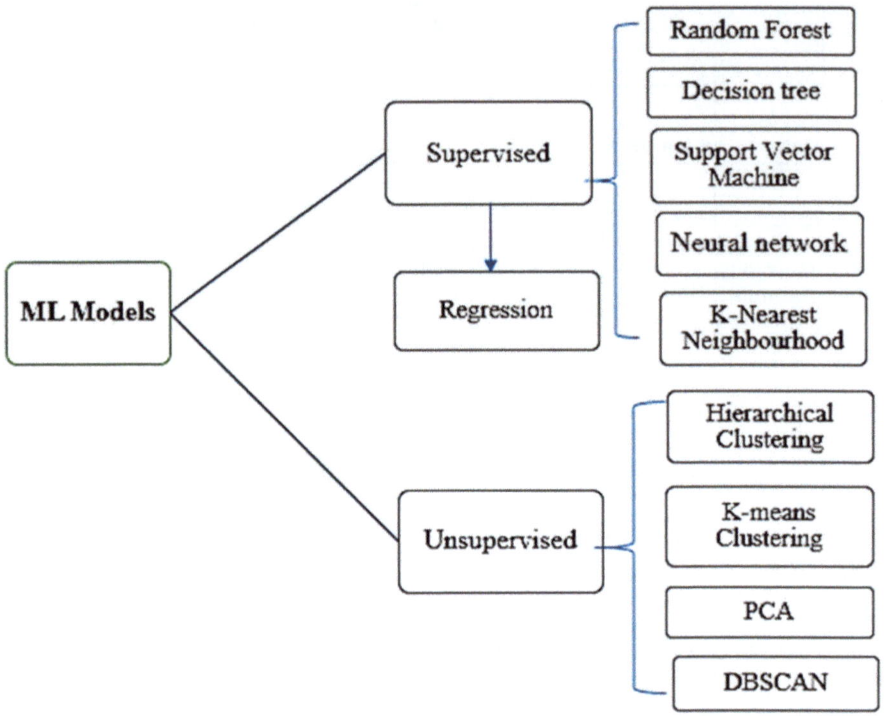

Fig. 3 Flowchart of ML models for surface and groundwater quality prediction

3.2 Decision Tree

A Decision Tree is a handy tool in ML applicable for both classifying data also making predictions. It simplifies complex decisions by breaking them down into a series of yes/no questions, more of creating a tree-like diagram having a root node and a branch node, in which each branch signifies a decision created on definite criteria, and each leaf shows the outcome. This clear structure makes Decision Trees easy to understand, with the Gini Index used for categoriztion and Mean Squared Error (MSE) for regression determining how the data is split (Breiman et al. 1986).

In terms of monitoring surface and groundwater quality, Decision Trees can be very effective. First, decide what you want to achieve—whether it's classifying water quality like "Safe" or "Unsafe," or predicting specific values such as contaminant levels. Then, identify your variables: the outcome you're interested in (like a water quality index or pollutant level) and the factors that might affect it (such as pH, turbidity, and other measurements). After collecting and preparing your data, you build your Decision Tree model using these inputs. You'll then check how well the model performs using metrics like accuracy for classification or MSE for regression.

A study found that Decision Trees (DT) worked better and were more efficient at predicting water quality categories compared to principal component analysis

(Saghebian et al. 2014). A study using Decision Trees (DT) came up with an earlier and more affordable way to calculate and predict Water Quality Indices (WQIs) (Ho et al. 2019). One of the strengths of Decision Trees is their simplicity and their ability to handle both numerical and categorical data, which is useful in environmental monitoring. The model proved to be accurate in forecasting WQI classes and is useful for comparing with other models (Kisi and Parmar 2016). Decision Trees are efficient, flexible, and handle missing data well. While other models might be quicker, Decision Trees are especially good at making short-term predictions. However, they can sometimes overfit, meaning they might work great with your specific data but not so well with new data. They can also be unbalanced, with minor variations in the data potentially important to diverse tree arrangements (Breiman et al. 1986; Zhang and Zhang 2021). Despite these drawbacks, Decision Trees are a valuable tool for water quality monitoring and can be even more powerful when used alongside other methods.

3.3 Random Forest (RF)

Random Forest (RF) develops several decision trees, each from a different sample of the original data using bootstrapping (sampling with replacement) (Voyant et al. 2017). Unlike traditional trees, which pick the best split from all available variables, RF adds some randomness by choosing a random subset of variables for each split. This randomness helps RF be more robust against overfitting compared to other methods. Random Forest (RF) is a technique that uses a bunch of decision trees to classify data. Rather than depending on a single tree, it combines the predictions from multiple trees to make a final decision. This method helps reduce errors and makes the model more accurate, especially as you add more trees. Each tree in the RF uses the CART procedure to sort the data based on what you're trying to predict.

Random Forests (RF) have proven to be quite effective for various surface and groundwater quality tasks. They've been used to predict levels of pollutants like nitrate and arsenic, assess the vulnerability of aquifers and track sources and flow paths of dissolved organic nitrogen (Wang et al. 2018). Combining RF with fuzzy methods has accurately determined both the Groundwater Quality Index (GQI) and Water Quality Index (WQI) (Norouzi and Moghaddam 2020). RF has also excelled in modeling fluoride in India and arsenic in Uruguay (Podgorski et al. 2018; Wu et al. 2021). Moreover, RF models paired with GIS have achieved high accuracy in tracking groundwater nitrate contamination over time and space (Judeh et al. 2022). A specific method for handling large datasets with RF has enabled national-scale predictions, such as in New Zealand. When combined with Shapley Additive exPlanations (SHAP), RF was found to be more accurate than traditional kriging for nitrate modeling (Li et al. 2022). RF is quite handy for monitoring surface and groundwater quality. It combines the results of many decision trees to give a more accurate picture of water quality, which helps in dealing with complex data from various indicators like pH, turbidity, and contaminants. It's good at handling missing data, which is often an issue in environmental studies, and generally avoids overfitting, meaning it can work well even with detailed datasets. On the other hand, RF can be a bit tricky to interpret. If you need clear explanations for water quality results, RF might not be the best choice since it's a bit of a "black box." It also requires a lot of computational power and memory, which can be a challenge with large datasets. RF might struggle with data that's imbalanced, where some issues are rare but important. Plus, if not set up carefully, it might not capture changes in water quality over time, like seasonal variations. Despite these challenges, RF remains a powerful tool for understanding and predicting the quality of water, as long as you're aware of its limitations.

3.4 K-Nearest Neighbor (KNN)

K-Nearest Neighbors (KNN) classifies a sample by looking at its nearest neighbors and picking the class that's most common among them. If there's a tie between classes, there are several ways to decide. However, KNN isn't great for large datasets because it must compare each new sample to every single piece of training data, which can be very slow and inefficient (Beyer et al. 1999). K-Nearest Neighbors (KNN) is a handy tool for keeping an eye on surface and groundwater quality. It works by comparing a new water sample to past samples that are similar, helping to figure out if the water is clean or if it has contaminants. For example, if you want to know if a new water sample meets safety standards, KNN looks at the most similar past samples to make a prediction.

One big advantage of KNN is that it's quite flexible and doesn't require you to set up a complicated model in advance. It can handle complex patterns in the data without needing a specific formula. Plus, KNN works well in real-time scenarios since it doesn't need a separate training phase; it uses the most recent data to make predictions as new samples come in. However, KNN does have some downsides. It can become slow and resource-intensive with large datasets because it must compare each new sample with every single past sample. This can be a problem if you need quick results or are dealing with lots of data. Also, KNN can struggle with noisy or irrelevant data, so it's important to clean and prepare the data carefully. Despite these challenges, KNN is still a useful method for water quality monitoring due to its ease of use and ability to handle complex data patterns.

3.5 Support Vector Machine (SVM)

Support Vector Machines (SVMs) can categorize data and predict number-based values in machine learning. They're commonly known for their ability to categorize data into different classes, but they also work well for tasks like forecasting trends (Irwan et al. 2023). SVMs distinguish different data categories by generating a hyperplane boundary. They try to make this boundary as wide as possible to reduce errors and differentiate between categories. Focusing on support vectors helps prevent overfitting. SVMs can also handle problems where there are more than two categories by adjusting parameters to accurately classify multiple classes. Support Vector Regression (SVR), a variation of SVM, is considered for more complex tasks like predicting values or evaluating functions. It's effective at solving difficult optimization problems and offers advantages such as avoiding local errors, having a solid mathematical foundation, and being scalable. However, SVR needs careful tuning of its parameters and a well-labeled dataset to perform well.

3.6 Artificial Neural Networks (ANN)

Artificial Neural Networks (ANNs) utilize artificial neurons to process information and solve complex problems, modeled after the human brain's neural network structure. These networks replicate the brain's learning and adapting abilities to tackle complex tasks. Imagine ANNs as a system that learns from examples: you give it lots of data, and it adjusts itself to make sure its predictions match what's been observed in the past. To be effective, the model uses a special function called the hyperbolic tangent sigmoid function to handle the data calculations. ANNs need a lot of examples to learn from. ANNs are modeled after the brain's neural networks and learn from data by automatically adjusting their settings. Each ANN has three main parts: an input layer that takes in data, hidden layers in the middle that do the heavy lifting, and an output layer that gives the result (Najah et al. 2013). The hidden layers work together to understand and process the data. For surface and groundwater quality monitoring the input layer has four neurons that take in data about turbidity, coliforms, temperature and pH. The hidden layers, which are adjustable, process this data. You can adjust the number of hidden layers and neurons per layer for optimal results. Finally, the output layer gives you the model's prediction (Azamathulla et al. 2011).

Nowadays, they're being used in environmental fields to predict things like groundwater levels, weather temperatures, water flow, and water quality. It's handy because it doesn't need a deep understanding of the system's inner workings and can handle problems with lots of variables easily. Artificial Neural Networks (ANNs) with Quick Propagation (QPANN) have been used to tackle groundwater salinity issues. By training the model with two years of real-time data, researchers developed a pumping strategy that keeps groundwater salinity below 2.5% of seawater for up to five years (Banerjee et al. 2011). In another practical case, combining a wireless water quality monitoring system with an ANN allowed for present updates on quality of groundwater. Newer ANN designs, like Radial Basis Function Neural Networks (RBFNN) combined with fuzzy clustering, have also been shown to improve the groundwater quality predictions accuracy (Azimi et al. 2019).

4 Unsupervised Learning Techniques

Unsupervised learning constitutes a part of machine learning that works with data that doesn't come with predefined labels or categories. The main objective is to uncover the underlying patterns, relationships, and structures within the data on its own, without being given explicit instructions on what to look for. Unsupervised learning algorithms are designed to group similar data points (clustering), reveal hidden structures (dimensionality reduction), or identify recurring patterns. In contrast to supervised learning, which relies on labeled data to teach the model what the correct outputs should be, Unsupervised learning processes data without any preconceived notions of the outcome. This approach is especially useful for handling complex datasets where labeling data is difficult, costly, or time-consuming. It helps in discovering insights and features directly from the data, making it invaluable for tasks where the structure of the data is unknown or where patterns need to be found without human intervention.

Various unsupervised learning algorithms have been used to predict and monitor surface and groundwater quality. These algorithms are effective at spotting patterns, grouping regions with similar water quality, and detecting unusual data points, which makes them valuable tools for environmental monitoring and helping guide decisions about water management. Here's a breakdown of some popular unsupervised algorithms and how they've been applied to water quality modeling:

4.1 K-means Clustering

K-means clustering, recognized for its simplicity, is frequently employed for clustering. It minimizes the variance within each cluster to group a dataset into K clusters. The procedure arbitrarily initializes K centroids, then iteratively allocates each observation to the nearest centroid and reassess the centroids built upon the new assignments. This procedure extends until the centroids no longer alteration significantly, resulting in distinct clusters where similar observations are grouped together. K-Means clustering has been applied in various studies to identify regions with similar groundwater quality profiles. For example, Xu et al. (2019) used K-Means clustering to classify groundwater in China's Guanzhong Basin based on chemical properties, grouping areas into low, moderate, and high contamination zones.

4.2 Hierarchical Clustering

In dendrograms, hierarchical clustering forms a tree-like hierarchy of clusters. Agglomerative hierarchical clustering and divisive hierarchical clustering respectively represent the bottom-up and top-down approaches. Agglomerative clustering begins with every datum as a separate cluster and subsequently merges the most similar clusters. In contrast, divisive clustering begins with one large cluster and successively splits it into smaller clusters. The result is a nested clustering structure that reveals relationships between clusters at different levels. Hierarchical clustering has been used to group groundwater samples based on their chemical compositions. Kshetrimayum et al. (2020) applied hierarchical clustering to classify water bodies in India, grouping them by their pollution levels based on parameters like pH and biochemical oxygen demand.

4.3 Principal Component Analysis (PCA)

Principal Component Analysis simplifies complex data by compressing it into a smaller set of variables, known as principal components. These components distill the essential patterns from the data, simplifying comprehension while retaining significance. PCA finds the main directions where the data varies the most and makes sure each new direction is independent of the others. This is especially useful in groundwater studies where there are many chemical indicators. For instance, Zhou et al. (2018) used PCA to make sense of groundwater data in the Pearl River Delta, helping to identify major contamination sources like industrial waste and urban runoff.

4.4 DBSCAN (Density-Based Spatial Clustering)

DBSCAN marks data points as part of a cluster if they meet a minimum density requirement and separates those that do not meet this requirement as noise. DBSCAN does not need the predetermined number of clusters like K-Means does. Instead, it groups data points based on the density of points in their neighborhood, making it particularly useful for detecting anomalies and outliers in datasets with complex structures. DBSCAN has been applied to identify dense clusters of contamination in urban groundwater systems. For example, Li et al. (2021) utilized DBSCAN to detect hotspots of surface water contamination, distinguishing between heavily polluted and sparsely polluted areas.

5 Evaluation Metrics

5.1 Accuracy

Accuracy measures how well a model's predictions usually align with the actual outcomes. It is a measure of overall correctness or how often the model gets the right answer.

$$Accuracy = \left[\frac{Number\ of\ correct\ predictions}{Total\ number\ of\ predictions}\right]$$

Suppose a model that predicts water quality, and evaluates 100 samples. If 85 of these predictions are correct, the accuracy is 85%. This shows that the model is generally reliable in making correct predictions, but it doesn't reveal how well it distinguishes between contamination and non-contamination. Accuracy is used to get a broad sense of how well a model performs. For example, Yin et al. evaluated models for classifying water samples into safe or contaminated categories, where high accuracy indicated that the model reliably predicted water quality overall.

Limitations

Accuracy can be misleading if the dataset is imbalanced. For instance, if only a small percentage of samples are contaminated and the model predicts all as clean, it might still show high accuracy, even though it's missing the contamination cases.

5.2 Precision

Precision is the percentage of correctly identified positive cases by the model. It's a way to check the reliability of the model when it estimates contamination. Essentially, it tells you how often the model's positive predictions are correct.

$$Precision = \frac{TP}{TP + FP}$$

- True Positives (TP): The model successfully recognizes positive cases.
- False Positives (FP): The model unsuccessfully recognizes negative cases as positive.
- True Negatives (TN): The model successfully recognizes negative cases.
- False Negatives (FN): The model unsuccessfully recognizes positive cases as negative.

If a model predicts 30 samples are contaminated, but only 20 of these predictions are correct, the precision is 67%. This indicates that when the model flags a sample is contaminated, it's correct 67% of the time. Precision is crucial when the cost of false positives is significant. Pereira et al. (2019) used precision to assess the performance of models detecting contaminants in groundwater. High precision means that the model's identifications are likely to be correct, which helps avoid unnecessary actions or treatments based on false alarms.

Limitations

Precision doesn't consider how many actual positive cases were missed. A model might have high precision but still miss many true contamination cases.

5.3 Recall

The model's ability to accurately identify positive cases is known as its recall or sensitivity. It's about capturing all possible positive cases and is crucial when it's important to detect every instance of contamination.

$$Recall = \frac{TP}{TP + FN}$$

- True Positives (TP): Correctly identified positives.
- False Negatives (FN): Missed positives (e.g., contaminated water incorrectly labeled as clean).

If there are 40 contaminated samples and the model correctly identifies 25, the recall is 62.5%. This means the model successfully detects 62.5% of the actual contamination cases, showing its effectiveness at finding positive instances. The recall is vital when it's critical to detect all instances of contamination. Smith et al. (2020) used recall to measure how well their model's identified pollution in surface water. A high recall indicates that the model is good at catching all instances of contamination, which is essential for protecting public health and ensuring environmental safety.

5.4 F1 Score

In scenarios with severe consequences for false positives and false negatives, precision and recall must be balanced. The balance between precision and recall is measured by the F1 score.

$$F1\,Score = \frac{2 \times Precision \times Recall}{Precision + Recall}$$

Studies such as Gonzalez et al. (2021) use the F1 Score to provide a comprehensive evaluation of model performance in water quality predictions, ensuring that both the accuracy of positive results and the completeness of positive case detection are considered.

By understanding these metrics, you can better evaluate the effectiveness of models used to predict water and groundwater quality, ensuring they are both accurate and reliable for practical applications.

6 Trends and Future Directions

Since 2008, machine learning techniques have been the go-to approach for groundwater quality modeling due to increased advanced computational power and data availability. 2022 witnessed a significant rise in research articles linked to this subject as a result of a stringent proposal deadline (Haggerty et al. 2023). Unsupervised and ensemble learning models provide less precise estimations than supervised learning models. Although supervised learning is mostly used, there's a need for more intricate modeling approaches.

Machine learning advancements and extensive water information are driving innovations in predictive modeling for both surface and groundwater quality. Recent AI developments include combining big data analytics with machine learning, employing hybrid models, and prioritizing model interpretability. The advancement of more robust and scalable models entails dealing with a variety of ecological circumstances, incorporating actual data from IoT devices, and developing AI-driven strategy frameworks for sustainable water management. The trends highlight addressing issues such as data shortages and ensuring applicability across several climatic and geographical situations. The combination of classical environmental models, deep learning, and ensemble approaches is expected to increase the accuracy and reliability of water quality forecasts.

6.1 Emerging Technologies in Machine Learning

The technologies are transforming predictive modeling for surface and groundwater quality. Utilizing IoT sensors, remote sensing, and GIS technology, these models analyze comprehensive datasets to discern patterns and predict water quality trends. These techniques-supervised learning, unsupervised learning, deep learning, and hybrid models-allow for precise predictions of contamination incidents and water quality parameters (Singha et al. 2021). These innovations improve environmental monitoring, offer valuable insights for sustainable water resource management and policy-making, and address data scarcity and model interpretability challenges.

6.2 Integration with IoT and Smart Sensors

IoT and smart sensors integrated with machine learning have transformed surface and groundwater quality predictive modeling. Inserting sensors into water bodies and their surroundings allows real-time, high-frequency collection of granular data on water quality parameters. The data collected through IoT sensors, handled by ML algorithms, advances ML models' predictive and analytical capabilities for detecting trends, and anomalies, and making informed decisions. IoT-ML synergy enhances water quality predictions' accuracy while empowering proactive management and intervention, minimizing risks. Figure 4 illustrates the key advantages of integrating IoT and smart sensors into water quality modeling.

IoT devices and sensors utilized in water quality modeling encompass various examples. YSI ProDSS is a transportable, multiparameter water quality meter that notices dissolved oxygen, turbidity, nutrient concentrations, and pH. Conductivity, pH, ORP, and additional variables can be measured by the Hach HQ40d, which is a multiparameter meter. The Aqua TROLL 600 instrument monitors pH, turbidity, dissolved oxygen, and conductivity. The Scan UV-V is a Spectrometer that monitors a variety of water quality indicators. Hach's HydroLAB Series contains sensors that measure turbidity, pH, and dissolved oxygen. SonTek-ARGOS measures and records water depth and velocity. Lovelace Smart Water Meters measure both water use and quality. The KROHNE OPTIFLEX 2200 C devices can detect both level and volume in liquids. NKE Marine Sensors monitor temperature and salinity in marine habitats. The Seabird SBE 63 device collects conductivity, temperature, and pressure data. Eureka Manta + monitors turbidity, pH, and dissolved oxygen levels. The OTT

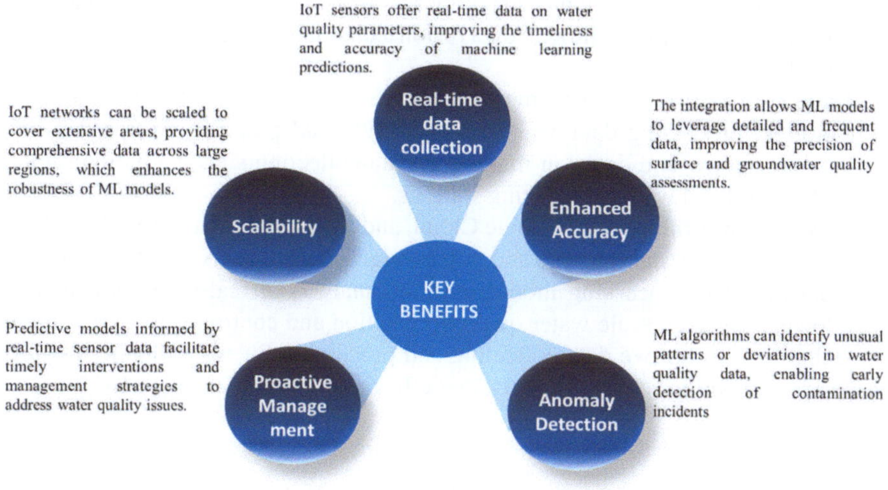

Fig. 4 Key advantages of integrating IoT and smart sensors in water quality modelling

Hydromet system provides water quality measurements and is interoperable with data management systems.

6.3 Enhancements in Data Collection and Processing

Emerging technologies are shifting the data collected and processed ways for ground and surface water quality prediction models. IoT and sensor technology advancements allow for real-time monitoring and high-resolution water quality indicators. These technologies include:

1. **Big Data Analytics**: Using big data analytics platforms improves the processing of large volumes of sensor data. Apache Hadoop and Apache Spark store, process, and analyze big datasets efficiently, providing the computing power essential for complex machine learning.
2. **Remote Sensing Technologies**: Satellite and aerial imagery advancements enable the collection of environmental data present on large scales. The satellite data of Copernicus Sentinel offer unique visions on change over time and surface water quality, enabling regional and global assessments.
3. **Advanced IoT Sensors**: These sensors are helpful to enhance sensitivity and accuracy in monitoring water quality parameters, including pollutants, turbidity, dissolved oxygen, and pH. Sensors such as the Aqua TROLL 600 and S::can UV–Vis Spectrometer offer continuous data for dynamic and thorough monitoring of water bodies.
4. **Data Fusion Techniques**: By combining remote sensing photography and ground-based sensors, we provide more accurate water conditions and enhanced forecasting capabilities through data fusion techniques that enhance water quality models.
5. **Edge Computing**: Edge Computing decreases latency and bandwidth application by processing data near its source. Edge AI processors offer real-time data processing and decision-making at data-collecting sites, including remote or decentralized water monitoring stations.
6. **Cloud Computing**: AWS, Google Cloud, and Microsoft Azure provide scalable infrastructure for data storage and processing. These systems enable the deployment of machine learning models and the analysis of real-time data streams, allowing for large-scale water quality prediction and control. These developing technologies improve data collecting and processing for water quality models, resulting in more accurate, timely, and effective resource management.

6.4 Challenges and Opportunities for Future Research

Future research in ML techniques for predicting surface and groundwater quality is characterized by both challenges and opportunities. Data scarcity and quality significantly impact the accuracy of ML models. Transfer learning, a new approach where models trained on data in one region can be applied to another region with limited data, is being used to tackle this issue. Explaining complex deep learning models, often viewed as black boxes, presents a significant challenge to decision-makers. SHAP, a technique being developed for greater transparency, offers insights into models' prediction mechanisms and produces interpretable and actionable results. ML models are increasingly being integrated into AI-driven policy-making for shaping water quality regulations. Data-driven understandings allow for supplementary adaptive and responsive managing strategies by enabling policymakers to make more informed decisions created in real-time environmental situations. These advancements demonstrate ML's ability to significantly improve predictive accuracy and promote sustainable water management practices.

7 Conclusion

Machine learning (ML) can make predicting water quality in real-time possible, helping water managers issue early warnings more effectively. Recent improvements in modeling, especially with ML and hybrid models, have made water quality index (WQI) predictions much more accurate. In this study, hybrid models were found to significantly enhance the accuracy of WQI estimates. However, when it comes to managing groundwater quality, there's still a lot of room for improvement. Using ML for groundwater assessments could save both time and money by reducing the need for extensive testing. Although many studies highlight the potential of ML for improving groundwater quality predictions, they often don't offer clear steps on how to bring these tools from research into real-world use by industries or government agencies.

To make the most of ML in groundwater quality management, data scientists and local authorities need to work more closely together. This could mean developing open-source ML tools, adapting water monitoring systems to better support ML, and increasing funding for ML projects. These efforts would help ensure that ML is effectively used to monitor and manage groundwater quality. In summary, this study shows that ML techniques, like Artificial Neural Networks (ANN) and Support Vector Machines (SVM), are very effective in predicting and monitoring surface and groundwater quality. ANN models are highly accurate and adaptable for complex water systems, while SVM models are great at simplifying these systems and making accurate predictions. These ML methods offer a strong way to improve how we monitor and manage water quality, leading to better and more informed decisions in water resource management.

References

APHA (2005) Standard methods for the examination of water and wastewater, 21st edn. American Public Health Association, Washington, DC

Azamathulla HM, Ghani AA, Leow CS, Chang CK, Zakaria NA (2011) Gene-expression programming for the development of a stage-discharge curve of the Pahang River. Water Resour Manage 25:2901–2916

Azimi S, Moghaddam MA, Monfared SH (2019) Prediction of annual drinking water quality reduction based on the groundwater resource index using the artificial neural network and fuzzy clustering. J Contam Hydrol 220:6–17

Banerjee P, Singh VS, Chatttopadhyay K, Chandra PC, Singh B (2011) Artificial neural network model as a potential alternative for groundwater salinity forecasting. J Hydrol 398(3–4):212–220

Beyer K, Goldstein J, Ramakrishnan R, Shaft U (1999) When is "nearest neighbor" meaningful? In: International conference on database theory. Springer, pp 217–235

Breiman L, Friedman J, Olshen RA, Stone CJ (1986) Classification and regression trees. Wadsworth Publishing Company, Belmont, CA

Elbeltagi A, Pande CB, Kouadri S, Islam ARMT (2022) Applications of various data-driven models for the prediction of groundwater quality index in the Akot basin, Maharashtra, India. Environ Sci Pollut Res 1–15

Gonzalez CA, Hernandez R, Lopez J (2021) Application of the F1 score in water quality model evaluation: A case study in groundwater contamination. Environ Model Softw 137:104957

Haggerty R, Sun J, Yu H, Li Y (2023) Application of machine learning in groundwater quality modeling—a comprehensive review. Water Res 233:119745

Ho JY, Afan HA, El-Shafie AH, Koting SB, Mohd NS, Jaafar WZB, El-Shafie A (2019) Towards a time and cost-effective approach to water quality index class prediction. J Hydrol 575:148–165

Irwan D, Ali M, Ahmed AN, Jacky G, Nurhakim A, Ping Han MC, El-Shafie A (2023) Predicting water quality with artificial intelligence: a review of methods and applications. Arch Comput Methods Eng 30(8):4633–4652

Judeh T, Almasri MN, Shadeed SM, Bian H, Shahrour I (2022) Use of GIS, statistics and machine learning for groundwater quality management: application to nitrate contamination. Water Resour 49(3):503–514

Kisi O, Parmar KS (2016) Application of least square support vector machine and multivariate adaptive regression spline models in long-term prediction of river water pollution. J Hydrol 534:104–112

Kshetrimayum B, Singh V, Bhatt D (2020) Application of hierarchical clustering to categorize lakes in India based on pollution levels. Environ Monit Assess 192(4):239

Li H, Yang J, Yu L, Zhang M (2021) Detection of urban water system pollution hotspots using DBSCAN. Water Res 194:116934

Li W, Ye X, Du X (2022) Imputation of ammonium nitrogen concentration in groundwater based on a machine learning method. Water 14(10):1595

Najah A, El-Shafie A, Karim OA, El-Shafie AH (2013) Application of artificial neural networks for water quality prediction. Neural Comput Appl 22:187–201

Norouzi H, Moghaddam AA (2020) Groundwater quality assessment using random forest method based on groundwater quality indices (case study: Miandoab plain aquifer, NW of Iran). Arab J Geosci 13(18):912

Omer NH (2019) Water quality—science, assessments and policy

Pereira SN, Ribeiro L, Pereira DA (2019) Evaluating the precision of groundwater contamination prediction models in urban areas. J Hydrol 573:207–216

Podgorski JE, Labhasetwar P, Saha D, Berg M (2018) Prediction modeling and mapping of groundwater fluoride contamination throughout India. Environ Sci Technol 52(17):9889–9898

Ramaraj M, Sivakumar R (2023) Integration of band regression empirical water quality (BREWQ) model with deep learning algorithm in spatiotemporal modeling and prediction of surface water quality parameters. Model Earth Syst Environ 9(3):3279–3304

Saghebian SM, Sattari MT, Mirabbasi R, Pal M (2014) Ground water quality classification by decision tree method in Ardebil region, Iran. Arab J Geosci 7:4767–4777

Sahour S, Khanbeyki M, Gholami V, Sahour H, Kahvazade I, Karimi H (2023) Evaluation of machine learning algorithms for groundwater quality modeling. Environ Sci Pollut Res 30(16):46004–46021

Singh S, Das A, Sharma P (2024a) Predictive modeling of water quality index (WQI) classes in Indian rivers: Insights from the application of multiple machine learning (ML) models on a decennial dataset. Stoch Environ Res Risk Assess 1–18

Singh V, Nema AK, Chouksey A, Suryawanshi A (2024b) Assessment of eco-environmental vulnerability using remote sensing and GIS tools in Maharashtra Region, India. Int J Environ Clim Change 14(4):119–129. https://doi.org/10.9734/ijecc/2024/v14i44103

Singha S, Pasupuleti S, Singha SS, Singh R, Kumar S (2021) Prediction of groundwater quality using efficient machine learning techniques. Chemosphere 276:130265

Smith A, Johnson K, Williams M (2020) Sensitivity analysis in surface water pollution detection: evaluating recall in predictive models. Water Res 182:115938

Spellman FR (2017) The drinking water handbook, 3rd edn. CRC Press, Boca Raton

Suryawanshi A, Nema AK, Jaiswal RK, Jain S, Kar SK (2021) Identification of soil erosion prone areas of Madhya Pradesh using USLE/RUSLE. J Agric Eng 58(2):177–191

Suryawanshi A, Dubey S, Sharma M (2023) Evaluating soil erosion through geospatial techniques: difficulties and prospects in the context of the Central Indian Chambal River Basin. Int J Environ Clim Change 13(11):4518–4533

Tchobanoglous G, Peavy HS, Rowe DR (1985) Environmental engineering. McGraw-Hill Interamericana, New York

Voyant C, Notton G, Kalogirou S, Nivet ML, Paoli C, Motte F, Fouilloy A (2017) Machine learning methods for solar radiation forecasting: a review. Renewable Energy 105:569–582

Wang B, Hipsey MR, Ahmed S, Oldham C (2018) The impact of landscape characteristics on groundwater dissolved organic nitrogen: insights from machine learning methods and sensitivity analysis. Water Resour Res 54(7):4785–4804

Wu B, Tian F, Zhang M, Piao S, Zeng H, Zhu W, Lu Y (2022) Quantifying global agricultural water appropriation with data derived from earth observations. J Clean Prod 358:131891

Wu R, Alvareda EM, Polya DA, Blanco G, Gamazo P (2021) Distribution of groundwater arsenic in Uruguay using hybrid machine learning and expert system approaches. Water 13(4):527

Xu X, Zhang L, Huang Q, Zhang S (2019) Assessment of groundwater contamination using the K-Means clustering algorithm in Guanzhong Basin, China. J Hydrol 572:343–355

Zhang H, Zhang L (2021) Decision tree for machine learning and data mining: a comprehensive review. Artif Intell Rev 54(3):2497–2524

Zhou Q, Chen Y, Zhao W, Sun H (2018) Application of PCA in identifying sources of groundwater contamination in the Pearl River Delta. Sci Total Environ 613–614:123–130

Surface Water and Groundwater Quality Forecasting Using Machine Learning Models

Salim Heddam

Abstract Fluorescent dissolved organic matter (FDOM) is one of the most critical water quality variables. Despite the significant improvement in water quality modelling and forecasting over the last few years, accurate forecasting of FDOM concentration is research challenge. In order to improves our knowledge skills, the present chapter presents a forecasting study based on the application of various machine learning models namely: (i) Random Forest Regression (RFR), (ii) Regularized Greedy Forest (RGF), (iii) Extremely Randomized Trees (ERT), (iv) Support Vector Regression (SVR), and (v) Relevance Vector Machine (RVM), for forecasting FDOM concentration at three different horizons: t, $t+1$ and $t+7$, measured at daily time scale. All models were developed using sig input lag from ($t-1$) to ($t-6$). All models were compared suing numerical and graphical comparison, i.e., the root mean squared error (RMSE), the mean absolute error (MAE), the coefficient of correlation (R), and the Nash–Sutcliffe efficiency (NSE), and graphical visualization, i.e., scatterplot, boxplot, violinplot and Taylor diagram. From the obtained results, we can draw the following conclusions. At the horizon (t), the best numerical performances were obtained by ERT equally with RFR with R, NSE, RMSE and MAE of approximately 0.936, 0.876, 1.367, and 0.733, respectively, while the RVM was found to bez the poorest model. At the horizon ($t+7$), the performances of the models were significantly decreased reaching the values of 0.464, 0.161, 3.555, and 2.269, in terms of R, NSE, RMSE and MAE, indices.

Keywords Forecasting · FDOM · ERT · RFR · RGF · RVM · SVR

1 Introduction

Dissolved organic matter (DOM) is the most important faction of the organic matter (OM) in the aquatic freshwater ecosystems (Niloy et al. 2023). Fluorescent dissolved organic matter (FDOM) is a fraction of the Chromophoric dissolved organic matter

S. Heddam (✉)
Agronomy Department, Faculty of Science, University 20 Août 1955 Skikda, Skikda, Algeria
e-mail: heddamsalim@yahoo.fr; s.heddam@univ-skikda.dz

Dimple et al. (eds.), *Surface Water and Groundwater Quality for Sustainable Utilization*, Springer Hydrogeology, https://doi.org/10.1007/978-981-96-8189-1_19

(CDOM) (Urban-Rich et al. 2006), and its presence in surface water has significant implications on water quality (Zhan et al. 2023). FDOM can be found in various natural aquatic environments and biochemical dyes, and it plays a critical role in the migration of some chemical components, i.e., heavy metals and organic contaminants (Fu et al. 2023). FDOM absorbs light and it occurs within the ultraviolet-blue range of the light (Daraei et al. 2024a). FDOM sensors can be used for detecting fluorescence signals from the fluorophore portion of the DOM (Daraei et al. 2024b). Few studies in the literature have reported an application of machine learning (ML) for modelling DOM, CDOM and FDOM. For example, Zounemat-Kermani et al. (2024) compared between various ML models for modelling FDOM concentration, namely, support vector regression (SVR), extreme learning machine (ELM), random forest regression (RFR), boosted trees (BT), and gene expression programming (GEP), using data collected at Caloosahatchee River in Florida, USA. The authors used various water quality variables namely, water pH, water temperature (T_w), dissolved oxygen (DO), specific conductance (SC), water turbidity (TU), nitrogen oxides (NO_x), and discharge (Q). From the obtained results it was found that, the RF and BT were the most accurate models with RMSE and MAE of approximately ≈0.84 and ≈24.84, respectively. In another study, Heddam (2014) applied the generalized regression neural network (GRNN) for modelling CDOM concentration using data from USGS station, and based on the water pH, T_w, TU and SC. The GRNN was found to be an accurate model having an R value of approximately 0.999.

Recently, the use of ML for water quality variables modeling has received great deal of attention and various published studies can be found in the literature. Parra et al. (2024) compared between various machine-learning models for modelling water turbidity (TU) based on the Red Green Blue (RGB) light source and a light detector. A total of 26 regression models were compared, among them the linear regression (LR), support vector regression (SVR), Gaussian process regression (GPR), and artificial neural networks (ANN), Boosted Trees (BTree) and Bagged Trees (Bagg), among others. High predictive accuracy was obtained, with correlation coefficient (R) reaching the value of 0.99. However, it is important to note that the models were calibrated based on high number of features, which makes the method impractical. Dehkordi et al. (2024) used fuzzy similarity analysis (FSA) algorithm for improving the performances of three ML models, namely SVR, partial least squares regression (PLS) and mixture density networks (MDN), applied and compared for modelling river water TU. The models were developed based on the Sentinel-2 remote sensing data and it was found that, the FSA significantly contributed to the improvement of ML models, and the FSA-MDN was more accurate with R-value of ≈0.959 compared to the values of ≈0.948 and ≈0.938, obtained by the FSA-SVR and FSA-PLSR, respectively. Liu and Li (2024) compared between four ML models namely, Bayesian ridge regression (BRR), decision tree regression (DTR), gradient boosting regression (GBR) and eXtreme Gradient Boosting (XGBoost) for modelling river water TU, based on the unmanned aerial vehicle (UAV) images. From the obtained results, it was found that, the hybrid fractional order derivation (FOD) and discrete wavelet transform (DWT) coupled with DTR (FOD-DWT-DTR) exhibited the best performances with R $\approx$ 0.948, root mean square error (RMSE $\approx$ 6.50), and mean absolute error (MAE $\approx$ 4.07), respectively.

Devi and Mamatha (2024) used the normalize difference turbidity index (NDTI), which is estimated using the spectral reflectance values of the water pixels derived from Landsat-8 images for modelling water TU. They compared between the LR, DTR, and the K-NN regression models. The DTR was found to be more accurate exhibiting an R, RMSE, MAE, of approximately $\approx$0.76, $\approx$3.802, $\approx$3.246, respectively. Zhang et al. (2024) investigated the feasibility and robustness and various ML models for water TU prediction in Nansi Lake, Shandong Province, China. They used Sentinel-2 remote sensing imagery with 20 and 17 feature bands. The proposed ML were respectively, random forest regression (RFR), the CatBoost, the XGBoost, the LightGBM, GBR and the stacking ensemble model (StEM). From the obtained results, the StEM guaranteed the best predictive accuracy with coefficient of determination (R^2) of approximately $\approx$0.812 compared to the values of $\approx$0.782, $\approx$0.792, $\approx$0.782, $\approx$0.791, and $\approx$0.791 obtained by the RFR, XGBoost, GBR, CatBoost, and LightGBM, respectively. Mousavi (2024) compared between ANN and adaptive neuro-fuzzy inference system (ANFIS) for modelling water TU. They used the threshold-based wavelet denoising method, as a data pre-processing technique coupled with the ANN and ANFIS. It was found that, ANFIS coupled with denoising method provided superior performances with R^2 and RMSE of approximately $\approx$0.937 and $\approx$10.360, respectively, compared to the values of $\approx$0.883 and $\approx$11.503 obtained using the ANN model. García-Nieto et al. (2024) applied four ML models namely, RFR, least absolute shrinkage and selection operator (Lasso), Ridge regression (RR), and ElasticNet regression (ENR) for modelling the outlet turbidity (TU) in different micro irrigation filters. The RFR exhibited the best numerical performances with R^2, R, RMSE, and MAE of approximately $\approx$0.8712, $\approx$0.9366, $\approx$0.3931, $\approx$0.2719, respectively. Hu et al. (2023) applied the long short-term memory (LSTM) for modelling water TU in West Taihu, the Lake Center, and the mouth of Gonghu Bay, China, and reported acceptable performances.

To our best knowledge, few studies have focused on modelling FDOM concentration using machine learning. Therefore, in the present chapter, we use five ML for forecasting FDOM concentration at three different horizons: t, $t + 1$ and $t + 7$. The proposed models were respectively: (i) Random Forest Regression (RFR), (ii) Regularized Greedy Forest (RGF), (iii) Extremely Randomized Trees (ERT), (iv) Support Vector Regression (SVR), and (v) Relevance Vector Machine (RVM). The performances of the models were compared based on the numerical and graphical comparisons.

2 Case Study and Dataset

In the present chapter, we use data collected form the United State Geological Survey (USGS) available at https://waterdata.usgs.gov/nwis/. We selected one station namely: (i) USGS 14211010, Clackamas River near Oregon City, Oregon, USA (Latitude 45° 22′ 46″, Longitude 122° 34′ 34″ NAD27). Fluorescent dissolved organic matter (FDOM) Data corresponds to the period ranging from 24/10/2014 to 15/09/2024. The FDOM is measured at daily time scale, and we select six lags' times as the input variables for forecasting the FDOM at time (t), ($t + 1$) and ($t + 7$) (Table 1). Location map of the FDOM station is depicted in Fig. 1. The correlation matrix is depicted in Fig. 2. Data was divided into training (70%) and validation (30%).

Fig. 1 Location map of FDOM USGS station

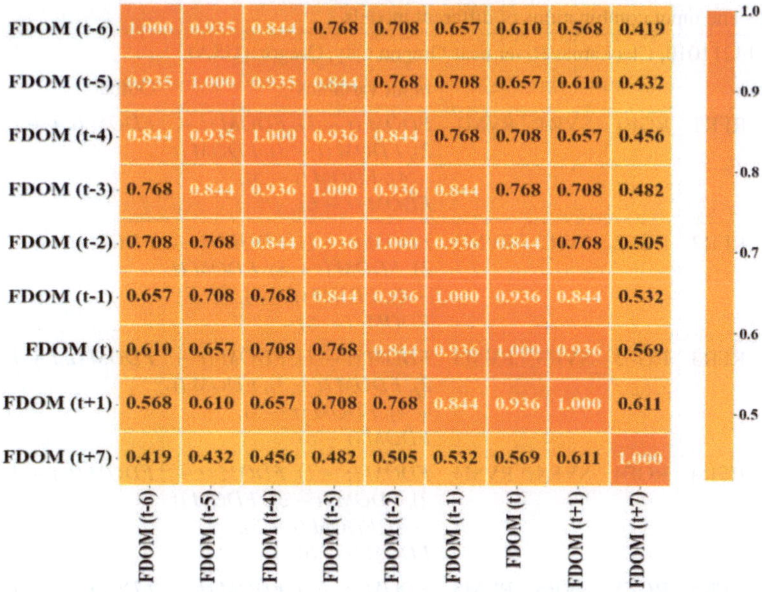

Fig. 2 The correlation matrix

3 Machine Learning Models

3.1 Random Forest Regression (RFR)

Random forest regression (RFR) proposed by Breiman (2001) was developed using the principals of classification and regression trees (CRT), and it is mainly composed from an ensemble of decision trees for improving the performances and prediction accuracy by the application of different algorithms: bootstrap aggregation (bagging) and random feature selection (Adebayo et al. 2024). Using the RFR model, the training algorithm randomly select only fewer variables which were presented as a candidate for the splitting strategy of the decision trees (DT) node. For any node, there are two possible cases: the splitting variables and the excluded variables (He et al. 2024). For solving a regression problem, the RFR use the mean value of the response of each DT as the final results (Yu et al. 2024). For the RFR architecture, each single tree starts by a so-called "*root node*", and in a second step, this "*root node*" is divided into "*child nodes*" using a "*splitting criterion*" which is in a general form a "*threshold set*" (Kamali et al. 2024). The RFR has been applied successfully for solving various technical and engineering problems: forecasting of flash flood susceptibility mapping (Wahba et al. 2024), optimization of biogas production (Pratap et al. 2024), meteorological prediction of $PM_{2.5}$ (Qi et al. 2024) and predicting financial distress in Latin American companies (Barboza and Altman 2024).

Table 1 The input combinations of different models

USGS 14211010, Clackamas River near Oregon City, Oregon (USA)						
Models					Input variables	Output
ERT1	RFR1	RGF1	SVR1	RVM1	*FDOM (t − 1), FDOM (t − 2), FDOM (t − 3), FDOM (t − 4), FDOM (t − 5), FDOM (t − 6)*	*FDOM (t, t + 1, t + 7)*
ERT2	RFR2	RGF2	SVR2	RVM2	*FDOM (t − 1), FDOM (t − 2), FDOM (t − 3), FDOM (t − 4), FDOM (t − 5), FDOM (t − 6)*	*FDOM (t, t + 1, t + 7)*
ERT3	RFR3	RGF3	SVR3	RVM3	*FDOM (t − 1), FDOM (t − 2), FDOM (t − 3), FDOM (t − 4), FDOM (t − 5), FDOM (t − 6)*	*FDOM (t, t + 1, t + 7)*
ERT4	RFR4	RGF4	SVR4	RVM4	*FDOM (t − 1), FDOM (t − 2), FDOM (t − 3), FDOM (t − 4), FDOM (t − 5), FDOM (t − 6)*	*FDOM (t, t + 1, t + 7)*
ERT5	RFR5	RGF5	SVR5	RVM5	*FDOM (t − 1), FDOM (t − 2), FDOM (t − 3), FDOM (t − 4), FDOM (t − 5), FDOM (t − 6)*	*FDOM (t, t + 1, t + 7)*
ERT6	RFR6	RGF6	SVR6	RVM6	*FDOM (t − 1), FDOM (t − 2), FDOM (t − 3), FDOM (t − 4), FDOM (t − 5), FDOM (t − 6)*	*FDOM (t, t + 1, t + 7)*

3.2 Regularized Greedy Forest (RGF)

Regularized greedy forest (RGF) was developed by Johnson and Zhang (2013). RGF uses the "*greedy search*" algorithm for a direct training of the DT (Ghosh et al. 2022), and by adding a regularization term in the loss function (Gargiulo et al. 2021). The forest handled by the RGF can be an ensemble of DT, and for the edge of each tree should be linked to a variable, threshold parameter, and decision rule (Ghosh et al. 2022). From it has been proposed, various application of the RGF can be found in the literature: short-term load forecasting (Yiu et al. 2024) and prediction of residential building occupancy (Bouyakhsaine et al. 2024).

3.3 Extremely Randomized Trees (ERT)

Extremely randomized trees (ERT) proposed by Geurts et al. (2006) works by building and ensemble of decision trees (DT) using the classical top-down algorithm

(Cheng et al. 2021). The ERT can be used for both classification and regression problems, and it is based on the idea of randomizing strongly both attribute and cut-point choice while splitting a tree node (Liang et al. 2021). The advantages of the ERT model can be summarized as follow: (i) the training process is conducted using all training data which can contribute significantly to the decrease of training error, (ii) the process of nodes splitting can be significantly enhanced by randomly selecting the best branching attributes and (iii) the final response of the ERT model is provided by the idea of voting decision (Cheng et al. 2021). Various application of the ERT can be found in the literature: land subsidence simulation (Wang et al. 2023), hierarchical brain tumour segmentation (Pinto et al. 2018), fault diagnosis in wireless sensor network (Saeed et al. 2021) and flood susceptibility mapping (Sachdeva and Kumar 2022).

3.4 Support Vector Regression (SVR)

Support vector regression (SVR) is a supervised ML model used for both regression or classification (SVM) problems. The objective of the SVM model is to construct the optimal separating hyperplane for categorizing the data composed from an output Y and an ensemble of input X. The standard SVR model use an λ insensitive loss function that have the sole objective, the identification of the approximation functions, that allows the calculation of the maximal value of the λ for the exact values of the training dataset (Kumari and Tanveer 2024). The SVR has been applied successfully for solving various engineering problems: solar radiation prediction (Adnan et al. 2024), fault diagnosis in photovoltaic systems (Lu et al. 2024), prediction of allowable storage time of finger millet grains (Joshi and Rao 2024), landslide susceptibility assessment (Prathom and Sujitapan 2024), and reliability estimation of complex structure (Zhu et al. 2024).

3.5 Relevance Vector Machine (RVM)

Relevance vector machine (RVM) is a supervised ML model mainly inspired from the Bayesian inference (Acosta et al. 2021). Based on Bayesian theory, the RVM uses a fast marginal likelihood maximization as the training algorithm and it is within a high capability of handling the imprecision of the predicted variable and reaching a high level of certainty. (Wang et al. 2024). The precision and robustness of the RVM are determined by the with factor, and it is supposed that, the parameters of the model will be with very low magnitude and with a sparse representation (Wang et al. 2024).

4 Different Result and Numerical Analysis

In this section, the performances of the five ML models, i.e., RFR, RGF, ERT, SVR, and RVM were evaluated and compared. The models were applied for forecasting FDOM concentration. The models were evaluated using four numerical indices namely, RMSE, MAE, R and NSE, calculated as follow:

$$RMSE = \sqrt{\frac{1}{N}\sum_{i=1}^{N}[(FDOM)o_i - (FDOM)_{Pi}]^2}, (0 \leq RMSE < +\infty) \quad (1)$$

$$MAE = \frac{1}{N}\sum_{i=1}^{N}|(FDOM)o_i - (FDOM)_{Pi}|, (0 \leq MAE < +\infty) \quad (2)$$

$$R = \left[\frac{\sum_{i=1}^{N}\left((FDOM)o_i - \overline{(FDOM)o}\right)\left((FDOM)_{Pi} - \overline{(FDOM)_P}\right)}{\sqrt{\sum_{i=1}^{N}((FDOM)o_i - \overline{(FDOM)o})^2 \sum_{i=1}^{N}((FDOM)_{Pi} - \overline{(FDOM)_P})^2}}\right] \quad (3)$$

$$NSE = 1 - \left[\frac{\sum_{i=1}^{N}[(FDOM)o_i - (FDOM)_{Pi}]^2}{\sum_{i=1}^{N}[(FDOM)o_i - \overline{(FDOM)o_i}]^2}\right], (-\infty < NSE \leq 1) \quad (4)$$

where $\overline{(FDOM)o}$ and $\overline{(FDOM)_P}$ are the mean measured and mean forecasted FDOM, respectively, and $(FDOM)o_i$ and $(FDOM)_{Pi}$ specify the observed and predicted FDOM for ith observations, and N shows the number of data points. The models were calibrated using 70% of the data and validated using the remaining 30% of the data. Further comparison between the models' performances was done using graphical comparison, i.e., the scatterplot, boxplot, violinplot and Taylor diagram. Obtained results are discussed hereafter for each station separately.

Results obtained are depicted in Table 2. In this section, we present the main results and we discuss the difference between different models at different forecasting horizons. The differences between models are evaluated tacking into account the goodness of fit, the predictive accuracy, and the predictive errors. The main finding from the Table 2 is that, for the first forecasting horizon a time (t) and during the training stage, excellent performances were obtained by all models. The best accuracy was obtained using ERT1 having an R, NSE, RMSE and MAE of ≈0.994, ≈0.990, ≈0.499 and ≈0.238, respectively, while the lowest accuracy was achieved using the SVR1 with R, NSE, RMSE and MAE of ≈0.940, ≈0.911 ≈1.522 and ≈0.658, respectively. Furthermore, the RGF1 was found to be more accurate compared to the RFR1, while the RVM1 model has provided superior performances compared to the RFR1, SVR1, and slightly superior to the RGF1. Furthermore, if all models were compared based on the NSE and the RMSE values we can see that, the ERT1 was the most accurate having the biggest NSE of ≈0.990 and 0.499, respectively. Moreover, when looking at the results obtained during the validation stage for the whole developed models, it seemed that the ERT1 was the most accurate exhibiting R, NSE, RMSE and MAE of ≈0.936, ≈0.876, ≈1.367 and ≈0.733, respectively. From the obtained results reported in Table 2, it was evident that, moving between the training and validation stages, the performances of all modes were remarkably decreased. The R, NSE, RMSE and MAE of the ERT1 were decreased by ≈5.835%, ≈11.515%, ≈63.497% and ≈67.531%, respectively. The performances of the SVR1 were decreased by ≈0.957%, ≈6.037%, ≈0.847% and ≈3.377%, respectively. Furthermore, among all models, the RVM1 was found to be the less accurate, exhibiting the lowest numerical performances with R, NSE, RMSE and MAE of ≈0.865, ≈0.744, ≈1.966 and ≈1.015, respectively. Scatterplot of measured and forecasted FDOM at the horizon (t) are depicted in Fig. 3, for which the superiority of the ERT1 was evident. The boxplot and violinplot for all models are depicted in Fig. 4.

For the forecasting horizon at ($t + 1$) as shown in Table 2, obtained results by the best models are depicted in Fig. 5 as a scatterplot between measured and predicted data. It is obvious from the graphical illustration that, the SVR2 was the most accurate model, while the RVM2 was the lowest one. During the validation stage, the SVR2 exhibited the best numerical performances with the biggest R (≈0.850) and NSE (≈0.703) and the lowest RMSE (≈2.115) and MAE (≈1.082) values, which illustrates that this model has very precise forecasting ability. The SVR2 improves the

Table 2 Results of FDOM concentration forecasting at different horizons

Models	Training				Validation			
	R	NSE	RMSE	MAE	R	NSE	RMSE	MAE
Scenario 01: forecasting FDOM at time (t)								
ERT1	0.994	0.990	0.499	0.238	0.936	0.876	1.367	0.733
RFR1	0.958	0.940	1.249	0.622	0.935	0.874	1.379	0.726
RGF1	0.981	0.972	0.850	0.450	0.929	0.863	1.437	0.792
SVR1	0.940	0.911	1.522	0.658	0.931	0.856	1.535	0.681
RVM1	0.984	0.978	0.762	0.443	0.865	0.744	1.966	1.015
Scenario 01: forecasting FDOM at time (t + 1)								
ERT2	0.987	0.980	0.731	0.386	0.839	0.702	2.119	1.228
RFR2	0.907	0.877	1.823	0.969	0.839	0.701	2.123	1.216
RGF2	0.957	0.941	1.266	0.727	0.827	0.676	2.211	1.286
SVR2	0.861	0.809	2.271	1.026	0.850	0.703	2.115	1.082
RVM2	0.956	0.940	1.274	0.739	0.728	0.519	2.693	1.523
Scenario 01: forecasting FDOM at time (t + 7)								
ERT3	0.971	0.950	1.163	0.757	0.540	0.275	3.304	2.139
RFR3	0.770	0.708	2.808	1.842	0.555	0.297	3.253	2.080
RGF3	0.872	0.820	2.203	1.485	0.487	0.188	3.496	2.345
SVR3	0.598	0.517	3.614	1.926	0.589	0.251	3.359	1.938
RVM3	0.805	0.755	2.573	1.705	0.464	0.161	3.555	2.269

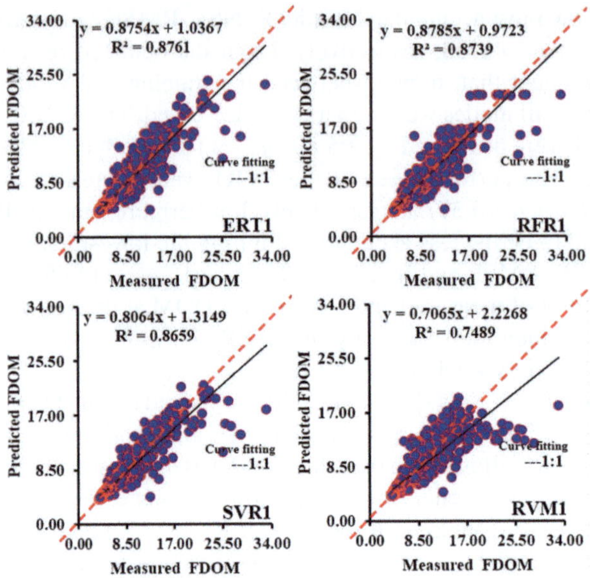

Fig. 3 Scatterplot of measured and forecasted FDOM concentration using different ML models for the validation dataset at the horizon (t)

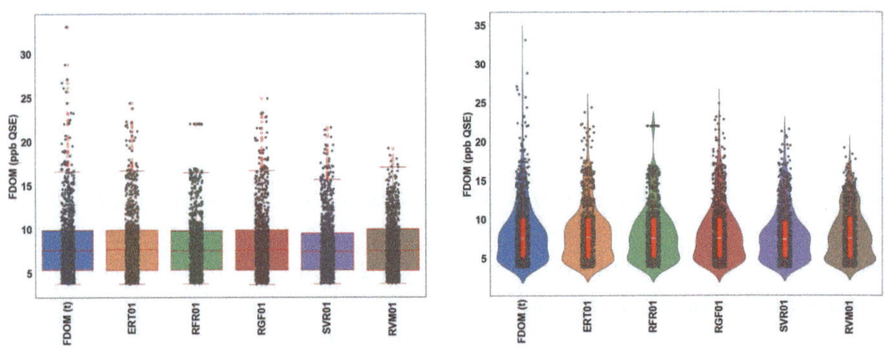

Fig. 4 Comparison between measured and forecasted FDOM concentration using different ML for the validation dataset at the horizon (t): boxplot (left panel) and violinplot (right panel)

RMSE and MAE of the ERT2 by 0.189% and 11.889%, respectively, and by 0.377% and 11.020%, compared to the RFR2 model. Similarly, the SVR2 improves the RMSE and MAE of RVM2 by 21.463% and 28.956%, and by 4.342% and 15.863% compared to the RGF2, respectively. Furthermore, it is clear from the results reported above that, the improvement gained using the SVR2 was more important compared to RVM2. Finally, all models were compared using Boxplot and Violinplot (Fig. 6). From Fig. 6, we can see that, the SVR2 was the only model that have close resemblance with the measured data, with high similarity compared to the ERT2 and RFR2.

It is clear that, the SVR2 was the most accurate model having the biggest R value. Beyond he numerical and graphical comparison, we can argue that all obtained results at the forecasting horizon (t + 1) were very acceptable, except the results obtained by the RVM2, which were less than all other models.

Results obtained at the forecasting horizon (t + 7) are reported in Table 2. Obtained results are depicted in Fig. 7 as a scatterplot and Fig. 8 as a boxplot and violinplot. The indices of RMSE, MAE, R and NSE achieved by the models on training and validation

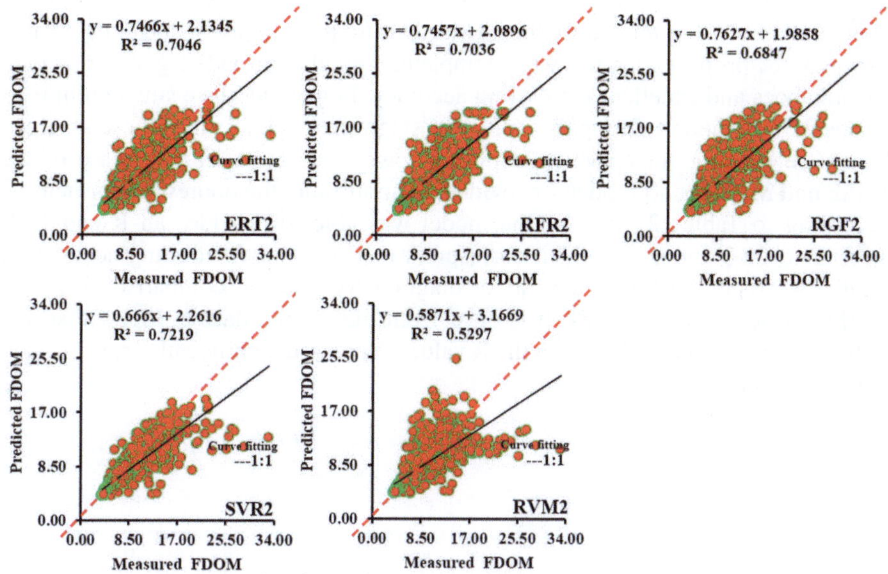

Fig. 5 Scatterplot of measured and forecasted FDOM concentration using different ML models for the validation dataset at the horizon (t + 1)

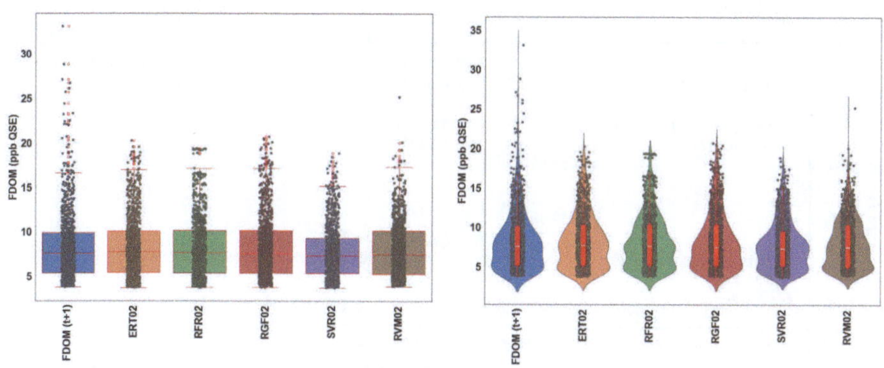

Fig. 6 Comparison between measured and forecasted FDOM concentration using different ML for the validation dataset at the horizon (t + 1): boxplot (left panel) and violinplot (right panel)

dataset are discussed hereafter. As shown in Tables 2, during the training stage the following results are worthy of discussion. First, by comparing all models, we found that the ERT3 obtains better results than the other, followed by the RGF3, RVM3, RFR3, while the SVR3 was ranked in the last place. This result demonstrates that the ERT3 significantly surpassed all other models. Second, by comparing the ERT3 model based on the calculated numerical performances, we found that ERT3 achieves significant improvement. The indices RMSE and MAE have deceased to around 58.85 and 58.90% compared to the RFR3, to around 47.208 and 49.024% compared to the RGF3, to around 67.82 and 60.69% compared to the SVR3, and to around 54.80 and 55.60% compared to the RVM3. During the validation stage, the results show that the proposed models were completely failed to provide a good numerical performances and excellent forecasting accuracy. In general, the evaluation indices achieved by all models during the validation stage are much less than those obtained during the training stage, which implies a lowest accuracy. The decrease of the RMSE and MAE for all models are evident in confirming the poorest performances. According to Tables 2, none of the model was able to provides an R and NSE superior to 0.60 and 0.30, which correspond to very poor forecasting accuracies. Finally, we report in Fig. 9 the Taylor diagram for models' comparison between the three forecasting horizons. It is clear from the plotted data in Fig. 9 that, the performances calculated based on the R values decreased significantly from horizon (t) to horizon ($t + 7$).

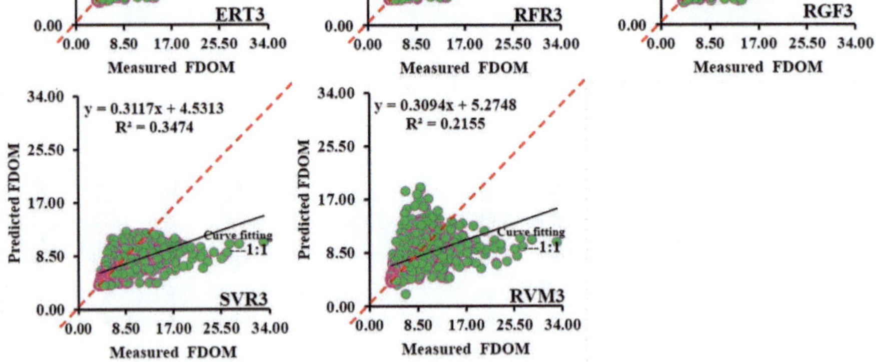

Fig. 7 Scatterplot of measured and forecasted FDOM concentration using different ML models for the validation dataset at the horizon (t + 7)

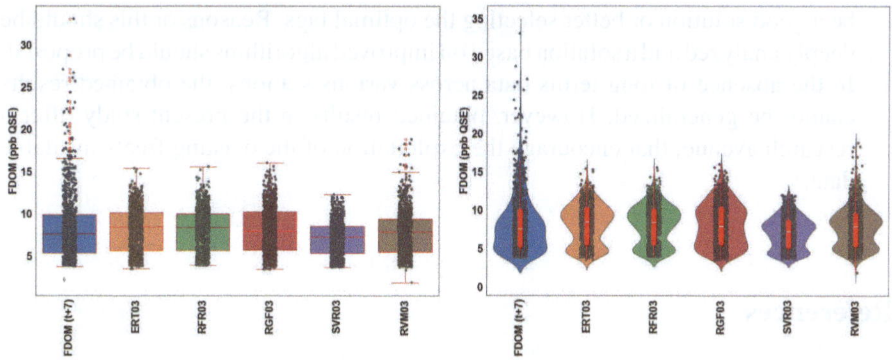

Fig. 8 Comparison between measured and forecasted FDOM concentration using different ML for the validation dataset at the horizon (t + 7): boxplot (left panel) and violinplot (right panel)

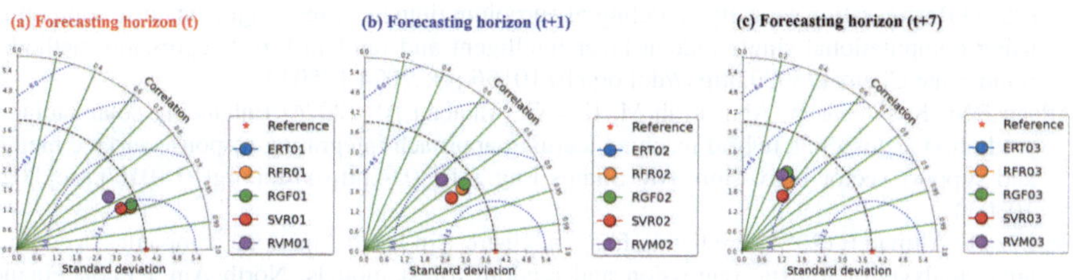

Fig. 9 Taylor diagram for models' comparison between measured and forecasted FDOM concentration for the validation dataset

5 Conclusion

This study comprehensively evaluated the forecasting capability of five machine learning models for FDOM forecasting. By comparing measured in situ FDOM from one USGS station versus calculated data, based on four assessment metrics (i.e. R, NSE, RMSE, and MAE), the following conclusions can be reported.

1. The forecasting accuracy was generally found to be better for shoot terms forecasting, i.e., at one day ahead, and significantly decreased at seven days ahead.
2. The RFR and ERT models were found to be the more accurate models, while the RVM was completely failed to correctly forecast the FDOM concentration. However, for more in depth analysis, the use of data from more stations is necessary for selecting the best lags times based on the site conditions.
3. Declining performances were observed with increasing forecasting horizon. Especially, significant degradations for all models based on all metrics. This decrease can be interpreted by the fact that, an increase in the forecasting horizon leads to an increase of the nonlinearity of the data, and the ACF, and PACF can

be a good solution or better selecting the optimal lags. Reasons or this should be deeply analyzed and a solution based on improved algorithms should be proposed.

4. In the absence of long terms data across various stations, the obtained results cannot be generalized. However, obtained results in the present study offer a research avenue, that encourage the exploitation of the existing freely available data.

References

Acosta SM, Amoroso AL, Anna AMOS, Junior OC (2021) Relevance vector machine with tuning based on self-adaptive differential evolution approach for predictive modelling of a chemical process. Appl Math Model 95:125–142. https://doi.org/10.1016/j.apm.2021.01.057

Adebayo BA, Owolabi TO, Taura U, Shamsah SMI, Alharbi FS, Alshaheen AM, Al-Jahwari F (2024) Refrigeration capacity modeling of europium titanate based magnetocaloric compounds using computational single hidden layer intelligent and random forest regression methods J Solid State Chem 125039.https://doi.org/10.1016/j.jssc.2024.125039

Adnan RM, Keshtegar B, Abusurrah M, Kisi O, Alkabaa AS (2024) Enhancing solar radiation prediction accuracy: a hybrid machine learning approach integrating response surface method and support vector regression. Ain Shams Eng J 103034.https://doi.org/10.1016/j.asej.2024.103034

Barboza F, Altman E (2024) Predicting financial distress in Latin American companies: a comparative analysis of logistic regression and random forest models. North Am J Econ Finance 72:102158. https://doi.org/10.1016/j.najef.2024.102158

Bouyakhsaine K, Brakez A, Draou M (2024) Prediction of residential building occupancy using machine learning with integrated sensor and survey data: insights from a living lab in Morocco. Energy Build 319:114519. https://doi.org/10.1016/j.enbuild.2024.114519

Breiman L (2001) Random forests. Mach Learn 45(1):5–32. https://doi.org/10.1023/A:1010933404324

Cheng H, Shi Y, Wu L, Guo Y, Xiong N (2021) An intelligent scheme for big data recovery in Internet of Things based on multi-attribute assistance and extremely randomized trees. Inf Sci 557:66–83. https://doi.org/10.1016/j.ins.2020.12.041

Daraei H, Bertone E, Awad J, Stewart RA, Chow CW, Duan J, Van Leeuwen J (2024a) A novel mathematical template for developing fDOM probe fluorescence signal correction models for freshwaters. J Environ Sci 146:103–117. https://doi.org/10.1016/j.jes.2023.06.011

Daraei H, Bertone E, Stewart RA, Awad J, Leavesley A, Gale M, Van Leeuwen J (2024b) Organic matter concentration and characteristic dynamics in surface waters post-bushfires and cyclones: fDOM sensors for environmental monitoring and control. Environ Sci Adv 3(6):950–963. https://doi.org/10.1039/d4va00036f

Dehkordi AT, Zoej MJV, Mehran A, Jafari M, Chegoonian AM (2024) Fuzzy similarity analysis of effective training samples to improve machine learning estimations of water quality parameters using Sentinel-2 remote sensing data. IEEE J Sel Top Appl Earth Observ Remote Sens. https://doi.org/10.1109/JSTARS.2024.3364020

Devi PD, Mamatha G (2024) Machine learning approach to predict the turbidity of Saki Lake, Telangana, India, using remote sensing data. Meas Sens 33:101139. https://doi.org/10.1016/j.measen.2024.101139

Fu Y, Li W, Liu T, Zhang Z, Li H, Xu J, Huang M (2023) CFSA-AGD: an accurate crosstalk fluorescence spectroscopic decomposition method for identifying and quantifying FDOMs in aquatic environments. Sci Total Environ 865:160950. https://doi.org/10.1016/j.scitotenv.2022.160950

García-Nieto PJ, García-Gonzalo E, Arbat G, Duran-Ros M, Pujol T, Puig-Bargués J (2024) Forecast of the outlet turbidity and filtered volume in different micro irrigation filters and filtration media by using machine-learning techniques. J Comput Appl Math 439:115606. https://doi.org/10.1016/j.cam.2023.115606

Gargiulo F, Duellmann D, Arpaia P, Schiano Lo Moriello R (2021) Predicting hard disk failure by means of automatized labeling and machine learning approach. Appl Sci 11(18):8293. https://doi.org/10.3390/app11188293

Geurts P, Ernst D, Wehenkel L (2006) Extremely randomized trees. Mach Learn 63(1):3–42. https://doi.org/10.1007/s10994-006-6226-1

Ghosh I, Chaudhuri TD, Alfaro-Cortés E, Gámez M, García N (2022) A hybrid approach to forecasting futures prices with simultaneous consideration of optimality in ensemble feature selection and advanced artificial intelligence. Technol Forecast Soc Change 181:121757. https://doi.org/10.1016/j.techfore.2022.121757

He Y, Wu C, Fan Y (2024) Exploring the drivers of local government budget coordination: a random forest regression analysis. Int Rev Econ Financ 93:1104–1113. https://doi.org/10.1016/j.iref.2024.04.004

Heddam S (2014) Generalized regression neural network (GRNN)-based approach for colored dissolved organic matter (CDOM) retrieval: case study of Connecticut River at Middle Haddam Station, USA. Environ Monit Assess 186:7837–7848. https://doi.org/10.1007/s10661-014-3971-7

Hu R, Xu W, Yan W, Wu T, He X, Cheng N (2023) Comparison between machine-learning-based turbidity models developed for different lake zones in a large shallow lake. Water 15(3):387. https://doi.org/10.3390/w15030387

Johnson R, Zhang T (2013) Learning nonlinear functions using regularized greedy forest. IEEE Trans Pattern Anal Mach Intell 36(5):942–954. https://doi.org/10.1109/TPAMI.2013.159

Joshi J, Rao PS (2024) Predictive modeling of allowable storage time of finger millet grains using artificial neural network and support vector regression approaches. J Food Eng 383:112224. https://doi.org/10.1016/j.jfoodeng.2024.112224

Kamali S, Mariani S, Hadianfard MA, Marzani A (2024) Inverse surrogate model for deterministic structural model updating based on random forest regression. Mech Syst Signal Process 215:111416. https://doi.org/10.1016/j.ymssp.2024.111416

Kumari A, Tanveer M (2024) LSTSVR+: least square twin support vector regression with privileged information. Eng Appl Artif Intell 136:108964. https://doi.org/10.1016/j.engappai.2024.108964

Liang Y, Zhang S, Qiao H, Yao Y (2021) IPromoter-ET: identifying promoters and their strength by extremely randomized trees-based feature selection. Anal Biochem 630:114335. https://doi.org/10.1016/j.ab.2021.114335

Liu B, Li T (2024) A machine-learning-based framework for retrieving water quality parameters in urban rivers using UAV hyperspectral images. Remote Sens 16(5):905. https://doi.org/10.3390/rs16050905

Lu SD, Liu HD, Wang M, Wu CC (2024) A novel strategy for multitype fault diagnosis in photovoltaic systems using multiple regression analysis and support vector machines. Energy Rep 12:2824–2844. https://doi.org/10.1016/j.egyr.2024.08.074

Mousavi S (2024) Conjugation of deep learning and de noising data methods for short-term water turbidity forecasting. J Hydro-Environ Res 52:26–37. https://doi.org/10.1016/j.jher.2023.12.002

Niloy NM, Habib SA, Islam MI, Haque MM, Shammi M, Tareq SM (2023) Distribution, characteristics and fate of fluorescent dissolved organic matter (FDOM) in the Bay of Bengal. Mar Pollut Bull 195:115467. https://doi.org/10.1016/j.marpolbul.2023.115467

Parra L, Ahmad A, Sendra S, Lloret J, Lorenz P (2024) Combination of machine learning and RGB sensors to quantify and classify water turbidity. Chemosensors 12(3):34. https://doi.org/10.3390/chemosensors12030034

Pinto A, Pereira S, Rasteiro D, Silva CA (2018) Hierarchical brain tumour segmentation using extremely randomized trees. Pattern Recogn 82:105–117. https://doi.org/10.1016/j.patcog.2018.05.006

Pratap V, Kumar S, Yadav BR (2024) Optimization of biogas production from thermal-alkali pretreated sludge using response surface methodology and random forest regressor model. J Taiwan Inst Chem Eng 105571.https://doi.org/10.1016/j.jtice.2024.105571
Prathom K, Sujitapan C (2024) Performance of logistic regression and support vector machine conjunction with the GIS and RS in the landslide susceptibility assessment: case study in Nakhon Si Thammarat, southern Thailand. J King Saud Univ Sci 103306.https://doi.org/10.1016/j.jksus.2024.103306
Qi L, Zheng H, Ding D, Wang S (2024) A comparison of meteorological normalization of PM2.5 by multiple linear regression, general additive model, and random forest methods. Atmos Environ 338:120854. https://doi.org/10.1016/j.atmosenv.2024.120854
Sachdeva S, Kumar B (2022) Flood susceptibility mapping using extremely randomized trees for Assam 2020 floods. Eco Inform 67:101498. https://doi.org/10.1016/j.ecoinf.2021.101498
Saeed U, Jan SU, Lee YD, Koo I (2021) Fault diagnosis based on extremely randomized trees in wireless sensor networks. Reliab Eng Syst Saf 205:107284. https://doi.org/10.1016/j.ress.2020.107284
Urban-Rich J, McCarty JT, Fernández D, Acuña JL (2006) Larvaceans and copepods excrete fluorescent dissolved organic matter (FDOM). J Exp Mar Biol Ecol 332(1):96–105. https://doi.org/10.1016/j.jembe.2005.11.023
Wahba M, Essam R, El-Rawy M, Al-Arifi N, Abdalla F, Elsadek WM (2024) Forecasting of flash flood susceptibility mapping using random forest regression model and geographic information systems. Heliyon 10(13). https://doi.org/10.1016/j.heliyon.2024.e33982
Wang Z, Guo L, Gong H, Li X, Zhu L, Sun Y, Zhu X (2023) Land subsidence simulation based on extremely randomized trees combined with Monte Carlo algorithm. Comput Geosci 178:105415. https://doi.org/10.1016/j.cageo.2023.105415
Wang C, Zhang X, Yun X, Fan X (2024) A novel hybrid optimized incremental relevance vector machine and filtering technique for state of charge estimation of lithium-ion batteries. J Energy Storage 90:111908. https://doi.org/10.1016/j.est.2024.111908
Yiu BWK, Zhang T, Lee CW (2024) Short-term load forecasting using regularized greedy forest-based ensemble model. IEEE Access. https://doi.org/10.1109/ACCESS.2024.3441642
Yu Y, Zhou Z, Sharma RP, Zhang L, Du M, Zhang H (2024) Comparing crown ratio models for spruce-fir broadleaved mixed forests using beta regression and random forest algorithm. Comput Electron Agric 225:109302. https://doi.org/10.1016/j.compag.2024.109302
Zhan Y, Zeng C, Chen X, Liu B, Li Y, Hur K, Dong S (2023) Hydrophile-lipophile balance solid phase extraction of surface water organics: fluorescent elution preference and overlooked fractions. Chemosphere 330:138770. https://doi.org/10.1016/j.chemosphere.2023.138770
Zhang J, Meng F, Fu P, Jing T, Xu J, Yang X (2024) Tracking changes in chlorophyll-a concentration and turbidity in Nansi Lake using Sentinel-2 imagery: a novel machine learning approach. Eco Inform 81:102597. https://doi.org/10.1016/j.ecoinf.2024.102597
Zhu H, Hao HK, Lu C (2024) Enhanced support vector machine-based moving regression strategy for response prediction and reliability estimation of complex structure. Aerosp Sci Technol 109634. https://doi.org/10.1016/j.ast.2024.109634
Zounemat-Kermani M, Kermani SG, Fadaee M, Aldallal A, Kisi O, Ramezani-Charmahineh A (2024) Modeling river water dissolved organic matter using ensemble computing and genetic programming techniques. Ecohydrol Hydrobiol. https://doi.org/10.1016/j.ecohyd.2024.04.003

Water Budgeting and Groundwater Utilization Assessment in Western Uttar Pradesh (India)

Kusum Pandey

Abstract In the last few decades, Uttar Pradesh has observed a notable rise in food production in agriculture. This was probable due to extensive fertilizer use, better varieties and more water resource utilization. This increased water resource need has led to extensive use of groundwater and surface water in Western Uttar Pradesh. Therefore, a quantitative groundwater assessment was carried out in the western Uttar Pradesh for the sustainable utilization of vulnerable groundwater resource. Ground Water Estimation Committee, 2015 (GEC-2015) methodology used for calculating the water budgeting and groundwater utilization. Of the twenty-four blocks of the study area, seven blocks of the critical category and two blocks of the semi-critical category was converted into over-exploited category; two blocks of the semi-critical category were converted into the critical category, whereas three blocks of the safe category were converted into the semi-critical category of the groundwater utilization development stage. Changing crop patterns, production of more high-water demanding crops, growing need for domestic and industrial use, resulting in over-exploitation of groundwater, could be the likely reasons for transformation of study region blocks from lower to higher groundwater utilization development stage.

Keywords Assessment · Groundwater · Overexploitation · Utilization

1 Introduction

Groundwater is the vital resource of freshwater in the globe. According to Shah et al. (2003), because of agricultural electricity subsidies, India is the world's biggest user of groundwater, drawing 150 km^3 per year. Due to uncontrolled population explosion, expending industries area, farming activities several countries face scarcity of water (Pandey et al. 2020). By understanding the balance of water in a system and the dynamics of groundwater, stakeholders can make informed decisions that

K. Pandey (✉)
Garhwal Regional Centre, G. B. Pant National Institute of Himalayan Environment, Srinagar, Uttarakhand, India
e-mail: khushipandey166@gmail.com

Dimple et al. (eds.), *Surface Water and Groundwater Quality for Sustainable Utilization*, Springer Hydrogeology, https://doi.org/10.1007/978-981-96-8189-1_20

promote sustainability and resilience (Pandey and Malik 2021). Therefore, water budgeting and groundwater utilization assessment are essential practices in the area for sustainable water management practices.

Varalakshmi et al. (2014) calculated groundwater recharge in the catchment areas of Osmansagar and Himayathsagar using GEC-2015 guidelines, consisting of basaltic and granitic terrains respectively and found that 22% of the rainfall contributes to the recharge of groundwater in the basin. While, Kumar and Seethapathi (2002) examined groundwater budgeting and confirmed different groundwater supplies in India's hard rock areas. Additionally, Kumar and Nachiappan (1995) developed an empirical equation for recharge estimation from precipitation at 25 points in Uttar Pradesh using the tritium tagging approach and based on the current hydrogeological conditions, that empirical equation can be applied in the data-scarce Uttar Pradesh district. Several groundwater recharge assessment techniques, including empirical, physical-based modelling, water level fluctuation, water balance approaches, and soil moisture approach, were compared by Walker et al. (2019). A large range of recharge values was discovered, owing to the ambiguity pertaining to each approach.

In Uttar Pradesh, agriculture is the main the occupation and farmers are generally reliant on groundwater for non-monsoon spell irrigation. This region lacks effective water resources for ecosystem services, crop production, and environmental preservation, resulting in an uneven water budget. In addition, due to rapidly population explosion, the need of freshwater is rising. Therefore, it is essential to manage and develop groundwater resources to overcome the issues associated with declining groundwater levels during the non-monsoon seasons. The aim of the study was to investigate the evaluation of the groundwater utilization development and estimation of water balance in the region for overall development.

2 Material and Methods

2.1 Study Area

The study area comprises total 24 blocks of four district of Uttar Pradesh in which 3 blocks of Saharanpur district, 6 blocks in Muzaffarnagar district, 12 blocks in Meerut district and 3 blocks in Ghaziabad district. The total geographical area includes 563,647 ha having latitude of 28° 66′ N to 29° 92′ N and longitude of 77° 46′ E to 78° 02′ E. The study area maximum and minimum temperature ranging between 43–45 °C and 3–4 °C respectively. The yearly rainfall in the region ranging from 933 to 1204 mm. The agricultural crops in the study area were wheat, paddy, sugarcane, jowar, bajara, maize, pulses, oilseeds and vegetables etc.

2.2 *Groundwater Balance*

The methodology used for ground water resource assessment is recommendation of ground water estimation committee, 2015 (GEC-2015) by the following equation.

$$\text{Recharge} - \text{Discharge} = \pm\text{Storage variation over a period}$$

2.2.1 Groundwater Recharge Assessment

The groundwater recharge in the region calculated by following equation,

$$P = P_r + P_c + P_i + P_g + P_s + P_t$$

where, P_r = Recharge from rainfall,
P_c = Recharge from seepage from canals,
P_i = Recharge from return flow of irrigation water,
P_g = Groundwater inflow into the zone,
P_s = Influent seepage from river, and
P_t = Recharge from seepage from tanks and ponds.
The above variables were assessed by the following procedures.

Recharge from Rainfall, P_r

The recharge from rainfall was estimated by Ministry of Water Resources, India (2015) given method:

$$P_r = 3.47(R - 38)^{0.4}$$

where, P_r and R were the rainfall penetration (cm), and annual rainfall (cm) respectively.

According to Ground Water Department et al. (1999) assumption, if rainfall was less than 10 mm during non-monsoon season then rainfall recharge was considered zero.

Recharge from Seepage from Canals, P_c

Recharge from seepage from canals were estimated by the Uttar Pradesh Groundwater Department norms.

Recharge from Return Flow of Irrigation Water, P_i

According to ARDC (1979) and CGWB (2009), 35% of the total water used for irrigation was expected to come from recharge from percolating irrigation water.

Groundwater Inflow into the Zone, P_g

According to ARDC (1979), the groundwater inflow from the area equals the groundwater outflow into the area were assumed under the region.

Influent Seepage from River, P_s

According to ARDC (1979), during monsoon the influent seepage to the rivers and effluent seepage from rivers in non-monsoon has considered equals under study.

Recharge from Seepage from Tanks and Ponds, P_t

Seepage from tanks and ponds were estimated by Uttar Pradesh Groundwater Department norms.

2.2.2 Groundwater Discharge Assessment

The groundwater discharge of the study area calculated by following equation.

$$\text{Discharge, } D = E_s + O_g + E_t + T_p$$

where, E_s = Effluent seepage to rivers,
O_g = Groundwater outflow from the area.
E_t = Evapotranspiration loss from groundwater reservoir, and
T_p = Groundwater pumpage from wells.

Effluent Seepage to Rivers, E_s

According to ARDC (1979), during non-monsoon the effluent seepage to the rivers and influent seepage from rivers in non-monsoon has considered equals under study.

Groundwater Outflow from the Area, O_g

According to ARDC (1979), the groundwater outflow from the area equals the groundwater inflow into the area were assumed under study.

Evapotranspiration Losses from Groundwater Reservoirs, E_t

In the region the depth to water table were equal or greater than 2 m, therefore evapotranspiration losses from the groundwater reservoir were assumed as zero.

Groundwater Pumpage from Wells, T_p

The groundwater pumpage was estimated via multiplying the amount of minor irrigation structures by their corresponding unit drafts which is gathered from the offices of the District Statistical Officers of the particular districts. Groundwater extraction for domestic and industrial use was assumed as 135 L/day/capita and 1% of pumpage from minor irrigation units respectively (ARDC 1979; CGWB 2009). The total groundwater demand for livestock consumption use was 10% of the overall water requirement (ARDC 1979; CGWB 2009). The groundwater consumption of livestock such as cattle, horse/donkey, pig, goat/sheep/poultry and others animals considered as 55 L/day, 30 L/day, 15 L/day, 5 L/day and 15 L/day respectively.

2.3 Groundwater Utilization Development Stage

The groundwater utilization development stage was proportion of net discharge and net recharge. In which, net recharge and net discharge was assessed by 85% of gross recharge and 70% of gross discharge respectively. GEC-2015 norms were used for categorization of groundwater development stage.

3 Results and Discussions

3.1 Groundwater Utilization Development Stage

Intensive groundwater development in the study areas can lead to over-exploitation leading to declining groundwater levels. Developing a groundwater structure is assumed ‘safe’ when the extraction rate from the groundwater does not surpass the natural recharge rate. According to the Central Ground Water Board (CGWB 2009), latest calculation of groundwater resources the evaluation units are classified as 'over-exploited,' 'critical,' 'semi-critical' and 'safe' based on the groundwater development stage. Therefore, the groundwater evaluation component is divided on four groups according to the amount of groundwater accessible and the groundwater draft. The study area category is determined by following the criteria recommended by GEC (1997).

In the Table 1 and Table 2, respectively, the groundwater inventory for the years 1994 and 2014 is given. From Table 1, in the year 1994, Baliakheri block of district Saharanpur; Charthawal and Muzaffarnagar blocks of district Muzaffarnagar; Daurala, Mawana, Macchra and Rohta blocks in district Meerut were in critical category, whereas Sarrorpur, Sardhana and Janikurd blocks of Meerut district; Muradnagar block in Ghaziabad district were under semi-critical category. However, Purkaji block in district Muzaffarnagar; Hastinapur and Paricchitgarh blocks in district Meerut were found under safe category. Ten remaining blocks of the area of study have been reported under the category of over-exploited. The maximum development stage for groundwater utilization was found as 182.09% in Shahpur block of district Muzaffarnagar and minimum as 45.19% in Hastinapur block in district Meerut. Since the development stage in the region was found to be 97.10% overall groundwater utilization, the study region can be placed under the critical category for the year 1994.

In the year 2014, from Table 2, the groundwater utilization development stage of Sardhana and Janikurd blocks in district Meerut was under critical category, while Purkaji block in district Muzaffarnagar; Hastinapur and Paricchitgarh blocks in district Meerut were found to be under semi-critical category. The remaining nineteen blocks of study area were under the development stage of overexploited groundwater utilization category. The maximum and minimum stage of groundwater utilization was found to be 208.0% and 73.70%, respectively, in the Jansath block of district Muzaffarnagar and district Meerut Paricchitgarh block. The entire study area was found to be under overexploited category, with 118.53% of groundwater utilization development stage.

Bar diagram for annual variation in the number of blocks under different development of groundwater use was drawn and is given in Fig. 1. From Fig. 1 and Table 3 it was found that 14 blocks out of 24 of the study area between 1994 and 2014 had transformed from the lower groundwater utilization development stage to the higher category. On the comparison of groundwater inventory for the years 1994 and 2014, it was found that during the year 1994 out of 24 blocks of the study area, 3 blocks were under safe category, 4 blocks under semi-critical category, 7 blocks under critical category and remaining 10 blocks were under overexploited category, where as in the year 2014, 19 blocks of study area were found to be under overexploited category, 3 blocks were under semi-critical category and 2 blocks were found to be under critical category.

The transformation of blocks from the development stage of the lower category of groundwater utilization to the higher category of groundwater utilization development stage for the period i.e. 1994–2014 is given in Table 3. Therefore, it can be concluded that during the study period of twenty-one year's (1994–2014) from 24 blocks of the study region 7 blocks from critical category and 2 blocks from

Table 1 Groundwater budgeting in year 1994

Block	Net recharge (ha m)	Net discharge (ha m)	Balance (ha m)	Stage of development (%)	Category
Baliakheri	7419.39	7190.81	228.57	96.91	Critical
Nagal	7456.96	7910.29	−453.33	106.07	Overexploited
Deoband	5712.39	6988.87	−1276.48	122.34	Overexploited
Charthawal	7332.98	6888.98	443.99	93.94	Critical
Purkaji	6891.71	4650.36	2241.34	67.47	Safe
Muzaffarnagar	7994.87	7689.42	305.44	96.17	Critical
Shahpur	4507.10	8207.30	−3700.21	182.09	Overexploited
Jansath	6237.05	6263.87	−26.82	100.43	Overexploited
Khatauli	7960.39	8300.81	−340.42	104.27	Overexploited
Saroorpur	5573.01	4531.42	1041.58	81.31	Semi-critical
Sardhana	5702.34	4541.87	1160.474	79.64	Semi-critical
Daurala	4943.51	4573.77	369.74	92.52	Critical
Mawana	6053.75	5885.12	168.62	97.21	Critical
Hastinapur	8869.64	4008.20	4861.43	45.19	Safe
Parichhitgarh	9859.69	5323.78	4535.90	53.99	Safe
Macchra	5767.52	5300.63	466.87	91.90	Critical
Rohta	4578.20	4402.28	175.91	96.15	Critical
Janikurd	4992.67	3528.80	1463.86	70.67	Semi-critical
Meerut	3001.03	3357.98	−356.95	111.89	Overexploited
Rajpura	5209.64	5338.9	−129.26	102.45	Overexploited
Kharkhoda	5040.76	5476.32	−435.56	108.64	Overexploited
Bhojpur	4190.28	5288.61	−1098.33	126.21	Overexploited
Muradnagar	5119.99	4270.58	849.40	83.40	Semi-critical
Razapur	3289.20	3937.68	−648.48	119.71	Overexploited
Overall				97.10	Critical

semi-critical category were converted to overexploited category; 2 blocks from semi-critical category were converted to critical category, additionally 3 blocks from safe category were converted to semi-critical category of groundwater utilization development stage. The 'Safe' areas represent the development potential of groundwater. The 'Semi-Critical' regions need the careful production of groundwater. Intensive monitoring, assessment and future groundwater development should be carried out in the 'Critical' and 'Over-exploited' regions, with appropriate water conservation measures. According to this method, the groundwater development stage is estimated as 118.53% signalling over-exploitation of the region and water conservation actions are required. The possible reasons for transformation of study area blocks from lower to higher groundwater utilization development stage may be as a result

Table 2 Groundwater budgeting in year 2014

Block	Net recharge (ha m)	Net discharge (ha m)	Balance (ha m)	Stage of development (%)	Category
Baliakheri	8192.79	8365.78	−172.99	102.11	Overexploited
Nagal	9086.89	10,516.39	−1429.51	115.73	Overexploited
Deoband	7003.00	7592.241	−589.24	108.41	Overexploited
Charthawal	8645.90	10,160.18	−1514.28	117.51	Overexploited
Purkaji	8647.17	6997.62	1649.55	80.92	Semi-critical
Muzaffarnagar	8746.44	9066.60	−320.17	103.66	Overexploited
Shahpur	7781.67	16,186.43	−8404.77	208.00	Overexploited
Jansath	2844.26	5755.69	−2911.43	202.36	Overexploited
Khatauli	8474.90	9743.33	−1268.44	114.96	Overexploited
Saroorpur	4438.32	5525.73	−1087.41	124.50	Overexploited
Sardhana	5472.47	5413.19	59.28	98.91	Critical
Daurala	5154.83	5253.11	−98.29	101.90	Overexploited
Mawana	5968.51	8770.45	−2801.94	146.94	Overexploited
Hastinapur	10,181.15	8533.19	1647.96	83.81	Semi-critical
Parichhitgarh	10,157.19	7486.59	2670.59	73.70	Semi-critical
Macchra	5658.55	5747.63	−89.09	101.27	Overexploited
Rohta	4245.49	4602.58	−357.10	108.41	Overexploited
Janikurd	4819.04	4798.41	20.62	99.57	Critical
Meerut	2355.83	2367.58	−11.76	100.49	Overexploited
Rajpura	4111.81	4582.03	−470.23	111.43	Overexploited
Kharkhoda	3260.13	4709.82	−1449.70	144.46	Overexploited
Bhojpur	7552.08	10,290.82	−2738.63	136.26	Overexploited
Muradnagar	7387.27	8337.863	−950.59	112.86	Overexploited
Overall				118.53	Overexploited

of changing in cropping patterns, more production of high-water demanding crops, growing needs for domestic and industrial uses resulting in groundwater overuse. Considering the current pattern of water table dropping, it is predicted that by 2025, over 150% of the study region would be combined with the region where water table has already risen above the critical level. Therefore, there is a requirement to monitor the water table decrease in this area by rising the groundwater draft and/or growing groundwater capacity. The groundwater draft can be minimized by implementing successful irrigation practices/technologies i.e. micro irrigation, bed planting, zero tillage laser levelling, crop diversification etc.

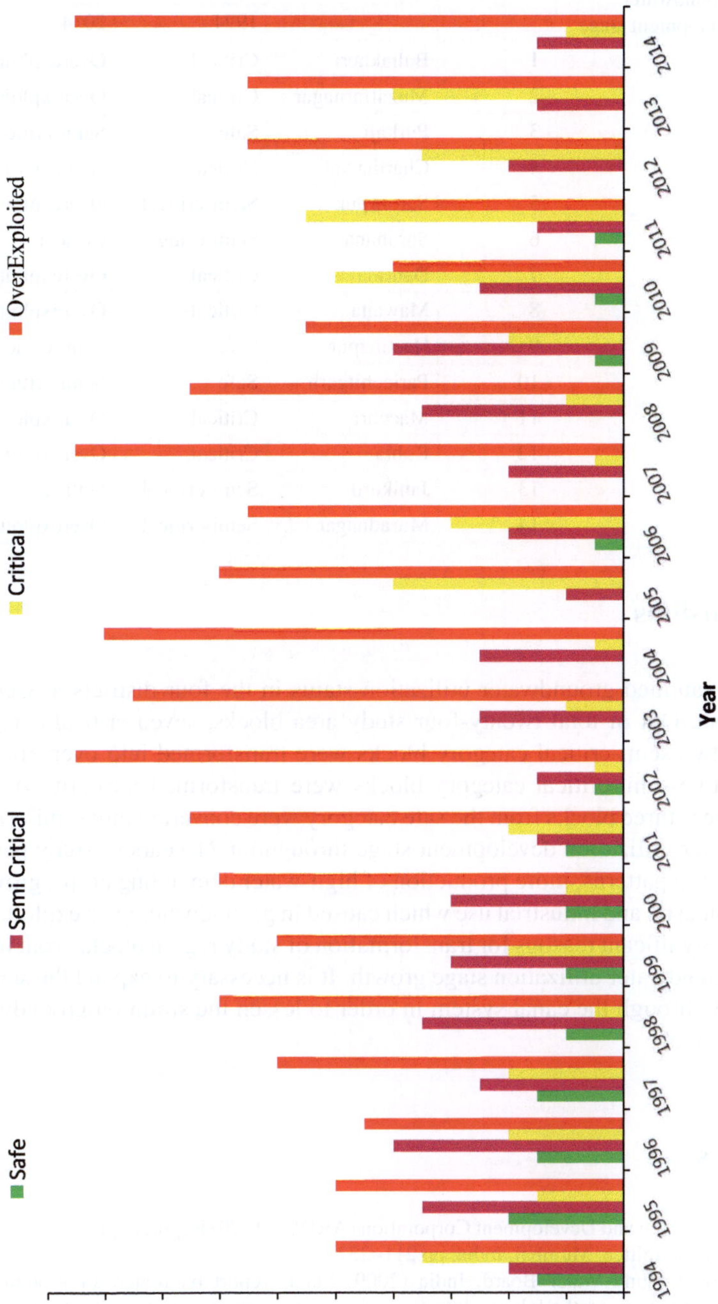

Fig. 1 Yearly variations in number of blocks under different groundwater utilization development stage categories

Table 3 Changes the blocks category of groundwater utilization development stage

S. no.	Block name	Year	
		1994	2014
1	Baliakheri	Critical	Overexploited
2	Muzaffarnagar	Critical	Overexploited
3	Purkaji	Safe	Semi-critical
4	Charthawal	Critical	Overexploited
5	Saroorpur	Semi-critical	Overexploited
6	Sardhana	Semi-critical	Critical
7	Daurala	Critical	Overexploited
8	Mawana	Critical	Overexploited
9	Hastinapur	Safe	Semi-critical
10	Paricchitgarh	Safe	Semi-critical
11	Macchra	Critical	Overexploited
12	Rohta	Critical	Overexploited
13	Janikurd	Semi-critical	Critical
14	Muradnagar	Semi-critical	Overexploited

4 Conclusions

The study examined groundwater utilization status in the four districts at western Uttar Pradesh. Out of total twenty-four study area blocks, seven critical category blocks and two semi-critical category blocks were transformed into overexploited categories; two semi-critical category blocks were transformed into critical categories, whereas three blocks from the safe category were converted into semi-critical of groundwater utilization development stage throughout 21 years of study period. Changes in crop patterns, more production of high-water demanding crops, growing needs for domestic and industrial use which caused in groundwater over-exploitation could be the significant reasons for transformation of study region blocks from lower to higher groundwater utilization stage growth. It is necessary to expand the surface water supply through the canal system in order to lessen the strain on groundwater in the study area.

References

Agricultural Refinance and Development Corporation (ARDC) (1979) Report of ground water over exploitation committee. Mumbai, India, pp 211–232

CGWB, Central Ground Water Board, India (2009) Status report on review of groundwater resources estimation methodology. http://www.indiaenvironmentportal.org.in/files/Statusreport_reviewmethodologycombined.pdf. Accessed 28 July 2020

GEC-Groundwater Resource Estimation Committee (2015) Ground water resource estimation methodology. Ministry of Water Resources report. Government of India, Delhi

Ground Water Department, Geohydrological Division, Uttar Pradesh (1999) Report on estimation of groundwater resources as per methodology-1997 in district Moradabad. Bareilly, T.M.No. 295/98-99

Ground Water Estimation Committee (GEC) (1997) Ground water estimation methodology—1997. Ministry of Water Resources, Govt of India

Kumar B, Nachiappan RP (1995) A mathematical approach based on tritium tagging technique to evaluate recharge to ground water due to monsoon rains. In: Proceedings of the international symposium on tracer technologies for hydrological systems, Colorado, USA, July, 1995, IAHS, vol 229, pp 237–248

Kumar CP, Seethapathi PV (2002) Assessment of natural groundwater recharge in upper Ganga canal command area. J Appl Hydrol Assoc Hydrol India XV 4:13–20

Ministry of Water Resources (MoWR), India (2015) A year of inclusive development in water resources sector. National Water Mission, Ministry of WR, RD&GR, Govt of India. http://www.nationalwatermission.gov.in/sites/default/files/COP%2021%20MoWR%20Booklet.pdf. Accessed 17 Jan 2017

Pandey K, Malik A (2021) Seasonal groundwater table depth prediction using fuzzy logic and artificial neural network in Gangetic plain, India. In: Advanced modelling and innovations in water resources engineering: select proceedings of AMIWRE 2021. Springer Singapore, Singapore, pp 549–564

Pandey K, Kumar S, Malik A, Kuriqi A (2020) Artificial neural network optimized with a genetic algorithm for seasonal groundwater table depth prediction in Uttar Pradesh, India. Sustainability 12(21):8932

Shah T, Scott C, Kishore A, Sharma A (2003) Energy–irrigation nexus in South Asia: approaches to agrarian prosperity with viable power industry (Research report No. 70). International Water Management Institute, Colombo

Varalakshmi V, Venkateswara Rao B, SuriNaidu L, Tejaswini M (2014) Groundwater flow modeling of a hard rock aquifer: case study. J Hydrol Eng 19(5):877–886

Walker D, Parkin G, Schmitter P, Gowing J, Tilahun SA, Haile AT, Yimam AY (2019) Insights from a multi-method recharge estimation comparison study. Groundwater 57(2):245–258

MIX
Papier aus verantwortungsvollen Quellen
Paper from responsible sources
FSC® C105338

If you have any concerns about our products,
you can contact us on
ProductSafety@springernature.com

In case Publisher is established outside the EU,
the EU authorized representative is:
Springer Nature Customer Service Center GmbH
Europaplatz 3, 69115 Heidelberg, Germany

Printed by Libri Plureos GmbH
in Hamburg, Germany